Advanced Nanomaterials

Edited by
Kurt E. Geckeler and
Hiroyuki Nishide

Further Reading

Tjong, Sie Chin

Carbon Nanotube Reinforced Composites

Metal and Ceramic Matrices

2009
ISBN: 978-3-527-40892-4

Zehetbauer, M. J., Zhu, Y. T. (eds.)

Bulk Nanostructured Materials

2009
ISBN: 978-3-527-31524-6

Vollath, D.

Nanomaterials

An Introduction to Synthesis, Properties and Applications

2008
ISBN: 978-3-527-31531-4

Astruc, D. (ed.)

Nanoparticles and Catalysis

2008
ISBN: 978-3-527-31572-7

Lee, Yoon S.

Self-Assembly and Nanotechnology

A Force Balance Approach

2008
ISBN: 978-0-470-24883-6

Eftekhari, Ali (Ed.)

Nanostructured Materials in Electrochemistry

2008
ISBN: 978-3-527-31876-6

Lazzari, M. / Liu, G. / Lecommandoux, S. (eds.)

Block Copolymers in Nanoscience

2006
ISBN: 978-3-527-31309-9

Kumar, Challa S. S. R. (Ed.)

Nanomaterials for the Life Sciences

10 Volume Set

2010
ISBN: 978-3-527-32261-9

Kumar, Challa S. S. R. (Ed.)

Nanotechnologies for the Life Sciences

10 Volume Set

2007
ISBN: 978-3-527-31301-3

Rao, C. N. R., Müller, A., Cheetham, A. K. (eds.)

Nanomaterials Chemistry

Recent Developments and New Directions

2007
ISBN: 978-3-527-31664-9

Advanced Nanomaterials

Edited by
Kurt E. Geckeler and Hiroyuki Nishide

Volume 1

WILEY-VCH Verlag GmbH & Co. KGaA

The Editors

Prof. Dr. Kurt E. Geckeler
Department of Nanobio Materials and
Electronics
World-Class University (WCU)
and
Department of Materials Science and
Engineering
Gwangju Institute of Science and
Technology (GIST)
1 Oryong-dong, Buk-gu
Gwangju 500-712
South Korea
E-mail: keg@gist.ac.kr

Prof. Hiroyuki Nishide
Department of Applied Chemistry
Waseda University
Ohkubo 3, Shinjuku
Tokyo 169-8555
Japan
E-mail: nishide@waseda.jp
and
Department of Nanobio Materials and
Electronics
World-Class University (WCU)
Gwangju Institute of Science and
Technology (GIST)
1 Oryong-dong, Buk-gu
Gwangju 500-712
South Korea

All books published by **Wiley-VCH** are carefully produced. Nevertheless, authors, editors, and publisher do not warrant the information contained in these books, including this book, to be free of errors. Readers are advised to keep in mind that statements, data, illustrations, procedural details or other items may inadvertently be inaccurate.

Library of Congress Card No.: applied for

British Library Cataloguing-in-Publication Data
A catalogue record for this book is available from the British Library.

Bibliographic information published by the Deutsche Nationalbibliothek
The Deutsche Nationalbibliothek lists this publication in the Deutsche Nationalbibliografie; detailed bibliographic data are available on the Internet at <http://dnb.d-nb.de>.

© 2010 WILEY-VCH Verlag GmbH & Co. KGaA, Weinheim

All rights reserved (including those of translation into other languages). No part of this book may be reproduced in any form – by photoprinting, microfilm, or any other means – nor transmitted or translated into a machine language without written permission from the publishers. Registered names, trademarks, etc. used in this book, even when not specifically marked as such, are not to be considered unprotected by law.

Composition   Toppan Best-set Premedia Limited
Printing and Bookbinding   Strauss GmbH, Mörlenbach
Cover Design   Schulz Grafik-Design, Fußgönheim

Printed in the Federal Republic of Germany
Printed on acid-free paper

ISBN: 978-3-527-31794-3

Contents

Preface *XV*
List of Contributors *XVII*

Volume 1

1 Phase-Selective Chemistry in Block Copolymer Systems *1*
Evan L. Schwartz and Christopher K. Ober
1.1 Block Copolymers as Useful Nanomaterials *1*
1.1.1 Introduction *1*
1.1.2 Self-Assembly of Block Copolymers *3*
1.1.3 Triblock Copolymers *4*
1.1.4 Rod–Coil Block Copolymers *7*
1.1.5 Micelle Formation *8*
1.1.6 Synthesis of Block Copolymers Using Living Polymerization Techniques *9*
1.1.6.1 Anionic Polymerization *10*
1.1.6.2 Stable Free Radical Polymerizations *11*
1.1.6.3 Reversible Addition–Fragmentation Chain Transfer (RAFT) Polymerization *12*
1.1.6.4 Atom Transfer Radical Polymerization *12*
1.1.6.5 Ring-Opening Metathesis Polymerization *13*
1.1.6.6 Group Transfer Polymerization *13*
1.1.7 Post-Polymerization Modifications *14*
1.1.7.1 Active-Center Transformations *14*
1.1.7.2 Polymer-Analogous Reactions *14*
1.2 Block Copolymers as Lithographic Materials *15*
1.2.1 Introduction to Lithography *15*
1.2.2 Block Copolymers as Nanolithographic Templates *17*
1.2.2.1 Creation of Nanoporous Block Copolymer Templates *20*
1.2.3 Multilevel Resist Strategies Using Block Copolymers *29*
1.3 Nanoporous Monoliths Using Block Copolymers *34*
1.3.1 Structure Direction Using Block Copolymer Scaffolds *34*
1.3.2 Nanopore Size Tunability *36*

Advanced Nanomaterials. Edited by Kurt E. Geckeler and Hiroyuki Nishide
Copyright © 2010 WILEY-VCH Verlag GmbH & Co. KGaA, Weinheim
ISBN: 978-3-527-31794-3

1.3.3	Functionalized Nanoporous Surfaces 38
1.4	Photo-Crosslinkable Nano-Objects 41
1.5	Block Copolymers as Nanoreactors 44
1.5.1	Polymer–Metal Solubility 44
1.5.2	Cluster Nucleation and Growth 46
1.5.3	Block Copolymer Micelle Nanolithography 47
1.6	Interface-Active Block Copolymers 48
1.6.1	Low-Energy Surfaces Using Fluorinated Block Copolymers 48
1.6.2	Patterning Surface Energies 49
1.6.3	Photoswitchable Surface Energies Using Block Copolymers Containing Azobenzene 51
1.6.4	Light-Active Azobenzene Block Copolymer Vesicles as Drug Delivery Devices 52
1.6.5	Azobenzene-Containing Block Copolymers as Holographic Materials 52
1.7	Summary and Outlook 54
	References 60

2 Block Copolymer Nanofibers and Nanotubes 67
Guojun Liu

- 2.1 Introduction 67
- 2.2 Preparation 69
- 2.2.1 Nanofiber Preparation 69
- 2.2.2 Nanotube Preparation 72
- 2.3 Solution Properties 74
- 2.4 Chemical Reactions 81
- 2.4.1 Backbone Modification 81
- 2.4.2 End Functionalization 85
- 2.5 Concluding Remarks 87
  Acknowledgements 88
  References 88

3 Smart Nanoassemblies of Block Copolymers for Drug and Gene Delivery 91
Horacio Cabral and Kazunori Kataoka

- 3.1 Introduction 91
- 3.2 Smart Nanoassemblies for Drug and Gene Delivery 92
- 3.3 Endogenous Triggers 93
- 3.3.1 pH-Sensitive Nanoassemblies 93
- 3.3.1.1 Drug Delivery 93
- 3.3.1.2 Gene Delivery 96
- 3.3.2 Oxidation- and Reduction-Sensitive Polymeric Nanoassemblies 99
- 3.3.3 Other Endogenous Triggers 101
- 3.4 External Stimuli 102
- 3.4.1 Temperature 102

3.4.2	Light	*105*
3.4.3	Ultrasound	*107*
3.5	Future Perspectives	*108*
	References	*109*

4 A Comprehensive Approach to the Alignment and Ordering of Block Copolymer Morphologies *111*
Massimo Lazzari and Claudio De Rosa

4.1	Introduction	*111*
4.1.1	Motivation	*111*
4.1.2	Organization of the Chapter	*112*
4.2	How to Help Phase Separation	*113*
4.3	Orientation by External Fields	*116*
4.3.1	Mechanical Flow Fields	*117*
4.3.2	Electric and Magnetic Fields	*118*
4.3.3	Solvent Evaporation and Thermal Gradient	*122*
4.4	Templated Self-Assembly on Nanopatterned Surfaces	*123*
4.5	Epitaxy and Surface Interactions	*126*
4.5.1	Preferential Wetting and Homogeneous Surface Interactions	*126*
4.5.2	Epitaxy	*128*
4.5.3	Directional Crystallization	*130*
4.5.4	Graphoepitaxy and Other Confining Geometries	*135*
4.5.5	Combination of Directional Crystallization and Graphoepitaxy	*138*
4.5.6	Combination of Epitaxy and Directional Crystallization	*140*
4.6	Summary and Outlook	*149*
	Acknowledgments	*150*
	References	*150*

5 Helical Polymer-Based Supramolecular Films *159*
Akihiro Ohira, Michiya Fujiki, and Masashi Kunitake

5.1	Introduction	*159*
5.2	Helical Polymer-Based 1-D and 2-D Architectures	*161*
5.2.1	Formation of Various 1-D Architectures of Helical Polysilanes on Surfaces	*162*
5.2.1.1	Direct Visualization of 1-D Rod, Semi-Circle and Circle Structures by AFM	*162*
5.2.1.2	Driving Force for the Formation of 1-D Architectures	*165*
5.2.2	Formation of Mesoscopic 2-D Hierarchical Superhelical Assemblies	*167*
5.2.2.1	Direct Visualization of a Single Polymer Chain	*167*
5.2.2.2	Formation of Superhelical Assemblies by Homochiral Intermolecular Interactions	*169*
5.2.3	Formation of 2-D Crystallization of Poly(γ-L-Glutamates) on Surfaces	*172*

5.2.3.1	Direct Visualization of 2-D Self-Organized Array by AFM	*173*
5.2.3.2	Orientation in 2-D Self-Organized Array	*174*
5.2.3.3	Intermolecular Weak van der Waals Interactions in 2-D Self-Organized Arrays	*175*
5.2.3.4	Comparison of Structures between a 2-D Self-Organized Array and 3-D Bulk Phase	*175*
5.2.4	Summary of Helical Polymer-Based 1-D and 2-D Architectures	*176*
5.3	Helical Polymer-Based Functional Films	*177*
5.3.1	Chiroptical Memory and Switch in Helical Polysilane Films	*178*
5.3.1.1	Memory with Re-Writable Mode and Inversion "−1" and "+1" Switch	*178*
5.3.1.2	Memory with Write-Once Read-Many (WORM) Mode	*182*
5.3.1.3	On-Off "0" and "+1" Switch Based on Helix–Coil Transition	*182*
5.3.2	Chiroptical Transfer and Amplification in Binary Helical Polysilane Films	*185*
5.3.3	Summary of Helical Polymer-Based Functional Films	*188*
	Acknowledgments	*189*
	References	*190*
6	**Synthesis of Inorganic Nanotubes**	*195*
	C.N.R. Rao and Achutharao Govindaraj	
6.1	Introduction	*195*
6.2	General Synthetic Strategies	*196*
6.3	Nanotubes of Metals and other Elemental Materials	*196*
6.4	Metal Chalcogenide Nanotubes	*206*
6.5	Metal Oxide Nanotubes	*214*
6.5.1	SiO_2 Nanotubes	*214*
6.5.2	TiO_2 Nanotubes	*216*
6.5.3	ZnO, CdO, and Al_2O_3 Nanotubes	*221*
6.5.4	Nanotubes of Vanadium and Niobium Oxides	*225*
6.5.5	Nanotubes of other Transition Metal Oxides	*228*
6.5.6	Nanotubes of other Binary Oxides	*230*
6.5.7	Nanotubes of Titanates and other Complex Oxides	*233*
6.6	Pnictide Nanotubes	*235*
6.7	Nanotubes of Carbides and other Materials	*240*
6.8	Complex Inorganic Nanostructures Based on Nanotubes	*240*
6.9	Outlook	*241*
	Referecnes	*241*
7	**Gold Nanoparticles and Carbon Nanotubes: Precursors for Novel Composite Materials**	*249*
	Thathan Premkumar and Kurt E. Geckeler	
7.1	Introduction	*249*
7.2	Gold Nanoparticles	*249*
7.3	Carbon Nanotubes	*251*

7.4	CNT–Metal Nanoparticle Composites	*254*
7.5	CNT–AuNP Composites	*255*
7.5.1	Filling of CNTs with AuNPs	*255*
7.5.2	Deposition of AuNPs Directly on the CNT Surface	*256*
7.5.3	Interaction Between Modified AuNPs and CNTs	*267*
7.5.3.1	Covalent Linkage	*268*
7.5.3.2	Supramolecular Interaction Between AuNPs and CNTs	*271*
7.6	Applications	*288*
7.7	Merits and Demerits of Synthetic Approaches	*289*
7.8	Conclusions	*291*
	Acknowledgments	*292*
	References	*292*

8 Recent Advances in Metal Nanoparticle-Attached Electrodes *297*
Munetaka Oyama, Akrajas Ali Umar, and Jingdong Zhang

8.1	Introduction	*297*
8.2	Seed-Mediated Growth Method for the Attachment and Growth of AuNPs on ITO	*298*
8.3	Electrochemical Applications of AuNP-Attached ITO	*300*
8.4	Improved Methods for Attachment and Growth of AuNPs on ITO	*302*
8.5	Attachment and Growth of AuNPs on Other Substrates	*306*
8.6	Attachment and Growth of Au Nanoplates on ITO	*308*
8.7	Attachment and Growth of Silver Nanoparticles (AgNPs) on ITO	*309*
8.8	Attachment and Growth of Palladium Nanoparticles PdNPs on ITO	*311*
8.9	Attachment of Platinum Nanoparticles PtNPs on ITO and GC	*312*
8.10	Electrochemical Measurements of Biomolecules Using AuNP/ITO Electrodes	*315*
8.11	Nonlinear Optical Properties of Metal NP-Attached ITO	*315*
8.12	Concluding Remarks	*316*
	References	*316*

9 Mesoscale Radical Polymers: Bottom-Up Fabrication of Electrodes in Organic Polymer Batteries *319*
Kenichi Oyaizu and Hiroyuki Nishide

9.1	Mesostructured Materials for Energy Storage Devices	*319*
9.2	Mesoscale Fabrication of Inorganic Electrode-Active Materials	*322*
9.3	Bottom-Up Strategy for Organic Electrode Fabrication	*323*
9.3.1	Conjugated Polymers for Electrode-Active Materials	*323*
9.3.2	Mesoscale Organic Radical Polymer Electrodes	*324*
9.4	Conclusions	*330*
	References	*330*

10 Oxidation Catalysis by Nanoscale Gold, Silver, and Copper *333*
Zhi Li, Soorly G. Divakara, and Ryan M. Richards

10.1	Introduction *333*	
10.2	Preparations *334*	
10.2.1	Silver Nanocatalysts *335*	
10.2.2	Copper Nanocatalysts *335*	
10.2.3	Gold Nanocatalysts *335*	
10.3	Selective Oxidation of Carbon Monoxide (CO) *337*	
10.3.1	Gold Catalysts *337*	
10.3.2	Silver Catalysts *342*	
10.3.3	Gold–Silver Alloy Catalysts *342*	
10.3.4	Copper Catalysts *343*	
10.4	Epoxidation Reactions *344*	
10.4.1	Gold Catalysts *344*	
10.4.2	Silver Catalysts *346*	
10.5	Selective Oxidation of Hydrocarbons *347*	
10.5.1	Gold Catalysts *349*	
10.5.2	Silver Catalysts *350*	
10.5.3	Copper Catalysts *350*	
10.6	Oxidation of Alcohols and Aldehydes *350*	
10.6.1	Gold Catalysts *351*	
10.6.2	Silver Catalysts *351*	
10.7	Direct Synthesis of Hydrogen Peroxide *353*	
10.8	Conclusions *354*	
	References *355*	

11 Self-Assembling Nanoclusters Based on Tetrahalometallate Anions: Electronic and Mechanical Behavior *365*
Ishenkumba A. Kahwa

- 11.1 Introduction *365*
- 11.2 Preparation of Key Compounds *366*
- 11.3 Structure of the $[(A(18C6))_4(MX_4)] [BX_4]_2 \cdot nH_2O$ Complexes *367*
- 11.4 Structure of the $[(Na(15C5))_4Br] [TlBr_4]_3$ Complex *368*
- 11.5 Spectroscopy of the *Cubic F23* $[(A(18C6))_4(MX_4)] [BX_4]_2 \cdot nH_2O$ *368*
- 11.6 Unusual Luminescence Spectroscopy of Some Cubic $[(A(18C6))_4(MnX_4)] [TlCl_4]_2 \cdot nH_2O$ Compounds *372*
- 11.7 Luminescence Decay Dynamics and 18C6 Rotations *374*
- 11.8 Conclusions *375*
  Acknowledgments *377*
  References *377*

12 Optically Responsive Polymer Nanocomposites Containing Organic Functional Chromophores and Metal Nanostructures *379*
Andrea Pucci, Giacomo Ruggeri, and Francesco Ciardelli

- 12.1 Introduction *379*
- 12.2 Organic Chromophores as the Dispersed Phase *380*

12.2.1	Nature of the Organic Dye *380*	
12.2.2	Polymeric Indicators to Mechanical Stress *381*	
12.2.2.1	Oligo(*p*-Phenylene Vinylene) as Luminescent Dyes *381*	
12.2.2.2	Bis(Benzoxazolyl) Stilbene as a Luminescent Dye *383*	
12.2.2.3	Perylene Derivatives as Luminescent Dyes *384*	
12.2.3	Polymeric Indicators to Thermal Stress *385*	
12.2.3.1	Oligo(*p*-Phenylene Vinylene) as Luminescent Dyes *385*	
12.2.3.2	Bis(Benzoxazolyl) Stilbene as Luminescent Dye *387*	
12.2.3.3	Anthracene Triaryl Amine-Terminated Diimide as Luminescent Dye *388*	
12.3	Metal Nanostructures as the Dispersed Phase *389*	
12.3.1	Optical Properties of Metal Nanoassemblies *389*	
12.3.2	Nanocomposite-Based Indicators to Mechanical Stress *391*	
12.3.2.1	The Use of Metal Nanoparticles *391*	
12.3.2.2	The Use of Metal Nanorods *395*	
12.4	Conclusions *397*	
	Acknowledgments *398*	
	References *398*	

13   Nanocomposites Based on Phyllosilicates: From Petrochemicals to Renewable Thermoplastic Matrices *403*

Maria-Beatrice Coltelli, Serena Coiai, Simona Bronco, and Elisa Passaglia

13.1	Introduction *403*	
13.1.1	Structure of Phyllosilicates *404*	
13.1.1.1	Clays *404*	
13.1.2	Morphology of Composites *408*	
13.1.3	Properties of Composites *411*	
13.2	Polyolefin-Based Nanocomposites *411*	
13.2.1	Overview of the Preparation Methods *412*	
13.2.2	Organophilic Clay and Compatibilizer: Interactions with the Polyolefin Matrix *414*	
13.2.3	The One-Step Process *426*	
13.3	Poly(Ethylene Terephthalate)-Based Nanocomposites *429*	
13.3.1	In Situ Polymerization *430*	
13.3.2	Intercalation in Solution *433*	
13.3.3	Intercalation in the Melt *434*	
13.4	Poly(Lactide) (PLA)-Based Nanocomposites *439*	
13.4.1	Overview of Preparation Methods *439*	
13.4.1.1	In Situ Polymerization *439*	
13.4.1.2	Intercalation in Solution *442*	
13.4.1.3	Intercalation in the Melt *443*	
13.5	Conclusions *447*	
	Acknowledgments *449*	
	References *450*	

Volume 2

14 Amphiphilic Poly(Oxyalkylene)-Amines Interacting with Layered Clays: Intercalation, Exfoliation, and New Applications *459*
Jiang-Jen Lin, Ying-Nan Chan, and Wen-Hsin Chang

15 Mesoporous Alumina: Synthesis, Characterization, and Catalysis *481*
Tsunetake Seki and Makoto Onaka

16 Nanoceramics for Medical Applications *523*
Besim Ben-Nissan and Andy H. Choi

17 Self-healing of Surface Cracks in Structural Ceramics *555*
Wataru Nakao, Koji Takahashi, and Kotoji Ando

18 Ecological Toxicology of Engineered Carbon Nanoparticles *595*
Aaron P. Roberts and Ryan R. Otter

19 Carbon Nanotubes as Adsorbents for the Removal of Surface Water Contaminants *615*
Jose E. Herrera and Jing Cheng

20 Molecular Imprinting with Nanomaterials *651*
Kevin Flavin and Marina Resmini

21 Near-Field Raman Imaging of Nanostructures and Devices *677*
Ze Xiang Shen, Johnson Kasim, and Ting Yu

22 Fullerene-Rich Nanostructures *699*
Fernando Langa and Jean-François Nierengarten

23 Interactions of Carbon Nanotubes with Biomolecules: Advances and Challenges *715*
Dhriti Nepal and Kurt E. Geckeler

24 Nanoparticle-Cored Dendrimers and Hyperbranched Polymers: Synthesis, Properties, and Applications *743*
Young-Seok Shon

25 Concepts in Self-Assembly *767*
Jeremy J. Ramsden

26 Nanostructured Organogels via Molecular Self-Assembly *791*
Arjun S. Krishnan, Kristen E. Roskov, and Richard J. Spontak

27 Self-assembly of Linear Polypeptide-based Block Copolymers *835*
Sébastien Lecommandoux, Harm-Anton Klok, and Helmut Schlaad

28 Structural DNA Nanotechnology: Information-Guided
Self-Assembly *869*
Yonggang Ke, Yan Liu, and Hao Yan

Index *881*

Preface

Nanotechnology has found an incredible resonance and a vast number of applications in many areas during the past two decades. The resulting deep paradigm shift has opened up new horizons in materials science, and has led to exciting new developments. Fundamentally, nanotechnology is dependent on the existence or the supply of new nanomaterials that form the prerequisite for any further progress in this new and interdisciplinary area of science and technology. Evidently, nanomaterials feature specific properties that are characteristic of this class of materials, and which are based on surface and quantum effects.

Clearly, the control of composition, size, shape, and morphology of nanomaterials is an essential cornerstone for the development and application of nanomaterials and nanoscale devices. The complex functions of nanomaterials in devices and systems require further advancement in the preparation and modification of nanomaterials. Such advanced nanomaterials have attracted tremendous interest during recent years, and will form the basis for further progress in this area. Thus, the major classes of novel materials are described in the twenty-eight chapters of this two-volume monograph.

The initializing concept of this book was developed at the *3rd IUPAC International Symposium on Macro- and Supramolecular Architectures and Materials (MAM-06): Practical Nanochemistry and Novel Approaches,* held in in Tokyo, Japan, 2006, within the framework of the biannual MAM symposium series. This monograph provides a detailed account of the present status of nanomaterials, and highlights the recent developments made by leading research groups. A compilation of state-of-the-art review chapters, written by over sixty contributors and well-known experts in their field from all over the world, covers the novel and important aspects of these materials, and their applications.

The different classes of advanced nanomaterials, such as block copolymer systems including block copolymer nanofibers and nanotubes, smart nanoassemblies of block copolymers for drug and gene delivery, aligned and ordered block copolymers, helical polymer-based supramolecular films, as well as novel composite materials based on gold nanoparticles and carbon nanotubes, are covered in the book. Other topics include the synthesis of inorganic nanotubes, metal nanoparticle-attached electrodes, radical polymers in organic polymer batteries, oxidation catalysis by nanoscale gold, silver, copper, self-assembling nanoclusters, optically responsive polymer nanocomposites, renewable thermoplastic matrices based on phyllosilicate nanocomposites, amphiphilic polymer–clay intercalation and applications, the synthesis and catalysis of mesoporous alumina, and nanoceramics for medical applications.

In addition, this book highlights the recent progress in the research and applications of structural ceramics, the ecological toxicology of engineered carbon nanoparticles, carbon nanotubes as adsorbents for the removal of surface water contaminants, molecular imprinting with nanomaterials, near-field Raman imaging of nanostructures and devices, fullerene-rich nanostructures, nanoparticle-cored dendrimers and hyperbranched polymers, as well as the interactions of carbon nanotubes with biomolecules. The book is completed with a series of chapters featuring concepts in self-assembly, nanostructured organogels via molecular self-assembly, the self-assembly of linear polypeptide-based block copolymers, and information-guided self-assembly by structural DNA nanotechnology.

The variety of topics covered in this book make it an interesting and valuable reference source for those professionals engaged in the fundamental and applied research of nanotechnology. Thus, scientists, students, postdoctoral fellows, engineers, and industrial researchers, who are working in the fields of nanomaterials and nanotechnology at the interface of materials science, chemistry, physics, polymer science, engineering, and biosciences, would all benefit from this monograph.

The advanced nanomaterials presented in this book are expected to result in commercial applications in many areas. As the science and technology of nanomaterials is still in its infancy, further research will be required not only to develop this new area of materials science, but also to explore the utilization of these novel materials. All new developments impart risks, and here also it is important to evaluate the risks and benefits associated with the introduction of such materials into the biosphere and ecosphere.

On behalf of all contributors to we thank the publishers and authors on behalf of all contributors for granting copyright permissions to use their illustrations in this book. It is also very much appreciated that the authors devoted their time and efforts to contribute to this monograph. Last, but not least, the major prerequisite for the success of this comprehensive book project was the cooperation, support, and understanding of our families, which is greatly acknowledged.

The Editors

List of Contributors

Kotoji Ando
Yokohama National University
Department of Material Science
and Engineering
79-1 Tokiwadai
Hohogaya-ku
Yokohama 240-8501
Japan

Besim Ben-Nissan
University of Technology
Faculty of Science
Broadway
P.O. Box 123
Sydney
NSW 2007
Australia

Simona Bronco
CNR-INFM-PolyLab c/o
Dipartimento di Chimica e
Chimica Industriale
Università di Pisa
Via Risorgimento 35
56126 Pisa
Italy

Horacio Cabral
The University of Tokyo
Department of Materials Engineering
Graduate School of Engineering
7-3-1 Hongo, Bunkyo-ku
Tokyo 113-8656
Japan

Ying-Nan Chan
National Taiwan University
Institute of Polymer Science and
Engineering
Taipei 10617
Taiwan

and

National Chung Hsing University
Department of Chemical Engineering
Taichung 40227
Taiwan

Wen-Hsin Chang
National Taiwan University
Institute of Polymer Science and
Engineering
Taipei 10617
Taiwan

Advanced Nanomaterials. Edited by Kurt E. Geckeler and Hiroyuki Nishide
Copyright © 2010 WILEY-VCH Verlag GmbH & Co. KGaA, Weinheim
ISBN: 978-3-527-31794-3

Jing Cheng
The University of Western
Ontario
Department of Civil and
Environmental Engineering
London, ON N6A 5B9
Canada

Andy H. Choi
University of Technology
Faculty of Science
Broadway
P.O. Box 123
Sydney
NSW 2007
Australia

Francesco Ciardelli
University of Pisa
CNR-INFM-PolyLab
c/o Department of Chemistry,
and Industrial Chemistry
Via Risorgimento 35
56126 Pisa
Italy

Serena Coiai
Centro Italiano Packaging and
Dipartimento di Chimica e
Chimica Industriale
Università di Pisa
Via Risorgimento 35
56126 Pisa
Italy

Maria-Beatrice Coltelli
Centro Italiano Packaging and
Dipartimento di Chimica e
Chimica Industriale
Università di Pisa
Via Risorgimento 35
56126 Pisa
Italy

Claudio De Rosa
University of Napoli "Federico II"
Department of Chemistry
Complesso Monte S. Angelo
Via Cintia
80126 Napoli
Italy

Soorly G. Divakara
Colorado School of Mines
Department of Chemistry and
Geochemistry
1500 Illinois St.
Golden, CO 80401
USA

Kevin Flavin
Queen Mary University of London
School of Biological and Chemical
Sciences
Mile End Road
London E1 4NS
UK

Michiya Fujiki
Nara Institute of Science and
Technology
Graduate School of Materials Science
8916-5 Takayama
Ikoma
Nara 630-0101
Japan

Kurt E. Geckeler
Gwangju Institute of Science and
Technology (GIST)
Department of Materials Science and
Engineering
1 Oryong-dong, Buk-gu
Gwangju 500-712
South Korea

List of Contributors

Achutharao Govindaraj
International Centre for
Materials Science
New Chemistry Unit and CSIR
Centre of Excellence in
Chemistry
Jawaharlal Nehru Centre for
Advanced Scientific Research
Jakkur P. O.
Bangalore 560 064
India

and

Solid State and Structural
Chemistry Unit
Indian Institute of Science
Bangalore 560 012
India

Jose E. Herrera
The University of Western
Ontario
Department of Civil and
Environmental Engineering
London, ON N6A 5B9
Canada

Ishenkumba A. Kahwa
The University of the West
Indies
Chemistry Department
Mona Campus
Kingston 7
Mona
Jamaica

Johnson Kasim
Nanyang Technological
University
School of Physical and
Mathematical Sciences
Division of Physics and Applied
Physics
Singapore 637371
Singapore

Kazunori Kataoka
The University of Tokyo
Department of Materials Engineering
Graduate School of Engineering
7-3-1 Hongo
Bunkyo-ku
Tokyo 113-8656
Japan

and

The University of Tokyo
Center for Disease Biology and
Integrative Medicine
Graduate School of Medicine
7-3-1 Hongo
Bunkyo-ku
Tokyo 113-0033
Japan

and

The University of Tokyo
Center for NanoBio Integration
7-3-1 Hongo
Bunkyo-ku
Tokyo 113-8656
Japan

Yonggang Ke
Arizona State University
Department of Chemistry and
Biochemistry & The Biodesign
Institute
Tempe, AZ 85287
USA

Harm-Anton Klok
Ecole Polytechnique Fédérale de
Lausanne (EPFL)
Institut des Matériaux, Laboratoire des
Polymères
STI-IMX-LP
MXD 112 (Bâtiment MXD), Station 12
1015 Lausanne
Switzerland

Arjun S. Krishnan
North Carolina State University
Department of Chemical &
Biomolecular Engineering
Raleigh, NC 27695
USA

Masashi Kunitake
Kumamoto University
Department of Applied
Chemistry and Biochemistry
2-39-1 Kurokami
Kumamoto 860-8555
Japan

Fernando Langa
Universidad de Castilla-La
Mancha
Facultad de Ciencias del
Medio Ambiente
45071 Toledo
Spain

Massimo Lazzari
University of Santiago de
Compostela
Department of Physical
Chemistry
Faculty of Chemistry and
Institute of Technological
Investigations
15782 Santiago de Compostela
Spain

Sebastien Lecommandoux
University of Bordeaux
Laboratoire de Chimie des
Polymères Organiques (LCPO)
UMR CNRS 5629
Institut Polytechnique de
Bordeaux
16 Avenue Pey Berland
33607 Pessac
France

Zhi Li
Colorado School of Mines
Department of Chemistry and
Geochemistry
1500 Illinois St.
Golden, CO 80401
USA

Jiang-Jen Lin
National Taiwan University
Institute of Polymer Science and
Engineering
Taipei 10617
Taiwan

Guojun Liu
Queens University
Department of Chemistry
50 Bader Lane
Kingston Ontario K7L 3N6
Canada

Yan Liu
Arizona State University
Department of Chemistry and
Biochemistry & The Biodesign
Institute
Tempe, AZ 85287
USA

Watoru Nakao
Yokohama National University
Department of Energy and Safety
Engineering
79-5 Tokiwadai
Hodogaya-ku
Yokohama 240-8501
Japan

Dhriti Nepal
Gwangju Institute of Science
and Technology (GIST)
Department of Materials Science
and Engineering
1 Oryong-dong, Buk-gu
Gwangju 500-712
South Korea

and

School of Polymer
Textile and Fiber Engineering
Georgia Institute of Technology
Atlanta, GA 30332
USA

Jean-François Nierengarten
Université de Strasbourg
Laboratoire de Chimie des
Matériaux Moléculaires
(UMR 7509)
Ecole Européenne de Chimie
Polymères et Matériaux
25 rue Becquerel
67087 Strasbourg, Cedex 2
France

Hiroyuki Nishide
Waseda University
Department of Applied
Chemistry
Tokyo 169-8555
Japan

Christopher K. Ober
Cornell University
Department of Materials Science
and Engineering
Ithaca, NY 14853
USA

Akihiro Ohira
National Institute of Advanced
Industrial Science and Technology
(AIST)
Polymer Electrolyte Fuel Cell Cutting-
Edge Research Center (FC-Cubic)
2-41-6 Aomi, Koto-ku
Tokyo 135-0064
Japan

Makoto Onaka
The University of Tokyo
Department of Chemistry
Graduate School of Arts and Sciences
Komaba, Meguro-ku
Tokyo 153-8902
Japan

Ryan R. Otter
Middle Tennessee State University
Department of Biology
Murfreesboro, TN 37132
USA

Kenichi Oyaizu
Waseda University
Department of Applied Chemistry
Tokyo 169-8555
Japan

Munetaka Oyama
Kyoto University
Graduate School of Engineering
Department of Material Chemistry
Nishikyo-ku
Kyoto 615-8520
Japan

Elisa Passaglia
University of Pisa
Department of Chemistry and
Industrial Chemistry
Via Risorgimento 35
56126 Pisa
Italy

Thathan Premkumar
Department of Materials Science
and Engineering
Gwangju Institute of Science
and Technology (GIST)
1 Oryong-dong, Buk-gu
Gwangju 500-712
South Korea

Andrea Pucci
University of Pisa
Department of Chemistry and
Industrial Chemistry
Via Risorgimento 35
56126 Pisa
Italy

Jeremy J. Ramsden
Cranfield University
Bedfordshire MK43 0AL
UK

and

Cranfield University at
Kitakyushu
2-5-4F Hibikino
Wakamatsu-ku
Kitakyushu 808-0135
Japan

C.N.R. Rao
International Centre for Materials
Science,
New Chemistry Unit and CSIR Centre
of Excellence in Chemistry
Jawaharlal Nehru Centre for Advanced
Scientific Research
Jakkur P. O.
Bangalore 560 064
India

and

Solid State and Structural Chemistry
Unit
Indian Institute of Science
Bangalore 560 012
India

Marina Resmini
Queen Mary University of London
School of Biological and
Chemical Sciences
Mile End Road
London E1 4NS
UK

Ryan M. Richards
Colorado School of Mines
Department of Chemistry and
Geochemistry
1500 Illinois St.
Golden, CO 80401
USA

Aaron P. Roberts
University of North Texas
Department of Biological Sciences &
Institute of Applied Sciences
Denton, TX 76203
USA

Kristen E. Roskov
North Carolina State University
Department of Chemical &
Biomolecular Engineering
Raleigh, NC 27695
USA

Giacomo Ruggeri
University of Pisa
CNR-INFM-PolyLab
c/o Department of Chemistry
and Industrial Chemistry
Via Risorgimento 35
56126 Pisa
Italy

Helmut Schlaad
Max Planck Institute of Colloids
and Interfaces
MPI KGF Golm
14424 Potsdam
Germany

Evan L. Schwartz
Cornell University
Department of Materials Science
and Engineering
Ithaca, NY 14853
USA

Tsunetake Seki
The University of Tokyo
Department of Chemistry
Graduate School of Arts and
Sciences
Komaba, Meguro-ku
Tokyo 153-8902
Japan

Ze Xiang Shen
Nanyang Technological University
School of Physical and Mathematical
Sciences
Division of Physics and Applied
Physics
Singapore 637371
Singapore

Young-Seok Shon
California State University, Long Beach
Department of Chemistry and
Biochemistry
1250 Bellflower Blvd
Long Beach, CA 90840
USA

Richard J. Spontak
North Carolina State University
Department of Chemical &
Biomolecular Engineering
Raleigh, NC 27695
USA

and

North Carolina State University
Department of Materials Science &
Engineering
Raleigh, NC 27695
USA

Koji Takahashi
Kyushu University
Hakozaki
Higashi-ku
Fukuoka 812-8581
Japan

Akrajas Ali Umar
Universiti Kebangsaan Malaysia
Institute of Microengineering and
Nanoelectronics
43600 UKM Bangi Selangor
Malaysia

Hao Yan
Arizona State University
Department of Chemistry and
Biochemistry & The Biodesign
Institute
Tempe, AZ 85287
USA

Ting Yu
Nanyang Technological
University
School of Physical and
Mathematical Sciences
Division of Physics and Applied
Physics
Singapore 637371
Singapore

Jingdong Zhang
Huazhong University of Science and
Technology
College of Chemistry and Chemical
Engineering
Wuhan 430074
China

1
Phase-Selective Chemistry in Block Copolymer Systems
Evan L. Schwartz and Christopher K. Ober

1.1
Block Copolymers as Useful Nanomaterials

1.1.1
Introduction

Despite our best efforts to chemically design functional nanomaterials, we cannot yet match the brilliance of Nature. One striking example of this fact comes from a tethering structure known as a byssus created by the bivalve, *Mytilus edulis*. Byssal threads are the highly evolved materials that *M. edulis* uses to provide secure attachments to rocks and pilings during filter feeding. The threads begin at the base of the mussel's soft foot and attach to a hard surface by an adhesive plaque. Under strong tidal forces, an ordinary material would not be able to withstand the contact stresses that would result from the meeting of such soft and hard surfaces. Recent studies have shown that *M. edulis* solves this materials design problem through the creation of a "fuzzy" interface that avoids abrupt changes in the mechanical properties by gradually changing the chemical composition of the thread [1]. The chemistry that it uses to accomplish this graded material involves the elegant use of collagen-based self-assembling block copolymers (BCPs) [2]. The ventral groove of the mussel's foot contains several pores that act as channels for a reaction–injection-molding process that creates the copolymer. For this, central collagen blocks are mixed with a gradient of either elastin-like (soft) blocks, amorphous polyglycine blocks (intermediate), or silk-like (stiff) threads to form "diblock" copolymers of gradually decreasing mechanical stiffness as *M. edulis* moves farther away from the rock interface. Spontaneous self-assembly of the biopolymer seems to occur by the metal-binding histidine groups found in between each block interface that may act as ligands for metal-catalyzed polymerizations. The transition metals used for these polymerizations, such as Zn and Cu, are extracted from the ambient water through filter feeding.

 M. edulis byssal thread is not the only example of a self-assembling chemical system found in Nature that seems perfectly suited to its environment. Self-assembly such as that found in *M. edulis* can be found in nearly every level of nature, from cellular structures such as lipid bilayers [3], the colonization of bacteria [4], and the formation of weather systems [5]. The concept of self-assembly is defined by the automatic organization of small components into larger patterns or structures [6]. As small components, nature often uses various molecular interactions, such as hydrophilic/hydrophobic effects and covalent, hydrogen, ionic and van der Waals bonds to construct nanomaterials with specific macroscale functionalities. As scientists, we have learned an extraordinary amount about how to construct better synthetic materials from careful studies of how structure fits function in natural materials [7].

In the field of soft matter, one type of self-assembling synthetic material that has already been introduced in the *M. edulis* example is the BCP. BCPs are composed of different types of polymer connected by a covalent bond [8]. Apart from their interesting physical properties that have resulted in their use in byssal threads, upholstery foam, box tape, and asphalt [9], BCPs are also interesting due to the ability of each polymer block, or *phase*, to physically separate on the nanometer scale into various self-assembled morphologies such as spheres, cylinders, and sheets. These structures are attractive to scientists for several reasons.

- First, if one of the phases is removed from the periodic, ordered lattice, then thin films of the material could be used as stencils to etch patterns into semiconductor substrates such as silicon or gallium arsenide. This application is of great interest to the semiconductor industry, which is currently searching for alternative technologies for sub-20 nm lithography.

- Second, chemists are interested in BCP templates because they provide the power to carry out chemical reactions within specific phases of the material. This ability opens up many new areas of chemistry for nanomaterial design, including the growth of functional nanoparticle arrays for catalytic applications, the selective sequestration of chemicals for drug delivery, and the creation of mesoporous monolithic structures as low-*k* dielectric materials.

- Third, chemical functionalities attached to one phase within BCPs can be driven to segregate to the surface, where they can be affected by external stimuli such as ultraviolet (UV) light. These *surface-responsive* materials could be lithographically patterned to control the selective adsorption of biomolecules for biosensor applications.

All of the above applications use *phase-selective chemistry* to effect changes to the BCP microstructure and create useful nanostructured materials. In this chapter, we will discuss not only the recent investigations in these areas but also many other new and interesting applications.

The chapter is organized into three sections. In the first section we will discuss the basics of BCP self-assembly, and include a more detailed analysis of the morphologies possible with this class of material, along with an overview on how they are made and modified. The second section will provide a literature review of relevant studies in the field, including descriptions of BCPs as lithographic materials, as *nanoreactors*, as photo-crosslinkable nanobjects, and as surface-responsive materials. The third section will conclude with a summary of the most important contributions, together with a few additional insights on the future direction of the field of phase-selective BCP systems.

1.1.2
Self-Assembly of Block Copolymers

The thermodynamics of polymer mixing plays a large role in the self-assembly of BCPs [10]. In typical binary polymer mixtures, it is entropically unfavorable for two dissimilar homopolymers to mix homogeneously, as both components feel repulsive forces that result in the formation of large "macrophases" of each component in the mixture, akin to the mixing of oil and water. In diblock copolymers, however, the two component polymer "blocks" are chemically attached with a covalent bond. Here, the covalent bond acts as an elastic restoring force that limits the phase separation to mesoscopic length scales, thus resulting in "microphase" separated structures. The size of these phases, which are also known as microdomains, scale directly as the two-thirds power of the copolymer molecular weight [11]. The specific shape of the microdomains relies on a number of factors that control how each of the blocks interacts with each other. In the simplest argument, if there are equal amounts of each polymer, the microdomains will form into distinct layers with planar interfaces. However, if there is more of one block than the other, then curved interfaces will result. This curvature minimizes the repulsive interfacial contact between the A and B block, which also minimizes the free energy of the system. The bend that forms can be characterized by the curvature radius, R, as shown in Figure 1.1. Therefore, the equilibrium morphology of the BCP can usually be predicted based on differential geometry.

Other, more complicated, 'self-consistent mean field' theoretical treatments can be used to calculate the equilibrium morphology of the BCP. These theories sum the free energy contributions between (i) the repulsive polymer–polymer interactions versus (ii) the elastic restoring force energy for a particular microphase structure. The microphase structure with the lowest free energy sum will be the final equilibrium morphology. These theoretical equilibrium morphologies can be mapped out on a phase diagram, as shown in Figure 1.2. A typical BCP phase diagram plots the product χN on the ordinate versus the volume ratio, f_A, on the independent axis. χ is known as the Flory–Huggins interaction parameter, which quantifies the relative incompatibility between the polymer blocks, and is inversely related to the temperature of the system. N is called the *degree of polymerization*, which is the total number of monomers per macromolecule. The volume fraction is represented by $f_A = N_A/N$, where N_A is the number of A monomers per molecule. For very low concentrations of A monomer, no phase separation will occur and the two polymers will mix homogeneously. However, at slightly higher compositions, where $f_A \ll f_B$, the A blocks form spherical microdomains in a matrix of B. The microdomains arrange on a body-centered cubic (BCC) lattice. Increasing the volume fraction to $f_A < f_B$ leads to an increase in the connectivity of the microdomains, triggering the spheres to coalesce into cylinders that arrange on a hexago-

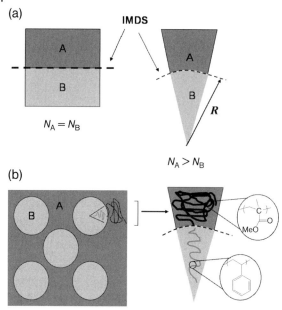

Figure 1.1 (a) Equal volume fractions of A and B blocks form layered structures called lamellae with curvature radius approaching infinity. Unequal volume fractions of A and B cause a curvature at the intermaterial dividing surface (IMDS) to minimize interfacial contact between the blocks and cause decrease of the curvature radius; (b) Schematic representing the application of this model in a sphere-forming (PS-b-PMMA) block copolymer system. Adapted from Ref. [29].

nal lattice. A roughly equal amount of both A and B blocks ($f_A \approx f_B$) will result in the formation of alternating layered sheets, or lamellae, of the A and B blocks. Any further increase in f_A ($f_A > f_B$), will cause the phases to invert, which means that the B block forms the microdomains in the matrix of A.

Thus, by tailoring the relative amount of A, the chemist can control the connectivity and dimensionality of the global BCP structure: spheres essentially represent zero-dimensional points in a matrix; cylinders represent one-dimensional lines; and lamellae represent two-dimensional sheets. Additionally, narrow regions of f_A exist in between the cylindrical and lamellar phase space where the two morphologies interpenetrate each other to form three-dimensional (3-D) "gyroid" [12, 13] network structures. Some reports of these morphologies have been published, and efforts have been put forth to take advantage of the added dimensionality with new applications [14, 15].

1.1.3
Triblock Copolymers

Adding extra polymer blocks to the BCP chain introduces additional levels of complexity into the self-assembled phase behavior. Core–shell morphologies [16], "knitting pattern" [17] and helical structures (Figure 1.3) are just a few of the exotic

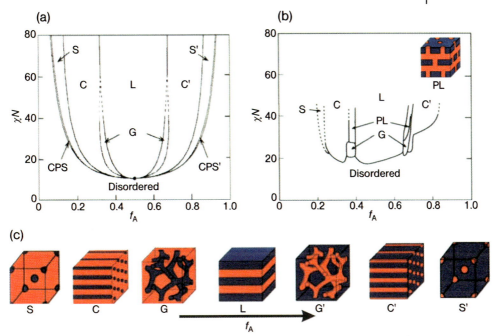

Figure 1.2 Phase diagram for linear AB diblock copolymers, comparing theory and experiment. (a) Self-consistent mean field theory predicts four equilibrium morphologies: spherical (S), cylindrical (C), gyroid (G), and lamellar (L), depending on the composition f and combination parameter χN. Here, χ is the Flory–Huggins interaction parameter (proportional to the heat of mixing A and B segments) and N is the degree of polymerization (number of monomers of all types per macromolecule); (b) Experimental phase portrait for poly(isoprene-*block*-styrene) diblock copolymers. Note the resemblance to the theoretical diagram. One difference is the observed perforated lamellae (PL) phase, which is actually metastable; (c) A representation of the equilibrium microdomain structures as f_A is increased for fixed χN. Reprinted with permission from Ref. [8]; © 2006, American Institute of Physics.

structures that have been found experimentally using triblock copolymers. Even more of the so-called "decorated phases" of tri-BCPs [18] have been predicted on a theoretical basis, but not yet found experimentally (Figure 1.4), offering a plethora of structures available to the chemist based on this template. In these cases, the phase behavior depends on two compositional variables and three relative incompatibility parameters (χ_{AB}, χ_{AC}, χ_{BC}), and thus the sequence of the components in the chain becomes important. For example, a poly(styrene-*block*-ethylene-*block*-butadiene) BCP may have completely different phase behavior than a poly(styrene-*block*-butadiene-*block*-ethylene) BCP at the same relative volume ratios. It is also possible to synthesize more than three blocks in the polymer chain – for example, a tetrablock terpolymer [19]. A more detailed look into the phase behavior and morphology of these complex systems is offered in a review by Abetz [20].

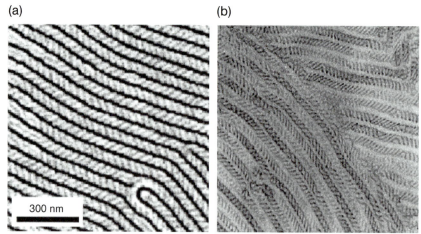

Figure 1.3 (a) Scanning electron microscopy image of the first layers of cylinders of a thin film of a triblock copolymer containing 17% styrene, 26% vinylpyridine, and 57% tert-butyl methacrylate after THF vapor exposure. The surface structures indicate a helix/cylinder morphology; (b) Transmission electron microscopy cross-section of a bulk sample of a triblock copolymer containing 26% styrene, 12% butadiene, and 62% tert-butyl methacrylate (MW = 218 000 g mol^{-1}).

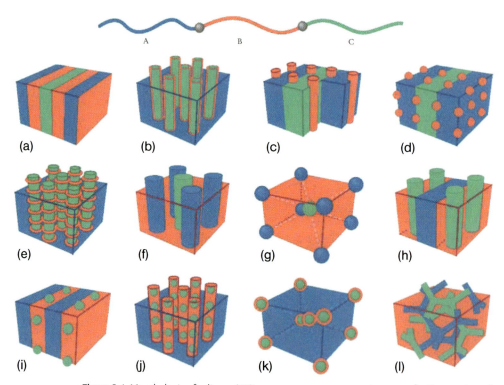

Figure 1.4 Morphologies for linear ABC triblock copolymers. A combination of block sequence (ABC, ACB, BAC), composition and block molecular weights provides an enormous parameter space for the creation of new morphologies. Reprinted with permission from Ref. [8]; © 2006, American Institute of Physics.

1.1.4
Rod–Coil Block Copolymers

There are essentially two types of BCP, both of which highlight interesting avenues for BCP self-assembly. "Coil–coil" BCPs, which are the most commonly studied, contain A and B blocks that can both be theoretically modeled as flexible chains. "Rod–coil" BCPs, on the other hand, have one polymer chain that is best represented as a rigid rod due to its stiff nature and anisotropic molecular shape. Rod-type molecules, also known as mesogens, can be incorporated into the main chain of a polymer backbone or appended from the polymer backbone as a side-chain substituent. Both types of rod–coil BCP have been shown to exhibit liquid crystalline (LC) behavior when placed in solution [21, 22]. These solutions are known as *lyotropic* solutions, which means that their phase behavior changes at different polymer concentrations. Initially, the polymers exhibit a disordered state called the *isotropic* phase; however, when the solution reaches a critical concentration, the molecular chains become locally packed and are forced to orient in a particular direction (*nematic* phase) due to the anisotropy of their shape. They can also arrange into several types of well-defined layers (*smectic* phases). By creating a BCP with a combination of a rod-like polymer block and a flexible polymer block, molecular level ordering characteristic of liquid crystals can be combined with the microphase-separated behavior typical of BCPs to produce hierarchical levels of self-assembly [23].

In a groundbreaking study on poly(hexylisocyanate-*block*-styrene) (PHIC-*b*-PS) – where the PHIC block represents the "rod" and the PS represents the "coil" block – it was found that, with increasing concentration of the polymer, isotropic, nematic and smectic LC phases each developed before the polymer adopted its final microphase-separated state [24]. As the PHIC chain was much longer than the PS chain in this case, the PHIC chain axis tilted with respect to the layer normal and interdigitated with the PS in order to accommodate the strain, resulting in wavy lamellae and never-before-seen zigzag and arrowhead morphologies. Electron diffraction experiments revealed ~1 nm spacings between the PHIC chains and a smectic layer repeat distance of approximately 200 nm. Furthermore, shearing a nematic solution of the polymer on a glass substrate induced over 10 μm of perfect long-range ordering of the layers, thus powerfully illustrating the multiple levels of ordering possible with LC-BCPs.

In further studies conducted by Mao and coworkers [25], a LC side group was attached as a pendent unit to a modified poly(styrene-*block*-isoprene) BCP. In this material, the phase transitions occurred in the opposite direction. The microphase separation of the classical lamellae and cylinders developed first, after which smectic layering of the LC blocks developed *within* the BCP microdomains due to constraint by the intermaterial dividing surface (IMDS) (Figure 1.5). Again, by incorporating a rigid block into a BCP framework, a hierarchy of ordering is observed. Unique chemical properties such as LC behavior may turn out to be crucial for future self-assembled synthetic materials.

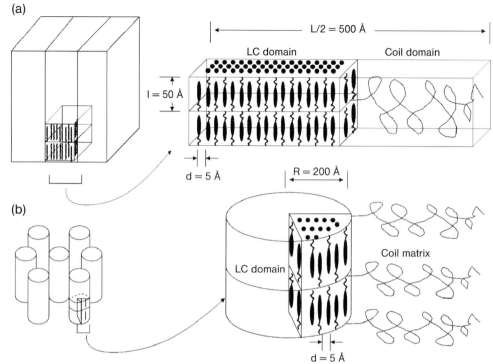

Figure 1.5 (a) A model showing the hierarchical levels of self-assembly using rod–coil block copolymers exhibiting liquid crystalline behavior in a lamellar morphology. (b) Structures can also form in the cylindrical morphology. Adapted from Ref. [25].

1.1.5
Micelle Formation

If the BCP is dissolved in a dilute solution with a solvent that dissolves only one of the blocks, the BCP will act as a surfactant molecule and micelle formation will occur. These materials are referred to as "amphiphilic" due to their dual polar/nonpolar chemical nature, and can thus dissolve partially in polar or nonpolar media. In dilute solutions, the soluble block "corona" will wrap itself around the insoluble "core" to minimize the repulsive contact forces between the insoluble block and the solvent, as illustrated in Figure 1.6. These micelles form structures with a defined size and shape, depending on the relative molecular weight of the blocks and the ionic strength of the solution. BCPs with large soluble blocks typically form spherical micelles due to small curvature radii, but smaller soluble block lengths can also form cylindrical micelles due to their greater curvature radii. The similarity of these micellar structures to biological cell vesicles [26] and liposomes

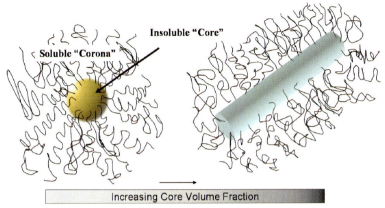

Figure 1.6 Schematic showing micelle structure. Amphiphilic block copolymers form micelles when dissolved in block-selective solvents. The soluble block "corona" stretches out into the solvent and masks the solvophobic "core". Cylindrical (and other) morphologies can be formed by tuning the relative volume fraction of the blocks.

has prompted many investigators to explore their use as templates [27], encapsulating agents [28], or drug delivery systems [29].

1.1.6
Synthesis of Block Copolymers Using Living Polymerization Techniques

BCPs are produced by the sequential addition of monomers into a "living" polymerization system [30]. Living polymerizations are characterized by a rapid initiation of the reactive chain end (e.g., carbanion, organometallic complex, etc.) and the lack of side reactions (e.g., chain termination or chain transfer) during growth of the polymer chain. In other words, a living polymer is a macromolecular species that will continue to grow as long as the monomer supply is replenished. The reactive chain end is then quenched to terminate further growth of the polymer during precipitation and purification.

A precise control of molecular weight is possible through living polymerization strategies. The degree of polymerization (N) is directly related to the molar amount of monomer (M) and the molar concentration of initiator, $[I]$, as shown in Equation 1.1:

$$N = \frac{[M]}{[I]} \qquad (1.1)$$

The living reaction will also be characterized by a narrow distribution of molecular weight, or polydispersity (M_w/M_n), of usually between 1.02 and 1.1, which means that that there is less than 30% standard deviation in the degree of polymerization of each of the chains. Many possible synthetic techniques are available to the

Table 1.1 Common living polymerization techniques for the preparation of block copolymers.

Polymerization technique	Monomers available
Anionic	Styrenes, vinylpyridines, methacrylates, acrylates, butadiene, isoprene, N-carboxyanhydrides (amino acids), ethylene oxide, lactones, hexamethylcyclotrisiloxane, 1,3-cyclohexadiene, isocyanates
Cationic ring-opening	Epoxides, siloxanes, tetrahydrofuran
Group transfer	Methacrylates, acrylates, nitriles, esters, butadienes, isoprenes
Ring-opening metathesis	Norbornenes
Stable free radical	Styrene, methacrylates, acrylates, acrylamides, dienes, acrylonitrile
Atom transfer radical	Styrenes, methacrylates, acrylates, acrylonitriles
Reversible addition – fragmentation chain transfer	Methacrylates, styrene, acrylates

$$Nu^{\ominus} + M \longrightarrow NuM^{\ominus} + M \longrightarrow Nu(P_n)^{\ominus} \xrightarrow{H^{\oplus}} Nu(P_n)\text{-}H$$

Scheme 1.1 The general mechanism of anionic polymerization.

chemist wishing to prepare a BCP, each with its own advantages and disadvantages. The types of polymerization suitable for each type of monomer are listed in Table 1.1. The most common techniques used to synthesize and modify BCPs are summarized briefly in the following paragraphs.

1.1.6.1 Anionic Polymerization

Anionic polymerization has become the most common technique in the synthesis of BCPs with narrow polydispersity [31]. The polymerization proceeds through the highly reactive carbanion chain end, usually created by an alkyl lithium initiator such as *sec*-BuLi or *n*-BuLi (Scheme 1.1). Due to the high reactivity of the chain end with other compounds, extremely stringent conditions must be met in order to avoid unwanted side reactions. Therefore, the polymerization must be carried out without any trace of oxygen or water, and all monomers and solvents must be extensively dried, degassed, and purified before use [32, 33]. The other main disadvantage to anionic polymerization is the limited range of monomers available for synthesis. Although the standard styrenes, methacrylates, butadiene, isoprene, ethylene oxide, vinylpyridines, and amino acids can all be synthesized using this technique, monomers with reactive functional groups cannot be used because they will interfere with the anionic chain end. Such monomers must therefore be protected before synthesis and then later deprotected. In some cases, such as the polymerization of acrylates, the reactions must be carried out at very low temperatures (−78 °C) in order to avoid terminating side reactions such as intrachain cyclization or "backbiting", caused by the reaction of the anionic center with a carbonyl group on the monomer.

For BCPs containing two distinctly different monomer types, such as the polymerization of polystyrene and polyethylene oxide, attention must be paid to the order of polymerization in order to maximize the efficiency of the reaction. For example, whilst a polystyryl lithium "macroinitiator" enables the rapid initiation of ethylene oxide, lithium-activated ethylene oxide will not efficiently initiate the polystyrene monomer and the reaction may not go to completion. This is due to the difference in relative reactivity between the oxyanion and carbanionic species.

There are, of course, many advantages to anionic polymerizations, besides the fact that they produce polymers with the lowest polydispersity. One advantage is that the chain end can be terminated with functional groups or coupling agents to produce telechelic polymers or complex macromolecular architectures, respectively. Examples of complex BCP architectures include ABA tri-BCPs, star, or graft BCPs [34].

1.1.6.2 Stable Free Radical Polymerizations

The amount of growth in the area of stable free radical polymerizations (SFRPs) during the past 20 years has been astounding. Although the model for SFRP was introduced by Otsu during the early 1980s [35], more recently, alkoxyamine initiators generated by the research group of Hawker [36] at IBM have led to dramatic improvements in the technique as introduced by Georges [37, 38] at Xerox during the early 1990s. These types of initiator contain a thermally cleavable C–O bond attached to a nitroxyl radical species. Running the reaction at high temperatures (80–90 °C) causes a reversible capping of the nitroxyl radicals, and allows monomer addition to the polymer chain only when the nitroxyl radical is in its detached state (Scheme 1.2). An advantage to SFRP is that the chemical rate of this detachment drops almost to zero at room temperature. Therefore, decreasing the temperature

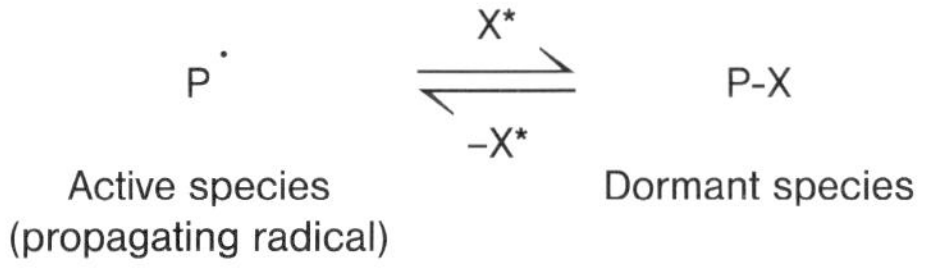

$$P^\bullet \underset{-X^*}{\overset{X^*}{\rightleftharpoons}} P\text{-}X$$

Active species   Dormant species
(propagating radical)

Scheme 1.2 The general mechanism of stable free radical polymerizations.

of the polymerization reactor essentially "switches off" the polymerization and allows the chemist to expose the first block to air, without terminating the reactive chain end. After precipitation, purification, and molecular weight characterization, the first block can be dissolved and heated in the presence of the second monomer to form the final product. There is a wide range of monomers available using SFRP, including styrenes, methacrylates, (meth)acrylonitriles, among others. Unfortunately, the polydispersities of the free radical polymerization process are not quite as low as anionic polymerization, and stereochemical control is not possible.

1.1.6.3 Reversible Addition–Fragmentation Chain Transfer (RAFT) Polymerization

The RAFT process is a variation of the living radical process that instead uses the thermal lability of a C–S bond to provide the insertion of monomer units [39]. A general scheme of the monomer addition/fragmentation step is shown in Scheme 1.3. RAFT is used for the polymerization of methacrylates, styrenes and acrylates, and can also be successfully applied to narrow polydispersity BCPs. Interestingly, in species with dithiocarbamate end groups, such as tetraethyldithiuram disulfide, Otsu and coworkers found that the C–S bond could photochemically dissociate, offering the possibility of initiating polymerizations purely with UV light [40]. This technique was subsequently used to produce several types of BCP [41–45]. The thiocarbonyl end group can be removed by aminolysis or reduction with tri-n-butylstannane to leave a saturated chain end, or by thermal treatment to leave an unsaturated chain end. It may also be functionalized with amino or carboxy-functionalized end groups [46].

1.1.6.4 Atom Transfer Radical Polymerization

Atom transfer radical polymerization (ATRP) is another rapidly maturing technology that easily allows the production of end-functionalized and low-polydispersity polymers. It has also been shown to be a highly versatile reaction for the production of a wide variety of polymer architectures such as stars, combs, and tapered BCPs [47]. The mechanism (Scheme 1.4) functions in similar manner to typical

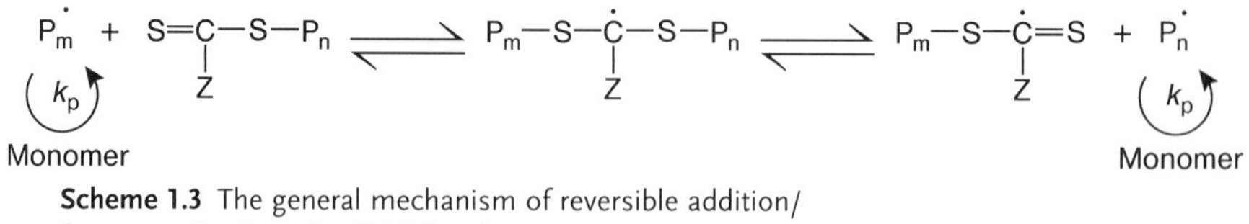

$$P_m^{\bullet} + S{=}C(Z){-}S{-}P_n \; \underset{}{\overset{}{\rightleftharpoons}} \; P_m{-}S{-}\overset{\bullet}{C}(Z){-}S{-}P_n \; \underset{}{\overset{}{\rightleftharpoons}} \; P_m{-}S{-}C(Z){=}S + P_n^{\bullet}$$

(with k_p, Monomer on both sides)

Scheme 1.3 The general mechanism of reversible addition/fragmentation/transfer (RAFT) polymerization.

$$R{-}X + Mt^n{-}Y/\text{ligand} \; \underset{k_{deact}}{\overset{k_{act}}{\rightleftharpoons}} \; \overset{\bullet}{R} \; (k_p, \text{Monomer}) + X{-}Mt^{n+1}{-}Y/\text{ligand}$$

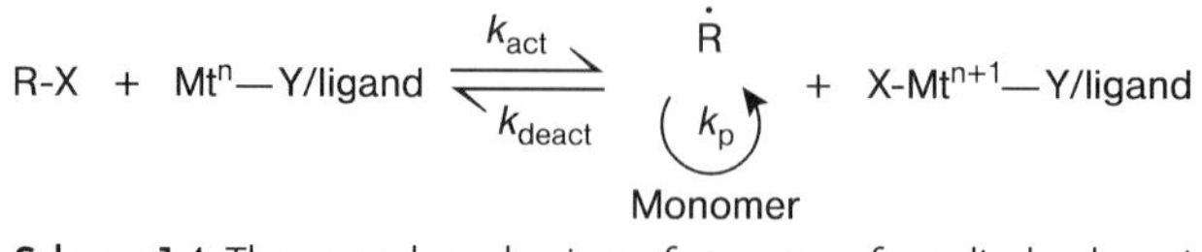

Scheme 1.4 The general mechanism of atom-transfer radical polymerization (ATRP).

living free radical processes, except that the active radical species undergoes a reversible redox process that is catalyzed by a transition metal complex attached to an amine-based ligand. The main disadvantage of ATRP is that these transition metals are difficult (if not impossible) to remove completely from the polymer after polymerization. ATRP has a wide range of monomers available for synthesis, however, including (meth)acrylates, (meth)acrylamides, styrenes, and acrylonitriles. Initiators for the process are usually alkyl halide species (R–X), and their presence at the end of the polymer chain allows for easy substitution reactions with functional groups.

1.1.6.5 Ring-Opening Metathesis Polymerization

Ring-opening metathesis polymerization (ROMP) is typically used for the ring-opening polymerization of cyclic olefins such as norbornenes and cyclooctadiene [48]. A general mechanism is presented in Scheme 1.5. ROMP also uses a metal catalyst that is usually composed of titanium, tungsten, or ruthenium attached to an aluminum ligand. Based on the results obtained by Robert Grubbs and coworkers, a selection of functional group-tolerant ruthenium catalysts has been synthesized, opening up new opportunities for structurally diverse BCPs, such as amphiphilic copolymers [49] used to coat chromatographic supports, water-soluble/conducting self-assembling materials [50], and flourescent BCPs for use in light-emitting devices [51].

1.1.6.6 Group Transfer Polymerization

Group transfer polymerization (GTP) is best suited for the polymerization of methacrylate and acrylate polymers [52]. A general mechanism is shown in Scheme 1.6. Esters, nitriles, styrenes, butadienes, isoprenes, and most other α,β-unsaturated

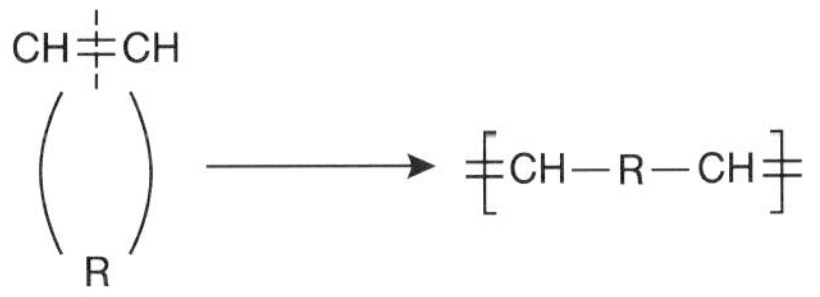

Scheme 1.5 The general mechanism of ring-opening metathesis polymerization (ROMP).

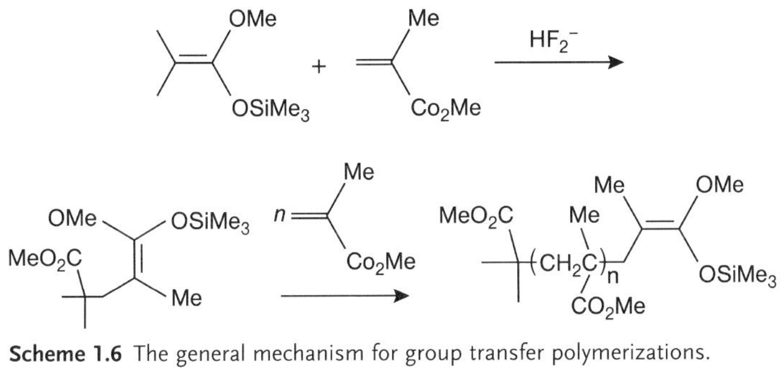

Scheme 1.6 The general mechanism for group transfer polymerizations.

compounds can also be prepared [53]. One interesting monomer that is typically prepared by GTP is poly(2-(dimethylamino)ethylmethacrylate) (PDAEMA). When polymerized with a hydrophobic methacrylate species, Billingham and coworkers found that the resulting amphiphilic BCP would easily form micelles in aqueous solution due to the water solubility of the PDAEMA block [54, 55]. Initiators often include silyl ketene acetal-type structures. Trace amounts of nucleophilic catalysts such as $TASHF_2$ are necessary to activate the silicon catalyst, along with large amounts of Lewis acids such as ZnX_2 (X = Cl, Br, I) to activate the monomer. A key advantage of GTP is that it can be performed at room temperature. Moreover, functionalized polymers can easily be added by the use of either: (i) a functionalized initiator or end-capping agents for functional groups attached to the end of the chain; or (ii) a functionalized monomer for functional groups evenly distributed throughout the polymer chain.

1.1.7
Post-Polymerization Modifications

Today, living polymerization techniques are available for a wide range of monomer types, and the possibilities are expanding daily. However, alternative routes are still necessary for the preparation of BCPs with highly specialized solubilities and functionalities, and this often requires post-polymerization modification steps such as active-center transformations and polymer-analogous reactions.

1.1.7.1 Active-Center Transformations

Often, one type of polymerization mechanism may not be suitable for both types of monomer used in the BCP. In this case, following formation of the first block, it is possible to alter the polymerization mechanism to suit the efficient addition of a second monomer to the chain. The active center can be modified either by *in situ* reactions or by isolation of the first block, followed by chemical transformation of the active center with a separate reaction; the polymerization can then continue after addition of the second monomer. For example, an SFRP mechanism can be transformed into an anionic ring-opening system for the polymerization of poly(styrene-*block*-ethylene oxide). First, the styrene undergoes SFRP in the presence of mercaptoethanol, a chain-transfer agent. The hydroxyl functionalized PS is then used as a "macroinitiator" for the anionic ring-opening polymerization of ethylene oxide [56]. Active center transformation has been used for the formation of poly(norbornene-*block*-vinylalcohol) BCPs through a combination of ROMP and aldol GTP [57], while a combination of cationic (not discussed) and anionic procedures have been used to polymerize poly(isobutylene-*block*- methyl methacrylate) BCPs [58]. Finally, each of the above mechanisms can also be transformed into an ATRP process, as described in a review by Matyjaszewski [59].

1.1.7.2 Polymer-Analogous Reactions

As we have seen, the creation of BCPs through living polymerization mechanisms restricts the number of monomers available for use. Additionally, functionalized polymers feature delicate protecting groups that may be unsuitable for the highly reactive initiators used in living polymerizations. Polymer-analogous reactions can create copolymers that could not have been synthesized within a living polymerization. These modifications are carried out on previously synthesized, or "precursor" BCPs such as poly(styrene-*block*-isoprene) with known molecular weights and narrow molecular weight distributions. Chemical transformation of the precursor polymer can be carried out selectively on individual blocks, on the entire copolymer, or on each block in sequence. Here, careful selection of the reaction conditions is vital to avoid any harmful side reactions such as degradation or crosslinking of the original polymer. If carried out successfully, the degree of polymerization, molecular weight distribution, and main chain architecture of the precursor polymer will remain the same, but the solubility and physical properties of the polymer may be altered completely. Examples of polymer analogous reactions include hydrogenation [60], epoxidation [61], hydrolysis [62], sulfonation [63], hydroboration/oxidation [64, 65], quaternization [66], hydrosilylation [67], and chloro/bromomethylation [68, 69]. Further details for each of these reactions is also available [71].

In this section, we have set the foundation for understanding how the chemistry of block copolymeric materials relates to the physics of their unique self-assembling properties. In the remainder of the chapter, it will become clear how these novel chemical strategies are used to effect practical physical applications.

1.2
Block Copolymers as Lithographic Materials

1.2.1
Introduction to Lithography

A modern integrated circuit is a complex, 3-D network of patterned wires, vias, insulators, and conductors. In order to transfer these patterns onto the silicon substrate it is first necessary to write the pattern, and for this the technique of photolithography is used, in which radiation-sensitive polymers play an integral role. Standard photolithography consists of essentially two steps: (i) writing of the pattern into a radiation-sensitive polymer thin film (resist); and (ii) transferring the pattern by etching into the underlying substrate [72]. A schematic of the traditional lithographic process is shown in Figure 1.7a. As photolithography and its associated technologies will become a recurring theme in this chapter, a brief primer on the subject will be useful at this point.

During the exposure process of a polymeric photoresist, UV radiation passes through a quartz photo mask that only allows a particular pattern of radiation to pass through and strike a thin film of photoresist. A chemical change is effected in the exposed regions of the polymer, which allows its subsequent development in a solvent, similar to the way in which photographic film is exposed and developed in a darkroom. The mechanism of the chemical change involves the forma-

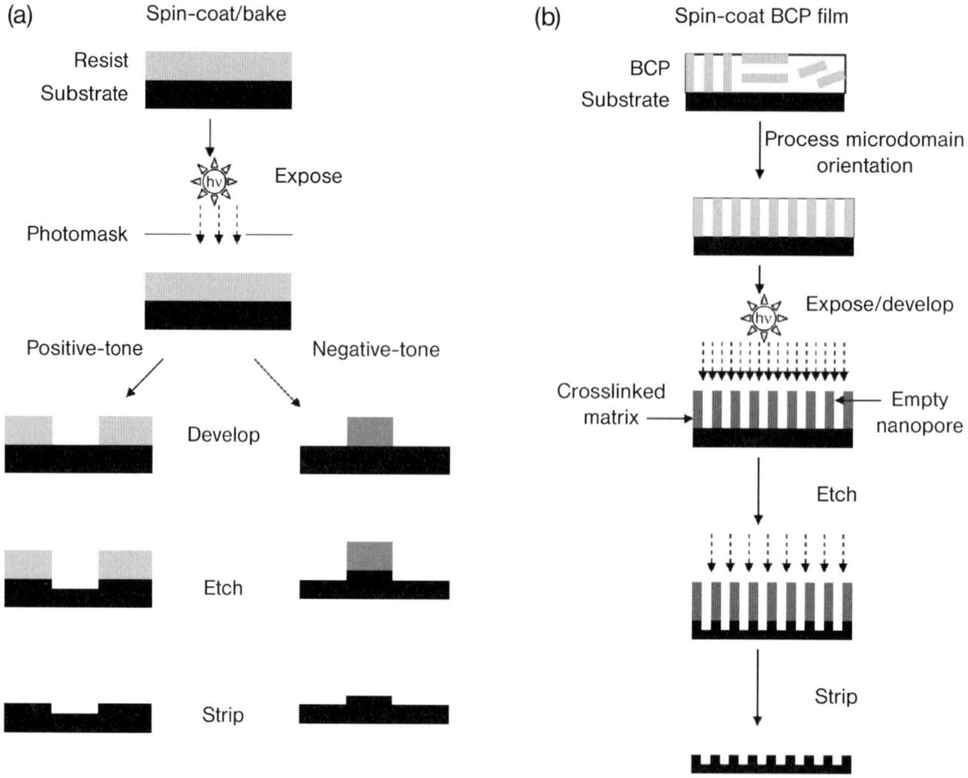

Figure 1.7 (a) Schematic of a traditional photolithography using positive-tone or negative-tone resist chemistry. (b) Schematic of block copolymer lithography, using a combination of positive-tone and negative-tone resist chemistry.

tion, rearrangement or breaking of bonds within the polymer chain. A resist's *sensitivity* is a measure of how efficiently it responds to a given amount of radiation, and might be compared to the ASA or ISO rating of a photographic film. A resist with a higher sensitivity will allow a satisfactory image to be produced for a smaller absorbed *dose* of radiation. The *resolution* of a photoresist is the size of the smallest structure that can be cleanly resolved after pattern development using standard microscopy techniques. This structure will then be used to efficiently pattern into the underlying substrate, assuming that the photoresist has sufficient *etch resistance* to withstand the harsh pattern transfer step. In this step, the patterned resist must withstand high-energy plasma sources that are designed to etch into silicon wafers. Polymers containing more carbon have stronger dry-etch resistance than those with lower amounts of carbon, whilst polymers containing high amounts of oxygen are etched easily. Even etch-resistant polymers may lose a small amount of material during the etching step, but maintaining the structure of the pattern is important for high resolution. If less than a ±10% change in the finest feature size of the polymer is etched into the substrate, the pattern transfer step is deemed successful [73]. This presents a huge challenge for the resist designer, who must tune the chemical components of the photoresist to satisfy two diametrically opposed requirements: to design a resist that is very responsive to ultraviolet radiation, but which, after the initial exposure, becomes highly resistant to the specific type of radiation and heat involved in the pattern transfer step. This is often achieved through the copolymerization of more than one type of monomer, each having its own functionality in the photoresist.

The realm of photolithography is split into two families of chemistry, based on the different physical properties possible for the exposed photoresist:

- *Positive-tone* resist chemistry refers to a photoresist that becomes *more soluble* after exposure to UV light. This can happen because of chemical deprotection, bond rearrangement, or chain-scission mechanisms.

- *Negative-tone* resist chemistry refers to a photoresist that becomes *less soluble* through the formation of crosslinked networks after exposure to UV light.

In this chapter, we will highlight the many approaches that have been used for one or both of these types of chemistries, in order to fabricate functional nanoscale-sized structures using BCPs.

1.2.2
Block Copolymers as Nanolithographic Templates

The lithography community has been extremely successful in its ability to pack progressively more circuit elements into a chip, as governed by the benchmark pace first predicted by Gordon Moore in 1989, which states that the transistor density of semiconductor chips will double roughly every 18 months [74]. Since then, new technologies have enabled this march down to smaller feature sizes. Photoresists with smaller pixel sizes such as molecular glass resists [75, 76], new processes such as nanoimprint lithography and step and flash lithography [77], as well as the development of smaller wavelength exposure sources [78], have catalyzed the production of feature sizes down below 50 nm.

The exposure wavelength, however, has become the rate-determining step in our ability to pattern small feature sizes. Extremely small wavelength sources such as electron beams (e-beams) and X-rays do not have wavelength limitations, but e-beam systems can only write features in a slow serial process that is not amenable to large-scale commercial processes. The high cost of the incorporation of these tools into a clean room is another disadvantage to their use. X-ray lithography requires the extremely high power of synchrotron sources, or electron storage rings, which are found in only a handful of locations around the world. Both, therefore, have proven to be impractical in a production setting. It seems that extreme ultraviolet (EUV) radiation sources can carry the lithography community down to 20 nm structures, but beyond that the semiconductor industry sees a need for innovative patterning strategies [79].

Design for modern integrated circuits usually requires a motif of several of different types of feature that may include – but are not limited to – regular patterns of straight and jogged lines and spaces, circular holes for contact openings, T-junctions, and columns of ferromagnetic media for data storage and memory applications. Interestingly, the shape of these features relate very well to the various geometries involved in BCP self-assembly (refer back to Figure 1.2). Furthermore, BCP microdomains form features that are much smaller than the current state-of-the-art standard photolithographic techniques. If one of the blocks in a BCP could be selectively degraded (i.e., positive tone chemistry), while the other block is crosslinked (i.e., negative tone chemistry) or has sufficient etch resistance, the result would be an ordered, nanoscale "stencil" that could be used to pattern circuit elements into a substrate. Theoretically, a density of over 10^{11} elements per square centimeter could be achieved over a large area with this technique [80]. Starting with the pioneering work of Lee in 1989 [81], the field of BCP lithography has exploded during the past ten years. Several excellent reviews on the subject are available [82–84].

Several obstacles stand in the way of BCP lithography becoming an industrially useful technology. First, when a BCP thin film is created, the microdomains will often seem disordered, appearing in randomly oriented grains along the sample surface. Long-range ordering of the BCP nanodomains is often necessary for a few of the possible applications of BCP lithography, such as in the creation of addressable, high-density information storage media. Second, the nanodomains tend to arrange parallel to the sample surface due to preferred interfacial interactions between one of the blocks and the substrate. However, in order to be lithographically useful as etch masks that are able to transfer patterns into a wafer, nanodomains such as lamellae and cylinders must be arranged perpendicularly so that they are physically and continuously connected from the polymer/air interface through to the substrate, as shown in the plan-view and cross-sectional scanning electron microscopy (SEM) images shown in Figure 1.8. Third, the thickness of the BCP must be carefully controlled. It has been shown that, for a lamellar PS-*b*-PMMA BCP, if the initial film thickness t is thicker than the natural period of the lamellae L_o and $t \neq (n + \frac{1}{2}) L_o$ (n an integer), then islands or holes of height L_o will form at the surface of the film and damage the homogeneity of the surface morphology. In the case where $t < L_o$, the situation becomes more complicated, as the lamellar chains will arrange perpendicular to the substrate due to the large entropic penalty associated with the chains having to compress to fit into the parallel orientation [85]. Other hybrid morphologies (Figure 1.9) have been found as the film thickness changes, due to the competition of several forces such as strong surface interactions, slow kinetics, and the thermodynamic driving force to arrange in layers commensurate with the height L_o. The physical complexity of block copolymer systems is staggering, and orientational control over these systems has developed into a field of its own [86]. Many research groups have achieved success in controlling the orientation of

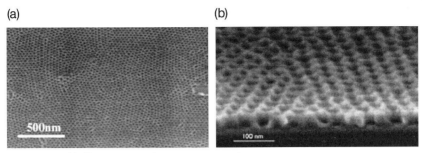

Figure 1.8 Field-effect scanning electron microscopy images obtained from a thin film of PS-*b*-PMMA after removal of the PMMA block. (a) Top view of the film; (b) A cross-sectional view. Reprinted with permission from Ref. [90]; © 2006, Wiley-VCH.

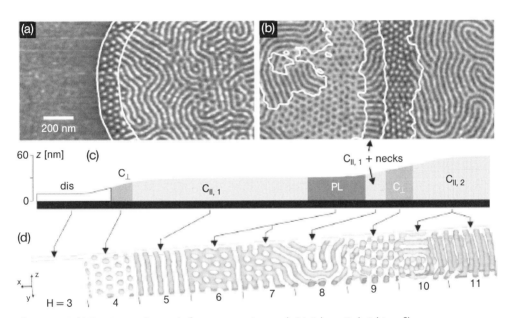

Figure 1.9 (a,b) Tapping-mode atomic force microscopy phase images of thin poly(styrene-*block*-butadiene-block-styrene) (PS-*b*-PB-*b*-PS) films on Si substrates after annealing in chloroform vapor. The surface is covered with an 10 nm-thick PB layer. Bright regions correspond to PS microdomains below the darker top PB layer. Contour lines calculated from the corresponding height images are superimposed; (c) Schematic height profile of the phase images shown in (a, b); (d) Simulation of a block copolymer film in one large simulation box with increasing film thickness. Reprinted with permission from Knoll A., Horvat, A., Lyakhova, K.S. et al. (2002) *Phys. Rev. Lett.*, **89**, 035501-1; © 2006, American Physical Society.

the self-assembled structures through special processing strategies; these include techniques such as thermal annealing [87], electric fields [88–90], mechanical shear [91–93], exposure to solvent vapor [18, 94, 95], physical confinement [96], neutral surfaces [97, 98], chemically nanopatterned surfaces [99–101], or a combination of any of the above [102].

1.2.2.1 Creation of Nanoporous Block Copolymer Templates

Once a high degree of long-range ordering has been achieved on a BCP thin film, a final processing step is necessary to create the nanolithographic template. A schematic of BCP lithography is shown in Figure 1.7b. Upon exposure to UV, chemical, or reactive ion etching (RIE), these systems are designed so that one of the blocks will be selectively degraded relative to the other block(s). In fact, the second block should ideally become photochemically crosslinked and thus highly immobile during the subsequent pattern transfer step in order to avoid distortion of the photo pattern. Both "wet" and "dry" chemical processes can be used during the pattern transfer step. Wet chemical etching involves the dissolution of the first block in an aggressive acid or base solvent, whereas "dry" chemical processes refer to exposure of the film to high-energy reactive ions and plasmas such as CF_4, O_2, SF_6, Cl_2, or argon gas. The etching process results from the combination of the kinetic energy of the ions (causing sputtering) and ion-induced chemical reactions that create volatile byproducts. The etching process affects polymers to different extents, depending on the chemical composition of the block. Several different types of BCP systems have been used as nanolithographic templates, such as poly(styrene-*block*-butadiene), poly(styrene-*block*-methyl methacrylate), poly(styrene-*block*- ferrocenyldimethylsilane), poly(styrene-*block*-lactic acid) and poly(α-methylstyrene-*block*-hydroxystyrene). The structures of the most commonly used sacrificial blocks are listed in Table 1.2.

Poly(Styrene-block-Butadiene) One of the first applications in this area also provides a model example of the concept of BCP lithography. Chaikin and coworkers [80, 111] created a thin film of microphase-separated poly(styrene-*block*-butadiene) (PS-*b*-PB), as shown in Figure 1.10a. Figure 1.10b shows how ozone was used to eliminate the PB spherical minority phase and open up windows in the PS matrix. In this instance, the minority PB block acts as a positive-tone resist due to its vulnerability to ozone chemical attack. The resulting spherical pores in the film provided less RIE resistance than the continuous PS matrix, thus creating a periodic array of 20 nm holes spaced 40 nm apart on a silicon nitride substrate. In the opposite strategy, illustrated in Figure 1.10c, the PB block was stained with osmium tetroxide vapor, which caused the PB block to exhibit a greater etch resistance than the PS block. In this case, the PB block acts like a negative-tone resist, resulting in removal of the PS matrix after RIE with CF_4. Therefore, the negative-tone system creates a pattern of raised dots on the substrate, instead of holes. This concept can easily be extended to create nanosized metal dots on any type of substrate for high-density information storage applications [127], or substrates such

Table 1.2 Commonly used sacrificial blocks for block copolymer templates.

Name	Structure	Method of removal	Reference(s)
Poly(methyl methacrylate)		Photolysis	[103–107]
		Etch selectivity	[108–110]
Poly(butadiene)		Ozonolysis	[80, 111, 112]
Poly(isoprene)		Reactive ion etching	[113, 114]
		Ozonolysis	[14, 111, 115]
Poly(α-methylstyrene)		Heat/Vacuum	[116, 117]
Poly(L-lactide)		Aqueous base dissolution	[118–121]
Poly(4-vinyl pyridine)		Reactive ion etching	[96]
Poly(ethylene oxide)		Water dissolution	[122]

Table 1.2 Continued.

Name	Structure	Method of removal	Reference(s)
Polystyrene	(polystyrene structure)	Reactive ion etching	[123, 124]
Poly(perfluorooctyl ethyl methacrylate)	(poly(perfluorooctyl ethyl methacrylate) structure with CH$_3$, CH$_2$, O, (CF$_2$)$_7$, CF$_3$)	Reactive ion etching	[125, 126]

as gallium arsenide can be patterned, opening up new avenues for the production of quantum dot (QD) structures [128].

Poly(Styrene-block-Methyl Methacrylate) Poly(styrene-*block*-methyl methacrylate) (PS-*b*-PMMA) has been the workhorse of the field of BCP lithography for several reasons. First, it is relatively simple to produce using anionic polymerization techniques, and is commercially available [129]. Second, PMMA acts as a readily degradable positive-tone resist on exposure to deep ultraviolet (DUV) or e-beam radiation. In fact, PMMA is already well established in the semiconductor industry as a positive-tone e-beam resist. The PMMA chain breaks up into oligomers through a chain scission mechanism (Scheme 1.7), and can then be removed from the matrix through dissolution in acetic acid. Third, in the same DUV exposure step, the polystyrene matrix acts as a weak negative-tone resist, becoming photochemically crosslinked through oxidative coupling, as shown in Scheme 1.8 [130]. Immobilizing the matrix phase through crosslinking strategies is very important in the creation of nanoporous materials. The huge increase in surface area that results from the removal of the minority domain creates a concomitant increase in surface free energy. A driving force for the minimization of this free energy creates a strong tendency for the nanopores to collapse, which would result in distorted etched patterns. A high glass transition temperature (T_g) relative to the processing temperature represents another means of stabilizing the nanoporous

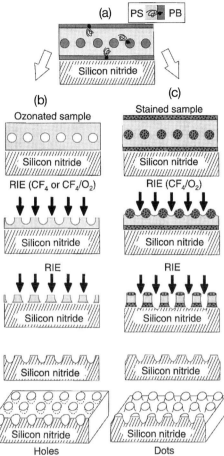

Figure 1.10 (a) Schematic cross-sectional view of a nanolithographic template consisting of a uniform monolayer of PB spherical microdomains on silicon nitride. PB wets the air and substrate interfaces; (b) Schematic of the processing flow when an ozonated copolymer film is used, which produces holes in silicon nitride; (c) Schematic of the processing flow when an osmium-stained copolymer film is used, which produces dots in silicon nitride. Reprinted with permission from Ref. [80]; © 1997, American Association for the Advancement of Science.

template; however, there must be a clear pathway in the film for the removal of degraded minority component (e.g., vertical cylinders). Otherwise, the low-molecular-weight products permeate through the matrix, leading to a decrease in T_g and collapse of the pores.

Stoykovich and colleagues in the Nealey group have substantially improved the directed self-assembly of lamellar microdomains of PS-*b*-PMMA. Traditional lithographic techniques were used to create a chemically nanopatterned surface that preferentially wets the PMMA domain. By using this preferential attraction,

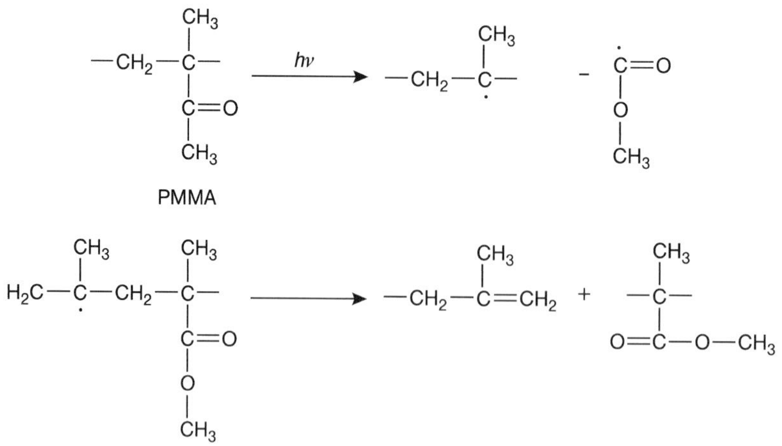

Scheme 1.7 Chain scission mechanism of PMMA.

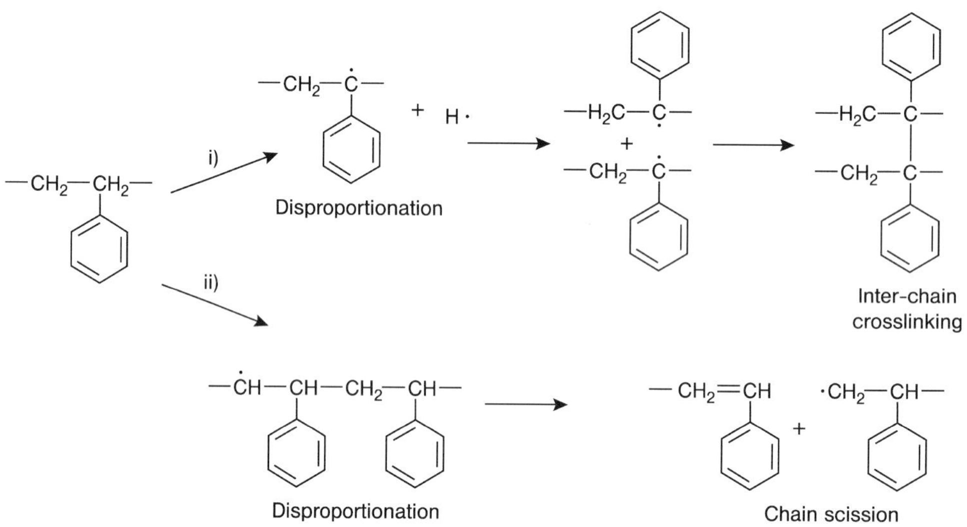

Scheme 1.8 Ultraviolet irradiation byproducts of polystyrene.

Stoykovich et al. were able to steer the vertically oriented lamellar morphologies through various bend angles, from 45° to 135° (Figure 1.11). The high curvature of these patterns induces a great deal of stress in the in the polymeric material, and leads to the formation of defects in the structure. It was found that, by blending small amounts of PMMA homopolymer, the homopolymer selectively swelled the PMMA block in the areas of high curvature (Figure 1.11b); this alleviated the stress in the material and in turn reduced the pattern defects [131].

Several successful applications of PS-b-PMMA as a nanolithographic template have already been achieved, with C. T. Black and coworkers at IBM being among the first to demonstrate the industrial feasibility of this technology. After optimization of the process window for maximum ordering of perpendicular-oriented nanodomains [132, 133], Black's group demonstrated the successful fabrication of metal-oxide-silicon (MOS) capacitors (Figure 1.12) [133–135], multinanowire

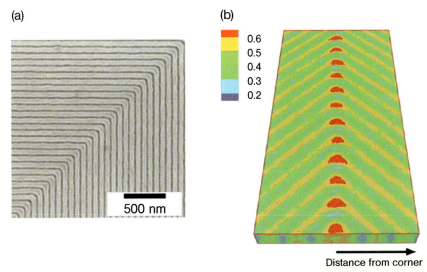

Figure 1.11 (a) A top-down SEM image of angled lamellae in a ternary PS-b-PMMA/PS/PMMA blend. The chemical surface pattern is fabricated with a 70 nm line spacing to match the natural period of the copolymer; (b) Theoretical concentration map of the distribution of the homopolymers on the surface. The homopolymers concentrate and swell the polymer at the bend area of the patterns to prevent the formation of defects. Reprinted with permission from Ref. [131]; © 2006, American Association for the Advancement of Science.

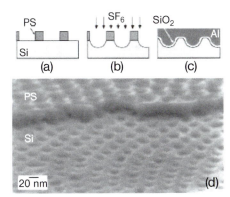

Figure 1.12 The process flow for MOS capacitors. (a) Polymer template formation on silicon surface; (b) RIE pattern transfer of PS template into silicon, followed by removal of the PS matrix; (c) SiO_2 growth followed by top Al gate electrode deposition; (d) SEM image at a 70° tilt after RIE etch. The remaining PS template is shown at the top. At the bottom, the nanoscale hexagonal array has been transferred into a Si counter electrode. Reprinted with permission from Ref. [134]; © 2006, American Institute of Physics.

Figure 1.13 Cross-sectional transmission electron microscopy image obtained from a PS-*b*-PMMA film (800 nm) annealed in an electric field of 25 V µm^{-1}. The block copolymer film is lying on top of a dark Au-film, which was used as the lower electrode. The upper electrode has been removed. Cylinders oriented normal to the substrate pass all the way through the sample. Reprinted with permission from Ref. [90]; © 2006, Wiley-VCH.

silicon field effect transistors [136], and FLASH memory devices [137]. The creation of high-aspect ratio patterns has always been problematic due to the difficulty of achieving a single microdomain orientation in thick BCP films. In this area, Thurn-Albrecht and coworkers reported that the application of a strong electric field, in combination with thermal annealing, creates 500 nm-long, vertically oriented PMMA cylinders that physically connect to the substrate, as shown in Figure 1.13. After PMMA removal, the underlying conducting substrate was used for the subsequent deposition of copper into the holes to form a matrix of continuous nanowires [90].

Asakawa and coworkers from the Toshiba Corporation were the first to pattern magnetic media for hard disk applications, by using a PS-*b*-PMMA template. In these investigations, the group took advantage of the large difference in etch resistance between the aromatic and acrylic polymer to produce the BCP template using dry etching techniques [108, 109]. Spiral-shaped circumferential grooves were imprinted into a hard-baked photoresist using a nickel master plate. The spherical PMMA microdomains then aligned within the walls of the grooves, and the PMMA was preferentially etched by oxygen plasma to create holes which connected to an underlying magnetic cobalt platinum film. The size of these holes could be adjusted by changing the molecular weight of the PS-*b*-PMMA. The holes were then filled with etch-resistant spin-on-glass, which acted as a mask while the remaining PS polymer and the underlying magnetic media were patterned by ion milling. After removal of the spin-on glass, the disk featured magnetic nanodots arranged in a spiral pattern.

PS-*b*-PMMA has proved to be an excellent system to perform studies on the ability of BCPs to act as nanolithographic stencils, although doubts persist regarding its potential to enter into industrial, high-volume production. For example, as noted previously, crosslinking the matrix phase of the BCP is necessary to prevent pore collapse during the rough pattern transfer and etching step. PS, however, cannot be crosslinked (or patterned) efficiently upon exposure to UV light. A combination of different photochemical processes, such as random chain scission, oxidative coupling and crosslinking, all occur at the same time during UV exposure, as shown previously in Scheme 1.8. Several groups have taken steps to correct this problem, with Hawker and coworkers [103] having randomly copolymerized thermally crosslinkable benzocyclobutene (BCB) groups with polystyrene (PS-*ran*-BCB). The degree of crosslinking can be tuned by increasing the amount of the BCB in the random copolymer. Microdomain ordering was induced by thermally annealing the matrix at 160 °C. Raising the temperature to 220 °C caused the matrix to be crosslinked, and the PMMA was exposed and developed in the normal manner.

Poly(Styrene-block-Lactic Acid)   Poly(styrene-*block*-lactic acid) (PS-*b*-PLA) is another copolymer that has been developed as a nanolithographic template. To synthesize this polymer, hydroxyl-terminated PS (prepared through anionic polymerization) was treated with triethylaluminum to form the corresponding aluminum alkoxide macroinitiator. This species was able to efficiently polymerize D,L-lactide through a ring-opening process [118]. The advantage of using this BCP as a nanolithographic template is that PLA undergoes main-chain cleavage simply by soaking it in an aqueous methanol mixture containing sodium hydroxide at 65 °C. The PS matrix is not affected at all by this treatment. Zalusky and coworkers have also published a complete phase diagram and characterization of PS-*b*-PLA [119]. Leiston-Balanger and coworkers used the benzocyclobutene crosslinking strategy explained in Section 1.2.2.1.2 but with PLA as the minority component, thus eliminating the photoprocessing step [138]. The group noted that, if a thermally degradable minority component were to be used, then a robust nanoporous template could be produced by using a completely thermal process.

Poly(Styrene-block-Ferrocenyldimethylsilane)   BCPs containing organometallic elements can also function very well as nanolithographic etch masks, as shown by the studies of Thomas and coworkers [113, 114, 139]. Thomas' group was able to synthesize a BCP of poly(styrene-*block*-ferrocenyldimethylsilane) (PS-*b*-PFS) that organized into lamellar and cylindrical microdomains by using an anionic ring-opening polymerization. The PFS block contained elemental iron and silicon, which made it highly resistant to dry-etching processes due to the creation of iron and silicon oxides during oxygen etching. The resulting etching ratio between PS and PFS was estimated to be as high as 50:1. Thus, by etching through the polystyrene, the BCP patterns were transferred into the underlying substrate in only one step. By using this polymer as a masking layer, Thomas *et al.* further demonstrated that an array of cobalt single-domain magnetic particles could be created through a tri-level etching strategy (Figure 1.14). Moreover, the magnetic properties and thermal stability of the dots could be tuned simply by changing the copolymer composition and etch depth into the cobalt [123, 140].

Poly(α-Methylstyrene–block-Hydroxystyrene)   Studies conducted by the present authors' group have focused on poly(α-methylstyrene-*block*-hydroxystyrene) (Pα-MS-*b*-HOST) that allows not only an efficient crosslinking of the hydroxystyrene matrix phase but also the ability to lithographically pattern the nanoporous tem-

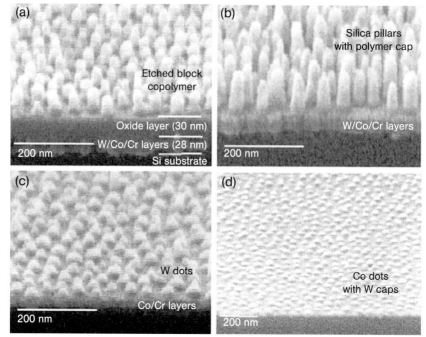

Figure 1.14 Tilted SEM images of the intermediate stages of lithographic processing. (a) An O$_2$-RIE-treated block copolymer thin film on a multilayer of silica, the metallic films, and the silicon substrate; (b) Pillars of silicon oxide capped with oxidized PFS after CHF$_3$-RIE; (c) Patterned tungsten film using CF$_4$ + O$_2$-RIE on top of a cobalt layer after removing the silica and residual polymer cap; (d) W-capped cobalt dot array produced by ion beam etching (note the different magnification). Reprinted with permission from Ref. [123]; © 2006, Wiley-VCH.

plates on select areas of the wafer [116, 117]. In these studies, traditional chemical amplification strategies [141], which have been recognized among the photoresist community for over 20 years, were used in combination with BCP lithography. An overview of the processing scheme is presented in Figure 1.15. The key to chemical amplification strategies is the use of cationic catalysts known as photoacid generators (PAGs); these produce a strong acid when exposed to DUV radiation (248 and 193 nm). Triflic acid generated by the PAG molecule (triphenylsulfonium triflate) was used to catalyze a condensation reaction [142] between the hydroxyl groups of the poly(hydroxystyrene) and a crosslinking species tetramethoxymethyl glycoluril (TMMGU; Powderlink 1174) to produce, in turn, a highly crosslinked network of the matrix phase (Scheme 1.9). The areas of the wafer where nanoporous templates are not needed are washed away with a solvent development step. The crosslinked regions of PHOST contained standing, 20 nm-diameter cylinders of poly(α-methylstyrene), a polymer that can be depolymerized and removed from the matrix with additional UV irradiation, heat,

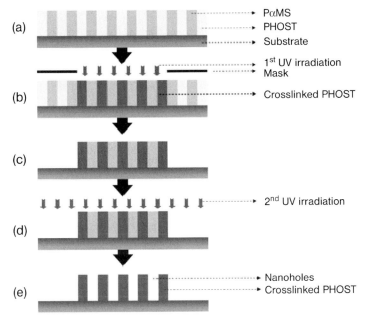

Figure 1.15 Novel nanofabrication process of obtaining spatially controlled nanopores. (a) Spin-coating of a PαMS-b-PHOST/PAG mixture onto a silicon wafer to form vertical cylinders of PαMS in the PHOST matrix; (b) Irradiate using a 248 nm stepper with a photomask and bake; (c) Develop with a mixed solvent to form micron-sized patterns on top of the substrate; (d) Irradiate using a 365 nm lamp under vacuum; (e) Form patterns with nanoporous channels. Reprinted with permission from Ref. [117]; © 2006, American Chemical Society.

and high vacuum. Thus, by using standard lithography procedures, it was possible to generate 450 nm resolution patterns of crosslinked PHOST containing 20 nm nanoporous substructures, as shown in Figure 1.16. Further processing of this polymer in order to maximize the degree of long-range ordering is currently under way [143]. It is believed that this combination of traditional "top-down" positive- and negative-tone lithography techniques, in combination with the power of "bottom-up" BCP self-assembly, holds much promise for the future of lithography.

1.2.3
Multilevel Resist Strategies Using Block Copolymers

The constant drive for smaller and smaller circuit device features with higher aspect ratios and more complex substrate topographies has caused many photolithographic engineers to rethink traditional single-level resist-processing strategies. As sub-100 nm feature sizes become the norm, artifacts arising from the lithographic process that were recently deemed insignificant have now become

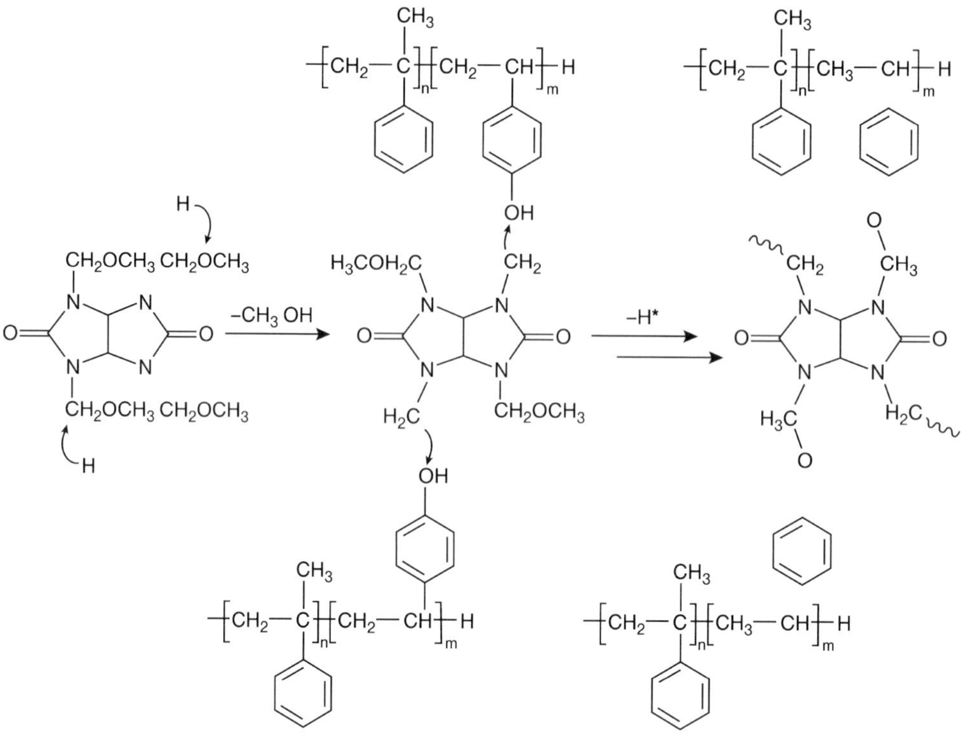

Scheme 1.9 Proposed crosslinking mechanism of PαMS-b-PHOST with TMMGU. After Roschert, H., Dammel, R., Eckes, C. et al. (1992), Proc. SPIE-Int. Soc. Opt. Eng., **1672**, 157.

major problems. One such artifact that limits resolution is the creation of the "standing wave effects" which occur when UV light reflects off the substrate surface after passing through the resist [73]. As a result, multilevel resist chemistries have been developed to incorporate polymeric planarizing layers to eliminate substrate topography variability and anti-reflection coatings to eliminate standing-wave patterns in photoresists. Commonly used multilevel strategies employ a photoresist imaging layer on top of the planarization layer. The resist used for the imaging layer is designed to provide high sensitivity to UV light exposure by providing a large number of photosensitive functional groups, and must be highly etch-resistant to the extremely harsh conditions imposed by the oxygen reactive ion-etching step. Organosilicon-containing polymers have been demonstrated to be excellent candidates as etch-resistant photoresists, due to their ability to form a protective SiO_2 ceramic upon exposure to oxygen plasma, as shown by Taylor and Wolf [144, 145]. The only disadvantage of these polymers is that their bulky side groups result in very low T_g-values, which make them susceptible to viscous flow and excess swelling during the development step. Photoresist engineers typically counter the problem of low T_g values by randomly copolymerizing the etch

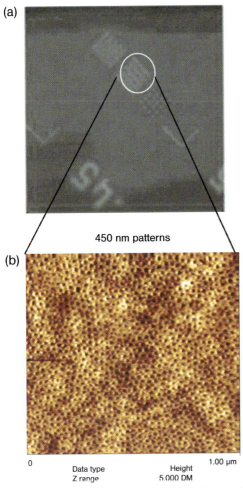

Figure 1.16 Atomic force microscopy height images showing: (a) the 450 nm photopattern produced through the crosslinking of the PHOST matrix; and (b) the 20 nm-diameter porous substructure generated after removal of the α-methylstyrene block.

resistant, low-T_g polymer with a photosensitive polymer that also has a high T_g. The copolymerization of the two monomers is necessary to avoid the inevitable macrophase separation that results when two homopolymers are mixed together. This composite style approach thereby simultaneously satisfies the requirements for high etch resistance, high sensitivity and high T_g, resulting in the best-performing photoresists.

BCPs seem to be a perfect fit for multilevel resist chemistries, due to their ability to segregate into chemically distinct levels and their intrinsic ability to avoid macrophase separation. It is well known that a BCP film will self-assemble such that the lower surface energy block is presented at the polymer/air interface. This principle can in turn be used to design layered systems of different functionality, as will be seen in the creation of semi-fluorinated BCPs for low-surface energy applications (see Section 1.6.1). For example, to create the imaging layer a polymer containing photosensitive *tert*-butyl functional groups may be used along with an etch-resistant organosilicon polymer (Figure 1.17). Conveniently, the *tert*-butyl groups contain three low-energy methyl groups that drive the segregation of the photoactive compound to the surface, where it is most effectively exposed to UV light. The surface-segregating properties are not limited to the *tert*-butyl group, however; many other protecting groups could be used. Moreover, the functional, chemically distinct polymers used in photoresists will exist in a microphase-separated state if they are incorporated into a BCP. The microdomains form with diameters within one radius of gyration of the polymer chain, which usually is about 5–30 nm. Therefore, a patterned lithographic feature on the BCP photoresist, currently approximately 100–200 nm, will contain numerous domains of the photosensitive and etch-resistant functionalities incorporated within it to ensure optimal performance of the photoresist.

The microphase separation of BCPs also means that each one of the polymer chains is confined to its own respective domain, and can be thought of as being artificially crosslinked. Thus, in negative-tone resists, less UV light is required to cause the polymers to become insoluble compared to a homopolymer of a similar molecular weight. BCPs have also been shown to be twice as sensitive to nonphase-separated random copolymers using the same monomer units and molecular weight, due to this confinement effect [146]. Confinement might also mean that the PAG moieties used in these resists might be clustering selectively inside one of the microdomains, and this effect might be magnified if the PAG is miscible in only one of the BCP microdomains [147]. Moreover, if the usually hydrophilic photoacid segregates inside the block containing the acid-labile functional groups, the effective concentration of the acid will be increased, which means that a higher percentage of the protecting groups will be converted to base-soluble –OH groups.

Hartney and coworkers [148] developed the first application for BCPs as bi-level e-beam resists in 1985, when they prepared a BCP of poly(chloromethylstyrene) for sensitivity and high T_g, and poly(dimethylsiloxane) (PDMS) that forms etch-resistant silicon oxide upon exposure to oxygen plasma and has a low T_g. Bowden *et al.* subsequently prepared a PDMS BCP grafted to PMMA that acted as a negative-tone resist [149, 150], while Jurek *et al.* created a novalac oligo-PDMS resist

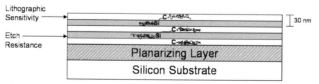

Figure 1.17 Multilevel resist strategy using block copolymers. Adapted from Ref. [148].

which showed resolutions of 500 nm and etch rates which were 36-fold slower than hard-baked novalac [151]. Gabor and coworkers attached a PDMS graft to the double bonds of an isoprene block in a styrene-*block*-isoprene BCP to create a negative-tone resist. In this way, Gabor et al. achieved line-space patterns of 200 nm using e-beam lithography, and demonstrated oxygen etching rates which were 42-fold slower than for polyimide [152]. Other resists which have incorporated PDMS for etch resistance in both block and graft copolymer architectures have also been reviewed [153].

Gabor and coworkers also prepared block and random copolymers of poly(tert-butyl methacrylate) and a silicon-containing methacrylate (poly(3-methacryloxy) propyl pentadimethyldisiloxane) (Figure 1.18) via group transfer polymerization for applications in 193 nm photolithography [154]. The BCP architecture allowed Gabor's group to incorporate a larger amount of the hydrophobic siloxane component for a high-oxygen RIE resistance, while maintaining solubility in an aqueous base developer. In fact, the BCPs were found to have a better development behavior in aqueous base than their random copolymer counterparts (Figure 1.19). Gabor et al. hypothesized that the exposed polymeric regions formed micelles in the aqueous base developer, with the silicon-containing block forming the core and the soluble methacrylic acid group forming the corona. In contrast, random copolymers would not have the ability to form these micellar structures; that is, the entire copolymer would need to be soluble in order for development to occur. However, the development behavior of these BCPs was far from perfect, presumably due to segregation of the PAG away from the surface of the film. Although their performance does not exceed commonly used industrial photoresists, the concept of using BCP architectures for multilevel photoresists is enticing, and there may be much more to learn from these types of systems.

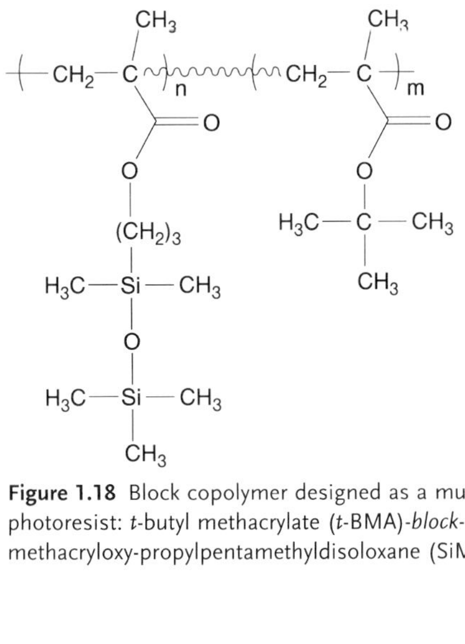

Figure 1.18 Block copolymer designed as a multilevel photoresist: *t*-butyl methacrylate (*t*-BMA)-*block*-3-methacryloxy-propylpentamethyldisoloxane (SiMA).

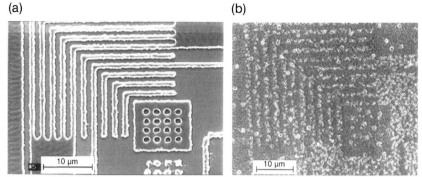

Figure 1.19 (a) A block copolymer photoresist poly(*tert*-butyl methacrylate-*block*-[3-(methacryloxy)propyl]pentamethyldisiloxane) with 10.0 wt% silicon, formulated with 5 wt% PAG. An exposure dose of 2 mJ cm^{-2} was used, and the resist was developed for 20 s in a mixture of 27% 0.21 N tetramethylammonium hydroxide (TMAH) and 73% iso-propyl alcohol; (b) Random copolymer with 9.7 wt% silicon, formulated with 5 wt% PAG, after 1.2 μm features are exposed to 19 mJ cm^{-2} and developed using the same conditions. While the random copolymer is insoluble, pinholes form during development and are thought to originate from phase-separated acid, which is removed by the developer. Reprinted with permission from Ref. [154]; © 2006, American Chemical Society.

1.3
Nanoporous Monoliths Using Block Copolymers

1.3.1
Structure Direction Using Block Copolymer Scaffolds

Besides the potential uses of BCPs as lithographic materials, their inherent microphase separation and chemical dissimilarity of the blocks also leads to other applications. For example, BCPs can serve as structure-directing agents for the formation of mesoporous monolithic materials through phase-selective chemistry. Monolithic nanoporous structures have a large surface:volume ratio, which means that there is a large amount of functionalizable surface area within a small volume of material. In one of the pioneering efforts in this area, Hashimoto and coworkers created a bicontinuous morphology with poly(styrene-*block*-isoprene) and selectively removed the minority isoprene domain with ozonolysis [115]. The group showed that coating the bulk structure with nickel metal does not completely block the nanopores. Rather, metal catalysts such as nickel are known for their ability to adsorb and split H_2 molecules for large-scale industrial reactions such as hydrogenation. The high surface:volume ratio of porous bicontinuous BCP morphologies make them excellent candidates for catalytic applications; indeed, one day they may even be used to make your margarine!

The BCPs used as structure-directing "scaffolds" to create mesoporous silicate structures could also be used as dielectric materials. Silicon dioxide (SiO_2) is typically used as a dielectric material to reduce undesired capacitive coupling between neighboring elements of an integrated circuit, due to its low dielectric constant (k)

of approximately 4.5. Air, however, has one of the lowest dielectric constants (~1); hence, the dielectric constant of SiO_2 can be dramatically decreased by filling it with voids, while maintaining good mechanical properties. So-called 'low-*k*' materials are highly desired by the semiconductor community because they allow faster switching speeds and lower heat dissipation in computer chipsets. Early studies conducted by Nakahama and coworkers led to the creation of monoliths of bicontinuous BCPs through a spin-coating process. This synthesis included a silyl-containing matrix block and an isoprene-based minority phase. Processing the film entailed hydrolytically crosslinking the silyl-containing block to prevent pore collapse, and ozonolysis to eliminate the isoprene minority domain [155]. Another group subsequently discovered a one-step, room-temperature UV irradiation/ozonolysis treatment to transform the matrix into a silicon oxycarbide ceramic and eliminate the polydiene minority phase. The silicon oxycarbide ceramic was stable at temperatures up to 400°C, and adjustment of the volume fraction of the BCP afforded an inverse bicontinuous phase to produce a nanorelief structure [14]. These mesoporous materials have also proved useful in the creation of photonic band gap materials [156], due to the possibility of tailoring the dielectric constant of the optical waveguides by sequestering optically active particles inside the matrix phase [157].

Watkins and coworkers have also demonstrated a novel technique to create mesoporous silicate structures by performing phase-selective chemistry inside one of the blocks [158]. In these studies, a tri-BCP of poly(ethylene oxide-*block*-propylene oxide-*block*-ethylene oxide) (PEO-*b*-PPO-*b*-PEO; also known as Pluronics®) was mixed with *p*-toluene sulfonic acid (pTSA) in an ethanol solution. Upon spin-casting the BCP onto a Si wafer, the BCP microphase separated into an ordered morphology containing spherical PPO microdomains. The pTSA catalyst segregated preferentially to the hydrophilic PEO matrix phase. The polymer was then placed in a chamber with humidified supercritical CO_2, so as to swell the polymer and allow the infiltration of a metal alkoxide, tetraethylorthosilicate (TEOS), into the polymer. The segregated acid in the hydrophilic domains then underwent a condensation reaction with the TEOS to form a silicon oxide network. Due to the phase selectivity of the acid segregation, no condensation reaction took place within the hydrophobic domains. The alcohol byproducts of the condensation reaction were quickly removed by the supercritical solvent, which rapidly pushed the condensation reaction to complete conversion. Finally, a calcination step in air at 400°C removed the organic block copolymer framework, leaving an inorganic silicon oxide replica of the original BCP (Figure 1.20). The process could also be carried out in standing cylindrical P(αMS-b-HOST) BCPs [159]. Eventually, the ability to pattern these monolithic silicate structures will lead to their use in future semiconductor fabrication paradigms.

Ulrich and coworkers reported the use of poly(isoprene-*block*-ethylene oxide) (PI-*b*-PEO) as a structure-directing agent for silica-type ceramic materials [160]. A mixture of prehydrolyzed (3-glycidyloxypropyl) trimethoxysilane (GLYMO) and aluminum *sec*-butoxide (Al(OBu)$_3$) was added to a solution of the PI-*b*-PEO and cast in a Petri dish. The Al(OBu)$_3$ triggered ring opening of the epoxy group that made the 3-glycidyloxypropyl ligand of the silane precursor compatible with the

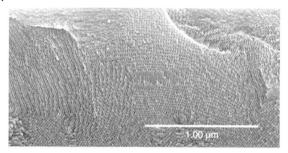

Figure 1.20 A scanning electron microscopy image showing the cross-section of a highly ordered mesoporous silicate film exhibiting a cylindrical morphology. The film was prepared by infusion and condensation of TEOS within a preorganized triblock PEO-*b*-PPO-*b*-PEO BCP film dilated with supercritical CO_2. The image reveals a preferential alignment of cylinders at the interfaces and grains of random orientation within the bulk of the film. Reprinted with permission from Ref. [158]; © 2006, American Association for the Advancement of Science.

PEO block [161]. Thus, addition of the inorganic material led to an increase in the volume fraction of the PEO block and the formation of multiple morphologies, depending on the concentration of the inorganic content in the solution. Spherical, cylindrical, lamellar and a novel type of bicontinuous morphology – called the "Plumber's Nightmare" morphology [162] – was found and characterized using transmission electron microscopy (TEM). Surprisingly, this phase did not occur in the neat PI-*b*-PEO BCP, indicating a radical transformation of the phase space through the addition of the inorganic content, possibly through a dramatic increase in the incompatibility (χ parameter) between the blocks. Additionally, calcination of the hybrid materials at 600 °C removed the organic phase and led to the creation of isolated nano-objects such as ceramic cylinders of the inorganic phases which ranged in size from 8.5 to 35 nm [160]. If combined with an organic ruthenium dye complex, these nano-objects could be used as fluorescent biomarkers in the field of nanobiotechnology [163].

1.3.2
Nanopore Size Tunability

Nanoporous BCP films can also be used as separation membranes or filtration devices [164, 165]. According to C.J. Hawker, "…the lateral density of pores in films prepared from block copolymers is nearly two times greater than that of aluminum oxide membranes and an order of magnitude greater than that of track-etched membranes" [82]. This increase in porosity corresponds to an ability to handle a higher flux of liquid, and thus a higher throughput filter. Furthermore, size-specific separation is possible because the nanopore size will be constant if a polymer with a low polydispersity is used. There is, however, a thermodynamic limit to how small the nanopores can be, this lower bound being based on the position of the order–disorder transition on the phase diagram relative to χN (refer back to Figure 1.2). Since small molecular weights are desired for smaller microdomains, a highly immiscible pair of polymers (corresponding to a greater χ value) is required to decrease their size. If χN falls below this lower bound, the copolymer will mix to form a single phase [166]. To date, structures below 12 nm have been difficult to achieve [167], although in order to tune the size of the nanopores to smaller dimensions, numerous processing methodologies have been developed.

Jeong and coworkers developed a method to alter nanopore size in PS-*b*-PMMA nanoporous films [168]. Their technique allowed the formation of two discrete sizes of nanopores, depending on the processing strategy used. A 10% blend of low-molecular-weight PMMA homopolymer into a PS-*b*-PMMA BCP resulted in the solubilization of the homopolymer into the center of the PMMA block. As the PMMA homopolymer was of a lower molecular weight than the PMMA block, it dissolved first when the film was washed with acetic acid (which is a selective solvent for PMMA). This led to the production of nanopores which were 6 nm in size, as determined by atomic force microscopy (AFM) measurements. DUV radiation, in combination with an acetic acid wash, allowed the removal of both the PMMA homopolymer and the PMMA block in the BCP matrix, which resulted in 22 nm-sized nanopores. Thus, two different size scales of nanoporous structures – 6 nm and 22 nm – were created using different processing schemes. Jeong *et al.* noted that the size of the smaller length-scale nanopores could be decreased if homopolymers of smaller molecular weights were used to infiltrate the PMMA block.

In an interesting approach to the preparation of nanoporous films, Ikkala and coworkers have incorporated alkylphenols, such as pentadecylphenol (PDP), into poly(styrene-*block*-4-vinylpyridine) (PS-*b*-P4VP) to create comb–coil supramolecular structures. Here, the phenol group of the PDP forms a strong hydrogen bond with the nitrogen donor on the pyridine group of the P4VP. The hydrogen bonding introduces an additional repulsive interaction that leads to the formation of "comb-like" layered structures of the PDP side chains inside the cylindrical P4VP microdomains. By using this approach, Ikkala *et al.* witnessed hierarchical structure formations such as lamellar-within-lamellar, lamellar-within cylinders, and lamellar-within-sphere morphologies [169, 170]. This was similar to the effects seen in liquid crystalline BCPs, except that the PDP side chains could easily be dissolved in methanol to create porous membranes [171].

The tunability of nanoporous monolithic structures has also been reported through the selective infusion of supercritical CO_2 inside the fluorinated block of poly(styrene-*block*- perfloro-octylethyl methacrylate) (PS-*b*-PFOMA) [125, 126]. Supercritical CO_2 has been shown to have a high affinity for fluorinated polymers [172]. After selective swelling of the CO_2 inside the PFOMA domain, quenching at 0 °C to lock in the PS matrix, and controlled depressurization at 0.5 MPa min^{-1}, nanocell formation was noted in the BCP film. It was also found that the nanocell diameter could be tuned from 10–30 nm by adjusting the saturation pressure of the CO_2 solvent, with low saturation pressures corresponding to small cell diameters, and *vice versa*. All of these approaches represent innovative means of decreasing the size scale of nanopores beyond that of the neat BCP.

1.3.3
Functionalized Nanoporous Surfaces

In nanoporous material, it may be very useful to control the chemical functionality of the nanopore wall for applications that require aqueous environments, such as microfluidic devices, water filtration, biocatalysis, or other chemical reactions that take place within the pore. In microfluidic devices, for example, compatibility between the substrate (pore wall) and the filler fluid (analyte) that passes through the nanoporous channels is necessary. Hydrophilic moieties such as hydroxyl groups attached to the pore wall will permit conduction of the aqueous solution through the membrane, as opposed to hydrophobic substrates that would hinder the flow of solution through the pores.

Several strategies are available for functionalization of the pore wall, three of which are shown in Figure 1.21. The first strategy (Figure 1.21a) is to introduce a functional group "spacer" between the covalent junction of the matrix and

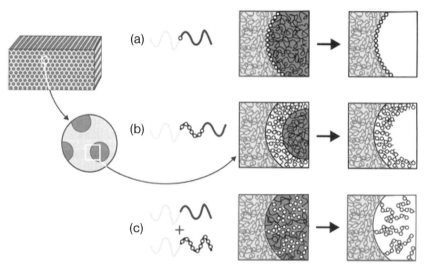

Figure 1.21 Possible routes to nanoporous materials with controlled pore-wall functionality. (a) A functional group is incorporated into the junction between the matrix and sacrificial blocks, and is exposed upon template degradation; (b) A functionalized mid-block is inserted between the matrix and sacrificial end block, producing a functional polymer brush at the pore wall upon removal of the template; (c) An AB/AC diblock copolymer blend is formed in which the common A block serves as the matrix, the B and C blocks are miscible, and only one of the two blocks is susceptible to degradative removal. In this manner, a functionalized nondegradable block can be introduced as a diffuse brush along the pore interior. Reproduced with permission from Ref. [176]; © 2006, Royal Society of Chemistry.

sacrificial blocks. This telechelic functionality can easily be introduced to the end of the living polymer chain during BCP synthesis, following addition of the first monomer and before addition of the second monomer. When the sacrificial block is removed through standard degradation procedures, the functional spacer will then line the pore wall. This approach has been demonstrated recently by Zalusky and coworkers [118, 119], Wolf and Hillmyer [120], and Hawker and coworkers [138]. Each of these groups made use of the hydroxyl-terminated poly(styrene) or poly(cyclohexylethylene) remaining when the PLA block had been removed. Degradation of the PLA left behind hydroxyl groups at the pore surfaces after removal, as proven by the reaction with trifluoroacetic anhydride and subsequent nuclear magnetic resonance (NMR) and infrared (IR) spectroscopic analyses. The only disadvantage to this relatively simple approach is the low density of functional groups that results, since only one –OH group exists per polymer chain. Zalusky et al. calculated the areal density of the hydroxyl groups to be approximately one –OH per $4\,nm^2$ for a sample with 22 nm-diameter pores [118].

Clearly, the solution to the low density of the functional groups would be to increase their number by polymerizing the functional unit, as shown in Figure 1.21b. In other words, a tri-BCP could be synthesized in which a functional center block (B) existed between the matrix (A) and the sacrificial (C) block. The success of this strategy, of course, would depend on the morphology of the tri-BCP, which can be difficult to predict. The ideal morphology, in this case, would be a hexagonally packed core–shell cylindrical morphology, with the functional block forming a shell around the sacrificial core, surrounded by a crosslinkable matrix. The first attempt at the formation of this morphology was demonstrated by Liu et al. with PI-b-PCEMA-b-PtBA (CEMA = cinnamoyloxyethyl methacrylate, tBA = tert-butyl acrylate) [173]. The tert-butyl group was removed by hydrolysis to form gas-permeable poly(acrylic acid) nanochannels. The research results of Liu and colleagues will be presented in greater detail in Section 1.4.

Recent investigations conducted by the group of Hillmyer have demonstrated great success in functionalizing nanoporous monoliths. A poly(styrene-block-polydimethylacrylamide-block-polylactide) (PS-b-PDMA-b-PLA) tri-BCP was synthesized by a combination of controlled ring-opening and free-radical polymerization techniques [174]. Aqueous base removal of the PLA minority phase and hydrolysis of the PDMA domain left carboxylic acid groups coating the pore wall, and these were used to chemically attach allylamine through carbodiimide coupling chemistry. ^{1}H NMR spectra of the dissolved polymer film verified a successful attachment of the N-allylamide group, and proved that the pore wall surface was chemically active. The same group was also successful in binding other functional groups such as a pyridine, a chiral hydroxyl, and an alkene in high yields. Other studies with PS-b-PI-b-PLA triblocks [175] by the Hillmyer group have produced an isoprene shell coating the pore wall following removal of the PLA phase by treatment with NaOH solution. These studies led to the production of a beautiful AFM image of a core–shell morphology (Figure 1.22) which was due to the differing mechanical contrasts between the PS matrix and PI shell. A tri-BCP containing

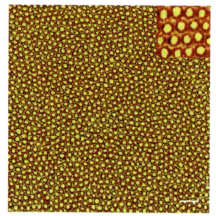

Figure 1.22 Tapping mode AFM phase image acquired from a PS-*b*-PI-*b*-PLA thin film. The scale bar at the lower right is 200 nm. The inset at upper right is 250 × 250 nm. Reprinted with permission from Ref. [175]; © 2006, American Chemical Society.

two selectively etchable blocks, such as PI and PLA, opens up the possibility of forming hollow "nanoring" donut-shaped structures through removal of the matrix and core-forming phases. The creation of coaxial nanowires would then be conceivable through the sequential electrodeposition of metals.

The third approach to add functionality to nanoporous BCP films, as shown in Figure 1.21c, is through the blending of two diblock copolymers. This idea is based on the use of two BCPs (AB and AC), in which the A blocks mix to form the matrix phase, and the B and C blocks must be selected to be miscible with one another to form a single cylindrical domain. If either one of the B or C blocks is degradable, it is possible to form nanopore walls coated with the nondegradable block. For example, if B is degraded by UV light, C will still be attached to its parent A molecules and will form a brush extending out towards the center of the nanopore containing the end-group functionality. This approach was championed in a recent investigation by Mao and coworkers, who blended a parent PS-*b*-PLA BCP with PS-*b*-PEO to form the blended PLA/PEO microdomains in a PS matrix [176]. The PLA was degraded through hydrolysis in an aqueous methanol/NaOH solution, leaving a PEO polymer brush (containing a hydroxyl end group) which extended into the nanopore. As a test to prove the success of their strategy, Mao *et al.* floated blended (PS-*b*-PEO/PS-*b*-PLA) and nonblended (PS-*b*-PLA) films on a water surface. The blended film proceeded to sink, which meant that the pores within the film had imbibed water. In contrast, the nonblended PS-b-PLA BCPs floated on the water surface for an indefinite period. Chemical functionalization strategies such as these will become increasingly important as the field of BCPs moves away from fundamental science and into the realm of practical applications.

1.4
Photo-Crosslinkable Nano-Objects

The design and fabrication of nanometer-sized structures with well-defined size and shape has recently aroused much interest, much of which has stemmed from the need for smaller electronic devices that are not possible to produce using conventional lithographic techniques. Indeed, small nano-objects could also be useful as biosensors capable of molecular recognition. A number of groups have emerged as front-runners in this field, with their use of photoactive BCP domains to fix unique microphase-separated structures. The Liu research group, for example, has been very prolific in this area, with the majority of their investigations lying in a variety of applications stemming from the unique photo-crosslinkable polymer, poly(2-cinnamoyloxyethyl methacrylate) (PCEMA). Through a dimerization process involving a [2 + 2] cycloaddition of two double bonds in neighboring chains (see Scheme 1.10), the PCEMA becomes photo-chemically crosslinked. An amphiphilic BCP containing PCEMA will form micelles when placed in a solvent that dissolves only one of the blocks. The shape and size of the micelles can then be changed by altering the relative chain lengths of the soluble and insoluble blocks, to produce spherical [177], cylindrical

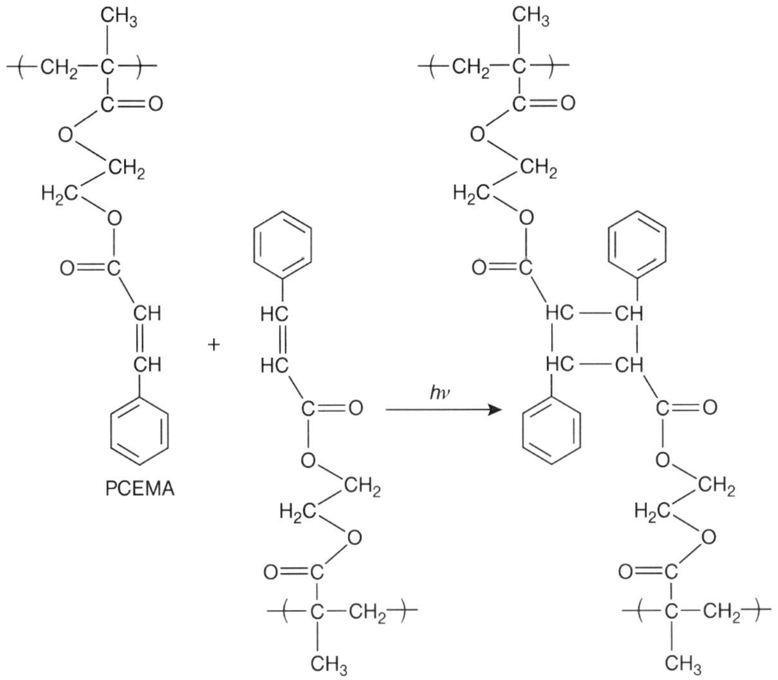

Scheme 1.10 Photocyclodimerization (photocyclo (2+2) addition) of poly(2-cinnamoylethyl methacrylate) (PCEMA).

[178, 179], vesicular [180], or donut-shaped [181] structures. These structures can then be fixed with UV light through the dimerization process to form permanent "nano-objects" that are stable in a wide variety of solvents. In an assortment of creative syntheses using this polymer, Liu's group has developed a wide range of structures, including nanotubes [182], porous [183], "shaved" and "hairy" nanospheres [184–187], nanospheres with crosslinked shells [188], crosslinked polymer brushes [189], nanofibers [184, 190–192], and nanochannels in polymer thin films [193, 194].

As an example of the power of their photocrosslinking strategy, nanotubes with hydrophilic nanochannels have been formed using the tri-BCP PbMA-b-PCEMA-b-PtBA (PbMA = poly(butyl methacrylate); see Figure 1.23) [195]. Bulk films (~0.2 mm thick) were produced by slow evaporation of the polymer in toluene. After an annealing step, the PtBA formed the hexagonally packed cylindrical core, surrounded by a shell of PCEMA in a PbMA matrix. The PCEMA is first photocrosslinked, after which the PtBA cylinders are hydrolyzed in methanol to yield nanochannels filled with poly(acrylic acid) (PAA) brushes. Subsequent dissolution of the PbMA matrix in tetrahydrofuran (THF) led to the production of free-floating, functionalized nanotubes. These rod-like nanotubes were also shown to exhibit LC properties in solution [196].

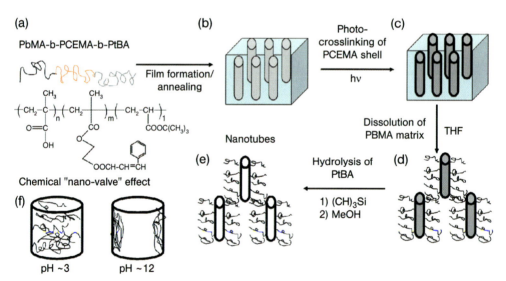

Figure 1.23 (a) A triblock copolymer of PbMA-b-PHEMA-TMS-b-PtBA was synthesized via anionic polymerization, deprotected, and then reacted with cinnamoyl chloride to form PbMA-b-PCEMA-b-PtBA; (b) Self-assembly of cylindrical morphologies developed upon spin-coating; (c) The PCEMA shells were crosslinked through a [2+2] cycloaddition upon exposure to UV light; (d) The PbMA matrix was dissolved in THF to free the cylinders; (e) The PtBA was hydrolyzed to form PAA-functionalized nanotubes; (f) Schematic of the chemical "nanovalve" effect. At low pH, low water permeability was found, whereas at high pH an increased permeability was found. Adapted from Ref. [195].

These studies also had ties to the creation of functionalized nanopores (see Section 1.3.3). In what was termed a "chemical valve effect," the permeability of the channels within the nanotubes could be tuned due to the swelling effect of the PAA brushes in water at different pH values. At low pH, the PAA chains formed a gel due to hydrogen bonding between the AA units, and prevented the flow of water through the pores. At high pH, the AA groups were converted to sodium acrylate, which did not form hydrogen bonds, and the ionized carboxyl groups were readily solvated by water. This allowed a high permeability of water through the tubes.

An obvious application for these type of functionalized material would be as a pH-sensitive membrane for filtration devices [70], although the hydrophilic PAA nanochannels may also act as hosts for the growth of inorganic nanoparticles. To demonstrate such capabilities, the authors were able to grow inorganic particles of CdS, an interesting semiconductor material, and a magnetic material, Fe_2O_3, by adding the appropriate metal salts and a reducing agent (see Section 1.5) [197, 198]. The Fe_2O_3 metal-impregnated nanofibers were found to be super-paramagnetic; that is, they were attracted to each other when a magnetic field was applied, but demagnetized when the field was turned off [199].

Taking their technology one step further, Liu's group has been successful in physically connecting the floating nanotubes to other structures, such as nanospheres. Here, a thick PS-b-PCEMA-b-PtBA film formed PtBA hexagonal cylindrical cores surrounded by a PCEMA shell and a PS matrix. After crosslinking the PCEMA, the PS was dissolved in THF to break the film up into nanotubes. Subsequent ultrasonication caused a shortening of the nanotubes and exposed the PtBA core chains at the ends, which were then hydrolyzed using trifluoroacetic acid to form PAA. The AA ends of the nanotubes were then grafted to polymeric spacers containing multiple amine groups on both ends. These amine-functionalized nanotubes were coupled with PCEMA-b-PAA nanospheres bearing surface carboxyl groups [183] through an amidation reaction, the result being the formation of "ball and chain" structures (see Figure 1.24) [182]. When a nanosphere

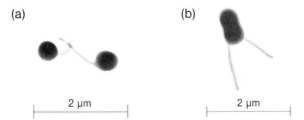

Figure 1.24 Transmission electron microscopy images of nanotube and nanosphere coupling products. (a) Singular nanotubes connected to nanospheres; (b) Two nanotubes connected to a nanosphere. Reprinted with permission from Ref. [182]; © 2006, American Chemical Society.

became attached to two nanotubes (see Figure 1.24b), Liu suggested that if the nanotubes were loaded with super-paramagnetic Fe_2O_3 particles they might act as "fingers," with an opening and closing motion induced by a magnetic field. In this way they could form the chemical basis of a magnetic "nano-hand!"

1.5
Block Copolymers as Nanoreactors

The directed self-assembly of inorganic nanoparticles into BCP templates can result in structures with interesting and useful electronic, optical, and magnetic properties. We have already seen the long-range ordering possible with BCP thin films. Yet, rather than selectively removing one of the blocks, it is also possible to use the ordered BCP pattern as a template for chemical reactions within one phase, leading to patterned clusters of functional nanoscale functional materials over large areas. Various types of functional nanostructures that could, in theory, result from these approaches include metal catalyst particles for fuel cell applications, ferromagnetic particles for high-density storage media [200], doped semiconductor clusters [201], QD structures [202, 203], and core–shell structures [204].

There are two approaches towards the use of BCPs as nanoreactors [205]. The first, most common approach uses an *"in situ"* production of metal nanoclusters inside BCP nanodomains [206]. Figure 1.25 shows a schematic of a cylindrical BCP that contains a functional group "receptor" which is located within the minority phase and is capable of selectively binding positively charged organometallic salts. Once the metal ions have been loaded, a subsequent reduction step can convert the metal ions into oxide, chalcogenide or zerovalent metal clusters, thus regenerating the receptor moiety [205]. The BCP can then accommodate repeated loading/reaction cycles [207] with the same or different metal salt.

In the second approach, the inorganic element is incorporated into an organometallic monomer, which can then be synthesized directly into the parent BCP. The previously mentioned PS-*b*-PFS (see Section 1.2.2.1.4) is an example of this type of polymer. This approach has been reviewed by Cummins *et al.* [208]. These BCPs self-assemble normally to leave an array of "pre-loaded" nanoreactors, ready and waiting for further chemical processing to form the inorganic clusters. This approach eliminates the metal salt loading step, which can take up to two weeks for bulk films, and may lead to nonselective sequestering of the reagents. However, the synthesis of new organometallic monomers for every application also presents huge challenges for the chemist.

1.5.1
Polymer–Metal Solubility

The solubility of inorganic compounds inside an organic matrix is made possible by careful adjustment of the polymer–metal bonding energy. Pearson's Hard–Soft, Acid–Base (HSAB) principle effectively describes the bonding behavior [28]. The

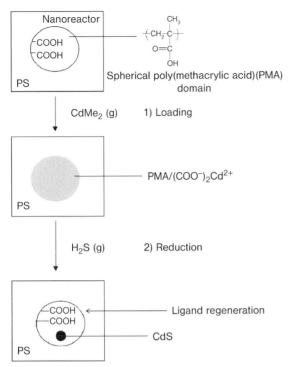

Figure 1.25 A typical scheme for the use of a block copolymer as a template for the selective growth of nanoparticles. Adapted from Ref. [205].

HSAB theory states that only chemicals of similar "hardness" will bond to each other; that is, soft acids will only bind to soft bases, and hard acids will only bind to hard bases. This hardness value is proportional to the difference between the materials' highest occupied molecular orbital (HOMO) energy and the lowest unoccupied molecular orbital (LUMO) energy, which also relates to the band gap of the material. Metals are conductors and have smaller band gaps, and so are classified as chemically "soft." Conversely, polymers are insulators with large band gaps and are classified as chemically "hard." Therefore, hard polymers cannot bind to soft metals unless functional groups are attached to the polymer that change the bonding energy and make it chemically softer. In block copolymers, poly(vinylpyridine) (PVP), containing the electron pair-donating nitrogen species, is most commonly used to effectively solubilize metal salts. The typical design for the loading step is to choose a weakly coordinated metal salt, such as $Pd(OAc)_2$, in which the transition metal acts as the soft acid bonded to a hard acetate base. According to the HSAB principle, this hard–soft bond is highly unstable, such that when the soft metal (Pd) encounters the functionalized soft polymer (PVP), a more stable soft–soft bond is created so as to encapsulate the metal within the block.

1.5.2
Cluster Nucleation and Growth

After block-selective loading of the metal salt, the next step is to reduce the salt to create the nanoparticles within each domain. Different applications may require different nanoparticle sizes and different numbers of clusters within each domain, as shown in Figure 1.26. For example, investigators in the field of electro-optics may only be interested in one nanoparticle per microdomain, whereas those in the field of catalysis might aim for a large number of nanoparticles to present a higher degree of functional surface area for subsequent chemical reactions. Formation of the nanoparticles proceeds through a nucleation and growth process. In typical nucleation and growth schemes, if the colloidal species has enough free energy to aggregate to sizes above a certain critical radius, R_c, then the particle will continue to grow; if not, the particle becomes unstable and falls apart to minimize the surface free energy. The critical radius is proportional to the interfacial tension of the polymer/particle interface (γ) and the degree of supersaturation (c/c_0), as shown in Equation 1.2 [28]:

$$R_c \propto \frac{\gamma}{\ln(c/c_0)} \tag{1.2}$$

Thus, one way to adjust the size and number of the particles is by adjustment of the degree of supersaturation of the reducing agent, which depends on the rate of chemical reaction of the reduction step [209]. Fast chemical reactions with strong

(a) 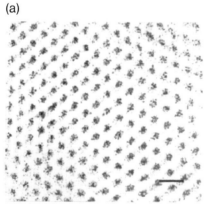 (b)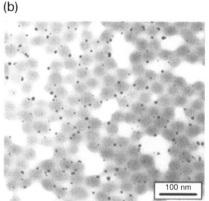

Figure 1.26 (a) Transmission electron microscopy (TEM) image showing multiple gold nanoclusters in each domain. Scale bar = 500 Angstroms. Reprinted with permission from Chan, Y., Ng Cheong, Schrock, R.R. and Cohen, R.E. (1992) *Chem. Mater.*, **4**, 885–894; (b) TEM image showing single nanoparticles in each domain. Reprinted with permission from Klingelhofer, S., Heitz, A., Greiner, S. et al. (1997) *J. Am. Chem. Soc.*, **119**, 10116; © 2006, American Chemical Society.

reducing agents such as lithium aluminum hydride (LiAlH$_4$) lead to high supersaturations, which in turn will mean a low critical radius for nucleation and a higher probability for particle nucleation within the microdomain. This would result in a large number of particles per domain, as shown in Figure 1.26a. In the reverse case, the use of weak reducing agents such as alkylsilanes would result in low supersaturation values. This will increase the critical radius for nucleation and decrease the probability that a nucleating event will occur, resulting usually in only one particle per microdomain, as shown in Figure 1.26b. Another way to control the critical radius is by changing the interfacial tension (γ) between the particle and the polymer, which can be achieved through the selection of different metal-binding blocks.

1.5.3
Block Copolymer Micelle Nanolithography

Building on the nanoreactor approach, nanoscale devices will require the precise placement of functional materials in one- or two-dimensional arrays on semiconductor interfaces. Increased levels of control over the exact location of these inorganic nanoparticles can come from a top-down approach using UV light or e-beam lithography, as shown from a series of reports from Spatz and coworkers. Their approach, which is known as "block copolymer micelle nanolithography" [210], is shown schematically in Figure 1.27a. As an example of their strategy, poly(styrene-*block*-2-vinylpyridine) (PS-*b*-P2VP) is dissolved in a selective solvent for PS to form

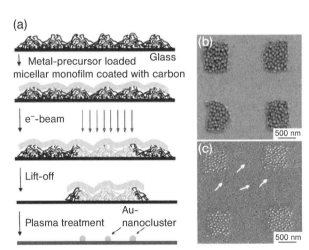

Figure 1.27 Application of monomicellar films as a negative e-beam resist on glass cover slips. (a) Schematic representation; (b) Monomicellar layers after lift-off; (c) Au-nanodots of 7 nm diameter after hydrogen plasma treatment. The white arrows point to holes which are characteristic for glass coverslips. Reprinted with permission from Glass, R., Moeller, M and Spatz, J.P. (2003) *Nanotechnology*, **14**, 1153–1160; © 2006, Institute of Physics.

micelles with P2VP cores and PS coronas. The diameter of these micelles is controlled by the molecular weight of the minority block and the interaction between the polymer blocks and the solvent. The micelles are then loaded with a metal salt of tetrachloroauric acid ($HAuCl_4$), which complexes with the P2VP core. The loaded micelles are deposited as a monolayer film onto a silicon wafer. The film then is exposed to a focused electron beam that oxidizes the polymer, creating carboxylic acids, ketones, aldehydes, and ethers on the polymer surface. These reactive groups bind to the silanol groups on the Si wafer, fixing the exposed area to the substrate, while the unexposed areas are washed away in a sonicating bath. Hydrogen plasma is then used to remove the underlying polymer layer and reduce the $HAuCl_4$, leaving gold dots in the same shape as the previous electron beam pattern (see Figure 1.27b). It was found that the size of the gold nanoparticles could be tuned by adjusting the concentration of metal salt in the BCP solution. These gold clusters are highly mechanically stable structures, and are being pursued for applications in immobilizing single proteins for biosensor applications, or as coatings for lenses [211].

1.6
Interface-Active Block Copolymers

1.6.1
Low-Energy Surfaces Using Fluorinated Block Copolymers

Molecular-level control over the surface properties of polymer films has become increasingly important in current research and development strategies. Polymers with tailored surface energies are used in a wide variety of applications, from planarizing or dielectric layers for semiconductor device fabrication to nonstick cookware and surfaces for combinatorial chemistry. In each of these applications, properties such as adhesion, wetting, lubrication and adsorption behavior must be carefully tuned so as to optimize performance. In the past, scientists toiled on the effects of individual molecular interactions such as entropic frustration between polymer blocks, hydrogen bonding, molecular shape anisotropy, coulombic interactions, and surface segregation. Now, we have learned how to combine these molecular interactions to produce powerful synergistic effects on materials structure.

Semi-fluorinated polymers attached to polymer backbones are often used as surface coatings for their hydrophobic and lipophobic behavior due to the highly chemically resistant nature of the C–F bond. They can also be used as surfactants, lubricating agents, emulsifiers, or photoresists. It has been found that the semi-fluorinated mesogens segregate to the polymer/air interface due to their low surface energy, and are often tilted with respect to the surface normal [212]. One problem with these highly hydrophobic materials, however, is the tendency for surface reconstruction to occur when placed in a highly polar solvent such as water (see Figure 1.28a). Surface reconstruction severely impairs the hydrophobic properties of the material.

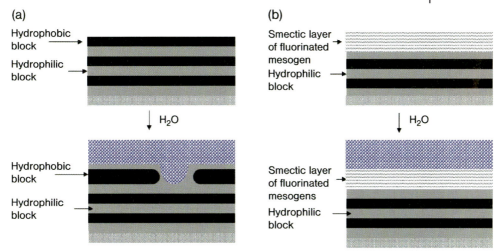

Figure 1.28 (a) Lamellar morphologies of a block copolymer consisting of hydrophobic and hydrophilic blocks self-assembled with the hydrophobic block situated at the air interface due to its lower surface energy. Exposure of the film to water causes a surface reconstruction that brings the hydrophilic block into contact with the water; (b) Block copolymers containing a hydrophobic fluorinated mesogen resist surface reconstruction due to the formation of a highly stable liquid crystalline smectic phase at the surface. Adapted from Gabor, A.H., Pruette, L.C. and Ober, C.K. (1996), *Chem. Mater.*, **8** (9), 2282–2290.

BCPs provide a solution to the problem of surface reconstruction. It has been found that the incorporation of semi-fluorinated chemical groups into microphase-separated BCPs produces materials that resist surface reconstruction due to the formation of a stable liquid crystalline smectic phase at the surface (as shown in Figure 1.28b). Since this discovery, semi-fluorinated BCPs have become very popular candidates as longlasting, nonbiofouling surface coatings for marine vessels, due to the inability of aquatic organisms to stick to their surfaces [65, 213]. A recent report showed that the surface hydrophobicity of fluorinated BCPs can be enhanced with supercritical CO_2 annealing due to a thickening of the smectic layer [214]. According to Langmuir's *Principle of Independent Surface Action*, the unique hydrophobic properties of a surface depend on: (i) the *nature*; and (ii) the *physical arrangement* of the atoms populating the surface of the material [215]. In the following examples, we will discuss how the surface energy of block copolymer films have been manipulated by changing these two variables.

1.6.2
Patterning Surface Energies

The facility to alter the hydrophobicity of a polymer in precise patterns would be a highly desirable trait for the formation of biologically active surfaces. This would allow the selective adsorption of biomolecules [216] or recombinant proteins [217] onto specific locations of a polymer film, which could then be used for biosensors or other "lab-on-a-chip" applications. By definition, the patterning of a photoresist readily accomplishes this solubility switch (see Section 1.2.1). Hayakawa and coworkers synthesized a hydroxylated poly(styrene-*block*-isoprene) BCP using polymer-analogous chemistry, followed by the grafting of a semi-fluorinated side chain onto the hydroxylated isoprene block [218]. The surface-segregated semi-fluorinated chains were capped with an acid-labile *tert*-butoxycarbonyl (TBOC) protecting group that masks a hydroxyl functionality at the end of the chain. To tailor the surface energy of the polymer surface layer, a chemical amplification strategy was used. A photoacid generator mixed into the polymer thin film produced a photoacid that deprotected the TBOC groups, such that the nonpolar methyl end group from the TBOC switched to a polar hydroxyl group during photoprocessing. This resulted in a decrease in the advancing and receding water contact angles by 14° and 15°, respectively. Annealing the film also induced a greater degree of surface ordering of the semi-fluorinated chain and increased the hydrophilicity of the exposed material by a small amount.

Other similar approaches have been used to accomplish the same effect. Böker et al. demonstrated that the highly hydrophobic perfluorinated side chains grafted to a hydroxylated poly(styrene-*b*-isoprene) BCP became completely removed after thermal annealing to dramatically alter the surface properties of the film [219]. Annealing at 340 °C for 15 min in a vacuum oven caused a thermal ester cleavage that resulted in decomposition of the perfluorodecanoyl side chains, but left the parent polymer backbone intact. This resulted in a considerable change of the advancing contact angle of the film, from 122° to 87°. As thermal heating could also be carried out locally on a polymer film, the author suggested that this approach could be used to pattern hydrophobic and hydrophilic regions on the master template of a printing press to control the dispersion of aqueous inks.

In a similar approach, Yang et al. used group transfer polymerization to synthesize a variety of methacrylate-based BCPs with semi-fluorinated chains functionalized with protecting groups, with the intent to use them as surface-active materials as well as photoresists [220]. Due to their transparency under 193 nm wavelength light, the semiconductor industry has shown great interest in fluorinated methacrylate polymers as 193 nm wavelength photoresists. Prior studies have also shown that fluorine-containing BCPs can outperform their random copolymer counterparts [154], and are able to develop in environmentally friendly supercritical CO_2 [221]. To investigate the effect of BCP microstructure on wetting behavior, an assortment of volume ratios for these copolymers were synthesized to provide a wide range of different microstructures and solubilities, but these did not have any effect on the surface energy of the films. The polymers with six $–CF_2–$ units and a $–CF_3$ end group showed the lowest critical surface tension, at approximately 7 mNewtons per meter. Rather than the commonly used *tert*-butyl protecting group, the acetal-type tetrahydropyranyl (THP) protecting group was used on the basis of its more polar and labile nature. Thermal deprotection of the THP groups formed acid and left –OH groups on the polymer chain ends that reduced the advancing water contact angle by 30°. After a period of annealing, it was also reported that that the free acid caused by the THP deprotection interacted with the Si–OH substrate and led to the formation of highly stable, nonreconstructing surfaces.

1.6.3
Photoswitchable Surface Energies Using Block Copolymers Containing Azobenzene

It is well known that polymers containing azobenzene groups attached to one of the blocks can exhibit light-responsive effects [222]. These chemical structures undergo a reversible *cis–trans* isomerization upon exposure to specific wavelengths of light (Figure 1.29), which also changes the molecular orientation of the LC azobenzene mesogens. It has also been found that the mesogens align perpendicular to the incident light polarization, which means that they will align in the film plane and along the direction of the propagation of radiation. Therefore, if the irradiation source is slanted at a given angle, the mesogens then align at the same angle within the film. This field is full of exciting potential for applications in the fields of holographic data storage, optical signal processing, and optical switching.

Light-induced molecular reorganization can also be used to tailor surface properties. Returning to the principle of independent surface action, there are two ways to change the wettability of a surface: (i) to change the nature of the end-group atoms; or (ii) to change the molecular orientation of the end-group atoms. Expanding on the latter point brings us to the topic of photoisomerizable fluorinated mesogens. Thin films of BCPs containing LC fluorinated mesogens have been shown to segregate into well-organized smectic layers on the surface, due to the low energy of the fluorinated block (see Section 1.6.1). Recently, the details have been reported of azobenzene BCPs with semi-fluorinated alkyl side chains that would allow structural modification of the fluorinated mesogens at the surface,

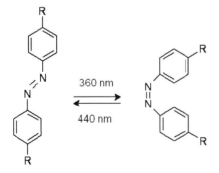

trans isomer   cis isomer

Figure 1.29 The chemical structure of an azobenzene chromophore and the reversible photoisomerization between *trans* and *cis* isomers.

and thus selective patterning of the films' wetting behavior [223]. It has also been shown that the copolymers with longer fluoroalkyl chain lengths resulted in a high degree of orientational order, but were highly resistant to molecular restructuring with photoisomerization. The copolymers with short fluoroalkyl segments showed a small change in advancing/receding contact angle measurements upon exposure to UV light, corresponding to changes from a hydrophobic to a slightly less hydrophobic surface.

Möller and coworkers have also investigated the behavior of a BCP consisting of a PHEMA block and a poly(methacrylate) block with 4-trifluoromethoxyazobenzene side groups for photoswitchable wetting applications [224]. The group found that the photoswitchable effect depended heavily on the packing density of the chromophores, which are higher in the *trans* state than in the *cis* state. Films that were switched to the *cis* state before film formation had a lower packing density and were more susceptible to photoinduced motions, due to the increased free volume in the film. Another group reported an 8° difference in water contact angles between the *cis* and *trans* states in 4-trifluoromethylazobenzene-containing BCPs [225].

1.6.4
Light-Active Azobenzene Block Copolymer Vesicles as Drug Delivery Devices

Tong and coworkers have shown that amphiphilic, azobenzene-containing BCP micelles are highly responsive to UV light [226]. In their ATRP synthesis, the hydrophobic block is a methacrylate-based, azobenzene-containing LC polymer, and the hydrophilic block is a random copolymer of poly(*tert*-butyl acrylate-*co*-acrylic acid). The BCP forms micelles in solution. Under UV light, the azobenzene groups in the core undergo a *trans–cis* photoisomerization that induces a change in the dipole moment in the BCP vesicle. This shift in the delicate hydrophilic/hydrophobic balance between the chains causes dissociation of the micelle. Subsequent exposure to visible light irradiation switches the azo molecule back to its *trans* state, restores the thermodynamic balance, and causes the micelles to reform (Figure 1.30). Tong *et al.* note that other groups [227] had found little effect of UV light irradiation on other azo-based amphiphilic BCPs, but hinted that the success of their system might be due to the thermodynamic lever made possible by the tunability of the hydrophilic random copolymer block. Combined with the ability for micelles to solubilize anticancer drugs [228] and to act as carriers for the site-specific transport of drugs [229], these UV light-responsive micelles are very exciting.

1.6.5
Azobenzene-Containing Block Copolymers as Holographic Materials

In the effort to store data on smaller and smaller length scales, volume holographic data storage has become an area of intense study in the scientific community. Recently, attention has turned to BCPs as a potential holographic material. In

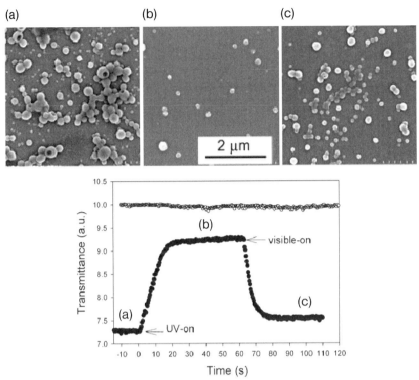

Figure 1.30 Changes in transmittance for a vesicle solution of PAzo74-b-(tBA46-AA22) exposed to UV (360 nm, 18 mW cm^{-2}) and visible (440 nm, 24 mW cm^{-2}) light irradiation. The vesicles are formed by adding 16% (v/v) of water in a dioxane solution with an initial polymer concentration of 1 mg ml^{-1}. (a–c) Typical SFM images for samples cast from the solution at different times: (a) before and (b) during UV light exposure, while (c) shows their reformation after visible light exposure. For comparison, also shown is the transmittance of the diblock copolymer solution in dioxane (no water added to induce the aggregation) subjected to the same conditions of UV and visible light irradiation. The abscissa of time is shifted to have the origin correspond to the application of UV irradiation. Reprinted with permission from Ref. [226]; © 2006, American Chemical Society.

order to understand where they fit in, some background into the technology will be necessary. Holography is a revolutionary technique to store and view data in which an optical interference pattern is produced by the intersection of two coherent laser beams. At the point of intersection, the phase and amplitude of the wave fields induce a chemical or physical change in the material, and are thus "recorded" onto the holographic material [230].

High-capacity holographic data storage requires precise, 3-D control over the index of refraction of the material. Among many potential candidates for holographic materials, azobenzene-containing polymers seem to be the best suited for holographic data storage for several reasons, including their high diffraction efficiency, resolution, and sensitivity, but mainly for the nematic–isotropic phase transition that occurs when the rod-like *trans* isomer is switched to the contracted *cis* isomer (refer to Figure 1.29). The disruption of LC ordering in the molecule takes place on a time scale of about 200 μs, which is a reasonable period for writing data. In 1995, in a landmark study conducted by Ikeda and coworkers [231], this phenomenon was used to record holographic gratings inside the bulk of a polymer film.

One drawback to the use of azobenzene-containing homopolymers and random copolymers is the astonishing formation of surface relief gratings [232]. Here, the polymer becomes physically displaced in the areas of the most intense illumination, due to a massive macroscopic motion of the azo-polymer chains. Despite an increased diffraction efficiency created by these photopatterned ridges, surface relief structures are permanent physical effects that are highly detrimental to the angular selectivity and rewritability required for volume holograms.

BCPs containing azobenzene side chains do not form surface relief gratings due to the confining effect of the microdomains on the azobenzene side chains [233]. This confinement effect, however, seems to have detrimental effects on the speed and magnitude of the *cis–trans* photoisomerization [234]. Häckel *et al.* synthesized a series of BCPs containing a polystyrene block and a polybutadiene block containing the photo-addressable azobenzene components [235]. The photo-addressable phase consisted of a statistical distribution of azobenzene side groups and benzoylbiphenyl side groups. The latter rod-type mesogen was introduced to increase the difference in refractive index between the illuminated and nonilluminated areas of the volume and to improve the stability of the orientation. Different azo:mesogen ratios were used to identify the polymer with the highest degree and stability of molecular reorientation. The polymer containing 35% of the mesogenic side groups showed a slight increase of the refractive index modulation over the period of a year. This strategy posted remarkable improvements over the stability of the recorded orientations.

1.7
Summary and Outlook

In this chapter we have brought to light many interesting chemical and physical applications possible using BCP-directed self-assembly (see Table 1.3). Unfortunately, however, many other fascinating applications have been omitted in the interest of space. An extraordinary amount of progress has been made in our ability to manipulate and optimize BCP structural ordering in various forms, including thin films, micelles, and monolithic bulk structures. The ordered microdomains may contain chemical functionality that can be used in a variety of fashions, such as templates for the nucleation and growth of inorganic nanoparticles, chemical valves inside cylindrical pores, monolithic nanoporous structures, or as removable components to form lithographic stencils. We have also seen how BCPs

Table 1.3 Block copolymers used for phase-selective chemistry.

Name	Structure	Function
Poly(styrene-*block*-isoprene) modified with 4-perfluoroalkyl azobenzene side groups		Photoswitchable surface engineering
Poly(styrene-*block*-methyl methacrylate)		Nanolithographic patterning
Poly(styrene-*block*-4-vinyl pyridine)		Micelle nanoreactors, nanolithographic patterning
Polystyrene-*block*-butadiene		Nanolithographic patterning

Table 1.3 Continued.

Name	Structure	Function
Poly(α-methylstyrene-*block*-hydroxystyrene)		Nanolithographic patterning
Poly(isoprene-*block*-ethylene oxide)		Structure-directing scaffolds for inorganic chemistry
Poly(styrene-*block*-L-lactide)		Functionalized nanoporous surfaces, nanolithographic patterning
t-butyl methacrylate (*t*-BMA)-*block*-3-methacryloxy-propylpentamethyldisoloxane (SiMA).		Bilevel block copolymer photoresists

1.7 Summary and Outlook

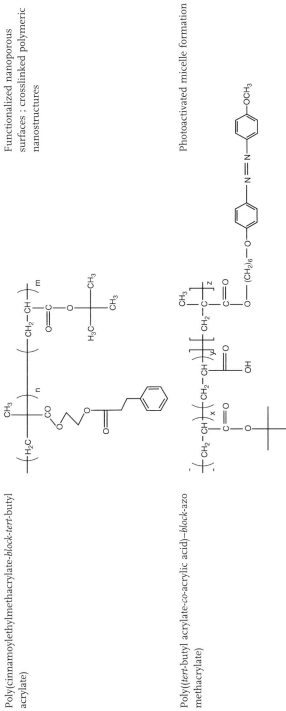

Functionalized nanoporous surfaces ; crosslinked polymeric nanostructures

Poly(cinnamoylethylmethacrylate-*block*-*tert*-butyl acrylate)

Photoactivated micelle formation

Poly((*tert*-butyl acrylate-*co*-acrylic acid)–*block*-azo methacrylate)

Table 1.3 Continued.

Name	Structure	Function
Modified poly(styrene-*block*-1,2-butadiene)		Holographic patterning

can behave as multilevel photoresists due to their ability to combine different polymers without macrophase separation. BCPs have also been used for their ability to segregate into distinct chemical levels for the tailoring of surface properties. Lastly, we have seen how the incorporation of LC moieties such as fluorinated mesogens or azobenzene chromophores into BCPs widens their repertoire to low-energy surface patterning and holographic data storage applications.

In all of these studies, novel chemical strategies have been used to generate functional results. Chemistry is at the heart of every physical effect arising from BCP self-assembly. Using myriad living polymerization strategies in combination with polymer-analogous reactions, BCPs with a vast range of functionalities can be synthesized with low polydispersity and well-defined molecular weights. Exquisite levels of control over the BCP can be achieved by altering the degree of polymerization of each of the blocks, through which the chemist can dial in any type of morphology from spheres, cylinders, bicontinuous networks, and lamellar phases. Even more morphologies are possible through micellization strategies, such as spherical, cylindrical, or onion-like structures. The physical and functional likeness of these structures to biological vesicles and lysosomes is uncanny. Moreover, these technologies are just starting to see the first of their applications as containers for drug delivery, and excellent reviews have been produced that highlight this area [29, 236].

In the field of BCP lithography, we have just begun to see the first practical uses emerge from all of the fundamental studies. The formation of cobalt magnetic dot arrays, flash memory devices, and MOS capacitors are only the "tip of the iceberg" of applications that can arise from this technology. Whilst industry has not overlooked these results, several key requirements are required in order for the semiconductor industry to take directed self-assembly seriously. Among these requirements are "improved long-range dimensional control … improved resolution and linear density by at least a factor of two over that achieved by top down lithography [~11 nm by 2010] … fabricated features with multiple sizes and pitches in the same layer in different regions of a chip …. " and so on [79]. If improvements in these areas are not made in the next four years, the field of directed self-assembly may miss its chance for widespread industrial use.

In the future, nothing stands in the way of our ability to pattern individual, isolated lines of perpendicular lamellar morphologies. To this end, the work of Ober and coworkers with the P(αMS-b-HOST) system, in combination with small molecular additives that undergo phase-selective photocrosslinking chemistry, seems very promising. One day, the rapidly growing field of organic electronics may converge with BCP lithography, and the ability to create patterned, well-ordered arrays of conducting nanowires may become possible [237–239]. In theory, the chemical nanopatterning approach mastered by the Nealey group should allow us to effectively control BCP self-assembly by forcing microdomains to organize into any type of lattice, such as simple cubic lattices. It should also be possible with chemical nanopatterning to pattern some of the more complicated features in semiconductor device components, such as nested and embedded jog structures, and t-junctions. Furthermore, if we consider the immense 3-D complexity of protein molecules, which are formed from various combinations of the 20 amino acids, we have only begun to scratch the surface of the dimensional control made possible by block copolymers. The incorporation of multiple blocks, hydrogen-bonded [240] and LC blocks [21] will be immensely rewarded by the ability to manipulate and exert control over multiple length scales of molecular self-assembly. Nor should we be constrained by the type of BCP that we use. PS-*b*-PMMA is a great lithographic system, but other, better-performing BCP systems may be on the horizon. New and improved chemistry will be vital for improvement in this field. It will certainly be exciting to follow the field of BCP research as it reaches its full maturity in the coming years.

References

1 Waite, J.H., Lichtenegger, H.C., Stucky, G.D. and Hansma, P. (2004) *Biochemistry*, **43**, 7653–7662.
2 Coyne, K.J., Qin, X. and Waite, J.H. (1997) *Science*, **277**, 1830.
3 Jones, M.N. and Chapman, D. (1995) *Micelles, Monolayers and biomembranes*, John Wiley & Sons, Inc., New York.
4 Shapiro, J.A. (1998) *Annu. Rev. Microbiol.*, **52**, 81.
5 Phillip, D. and Stoddart, J.F. (1996) *Angew. Chem. Int. Ed. Engl.*, **35**, 1155.
6 Whitesides, G.M., Mathias, J.P. and Seto, C.T. (1991) *Science*, **254**, 1312–1319.
7 Ball, P. (2001) *Nature*, **413**, 667–668.
8 Bates, F.S. and Fredrickson, G.H. (1999) *Phys. Today*, **52**, 32–38.
9 Holden, G., Legge, N.R., Quirk, R. and Schroeder, H.E. (1996) *Thermoplastic Elastomers*, 2nd edn, Schoder Druck GmbH & Co KG, Gersthofen.
10 Bates, F.S. and Fredrickson, G.H. (1990) *Annu. Rev. Phys. Chem.*, **41**, 525–557.
11 Hashimoto, T., Shibayama, M. and Kawai, H. (1980) *Macromolecules*, **13**, 1237.
12 Hajduk, D.A., Harper, P.E., Gruner, S.M., Honeker, C.C., Kim, G., Thomas, E.L. and Fetters, L.J. (1994) *Macromolecules*, **27**, 4063.
13 Zhao, J., Majumdar, B., Schulz, M.F., Bates, F.S., Almdal, K., Mortensen, K., Hajduk, D.A. and Gruner, S.M. (1996) *Macromolecules*, **29**, 1024.
14 Chan, V.Z.H., Hoffman, J., Lee, V.Y., Latrou, H., Avgeropoulos, A., Hadjichristidis, N. and Miller, R.D. (1999) *Science*, **286**, 1716–1719.
15 Urbas, A.M., Maldovan, M., DeRege, P. and Thomas, E.L. (2002) *Adv. Mater.*, **14**, 1850–1853.
16 Blomberg, S., Ostberg, S., Harth, E., Bosman, A.W., Van Horn, B. and Hawker, C.J. (2002) *J. Polym. Sci., Part A: Polym. Chem.*, **40**, 1309–1320.
17 Stadler, R., Auschra, C., Beckmann, J., Krappe, U., Voight-Martin, I. and Leibler, L. (1995) *Macromolecules*, **28**, 3080–3097.
18 Zheng, W. and Wang, Z.G. (1995) *Macromolecules*, **28**, 7215–7223.
19 Takahashi, K., Hawegawa, H., Hashimoto, T., Bellas, V., Iatrou, H. and Hadjichristidis, N. (2002) *Macromolecules*, **36**, 4859.
20 Abetz, V. and Simon, P.F.W. (2005) *Adv. Polym. Sci.*, **189**, 125–212.
21 Adams, J. and Gronski, W. (1989) *Makromol. Chem. Rapid Commun.*, **10**, 553–557.
22 Gallot, B. (1996) *Prog. Polym. Sci.*, **21**, 1035–1088.
23 Muthukumar, M., Ober, C.K. and Thomas, E.L. (1997) *Science*, **277**, 1225–1232.
24 Chen, J.T., Thomas, E.L., Ober, C.K. and Mao, G.P. (1996) *Science*, **273**, 343–346.
25 Mao, G.P., Wang, J., Clingman, S.R., Ober, C.K., Chen, J.T. and Thomas, E.L. (1997) *Macromolecules*, **30**, 2556–2567.
26 Discher, B.M., Won, Y.Y., Ege, D.S., Lee, J.C.M., Bates, F.S., Discher, D.E. and

Hammer, D.A. (1999) *Science*, **284**, 1143–1146.

27 Foerster, S. (2003) *Top. Curr. Chem.*, **226**, 1–28.

28 Forster, S. and Antonietti, M. (1998) *Adv. Mater.*, **10**, 195–217.

29 Forster, S. and Plantenberg, T. (2002) *Angew. Chem. Int. Ed. Engl.*, **41**, 688–714.

30 Hadjichristidis, N., Pitsikalis, M. and Iatrou, H. (2005) *Adv. Polym. Sci.*, **189**, 124.

31 Hsieh, H. and Quirk, R. (1996) *Anionic Polymerization: Principles and Practical Applications*, Vol. **34**, Marcel Dekker, Inc., New York.

32 Hadjichristidis, N., Pitsikalis, M., Iatrou, H. and Pispas, S. (2000) *J. Polym. Sci., Part A: Polym. Chem.*, **38**, 3211–3234.

33 Ndoni, S., Papadakis, C.M., Bates, F.S. and Almdal, K. (1995) *Rev. Sci. Instrum.*, **66**, 1090–1095.

34 Hadjichristidis, N., Pitsikalis, M., Pispas, S. and Iatrou, H. (2001) *Chem. Rev.*, **101**, 3747–3792.

35 Otsu, T., Yoshida, M. and Tazaki, T. (1982) *Makromol. Chem., Rapid Commun.*, **3**, 133.

36 Hawker, C.J., Bosman, A.W. and Harth, E. (2001) *Chem. Rev.*, **101**, 3661–3688.

37 Georges, M.K., Veregin, R.P.N., Kazmaier, P.M., Hamer, G.K. and Saban, M. (1994) *Macromolecules*, **27**, 7228–7229.

38 Georges, M.K., Veregin, R.P.N., Kazmaier, P.M. and Hamer, G. (1993) *Macromolecules*, **26**, 2987–2988.

39 Otsu, T. and Matsumoto, A. (1998) *Adv. Polym. Sci.*, **136**, 75–136.

40 Otsu, T., Yoshida, M. and Kuriyama, A. (1982) *Polym. Bull.*, **7**, 45.

41 Otsu, T., Kuriyama, A. and Yoshida, M. (1983) *J. Polym. Sci. Tech. (Japan)*, **40**, 583.

42 Otsu, T. and Yoshida, M. (1982) *Polym. Bull.*, **7**, 197.

43 Haque, S.A. (1994) *J. Macromol. Sci., Pure Appl. Chem. A*, **31**, 827.

44 Liu, F.T., Cao, S.Q. and Yu, X.D. (1993) *J. Appl. Polym. Sci.*, **48**, 425.

45 Opresnik, M. and Sebenik, A. (1995) *Polym. Int.*, **36**, 13.

46 Moad, G., Chong, Y.K., Postma, A., Rizzardo, E. and Thang, S.H. (2005) *Polymer*, **46**, 8458–8468.

47 Matyjaszewski, K. and Xia, J. (2001) *Chem. Rev.*, **101**, 2921–2990.

48 Gibson, V.C. (1994) *Adv. Mater.*, **6**, 37–42.

49 Buchmeiser, M.R., Seeber, G., Mupa, M. and Bonn, G.K. (1999) *Chem. Mater.*, **11**, 1533–1540.

50 Royappa, A.T., Saunders, R.S., Rubner, M.F. and Cohen, R.E. (1998) *Langmuir*, **14**, 6207–6214.

51 Fogg, D.E., Radzilowski, L.H., Blanski, R., Schrock, R.R. and Thomas, E.L. (1997) *Macromolecules*, **30**, 417–426.

52 Brittain, W. (1992) *Rubber Chem. Technol.*, **65**, 580.

53 Sogah, D.Y., Hertler, W.R., Webster, O.W. and Cohen, G.M. (1987) *Macromolecules*, **20**, 1473–1488.

54 Mykytiuk, J., Armes, S.P. and Billingham, N.C. (1992) *Polym. Bull.*, **29**, 139.

55 Rannard, S.P., Billingham, N.C., Armes, S.P. and Mykytiuk, J. (1993) *Eur. Polym. J.*, **29**, 407.

56 AG and Goldschmidt, T. (1992), Germany.

57 Risse, W. and Grubbs, R.H. (1989) *Macromolecules*, **22**, 1558–1562.

58 Feldthusen, J., Ivan, B. and Muller, A.H.E. (1998) *Macromolecules*, **31**, 578–585.

59 Matyjaszewski, K. and Xia, J. (2001) *Chem. Rev.*, **101**, 2963.

60 McGrath, M.P., Sall, E.D. and Tremont, S.J. (1995) *Chem. Rev.*, **95**, 381.

61 Jian, X. and Hay, A.S. (1991) *J. Polym. Sci., Part A: Polym. Chem.*, **29**, 1183.

62 Ramireddy, C., Tuzar, Z., Prochazka, K., Webber, S.E. and Munk, P. (1992) *Macromolecules*, **25**, 2541.

63 Valint, P.L. and Bock, J. (1988) *Macromolecules*, **21**, 175.

64 Chung, T.C., Raate, M., Berluche, E. and Schulz, D.N. (1988) *Macromolecules*, **21**, 1903–1907.

65 Krishnan, S., Wang, N., Ober, C.K., Finlay, J.A., Callow, M.E., Callow, J.A., Hexemer, A., Sohn, K., Kramer, E.J. and Fischer, D. (2006) *Biomacromolecules*, **7**, 1449–1462.

66 Selb, J. and Gallot, Y. (1985) In *Developments in Block Copolymers*, Vol. 2 (ed. I. Goodman), Elsevier, London, p. 27.

67 Cameron, G.G. and Qureshi, M.Y. (1981) *Makromol. Chem.*, **2**, 287.

68 Pepper, K.W., Paisley, H.M. and Young, M.A. (1953) *J. Am. Chem. Soc.*, 4097.

69 Rahlwes, D., Roovers, J.E.L. and Bywater, S. (1977) *Macromolecules*, **10**, 604.

70 Liu, G., Ding, J., Guo, A., Herfort, M. and Bazett-Jones, D. (1997) *Macromolecules*, **30**, 1851–1853.

71 Hadjichristidis, N., Pispas, S. and Floudas, G. (2003) *Block Copolymers: Synthetic Strategies, Physical Properties, and Applications*, John Wiley & Sons, Inc., Hoboken, NJ.

72 Reichmanis, E. (1993) In *Polymers for Electronic and Photonic Applications* (ed. C.P. Wong), Academic Press, pp. 67–116.

73 Thompson, L.F. and Bowden, M.J. (1983) In *Introduction to Microlithography*, ACS Symposium Series, Vol. **412**, 1st edn, American Chemical Society, Washington, DC, pp. 162–214.

74 Moore, G.E. (1965) *Electronics*, **38**. ftp://download.intel.com/museum/Moores_Law/Articles-Press_Releases/Gordon_Moore_1965_Article.pdf

75 Chang, S.W., Ayothi, R., Bratton, D., Yang, D., Felix, N., Cao, H.B., Deng, H. and Ober, C.K. (2006) *J. Mater. Chem.*, **16**, 1470–1474.

76 Yang, D., Chang, S.W. and Ober, C.K. (2006) *J. Mater. Chem.*, **16**, 1693–1696.

77 Xia, Y., Rogers, J.A., Paul, K.E. and Whitesides, G.M. (1999) *Chem. Rev.*, **99**, 1823–1848.

78 McCord, M. and Rooks, M. (1997) In *Handbook of Microlithography, Micromachining and Microfabrication*, Vol. **1**, Ch. 2 (ed. P. Rai-Choudhury), SPIE Press, Bellingham, WA.

79 Garner, M.C., Herr, D. and Krautschik, C. (2006) Semiconductor Research Corporation Internal Document.

80 Park, M., Harrison, C., Chaikin, P.M., Register, R.A. and Adamson, D.H. (1997) *Science*, **276**, 1401–1404.

81 Lee, J.S., Hirao, A. and Nakahama, S. (1989) *Macromolecules*, **22**, 2602.

82 Hawker, C.J. and Russell, T.P. (2005) *MRS Bull.*, **30**, 952–966.

83 Harrison, C.H., Dagata, J.A. and Adamson, D.H. (2004) In *Developments in Block Copolymer Science and Technology* (ed. I.W. Hamley), John Wiley & Sons, pp. 295–323.

84 Li, M., Coenjarts, C.A. and Ober, C.K. (2005) *Adv. Polym. Sci.*, **190**, 183–226.

85 Fasolka, M.J. and Mayes, A.M. (2001) *Annu. Rev. Mater. Res.*, **31**, 323–355.

86 Segalman, R.A. (2005) *Mater. Sci. Eng.*, R, **48**, 191–226.

87 Hahm, J. and Sibener, S.J. (2001) *J. Chem. Phys.*, **114**, 4730–4740.

88 Amundson, K., Helfand, E., Quan, X., Hudson, S.D. and Smith, S.D. (1994) *Macromolecules*, **27**, 6559–6570.

89 Thurn-Albrecht, T., DeRouchey, J., Russell, T.P. and Kolb, R. (2002) *Macromolecules*, **35**, 8106–8110.

90 Thurn-Albrecht, T., Steiner, R., DeRouchey, J., Stafford, C.M., Huang, E., Bal, M., Tuominen, M.T., Hawker, C.J. and Russell, T.P. (2000) *Adv. Mater.*, **12**, 787–790.

91 Daniel, C., Hamley, I.W., Mingvanish, W. and Booth, C. (2000) *Macromolecules*, **33**, 2163–2170.

92 Fredrickson, G.H. (1994) *J. Rheol.*, **38**, 1045–1067.

93 Hamley, I.W. (2000) *Curr. Opin. Colloid Interface Sci.*, **5**, 342–350.

94 Xuan, Y., Peng, J., Cui, L., Wang, H., Li, B. and Han, Y. (2004) *Macromolecules*, **37**, 7301–7307.

95 Fukunaga, K., Elbs, H., Magerle, R. and Krausch, G. (2000) *Macromolecules*, **33**, 947–953.

96 Segalman, R.A., Yokoyama, H. and Kramer, E.J. (2001) *Adv. Mater.*, **13**, 1152–1155.

97 Huang, E., Pruzinsky, S., Russell, T.P., Mays, J. and Hawker, C.J. (1999) *Macromolecules*, **32**, 5299–5303.

98 Huang, E., Russell, T.P., Harrison, C., Chaikin, P.M., Register, R.A., Hawker, C.J. and Mays, J. (1998) *Macromolecules*, **31**, 7641–7650.

99 Peters, R.D., Yang, X.M., Wang, Q., de Pablo, J.J. and Nealey, P.F. (2000) *J. Vac. Sci. Tech.*, B, **18**, 3530–3534.

100 Kim, S.O., Solak, H.H., Stoykovich, M.P., Ferrier, N.J., de Pablo, J.J. and Nealey, P.F. (2003) *Nature*, **424**, 411–414.

101 Yang, X.M., Peters, R.D., Nealey, P.F., Solak, H.H. and Cerrina, F. (2000) *Macromolecules*, **33**, 9575–9582.
102 Kim, S.H., Misner, M.J., Xu, T., Kimura, M. and Russell, T.P. (2004) *Adv. Mater.*, **16**, 226–231.
103 Drockenmuller, E., Li, L.Y.T., Ryu, D.Y., Harth, E., Russell, T.P., Kim, H.C. and Hawker, C.J. (2005) *J. Polym. Sci., Part A: Polym. Chem.*, **43**, 1028–1037.
104 Russell, T.P., Thurn-Albrecht, T., Tuominen, M., Huang, E. and Hawker, C.J. (2000) *Macromol. Symp.*, **159**, 77–88.
105 Xu, T., Stevens, J., Villa, J.A., Goldbach, J.T., Guarini, K.W., Black, C.T., Hawker, C.J. and Russell, T.P. (2003) *Adv. Funct. Mater.*, **13**, 698–702.
106 Kim, H.C., Jia, X., Stafford, C.M., Kim, D.H., McCarthy, T.J., Tuominen, M., Hawker, C.J. and Russell, T.P. (2001) *Adv. Mater.*, **13**, 795–797.
107 Shin, K., Leach, K.A., Goldbach, J.T., Kim, D.H., Jho, J.Y., Tuominen, M., Hawker, C.J. and Russell, T.P. (2002) *Nano Lett.*, **2**, 933–936.
108 Asakawa, K. and Hiraoka, T. (2002) *J. Appl. Phys., 1 (Japan)*, **41**, 6112–6118.
109 Asakawa, K., Hiraoka, T., Hieda, H., Sakurai, M., Kamata, Y. and Naito, K. (2002) *J. Photopolym. Sci. Technol.*, **15**, 465–470.
110 Naito, K., Hieda, H., Sakurai, M., Kamata, Y. and Asakawa, K. (2002) *IEEE Trans. Magn.*, **38**, 1949–1951.
111 Harrison, C., Park, M., Chaikin, P.M., Register, R.A. and Adamson, D.H. (1998) *J. Vac. Sci. Tech., B*, **16**, 544–552.
112 Mansky, P., Harrison, C.K., Chaikin, P.M., Register, R.A. and Yao, N. (1996) *Appl. Phys. Lett.*, **68**, 2586–2588.
113 Lammertink, R.G.H., Hempenius, M.A., Van Den Enk, J.E., Chan, V.Z.H., Thomas, E.L. and Vancso, G.J. (2000) *Adv. Mater.*, **12**, 98–103.
114 Lammertink, R.G.H., Hempenius, M.A., Chan, V.Z.H., Thomas, E.L. and Vancso, G.J. (2001) *Chem. Mater.*, **13**, 429–434.
115 Hashimoto, T., Tsutsumi, K. and Funaki, Y. (1997) *Langmuir*, **13**, 6869–6872.
116 Du, P., Li, M., Douki, K., Li, X., Garcia, C.B.W., Jain, A., Smilgies, D.M., Fetters, L.J., Gruner, S.M., Wiesner, U. and Ober, C.K. (2004) *Adv. Mater.*, **16**, 953–957.
117 Li, M., Douki, K., Goto, K., Li, X., Coenjarts, C., Smilgies, D.M. and Ober, C.K. (2004) *Chem. Mater.*, **16**, 3800–3808.
118 Zalusky, A.S., Olayo-Valles, R., Taylor, C.J. and Hillmyer, M.A. (2001) *J. Am. Chem. Soc.*, **123**, 1519–1520.
119 Zalusky, A.S., Olayo-Valles, R., Wolf, J.H. and Hillmyer, M.A. (2002) *J. Am. Chem. Soc.*, **124**, 12761–12773.
120 Wolf, J.H. and Hillmyer, M.A. (2003) *Langmuir*, **19**, 6553–6560.
121 Rzayev, J. and Hillmyer, M.A. (2005) *Macromolecules*, **38**, 3–5.
122 Lin, Z., Kim, D.H., Wu, X., Boosahda, L., Stone, D., LaRose, L. and Russell, T.P. (2002) *Adv. Mater.*, **14**, 1373–1376.
123 Cheng, J.Y., Ross, C.A., Chan, V.Z.H., Thomas, E.L., Lammertink, R.G.H. and Vancso, G.J. (2001) *Adv. Mater.*, **13**, 1174–1178.
124 Cheng, J.Y., Ross, C.A., Thomas, E.L., Smith, H.I. and Vancso, G.J. (2002) *Appl. Phys. Lett.*, **81**, 3657–3659.
125 Yokoyama, H., Li, L., Nemoto, T. and Sugiyama, K. (2004) *Adv. Mater.*, **16**, 1542–1546.
126 Li, L., Yokoyama, H., Nemoto, T. and Sugiyama, K. (2004) *Adv. Mater.*, **16**, 1226–1229.
127 Park, M., Chaikin, P.M., Register, R.A. and Adamson, D.H. (2001) *Appl. Phys. Lett.*, **79**, 257–259.
128 Li, R.R., Dapkus, P.D., Thompson, M.E., Jeong, W.G., Harrison, C., Chaikin, P.M., Register, R.A. and Adamson, D.H. (2000) *Appl. Phys. Lett.*, **76**, 1689–1691.
129 Poly(styrene-block-methylmethacrylate). Polymer Source, Inc. Available at: http://www.polymersource.com/shoppingCart/product.asp?ID=408
130 Ranby, B. and Rabek, J.F. (1975) *Photodegradation, Photo-oxidation and Photostabilization of Polymers*, John Wiley & Sons, Inc., New York.
131 Stoykovich, M.P., Mueller, M., Kim, S.O., Solak, H.H., Edwards, E.W., de Pablo, J.J. and Nealey, P.F. (2005) *Science*, **308**, 1442–1446.
132 Guarini, K.W., Black, C.T. and Yeung, S.H.I. (2002) *Adv. Mater.*, **14**, 1290–1294.
133 Guarini, K.W., Black, C.T., Milkove, K.R. and Sandstrom, R.L. (2001) *J. Vac. Sci. Tech., B*, **19**, 2784–2788.
134 Black, C.T., Guarini, K.W., Milkove, K.R., Baker, S.M., Russell, T.P. and

Tuominen, M.T. (2001) *Appl. Phys. Lett.*, **79**, 409–411.

135 Guarini, K.W., Black, C.T., Zhang, Y., Kim, H., Sikorski, E.M. and Babich, I.V. (2002) *J. Vac. Sci. Tech., B*, **20**, 2788–2792.

136 Black, C.T. (2005) *Appl. Phys. Lett.*, **87**, 163116.

137 Guarini, K.W., Black, C.T., Zhang, Y., Babich, I.V., Sikorski, E.M. and Gignac, L.M. (2003) Low voltage, scalable nanocrystal FLASH memory fabricated by templated self assembly. *Technical Digest – International Electron Devices Meeting*, pp. 541–544.

138 Leiston-Belanger, J.M., Russell, T.P., Drockenmuller, E. and Hawker, C.J. (2005) *Macromolecules*, **38**, 7676–7683.

139 Lammertink, R.G.H., Hempenius, M.A., Thomas, E.L. and Vancso, J. (1999) *J. Polym. Sci., Part B: Polym. Phys.*, **37**, 1009–1021.

140 Cheng, J.Y., Ross, C.A., Thomas, E.L., Smith, H.I., Lammertink, R.G.H. and Vancso, G.J. (2002) *IEEE Trans. Magn.*, **38**, 2541–2543.

141 Ito, H. and Willson, C.G. (1982) *Polym. Eng. Sci.*, **23**, 1021.

142 Shaw, J.M. and Gelorme, J.D. (1997) *IBM J. Res. Dev.*, **41**, 81.

143 Bosworth, J.K., Paik, M.Y., Ruiz, R. et al. (2008) *ACS Nano*, **2** (7), 1396.

144 Taylor, G.N. and Wolf, T.M. (1980) *Polym. Eng. Sci.*, **20**, 1087.

145 Taylor, G.N., Wolf, T.M. and Moran, J.M. (1981) *J. Vac. Sci. Tech.*, **19**, 872.

146 Gabor, A.H., Pruette, L.C. and Ober, C.K. (1996) *Chem. Mater.*, **8**, 2282–2290.

147 Uhrich, K.E., Reichmanis, E. and Baiocchi, F.A. (1994) *Chem. Mater.*, **6**, 295–301.

148 Hartney, M.A., Novembre, A.E. and Bates, F.S. (1985) *J. Vac. Sci. Tech., B*, **3**, 1346–1351.

149 Bowden, M.J., Gozdz, A.S., Desimone, J.M., McGrath, J.E., Ito, S. and Matsuda, M. (1992) *Makromol. Chem., Macromol. Symp.*, **53**, 125.

150 DeSimone, J.M., York, G.A., McGrath, J.E., Gozdz, A.S. and Bowden, M.J. (1991) *Macromolecules*, **24**, 5330.

151 Jurek, M.J. and Reichmanis, E. (1989) In *Polymers in Microlithography*, ACS Symposium Series, Vol. **412** (eds E. Reichmanis, S.A. MacDonald and T. Iwayanagi), American Chemical Society, Washington DC, p. 158.

152 Gabor, A.H., Lehner, E.A., Mao, G.P., Schneggenburger, L.A. and Ober, C.K. (1994) *Chem. Mater.*, **6**, 927–934.

153 Gabor, A.H. and Ober, C.K. (1995) In *Microelectronics Technology: Polymers in Advanced Imaging and Packaging*, ACS Symposium Series, Vol. **614** (eds E. Reichmanis, S.A. MacDonald, T. Iwayanagi, C.K. Ober and T. Nishikubo), American Chemical Society, Washington, DC, pp. 281–298.

154 Gabor, A.H., Pruette, L.C. and Ober, C.K. (1996) *Chem. Mater.*, **8**, 2282–2290.

155 Lee, J.S., Hirao, A. and Nakahama, S. (1989) *Macromolecules*, **22**, 2602–2606.

156 Fink, Y., Urbas, A.M., Bawendi, M.G., Joannopoulos, J.D. and Thomas, E.L. (1999) *J. Lightwave Technol.*, **17**, 1963–1969.

157 Maldovan, M., Bockstaller, M.R., Thomas, E.L. and Carter, W.C. (2003) *J. Appl. Phys., B*, **76**, 877–889.

158 Pai, R.A., Humayun, R., Schulberg, M.T., Sengupta, A., Sun, J.N. and Watkins, J.J. (2004) *Science*, **303**, 507–511.

159 Nagarajan, S., Pai, R.A., Russell, T.P., Watkins, J.J., Li, M., Bosworth, J.K., Busch, P., Smilgies, D.M. and Ober, C.K. (2008) *Adv. Mater.*, **20** (2), 246.

160 Ulrich, R., Du Chesne, A., Templin, M. and Wiesner, U. (1999) *Adv. Mater.*, **11**, 141–146.

161 Simon, P.F.W., Ulrich, R., Spiess, H.W. and Wiesner, U. (2001) *Chem. Mater.*, **13**, 3464–3486.

162 Finnefrock, A.C., Ulrich, R., Toombes, G.E.S., Gruner, S.M. and Wiesner, U. (2003) *J. Am. Chem. Soc.*, **125**, 13084–13093.

163 Ulrich, R. (2000) *Morphologien und Eigenschaften strukturierter organisch-anorganischer Hybridmaterialien*, Logos-Verlag, Berlin.

164 Li, M. (2004) In *Materials Science and Engineering*, Cornell University, Ithaca, NY, p. 216.

165 Li Minqi, Development of functional block copolymers for nanotechnology. PhD Dissertation (2004), Cornell

University Materials Science and Engineering.
166 Hamley, I.W. (1998) *The Physics of Block Copolymers*, Oxford University Press, Oxford.
167 Xu, T., Kim, H.-C., DeRouchey, J., Seney, C., Levesque, C., Martin, P., Stafford, C.M. and Russell, T.P. (2001) *Polymer*, **42**, 9091–9095.
168 Jeong, U., Kim, H.C., Rodriguez, R.L., Tsai, I.Y., Stafford, C.M., Kim, J.K., Hawker, C.J. and Russell, T.P. (2002) *Adv. Mater.*, **14**, 274–276.
169 Ruokolainen, J., ten Brinke, G. and Ikkala, O. (1999) *Adv. Mater.*, **11**, 777–780.
170 Ikkala, O. and ten Brinke, G. (2002) *Science*, **295**, 2407–2409.
171 Maki-Ontto, R., de Moel, K., de Odorico, W., Ruokolainen, J., Stamm, M., ten Brinke, G. and Ikkala, O. (2001) *Adv. Mater.*, **13**, 117–121.
172 DeSimone, J.M., Guan, Z. and Elsbernd, C.S. (1992) *Science*, **257**, 945.
173 Liu, G., Ding, J. and Stewart, S. (1999) *Angew. Chem. Int. Ed. Engl.*, **38**, pp. 835–838.
174 Rzayev, J. and Hillmyer, M.A. (2005) *J. Am. Chem. Soc.*, **127**, 13373–13379.
175 Guo, S., Rzayev, J., Bailey, T., Zalusky, A., Olayo-Valles, R. and Hillmyer, M.A. (2006) *Chem. Mater.*, **18**, 1719–1721.
176 Mao, H., Arrechea, P.L., Bailey, T.S., Johnson, B.J.S. and Hillmyer, M.A. (2004) *Faraday Discuss.*, **128**, 149–162.
177 Tuzar, Z. and Kratochvil, P. (1993) *Colloids Surf. A*, **15**, 1–83.
178 Canham, P.A., Lally, T.P., Price, C. and Stubbersfield, R.B. (1980) *Faraday Trans.*, **55**, 1857.
179 Jenekhe, S.A. and Chen, X.L. (1998) *Science*, **279**, 1903.
180 Ding, J. and Liu, G. (1997) *Macromolecules*, **30**, 655.
181 Ding, J., Liu, G. and Yang, M. (1997) *Polymer*, **38**, 5497.
182 Liu, G., Yan, X., Li, Z., Zhou, J. and Duncan, S. (2003) *J. Am. Chem. Soc.*, **125**, 14039–14045.
183 Henselwood, F. and Liu, G. (1998) *Macromolecules*, **31**, 4213.
184 Ding, J. and Liu, G. (1998) *J. Phys. Chem. B*, **102**, 6107–6113.
185 Stewart, S. and Liu, G. (1999) *Chem. Mater.*, **11**, 1048–1054.
186 Guo, A., Liu, G. and Tao, J. (1996) *Macromolecules*, **29**, 2487–2493.
187 Tao, J., Liu, G., Ding, J. and Yang, M. (1997) *Macromolecules*, **30**, 4084–4089.
188 Ding, J. and Liu, G. (1998) *Macromolecules*, **31**, 6554.
189 Ding, J. and Liu, G. (1999) *Langmuir*, **15**, 1738–1747.
190 Liu, G. (1997) *Adv. Mater.*, **9**, 437–439.
191 Liu, G., Ding, J., Qiao, L., Guo, A., Dymov, B.P., Gleeson, J.T., Hashimoto, T. and Saijo, K. (1999) *Chem. Eur. J.*, **5**, 2740–2749.
192 Liu, G., Qiao, L. and Guo, A. (1996) *Macromolecules*, **29**, 5508–5510.
193 Liu, G. and Ding, J. (1998) *Adv. Mater.*, **10**, 69.
194 Liu, G., Ding, J., Guo, A., Herfort, M. and Bazett-Jones, D. (1997) *Macromolecules*, **30**, 1851.
195 Yan, X., Liu, F., Li, Z. and Liu, G. (2001) *Macromolecules*, **34**, 9112–9116.
196 Stewart, S. and Liu, G. (2000) *Angew. Chem. Int. Ed. Engl.*, **39**, 340–344.
197 Liu, G., Ding, J., Hashimoto, T., Kimishima, K., Winnik, F.M. and Nigam, S. (1999) *Chem. Mater.*, **11**, 2233–2240.
198 Underhill, R.S. and Liu, G. (2000) *Chem. Mater.*, **12**, 2082–2091.
199 Xiaohu, Y., Liu, G., Liu, F., Tang, B.Z., Peng, H., Pakhomov, A.B. and Wong, C.Y. (2001) *Angew. Chem. Int. Ed. Engl.*, **40**, 3593–3596.
200 Abes, J.I., Cohen, R.E. and Ross, C.A. (2003) *Chem. Mater.*, **15**, 1125–1131.
201 Kane, R.S., Cohen, R.E. and Silbey, R. (1999) *Chem. Mater.*, **11**, 90–93.
202 Fogg, D.E., Radzilowski, L.H., Dabbousi, B.O., Schrock, R.R., Thomas, E.L. and Bawendi, M.G. (1997) *Macromolecules*, **30**, 8433–8439.
203 Fogg, D.E., Radzilowski, L.H., Blanski, R., Schrock, R.R. and Thomas, E.L. (1997) *Macromolecules*, **30**, 417–426.
204 Aizawa, M. and Buriak, J. (2006) *J. Am. Chem. Soc.*, **128**, 5877–5886.
205 Cohen, R.E. (2000) *Curr. Opin. Solid State Mater. Sci.*, **4**, 587–590.
206 Ciebien, J.F., Clay, R.T., Sohn, B.H. and Cohen, R.E. (1998) *New J. Chem.*, **22**, 685–691.

207 Kane, R.S., Cohen, R.E. and Silbey, R. (1996) *Chem. Mater.*, **8**, 1919–1924.
208 Cummins, C.C., Beachy, M.D., Schrock, R.R., Vale, M.G., Sankaran, V. and Cohen, R.E. (1991) *Chem. Mater.*, **3**, 1153–1163.
209 Kane, R.S., Cohen, R.E. and Silbey, R.J. (1999) *Langmuir*, **15**, 39–43.
210 Spatz, J.P., Herzog, T., Moessmer, S., Ziemann, P. and Moeller, M. (1999) *Adv. Mater.*, **11**, 149–153.
211 Arnold, M., Calvacanti, A.A., Glass, R., Blummel, J., Eck, W., Kessler, H. and Spatz, J.P. (2004) *ChemPhysChem*, **5** (3), 383.
212 Genzer, J. and Kramer, E.J. (1997) *Phys. Rev. Lett.*, **78**, 4946–4949.
213 Youngblood, J.P., Andruzzi, L., Ober, C.K., Hexemer, A., Kramer, E.J., Callow, J.A., Finlay, J.A. and Callow, M.E. (2003) *Biofouling*, **19**, 91–98.
214 Yokoyama, H. and Sugiyama, K. (2004) *Langmuir*, **20**, 10001–10006.
215 Langmuir, I. (1916) *J. Am. Chem. Soc.*, **38**, 2221.
216 Pan, F., Wang, P., Lee, K., Wu, A., Turro, N.J. and Koberstein, J.T. (2005) *Langmuir*, **21**, 3605–3612.
217 Cresce, A.V., Silverstein, J.S., Bentley, W.E. and Kofinas, P. (2006) *Macromolecules*, **39**, 5826–5829.
218 Hayakawa, T., Wang, J., Sundararajan, N., Xiang, M., Li, X., Glusen, B., Leung, G.C., Ueda, M. and Ober, C.K. (2000) *J. Phys. Org. Chem.*, **13**, 787–795.
219 Boker, A., Reihs, K., Wang, J., Stadler, R. and Ober, C.K. (2000) *Macromolecules*, **33**, 1310–1320.
220 Yang, S., Wang, J., Valiyaveetil, S. and Ober, C.K. (2000) *Chem. Mater.*, **12**, 33–40.
221 Sundararajan, N., Yang, S., Ogino, K., Valitaveetil, S., Wang, J., Zhou, X. and Ober, C.K. (2000) *Chem. Mater.*, **12**, 41–48.
222 Natansohn, A. and Rochon, P. (2002) *Chem. Rev.*, **102**, 4139.
223 Paik, M.Y., Krishnan, S., You, F., Li, X., Hexemer, A., Ando, Y., Ho Kang, S., Fischer, D.A., Kramer, E.J. and Ober, C.K. (2007) *Langmuir*, **23** (9), 5110.
224 Moller, G., Harke, M., Motschmann, H. and Prescher, D. (1998) *Langmuir*, **14**, 4955.
225 Feng, C.L., Jin, J., Zhang, Y.J., Song, Y.L., Xie, L.Y., Qu, G.R., Xu, Y. and Jiang, L. (2001) *Surf. Interface. Anal.*, **32**, 121.
226 Tong, X., Wang, G., Soldera, A. and Zhao, Y. (2005) *J. Phys. Chem.*, **109**, 20281–20287.
227 Ravi, P., Sin, S.L., Gan, L.H., Gan, Y.Y., Tam, K.C., Xia, X.L. and Hu, X. (2005) *Polymer*, **46**, 137.
228 Yokoyama, M., Inoue, S., Kataoka, K., Yui, N. and Sakurai, Y. (1987) *Makromol. Chem. Rapid Commun.*, **8**, 431–435.
229 Kwon, G.S. (1998) *Crit. Rev. Therap. Drug Carr. Sys.*, **15**, 481–512.
230 Ashley, J., Bernal, M.P., Burr, G.W., Coufal, H., Guenther, H., Hoffnagle, J.A., Jefferson, C.M., Marcus, B., Macfarlane, R.M., Shelby, R.M. and Sincerbox, G.T. (2000) *IBM J. Res. Dev.*, **44**, 341–368.
231 Ikeda, T. and Tsutsumi, O. (1995) *Science*, **268**, 1873–1875.
232 Yamamoto, T., Hasegawa, M., Kanazawa, A., Shiono, T. and Ikeda, T. (1999) *J. Phys. Chem. B*, **103**, 9873–9878.
233 Frenz, C., Fuchs, A., Schmidt, H.W., Theissen, U. and Haarer, D. (2004) *Macromol. Chem. Phys.*, **205**, 1246–1258.
234 Tong, X., Cui, L. and Zhao, Y. (2004) *Macromolecules*, **37**, 3101–3112.
235 Hackel, M., Kador, L., Kropp, D., Frenz, C. and Schmidt, H.W. (2005) *Adv. Funct. Mater*, **15**, 1722–1727.
236 Kita-Tokarczyk, K., Grumelard, J., Haefele, T. and Meier, W. (2005) *Polymer*, **46**, 3540–3563.
237 Liu, J., Sheina, E., Kowalewski, T. and McCullough, R.D. (2002) *Angew. Chem. Int. Ed. Engl.*, **41**, 329–332.
238 Li, M., Li, X. and Ober, C.K. (2001) *Polym. Mater. Sci. Eng.*, **84**, 715.
239 Hempenius, M.A., Langeveld-Voss, B.M.W., van Haare, J.A.E.H., Janssen, R.A.J., Sheiko, S.S., Spatz, J.P., Moller, M. and Meijer, E.W. (1998) *J. Am. Chem. Soc.*, **120**, 2798–2804.
240 Ruokolainen, J., Makinen, R., Torkkeli, M., Makela, T., Serimaa, R., ten Brinke, G. and Ikkala, O. (1998) *Science*, **280**, 557–560.

2
Block Copolymer Nanofibers and Nanotubes
Guojun Liu

2.1
Introduction

Block copolymer nanofibers in this chapter refer to cross-linked cylindrical structures that are made from block copolymers with diameters below 100 nm and lengths up to hundreds of micrometers. Figure 2.1 depicts the structures of nanofibers prepared from an A-B diblock and an A-B-C triblock copolymer, respectively. In the case of diblock nanofibers, either the core or the corona [1, 2] can be cross-linked. Once dried, nanofibers with cross-linked coronas may not re-disperse readily in solvents, the same as for block copolymer nanospheres with cross-linked shells [3]. Hence, our discussion in this chapter will be mainly on diblock nanofibers with cross-linked cores. For nanofibers with cross-linked cores, the soluble corona chains stretch into the solvent phase helping disperse the fibers, and the cross-linked core provides the structural stability. Although the nanofibers are depicted in Figure 2.1 as being rigid and straight, in reality, they can bend or contain kinks.

Three scenarios can be differentiated for triblock nanofibers. In scenario one, the corona block is cross-linked. Again because of dispersibility considerations we have not prepared and studied such fibers. Our focus so far has been on scenario two where the middle layer of the nanofiber is cross-linked. In scenario three, the innermost block is cross-linked. The structure of such a nanofiber bears close resemblance to block copolymer cylindrical brushes [4] when the inner most block is short relative to the other blocks. Block copolymer cylindrical brushes can be obtained by polymerizing a diblock macromer. There have been a number of reports on the properties and applications of block copolymer cylindrical brushes [4]. While this chapter will not go beyond triblock nanofibers, except for one case when nanofibers of a tetrablock copolymer are discussed, the methodologies developed for di- and triblock copolymer nanofiber preparations should apply equally well to a preparation of nanofibers from tetra- and pentablock copolymers and to more complex copolymers.

Figure 2.1 Structural illustration of a diblock nanofiber (top) and a triblock nanofiber (bottom).

Figure 2.2 Two configurations for a nanotube.

In principle, nanotubes can be prepared by cross-linking tubular micelles [5–8] of diblock copolymers. The focus of discussion here is, however, on nanotubes derived from triblock copolymers. Two possible configurations for such nanotubes are depicted in Figure 2.2.

In case one, the tubular core is void. Such a nanotube is derived from a triblock nanofiber with a cross-linked intermediate layer by degrading the innermost block [9]. In case two, the tubular core is lined by a polymer. In this instance, a nanotube is derived from a triblock nanofiber with a cross-linked intermediate layer by cleaving pendant groups off the core block [10].

The preparation of block copolymer nanofiber [11] and nanotube [9] structures were only reported for the first time a few years ago. Over the years, there have been reports on the use of block copolymer nanofibers and nanotubes as vehicles for drug delivery [12], as scaffolds for cell growth [13, 14], as precursors for ceramic magnetic nanowires [15, 16] and as precursors for carbon nanofibers [17, 18], etc. While finding novel applications for such structures is of paramount importance, the emphasis of this chapter will be on research undertaken in my group aimed at achieving a fundamental understanding of the physical and chemical properties of these materials. In Section 2.2, the preparation of block copolymer nanofibers and nanotubes will be described and the solution properties of the nanofibers and nanotubes will be discussed in Section 2.3. The different reaction patterns of nanofibers and nanotubes will be examined in Section 2.4. In Section 2.5, some conclusions will be drawn and my perspectives on where block copolymer nanofiber and nanotubes research is going will be presented.

2.2
Preparation

2.2.1
Nanofiber Preparation

Nanofibers can be prepared from chemically processing either a block-segregated copolymer solid [19] or block copolymer cylindrical micelles [20] formed in a block-selective solvent. Figure 2.3 depicts the processes involved to obtain nanofibers from a block-segregated diblock solid.

The first step involves casting a film from a diblock with an appropriate composition so that the minority block segregates from the majority block forming hexagonally packed cylinders. This is followed by cross-linking the cylindrical domains. The cross-linked cylinders are then levitated from the film or separated from one another by stirring the film in a solvent that solubilizes the uncross-linked diblock.

Our literature search revealed that the first report on block copolymer nanofiber synthesis appeared in 1996 [11]. In that report, block-segregated solids of two polystyrene-*block*-poly(2-cinnamoyloxyethyl methacrylate), or PS-PCEMA, samples were used as the precursor. The two PS-PCEMA samples used had 1250 and 160 units of PS and PCEMA for polymer 1 and 780 and 110 for polymer 2, respectively. Thus, the volume fraction of PCEMA was ≈26% in both samples. After film formation from a diblock by evaporating a toluene solution slowly, the film was annealed at 90 °C for days to ensure clean phase segregation between PS and PCEMA. This film was then sectioned by ultra-microtoming to yield thin slices for transmission electron microscopic (TEM) examination. Shown in the left panel of Figure 2.4 is

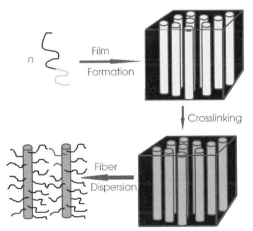

Figure 2.3 Schematic illustration of the steps involved in the preparation of diblock copolymer nanofibers from a block-segregated solid.

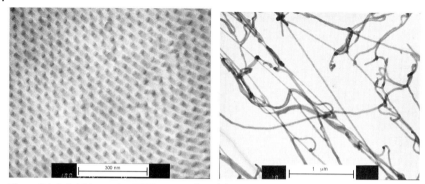

Figure 2.4 Left: Thin-section TEM image of a solid PS$_{1250}$-PCEMA$_{160}$ sample. The dark PCEMA cylinders are slanted diagonally (from bottom right to top left). Right: TEM image of the PS$_{1250}$-PCEMA$_{160}$ nanofibers on a carbon-coated copper grid aspirated from THF.

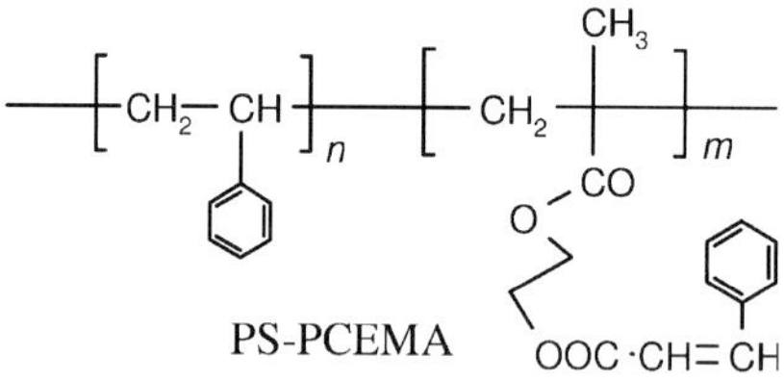

Scheme 2.1

a TEM image of a thin section for polymer 1 or PS$_{1250}$-PCEMA$_{160}$, where the subscripts denote the numbers of styrene and CEMA units, respectively (Scheme 2.1). It can be seen that the OsO$_4$-stained dark PCEMA phase consist of hexagonally packed cylinders dispersed in the PS matrix and that the cylinders slant, aligned along a diagonal direction of the image. A similar PCEMA block segregation pattern was found for polymer 2. Such films were then irradiated by UV light to cross-link the PCEMA cylindrical domains. Stirring the irradiated samples in THF helped levitate the cross-linked PCEMA cylinders from the films to yield solvent-dispersible nanofibers. In the right panel, a TEM image of the resultant nanofibers is shown. It can be observed that the fibers in this sample are still entangled.

The groups working with Wiesner [21], Ishizu [22], Muller [23] and my group [24], have also used chemical rather than photochemical methods to cross-link the cylindrical domains of block copolymer solids to prepare nanofibers. To prepare nanofibers from polystyrene-*block*-polyisoprene (PS-PI), we utilized a diblock that had 220 styrene and 140 isoprene units, respectively, which corresponded to a volume fraction of ≈30% for PI. The cylindrical domains were then cross-linked by exposing the film to sulfur monochloride (Scheme 2.2).

Isolated nanofibers were obtained by separating the cross-linked cylindrical domains in THF. The left panel of Figure 2.5 shows a TEM image of such nanofibers.

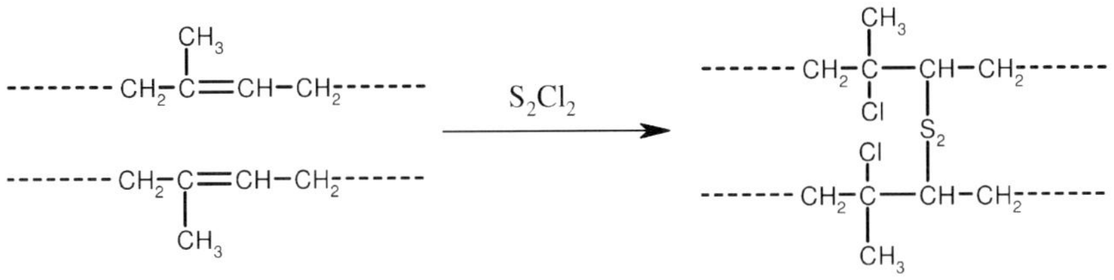

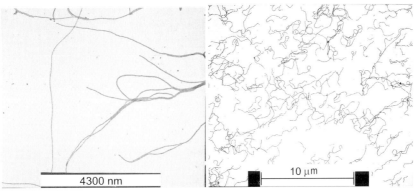

Scheme 2.2

Figure 2.5 Left: TEM image of PS-PI nanofibers prepared from the cross-linking of block-segregated PS_{220}-PI_{140} solid. The length of the white box is 4300 nm. Right: TEM image of PS-PI nanofibers prepared from the cross-linking of PS_{130}-PI_{370} cylindrical micelles.

A drawback with chemical cross-linking of block-segregated solids is the long diffusion time required for the cross-linker to penetrate the film. Insufficient reaction time leads to non-uniform cross-linking with higher degrees of cross-linking found close to the surfaces. Such non-uniform cross-linking can occur with photo-cross-linking for the short penetration distance of the light. These complications can be avoided by preparing nanofibers starting from block copolymer cylindrical micelles formed in a block-selective solvent.

Figure 2.6 depicts processes involved in preparation of diblock nanofibers starting from cylindrical micelles. Firstly, this requires the preparation of a diblock with an appropriate composition. Then, a selective solvent has to be found that solubilizes only one block of the diblock copolymer. In such a block-selective solvent, the insoluble blocks of different chains aggregate to form a cylindrical core stabilized by chains of the soluble block. Nanofibers are obtained by cross-linking the core chains.

A report on the preparation of block copolymer nanofibers from the cross-linking of cylindrical micelles first appeared in 1997 [25]. The cylindrical micelles were prepared also from PS-PCEMA in refluxing cyclopentane, which solubilized PS and not PCEMA. Nanofibers were obtained after the photocross-linking of the

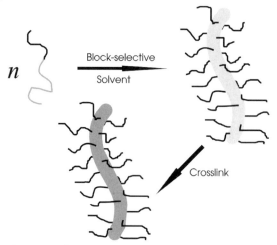

Figure 2.6 Schematic illustration of the steps involved in the preparation of diblock copolymer nanofibers from cylindrical micelles formed in a block-selective solvent.

PCEMA cores. Aside from photocross-linking, the groups working with Bates [26], Discher [27], Manner [2], Wooley [28] and Stupp [13], in addition to my group [29, 30], have reported on the chemical cross-linking of cylindrical micelles to prepare nanofibers. For example, recently we prepared nanofibers by cross-linking PS-PI cylindrical micelles formed in a block-selective solvent, N,N-dimethyl acetamide, for PS [24, 31]. For cylindrical micelle formation with PI as the core, the PI weight fraction should be relatively high. We followed the recipe of Price [32] and used a sample consisting of 130 styrene units and 370 isoprene units. The cylindrical fibers were cross-linked using S_2Cl_2. The right-hand panel of Figure 2.5 shows a TEM image of such nanofibers. Compared with fibers prepared from the solid-state syntheses described above, the fibers in the right-hand panel of Figure 2.5 are shorter. A more quantitative analysis of the length distributions revealed that the fibers from the solution preparation approach are more monodisperse. Furthermore, the preparation yield can be high, approaching 100%.

2.2.2
Nanotube Preparation

Nanotubes have been prepared by my group mainly via the derivatization of triblock nanofibers. The first block copolymer nanotubes were prepared from PI_{130}-$PCEMA_{130}$-$PtBA_{800}$ [9], where PtBA denotes poly(tert-butyl acrylate). This involved firstly dispersing the triblock in methanol. As only the PtBA block was soluble in methanol, the triblock self-assembled into cylindrical micelles consisting of a PI core encapsulated in an insoluble PCEMA intermediate layer and a PtBA corona. After photocross-linking the PCEMA intermediate layer, the PI core block was

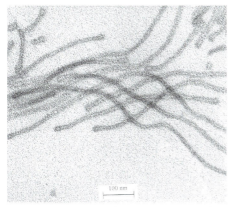

Figure 2.7 TEM image of PCEMA-P*t*BA nanotubes where the PI core has been degraded.

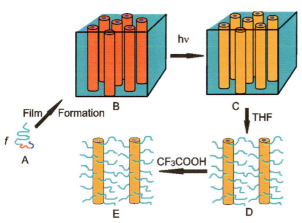

Figure 2.8 Schematic illustration of processes involved in preparing PS-PCEMA-PAA nanotubes.

degraded by ozonolysis to yield nanotubes. The removal of the PI block was demonstrated by infrared absorption and TEM analyses. More importantly, Rhodamine B could be loaded into the tubular core. Figure 2.7 shows a TEM image of such nanotubes stained by OsO$_4$. The center of each tube appears lighter than the PCEMA intermediate layer because the PI block was decomposed.

To facilitate the incorporation of inorganic species into the tubular core, nanotubes containing PAA-lined cores were more desirable. Figure 2.8 depicts the steps involved in the preparation of PS-PCEMA-PAA nanotubes from PS$_{690}$-PCEMA$_{170}$-P*t*BA$_{200}$ via a solid-state precursor approach [10].

Step 1 (A → B in Figure 2.8) involved casting films from the triblock containing concentric P*t*BA and PCEMA core-shell cylinders dispersed in the matrix of PS. This required the PS volume fraction to be ≈70% [33] and was achieved by mixing some PS homopolymers (hPS) with the triblock to increase the PS volume fraction.

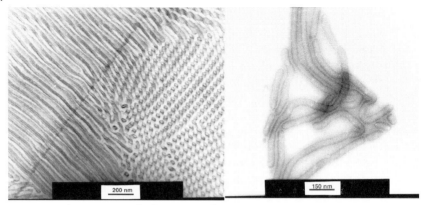

Figure 2.9 Left: TEM image of a thin section of the PS_{690}-$PCEMA_{170}$-$PtBA_{200}$ solid. Right: TEM image of the PS-PCEMA-PAA nanotubes.

The left-hand panel of Figure 2.9 shows a TEM image of a thin section of an hPS/PS-PCEMA-PtBA solid. On the right-hand side of this image we can see numerous concentric light and dark ellipses with short stems. These represent projections of cylinders with PCEMA shells and PtBA cores aligned slightly off the normal direction of the image. The PCEMA shells appear darker, because OsO_4 stained the PCEMA selectively. The diameter of the PtBA core is ≈20 nm. On the left of this image we see cylinders lying in the plane of the picture. Thus, the orientation of the cylindrical domains varied from grain to grain, in the micrometer size range, because we did not take special measures to effect their macroscopic alignment.

In step 2 (B → C, in Figure 2.8), the block-segregated copolymer film was irradiated with UV light to cross-link the PCEMA shell cylinder. The cross-linked cylinders were levitated from the film by stirring in THF (C → D). PS-PCEMA nanotubes containing PAA-lined tubular cores were prepared by hydrolyzing the PtBA block in methylene chloride and trifluoroacetic acid (D → E). The right-hand panel of Figure 2.9 shows a TEM image of the intestine-like nanotubes. The stained PCEMA layer does not have a uniform diameter across the nanotube length because of its uneven collapse during solvent evaporation. The presence of PAA groups inside the tubular core was demonstrated by our ability to carry out various aqueous reactions inside the tubular core, as will be discussed later.

2.3
Solution Properties

Figure 2.10 shows a comparison between the structures of a PS-PI nanofiber and a poly(n-hexyl isocyanate), PHIC, chain. In a PHIC chain, the backbone is made of imide units joined linearly and the hairs are the hexyl groups. Their counterparts in a PS-PI nanofiber are the cross-linked PI cylinder and PS chains, respec-

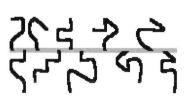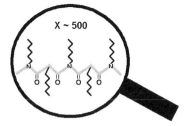

Figure 2.10 Structural comparison between a PS-PCEMA nanofiber (left) and a PHIC chain (right) at different magnifications.

tively. Other than a large size difference, a nanofiber bears remarkable structural resemblance to PHIC. Thus, block copolymer nanofibers can be viewed as a macroscopic counterpart of a polymer chain or a "suprapolymer" chain or "giant" polymer chain [34]. In this subsection, preliminary results showing the similarities and dissimilarities between the solution properties of polymer chains and diblock nanofibers will be reviewed.

To study the dilute solution properties of nanofibers and polymer chains, the fibers should be made sufficiently short, so that they remain dispersed in the solvent for a long, or even an infinitely long, period of time. The use of relatively short nanofibers also ensured their characterization by classic techniques, such as light scattering (LS) and viscometry. While we have studied nanofibers prepared from several block copolymer families, for clarity, the discussion will be restricted to PS-PI nanofibers obtained by cross-linking cylindrical micelles of PS_{130}-PI_{370} formed in N,N-dimethyl acetamide [24]. The preparation of such fibers has been discussed previously and the right-hand panel of Figure 2.5 shows a TEM image of the nanofibers thus prepared in THE after aspiration onto a carbon-coated copper grid. As the magnification was known for such images, we were able to measure manually the lengths of more than 500 fibers for this sample. The data from such measurements allowed us to construct the length distribution function of this sample denoted as fraction 1 or F1 in Figure 2.11. From the length distribution function, we obtained the weight- and number-average lengths and L_w and L_n. The L_w and L_w/L_n values are 3490 nm and 1.35 for this sample.

While ultracentrifugation [34] or density gradient centrifugation could have been used, in principle, to separate the fibers into fractions of different lengths, we obtained nanofiber fractions with shorter lengths by breaking up the longer nanofibers by ultrasonication [26]. By adjusting the ultrasonication time, we produced fibers of different lengths. Also shown in Figure 2.11 are the length distribution functions for samples denoted as F3 and F5, which were ultrasonicated for 4 and 20 h, respectively. As ultrasonication time increased, the distribution shifted to shorter lengths.

These fiber fractions were sufficiently short and allowed us to determine their weight-average molar, M_w, by light scattering. Figure 2.12 shows a Zimm plot

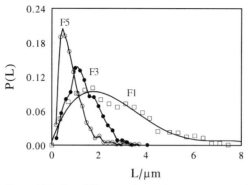

Figure 2.11 Plot of fiber population density P(L) versus length L for PS_{130}-PI_{370} nanofiber fractions 1 (□), 3 (●) and 5 (○) generated from TEM image analysis.

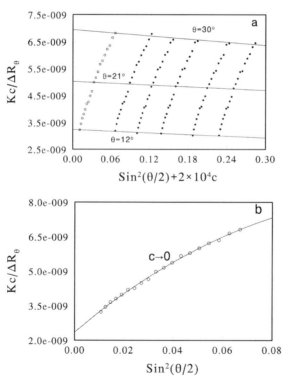

Figure 2.12 Zimm plot for the light scattering data of F3 in the scattering angle range of 12 to 30°. The solid circles represent the experimental data. The hollow circles represent the extrapolated $Kc/\Delta R_\theta|_{c\to 0}$ data. (a) Linear extrapolation of data to zero concentration at the highest and lowest scattering angles of 30 and 12° is illustrated. (b) The result of curve fitting of the $Kc/\Delta R_\theta|_{c\to 0}$ data using Equation (2.1).

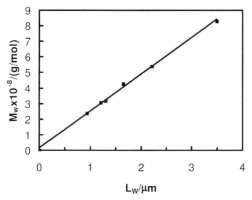

Figure 2.13 Increase in LS M_w with TEM L_w for PS_{130}-PI_{370} nanofiber fractions.

for the light scattering data for sample F3 in the scattering angle, θ, range of 12 to 30°.

The data quality appears high. Multiple runs of the same sample indicated that the data precision was high.

For the large-sized fibers, the $Kc/\Delta R_\theta$ data varied with $\sin^2(\theta/2)$ or the square of the scattering wave vector q non-linearly, despite the low angles used. We fitted the data using Equation (2.1):

$$\frac{Kc}{\Delta R_\theta} = \frac{1}{M_w}[1+(1/3)q^2 R_G^2 - kq^4 R_G^4] + 2A_2 c \qquad (2.1)$$

and obtained M_w, the radii of gyration R_G and the second Virial coefficient A_2 for the different fractions. Figure 2.13 plots the resultant M_w, versus L_w, where the values for L_w were obtained from TEM length distribution functions $P(L)$. The linear increase in M_w with L_w suggests the validity of the M_w values determined. The validity of the M_w value for F3 was further confirmed recently by Professor Chi Wu's group at the Chinese University of Hong Kong, who performed a light scattering analysis of a nanofiber sample down to $\theta = 7°$. At such low angles, the $kq^4 R_G^4$ term in Equation (2.1) was not required for curve fitting and data analysis by the Zirnm method should yield accurate M_w and R_G values.

After nanofiber characterization, we then proceeded to check the dilute solution viscosity properties. Our experiments indicated that the nanofiber solutions were analogous to polymer solutions and were shear thinning, i.e., the viscosity of a sample decreased with increasing shear rate. This occurred for the alignment of the nanofibers along the shearing direction above a shear rate γ of $\approx 0.1\,s^{-1}$ [35]. While both nanofiber and polymer solutions are shear thinning, the fields required for shear thinning are dramatically different. Polymers of ordinary molar mass, e.g. $<10^6\,g\,mol^{-1}$, would experience shear thinning only if $\gamma \gtrsim 10^{-4}\,s^{-1}$ [36].

The huge difference should be a direct consequence of the drastically different sizes between the two.

To minimize the shear-thinning effect, we measured the viscosities of dilute solutions of the nanofiber fractions in THF using a laboratory-built rotating cylinder viscometer at $\gamma = 0.082\,\text{s}^{-1}$ [37]. Figure 2.14 shows the $(\eta_r - 1)/c$ data plotted against nanofiber concentrations c, where η_r, the relative viscosity, is defined as the ratio between the viscosities of the nanofiber solution and solvent THF. The solid lines represent the best fit to the experimental data by Equation (2.2):

$$(\eta_r - 1)/c = [\eta] + k_h [\eta]^2 c \tag{2.2}$$

where $[\eta]$ is the intrinsic viscosity and k_h is the Huggins coefficient. The linear dependence between $(\eta_r - 1)/c$ and c is in striking agreement with the behavior of polymer solutions. Even more interesting, k_h took values mostly between 0.20 and 0.60 in agreement with those found for polymers [36].

We further treated the $[\eta]$ data with the Yamakawa–Fujii–Yoshizaki (YFY) theory, originally developed for wormlike chains [38, 39]. According to Bohdanecky [40], the YFY theory could be cast in a much simpler form, Equation (2.3):

$$\left(M_w^2/[\eta]\right)^{1/3} = A + BM_w^{1/2} \tag{2.3}$$

for chains with a wide range of reduced chain lengths. In Equation (2.3), A and B are fitting parameters that are related to the persistence length l_p and the hydrodynamic diameter d_h of the chains, respectively. Figure 2.15 shows the data that we obtained for the PS_{130}-PI_{370} nanofibers in THF plotted following Equation (2.3). From the intercept A and slope B of the straight line, we calculated l_p and d_h for the nanofibers to be (1040 ± 150) and (69 ± 18) nm, respectively.

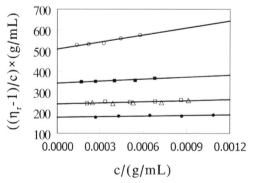

Figure 2.14 From top to bottom, plot of $(\eta_r - 1)/c$ versus c for PS_{130}-PI_{370} nanofiber fractions 2, 3, 4 and 6 in THF. All the η_r data were obtained using the viscometer at a shear rate of $0.082\,\text{s}^{-1}$ with the exception of those denoted by (Δ), which were obtained at a shear rate of $0.047\,\text{s}^{-1}$.

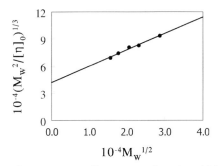

Figure 2.15 Nanofiber viscosity data plotted following the Bohdanecky method.

Table 2.1 Persistence length l_p, and hydrodynamic diameter d_h of the nanofibers calculated from the viscosity data for PS_{130}-PI_{370} nanofiber fractions in various solvents.

Solvent	d_h/nm	l_p/nm
THF	69 ± 18	1040 ± 150
THF – DMF = 50/50	61±	850 ± 90
THF – DMF = 30/70	51 ± 12	830 ± 60

This procedure was repeated for the nanofibers in different solvents. Table 2.1 summarizes the l_p and d_h values that we determined in three different solvents for the PS_{130}-PI_{370} nanofibers. The d_h value in THF compares well with what we estimated from the sum between the diameter of the cross-linked PI core determined from TEM and the root-mean-square end-to-end distance of the PS coronal chains, and thus suggests the applicability of the FYF theory to the nanofiber solutions. What is more convincing is the decreasing trend for the determined d_h values with increasing DMF content in THF–DMF mixtures. While both THF and DMF solubilize PS, d_h decreased with increasing DMF content because the extent of swelling for the cross-linked PI core decreased with increasing DMF content.

The l_p values reported in Table 2.1 are comparable to those reported by Discher and coworkers [41, 42] and by Bates and coworkers [43, 44] for PEO-PI cylindrical micelles with a core diameter of ≈20 nm prepared in water, where PEO denotes poly(ethylene oxide). While Bates and coworkers deduced the l_p values from small-angle neutron scattering, Discher and coworkers determined the l_p values from fluorescence microscopy. In the latter case, they compared the dynamic behavior of single cylindrical micelles before and after PI core cross-linking. After micelle cross-linking, the micelles became much more rigid dynamically, which means that the contour or conformation of the fibers, in contrast to the micelles, changed or flexed very little with time, despite their rotation in space as approximate rigid rotors. By performing a dynamic analysis of the flexion motion by subtracting off the spontaneously curved average shape of the fibers, they concluded that the dynamic l_p values of the fibers were about 50 times higher than those of the cylindrical micelles. From viscometry, one deduces the *static* l_p values of the nanofibers, which measure on average how much an ensemble of fibers bends. Therefore, one should not compare the viscometry l_p values determined by us with those determined by Discher and coworkers, who totally ignored the locked-in curvatures of the fibers in their analysis.

To get a clue to the static l_p values of the PEO-PI fibers studied by Discher and coworkers, from their fluorescence microscopy images we noticed that the kinks in the original cylindrical micelles were locked in after micelle cross-linking and the fibers assumed conformations similar to those before micelle cross-linking. Thus, the static l_p values of the nanofibers should be similar to those of the cylindrical micelles. The fact that the l_p values that we determined from viscometry are comparable to those of the PEO-PI cylindrical micelles with similar core diameters again suggests the validity of the YFY theory in treating the nanofiber viscosity data.

The above study demonstrates that block copolymer nanofibers have dilute solution properties similar to those of polymer chains. In an earlier report [45], we also demonstrated that block copolymer nanofibers have concentrated solution properties similar to those of polymer chains. According to the theories of Onsager [46] and Flory [47], polymer chains with $l_p/d_h > 6$ would form a liquid crystalline phase above a critical concentration. We did show the presence of such a liquid crystalline phase by polarized optical microscopy for PS-PCEMA nanofibers dissolved in bromoform at concentrations above ≈25 wt-% [45]. Furthermore, we observed that such liquid crystalline phases disappeared as the temperature was raised and the liquid crystalline to disorder transition was fairly sharp.

While block copolymer nanofibers behave similarly to polymer chains in many aspects, the drastic size difference between the two dictates that they have substantial property differences. Because of the large size of the nanofibers, they obviously move more sluggishly. Hence, we observed that a liquid crystalline phase was formed only after the PS-PCEMA nanofiber solution was sheared mechanically. Also, because of their sluggishness, the liquid crystalline phase could not reform spontaneously after cooling a system if it had been heated above the liquid crystalline to disorder transition temperature. Thus, we can predict, without performing any sophisticated experiments, that the analogy between nanofibers and polymer chains will fail after the molar mass or the size of the nanofibers exceeds a critical value. As the size of the nanofiber increases, the gravitational force driving the settling of the nanofiber increases and the dispersibility of the nanofiber decreases. Furthermore, the van der Waals forces between different nanofibers increase [48], which can cause different nanofibers to cluster and settle.

We recently examined the stability of nanofibers dispersed in THF prepared from PS_{130}-PI_{370}. This particular nanofiber sample had $L_w = 1650$ nm, $L_W/L_n = 1.21$ and $M_w = 4.3 \times 10^8$ g mol^{-1}, respectively. At a concentration of $\approx 8 \times 10^{-3}$ g mL^{-1} and under gentle stirring, no nanofiber settling was observed during 4 days of observation by light scattering. Without stirring, we noticed a 10% decrease in the light scattering intensity of the solution, which corresponded to ≈10 wt-% settling of the nanofibers in the first 4 days. No noticeable further settling was observed in another 8 days [24]. This could indicate that the longer fibers in this sample exceeded the critical size for settling. Our light scattering and centrifugation experiments suggested that the longer fibers first clustered and then settled. The fact that the clustering could be prevented by gentle stirring suggests that only a very shallow attraction potential existed between the fibers.

Although the critical length for settling was short for this sample, several rnicrometers, the critical length depends on many factors including the relative length between the core and soluble block and the absolute diameter of the cores. Methods of increasing the nanofiber dispersity may include increasing the length of the soluble block relative to the core block and decreasing the core diameter.

Because of the differences between the polymer chains and the nanofibers, we expect differences in the performances of these two classes of bulk materials. Unfortunately, the mechanical properties of block copolymer nanofibers or nanofiber composites have not been studied so far. We have not performed any detailed studies of solution properties of block copolymer nanotubes. As a result of the structural similarities between the two, we expect the nanofibers and nanotubes to have many similar solution properties.

2.4
Chemical Reactions

The similarities between the structure and the properties of the solutions between nanofibers and polymer chains prompted us to ask the question as to whether nanofibers and nanotubes would have chemical reaction patterns similar to those of polymer chains. A PI chain can be hydrogenated via "backbone modification" to yield a polyolefin chain. Through techniques such as anionic polymerization, etc., one can readily prepare "end-functionalized" polymers. The end-groups can be further derivatized or used for additional end-fiinctionalization. This section will show that block copolymer nanotubes can also undergo backbone modification and end-functionalization.

2.4.1
Backbone Modification

Backbone modification has already been involved to convert triblock nanofibers into nanotubes. Apart from the performance of organic reactions to the nanofibers and nanotubes, this sub-section discusses the performance of inorganic reactions in the cores of the nanotubes to convert them into polymer–inorganic hybrid nanofibers. Block copolymer nanofibers and nanotubes are soft materials. They will most probably find applications in bio-related disciplines, such as in the medical, pharmaceutical and cosmetic industries. For applications in nanoelectronic devices, polymer–inorganic hybrid nanofibers would be more desirable [49,

50]. The first report on the preparation of block copolymer–inorganic hybrid nanofibers appeared in 2001, which dealt with filling of the core of the PS-PCEMA-PAA nanotubes by γ-Fe$_2$O$_3$ [10].

The preparation first involved the equilibration between the nanotubes and FeCl$_2$ in THF. Fe(II) entered the nanotube core to bind with the core carboxyl groups. The extraneous FeCl$_2$ was then removed by precipitating the Fe(II)-containing nanotubes into methanol. Adding NaOH dissolved in THF containing 2 vol-% of water precipitated Fe(II) trapped in the nanotube core as ferrous oxide. The ferrous oxide was subsequently oxidized to γ-Fe$_2$O$_3$ via the addition of hydrogen peroxide [51]. The top panel in Figure 2.16 shows a TEM image of the hybrid nanofibers. The γ-Fe$_2$O$_3$ particles can be seen to be produced exclusively inside the nanotube cores.

The production of γ-Fe$_2$O$_3$ in the confined space of the "nanotest-tubes" resulted in particles that were nanometer-sized. Hence the particles were superparamag-

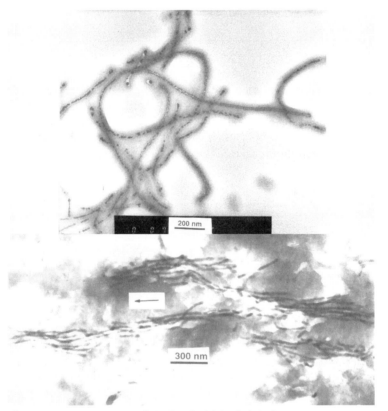

Figure 2.16 Top: TEM image of PS-PCEMA-PAA/Fe$_2$O$_3$ hybrid nanofibers. Bottom: Bundling and alignment of the nanofibers in a magnetic field. The arrow indicates the magnetic field direction.

netic, as demonstrated by the results of our magnetic property measurement [10]. This meant that they were magnetized only in the presence of an external magnetic field and were demagnetized when the field was removed. To see how such fibers behaved in a solvent in a magnetic field, we dispersed the fibers in a solvent mixture consisting of THF, styrene, divinylbenzene, and a free radical initiator AIBN. The fiber dispersion was then dispensed into an NMR tube and mounted in the sample holder of an NMR instrument. In the 4.7-T magnetic field of the NMR, the solvent phase was gelled by raising the temperature to 70 °C to polymerize styrene and divinylbenzene. Thin sections were obtained from the gelled sample by ultramicrotoming. Shown in the bottom panel of Figure 2.16 is a TEM image of nanofibers in a gelled sample. One consequence of the induced magnetization of the fibers is that they attracted one another and bundled in a magnetic field. Also clear from this image is that the fibers aligned along the magnetic field direction.

The bundling and alignment of the hybrid nanofibers in a magnetic field have important practical implications. For example, the controlled bundling of several nanofibers may form the basis of magnetic nanomechanical devices. For the construction of water-dispersible magnetic nanomechanical devices, the superparamagnetic nanofibers need to be water dispersible. We recently prepared water-dispersible polymer–Pd hybrid catalytic nanofibers from a tetrablock copolymer [52] and more recently polymer–Pd–Ni superparamagnetic nanofibers from a triblocic copolymer [53]. The tetrablock that we used was PI-PtBA-P(CEMA-HEMA)-PGMA, where PGMA, being water soluble, denotes poly(glyceryl methacrylate) and P(CEMA-HEMA) denotes a random copolymer of CEMA and 2-hydroxyethyl methacrylate. The hydroxyl groups of the precursory PHEMA block was not fully cinnamated because P(CEMA-HEMA) facilitated the transportation of Pd^{2+} and Ni^{2+}.

We prepared the polymer–Pd hybrid nanofibers following the scheme depicted in Figure 2.17 [52]. This involved first dispersing freshly-prepared PI-PtBA-P(CEMA-HEMA)-PGMA in water to yield cylindrical aggregates (A → B in Figure 2.17). Such aggregates consisted of a PGMA corona and a PI core. Sandwiched between these two layers are a thin PtBA layer and a P(CEMA-HEMA) layer. Such cylindrical aggregates were then irradiated to cross-link the (CEMA-HEMA) layer (B → C). The PI core was degraded by ozonolysis (C → D). By controlling the ozonolysis time, we could control the degree of PI degradation. When not fully degraded, the residual double bonds of the PI fragments trapped inside the nanotubular core were able to sorb Pd(II), most probably via π-allyl complex formation, Scheme 2.3.

The complexed Pd(II) was then reduced by $NaBH_4$ to Pd (D → E). The left panel of Figure 2.18 is a TEM image for such nanotubes containing 4.0 wt-% reduced Pd nanoparticles. The Pd-loaded nanofibers were dispersible in water where many water-based electroless plating reactions occur. Thus, Pd could serve as a catalyst for the further electroless deposition of other metals. We, for example, loaded more Pd into the tubular core via electroless Pd plating onto the initially formed Pd nanoparticles to yield essentially continuous Pd nanowires (E → F,

2 Block Copolymer Nanofibers and Nanotubes

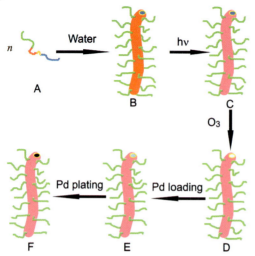

Figure 2.17 Schematic illustration of the processes involved to produce water-dispersible polymer–Pd hybrid nanofibers.

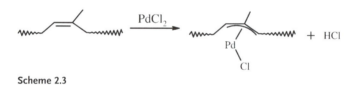

Scheme 2.3

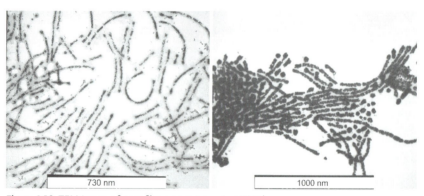

Figure 2.18 TEM image of nanofibers containing 4 wt-% Pd (left); TEM image of nanofibers containing 18.4 wt-% Pd (right). The scale bars in the form of white boxes are 730 and 1000 nm long, respectively.

Figure 2.17). The right-hand image in Figure 2.18 shows a TEM image of such hybrid nanofibers after the incorporation of Pd to a total of 18.4 wt-%. In fact, we could tune the amount of Pd loaded into the nanotubes by adjusting the relative amounts of the Pd-loaded nanotubes and Pd^{2+} in a plating bath. As the Pd content increased, eventually the hybrid fibers could not be dispersed in water. In addition to Pd plating, we have also succeeded in the plating of Ni into the core of such nanotubes, as evidenced by our recent success in preparing water-dispersible triblock–Pd–Ni magnetic nanofibers [53].

The use of a tetrablock copolymer for the above project seems to be an over kill, as in the end the PtBA block was not used at all. This was, however, not by design. Our initial plan was to fully degrade the PI block and then to hydrolyze the PtBA block. We planned to introduce Pd(II) via its binding with the carboxyl groups of PAA. The binding between Pd(II) and the residual double bonds of PI was a surprise to us and should be useful in the future as a staining method for PI in the elucidation of complex segregation patterns of block copolymers.

2.4.2
End Functionalization

In the "supramolecular chemistry" of nanotubes and nanofibers, the end-functional groups should not be interpreted as the traditional carboxyl or amino groups, etc. Rather, they should be other nano "building blocks", including nanospheres and nanofibers or nanotubes of a different composition. We first end-functionalized PS-PCEMA-PAA nanotubes by attaching to them water-dispersible PAA-PCEMA nanospheres and spheres bearing surface carboxyl groups that were prepared from emulsion polymerization [54]. Figure 2.19 shows the steps involved in coupling the nanotubes and PCEMA-PAA nanospheres.

To ensure that the PAA core chains were exposed at the ends, we ultrasonicated the pristine nanotubes to shorten them. The carboxyl groups of the nanotubes were then reacted with the amino groups of a triblock PAES-PS-PAES, where PAES denotes poly[4(2-aminoethyl)styrene], in the presence of catalyst 1-[3-

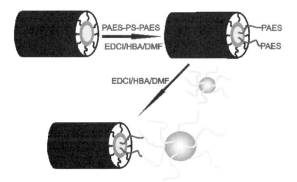

Figure 2.19 End-functionalization of PS-PCEMA-PAA nanotubes by PAA-PCEMA nanospheres.

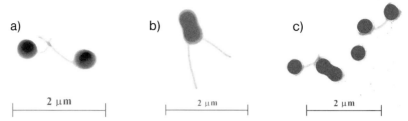

Figure 2.20 TEM images of nanotube and emulsion nanosphere coupling products.

(dimethyl-amino)propyl]-3-ethylcarbodiimide hydrochloride (EDCI) and co-catalyst 1-hydroxy-benzotriazole (HBA). As PAES-PS-PAES was used in excess, it was grafted onto the nanotubes mainly via one end only. We then purified the sample. Nanotubes bearing terminal PAES chains were subsequently reacted with the carboxyl groups on the surfaces of the nanospheres again using EDCI and HBA as the catalysts. Coupling between the nanotubes and emulsion spheres containing surface carboxyl group was achieved in a similar manner.

Figure 2.20 shows the typical products obtained from coupling the PAES-PS-PAES-treated nanotubes with a batch of emulsion nanospheres bearing surface carboxyl groups. The product in Figure 2.20a resulted from the coupling between one tube and one sphere. As the spherical "head" is water-dispersible and the tube "tail" is hydrophobic, this structure may be viewed as a macroscopic counterpart of a surfactant molecule or a "super-surfactant". Figure 2.20b shows the attachment of two tubes to one sphere, which had fused with another sphere probably during TEM specimen preparation. "Dumbbell-shaped molecules" were formed from the attachment of one tube to two spheres at the opposite ends, as seen in Figure 2.20c. The products depicted in Figure 2.20a–c co-existed regardless of whether we changed the tube to microsphere mass ratio from 20/1 to 1/20. At the high tube to emulsion sphere mass ratio of 20/1, the super-surfactant and dumbbell-shaped species were the major products. At a mass ratio of 1/1 and 1/20, the dumbbell-shaped product dominated. Other than product control by adjusting the stoichiometry, an effective method to eliminate the dumbbell-shaped product was to use nanotubes labeled at only one end by PAES-PS-PAES. These tubes were obtained by using ultrasonication to break up nanotubes that contained end-grafted PAES-PS-PAES chains. For example, the ultrasonication of the PAES-PS-PAES-grafted nanotubes for 8 h reduced the L_w of a nanotube sample from 701 to 252 nm and L_n from 515 to 187 nm. The reaction between the shortened tubes and the nanospheres at a tube to sphere mass ratio of 1/20 yielded almost exclusively the supersurfactant structure with unreacted nanospheres. The content of the multi-armed structure increased as the nanotube to microsphere mass ratio increased.

More recently, we have used the same chemistry to couple PS-PCEMA-PAA nanotubes with PGMA-PCEMA-PAA nanotubes [55]. Figure 2.21 shows the nano-

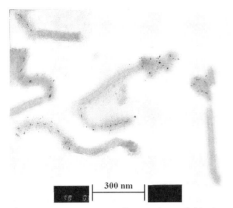

Figure 2.21 TEM image of nanotube multiblocks.

tube multiblocks that we prepared. To facilitate the easy differentiation between the two types of nanotubes, we loaded Pd nanoparticles into the PGMA-PCEMA-PAA nanotubes. Using this method, we succeeded in preparing both nanotube di- and triblocks with structures similar to di- and triblock copolymers. The nanotube multiblocks should self-assemble in a manner similar to the block copolymers.

2.5
Concluding Remarks

Block copolymer nanofibers and nanotubes can now be readily prepared with high yields. Such nanofibers have interesting chemical and physical properties. More pressing challenges in this area of research are to find and to realize the commercial applications for these nanostructures. The latter will be greatly facilitated with more participation from industrial partners.

On the fundamental research side, the construction and study of superstructures prepared from the coupling of different nano- and micro-components are a very interesting and promising area of frontier research. The super-surfactants of Figure 2.20a may, for example, self-assemble as do surfactant molecules to form supermicelles or artificial cells, which contain structural order to several length scales. The multi-armed structure of Figure 2.20b and its analogues will be of particular irrterest if the nanotubes are replaced by PS-PCEMA-PAA/γ-Fe$_2$O$_3$ hybrid nanofibers. Figure 2.22 shows, for example, just such an interesting structure, which can be prepared by attaching three PS-PCEMA-PAA/γ-Fe$_2$O$_3$ hybrid nanofibers to one microsphere.

These fibers will attract one another in a magnetic field due to their magnetization and thus grab a nanoobject. Such a nanoobject can be moved around by focussing a laser beam on the microsphere that is trapped through an optical

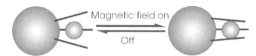

Figure 2.22 Schematic illustration of the operation of an optical magnetic nanohand.

tweezing mechanism. The nanoobject can then be released at a desired spot by turning off the magnetic field. Thus, such a device can function as an "optical magnetic nanohand". Last but not the least, studies on the concentrated solution properties and nanofiber composite properties should be performed to see if there are any novel desirable applications for these materials.

Acknowledgements

G. L. is very grateful for financial support from the Natural Sciences and Engineering Research Council of Canada, Defense Research and Development Canada, Canada Research Chairs Program, Canada Foundation for Innovations, Ontario Innovation Trust. G. L. would also like to thank Drs. Xiaohu Yan, Zhao Li, Sean Stewart, Jianfu Ding, and Lijie Qiao for carrying out the work reviewed in this chapter.

This Chapter has been published previously in:
Lazzari, Massimo / Liu, Guojun / Lecommandoux, Sebastién (eds.)
Block Copolymers in Nanoscience
2006
ISBN-13: 978-3-527-31309-9-Wiley-VCH, Weinheim

References

1 Wang, X.S., Wang, H., Coombs, N., Winnik, M.A. and Manners, I. (2005) *J. Am. Chem. Soc.*, **127**, 8924–8925.
2 Wang, X.S., Arsenault, A., Ozin, G.A., Winnik, M.A. and Manners, I. (2003) *J. Am. Chem. Soc.*, **125**, 12686–12687.
3 Ding, J.F. and Liu, G.J. (1998) *Macromolecules*, **31**, 6554–6558.
4 Zhang, M.F., Muller, A.H.E. (2005) *J. Polym. Sci. A: Polym. Chem.*, **43**, 3461–3481.
5 Yu, K., Zhang, L.F., Eisenberg, A. (1996) *Langmuir*, **12**, 5980–5984.
6 Raez, J., Manners, I. and Winnik, M.A. (2002) *J. Am. Chem. Soc.*, **124**, 10381–10395.
7 Jenekhe, S.A. anéd Chen, X.L. (1999) *Science*, **283**, 372–375.
8 Grumelard, J., Taubert, A. and Meier, W. (2004) *Chem. Commun.*, 1462–1463.
9 Stewart, S. and Liu, G. (2000) *Angew. Chem., Int. Ed. Engel.*, **39**, 340–344.
10 Yan, X.H., Liu, G.J., Liu, F.T., Tang, B.Z., Peng, H., Pakhomov, A.B. and Wong, C.Y. (2001) *Angew Chem., Int. Ed. Engl.*, **40**, 3593–3596.

11 Liu, G.J., Qiao, L.J. and Guo, A. (1996) *Macromolecules*, **29**, 5508–5510.
12 Kim, Y., Dalhaimer, P., Christian, D.A. and Discher, D.E. (2005) *Nanotechnology*, **16**, S484–S491.
13 Silva, G.A., Czeisler, C., Niece, K.L., Beniash, E., Harrington, D.A., Kessler, J.A. and Stupp, S.I. (2004) *Science*, **303**, 1352–1355.
14 Stupp, S.I. (2005) *MRS Bull.*, **30**, 546–553.
15 Massey, J.A., Winnik, M.A., Manners, I., Chan, V.Z.H., Ostermann, J.M., Enchelmaier, R., Spatz, J.P. and Moller, M. (2001) *J. Am. Chem. Soc.*, **123**, 3147–3148.
16 Garcia, C.B.W., Zhang, Y.M., Mahajan, S. DiSalvo, F. and Wiesner, U. (2003) *J. Am. Chem. Soc.*, **125**, 13310–13311.
17 Kowalewski, T., Tsarevsky, N.V. and Matyjaszewski, K. (2002) *J. Am. Chem. Soc.*, **124**, 10632–10633.
18 Tang, C.B., Tracz, A., Kruk, M., Zhang, R., Smilgies, D.M., Matyjaszewski, K. and Kowalewski, T. (2005) *J. Am. Chem. Soc.*, **127**, 6918–6919.
19 Bates, F.S. and Fredrickson, G.H. (1999) *Phys. Today*, **52**, 32–38.
20 Cameron, N.S., Corbierre, M.K. and Eisenberg, A. (1999) *Can. J. Chem.*, **77**, 1311–1326.
21 Templin, M., Franck, A., DuChesne, A., Leist, H., Zhang, Y.M., Ulrich, R., Schadler, V. and Wiesner, U. (1997) *Science*, **278**, 1795–1798.
22 Ishizu, K., Ikemoto, T. and Ichimura, A. (1999) *Polymer*, **40**, 3147–3151.
23 Liu, Y.F., Abetz, V. and Muller, A.H.E. (2003) *Macromolecules*, **36**, 7894–7898.
24 Yan, X., Liu, G. and Li, H. (2004) *Langmuir*, **20**, 4677–4683.
25 Tao, J., Stewart, S., Liu, G.J. and Yang, M.L. (1997) *Macromolecules*, **30**, 2738–2745.
26 Won, Y.Y., Davis, H.T. and Bates, F.S. (1999) *Science*, **283**, 960–963.
27 Dalhaimer, P., Bermudez, H. and Discher, D.E. (2004) *J. Polym. Sci. B: Polym. Phys.*, **42**, 168–176.
28 Ma, Q.G., Remsen, E.E., Clark, C.G., Kowalewski, T. and Wooley, K.L. (2002) *Proc. Natl. Acad. Sci. USA*, **99**, 5058–5063.
29 Liu, G.J., Li, Z. and Yan, X.H. (2003) *Polymer*, **44**, 7721–7727.
30 Yan, X.H. and Liu, G.J. (2004) *Langmuir*, **20**, 4677–4683.
31 Liu, G.J. and Zhou, J.Y. (2003) *Macromolecules*, **36**, 5279–5284.
32 Price, C. (1983) *Pure Appl. Chem.*, **55**, 1563–1572.
33 Breiner, U., Krappe, U., Abetz, V. and Stadler, R. (1997) *Macromol. Chem. Phys.*, **198**, 1051–1083.
34 Liu, G.J., Yan, X.H., Qiu, X.P. and Li, Z. (2002) *Macromolecules*, **35**, 7742–7747.
35 Liu, G.J., Yan, X.H. and Duncan, S. (2003) *Macromolecules*, **36**, 2049–2054.
36 Moore, W.R. (1969) *Progr. Polym. Sci.*, **1**, 3–43.
37 Zimm, B.H. and Crothers, D.M. (1962) *Proc. Natl. Acad. Sci. USA*, **48**, 905.
38 Yamakawa, H. and Fujii, M. (1974) *Macromolecules*, **7**, 128–135.
39 Yamakawa, H. and Yoshizaki, T. (1980) *Macromolecules*, **13**, 633–643.
40 Bohdanecky, M. (1983) *Macromolecules*, **16**, 1483–1492.
41 Dalhaimer, P., Bates, F.S. and Discher, D.E. (2003) *Macromolecules*, **36**, 6873–6877.
42 Geng, Y., Ahmed, F., Bhasin, N. and Discher, D.E. (2005) *J. Phys. Chem. B*, **109**, 3772–3779.
43 Won, Y.Y., Davis, H.T., Bates, F.S., Agamalian, M. and Wignall, G.D. (2000) *J. Phys. Chem. B*, **104**, 9054.
44 Won, Y.Y., Paso, K., Davis, H.T. and Bates, F.S. (2001) *J. Phys. Chem. B*, **105**, 8302–8311.
45 Liu, G.J., Ding, J.F., Qiao, L.J., Guo, A., Dymov, B.P., Gleeson, J.T., Hashimoto, T. and Saijo, K. (1999) *Chem. Eur. J.*, **5**, 2740–2749.
46 Onsager, L. (1949) *Ann. N.Y. Acad. Sci.*, **51**, 627–659.
47 Flory, P.J. (1956) *Proc. R. Soc. London, Ser. A: Math. Phys. Sci.*, **234**, 73–89.
48 Hunter, R.J. (1989) *Foundations of Colloid Science*, Vol. 1, Oxford University Press, Oxford.
49 Xia, Y.N., Yang, P.D., Sun, Y.G., Wu, Y.Y., Mayers, B., Gates, B., Yin, Y.D., Kim, F. and Yan, Y.Q. (2003) *Adv. Mater.*, **15**, 353–389.
50 Lieber, C.M. (2003) *MRS Bult.*, **28**, 486–491.

51 Ziolo, R.F., Giannelis, E.P., Weinstein, B.A., Ohoro, M.P., Ganguly, B.N., Mehrotra, V., Russell, M.W. and Huffman, D.R. (1992) *Science*, **257**, 219–223.

52 Li, Z. and Liu, G.J. (2003) *Langmuir*, **19**, 10480–10486.

53 Yan, X.H., Liu, G.J., Haeussler, M. and Tang, B.Z. (2005) *Chem. Mater.*, **17**, 6053–6059.

54 Liu, G., Yan, X., Li, Z., Zhou, J. and Duncan, S. (2003) *J. Am. Chem. Soc.*, **125**, 14039–14045.

55 Yan, X., Liu, G. and Li, Z. (2004) *J. Am. Chem. Soc.*, **126**, 10059–10066.

3
Smart Nanoassemblies of Block Copolymers for Drug and Gene Delivery

Horacio Cabral and Kazunori Kataoka

3.1
Introduction

In Nature, various complex architectures are formed from a limited choice of building units such as lipids or amino acids. These natural assemblies form a large variety of biological devices with specific cellular functions. Conversely, synthetic polymers can be prepared from a wide range of monomers, and these macromolecules can construct singular architectures and shapes. The interest in these structures relies on their characteristic sizes in the mesoscopic range (<100 nm), and that their constitution and shape lead to materials with particular properties and functions. Nevertheless, these artificial structures could not yet attain the complexity achieved by the natural assemblies.

One particular example of man-made nanoassemblies is that of block copolymers. The reason for using block copolymeric systems to prepare nanostructures is their simplicity to form nanoscale objects with expected shapes and sizes, without any additional trigger. Block copolymers consist of two or more covalently bonded blocks with different physical and chemical properties. In addition, each block should have the features to control the self-assembly process. The driving force for core–shell-structured self-assembly consists of repulsive interactions between incompatible domains, such as the case of amphiphilic diblock copolymers, which contain a hydrophobic and a hydrophilic block. Moreover, the covalent link between the blocks is responsible for the microphase separation, and prevents the system from further separation on a macroscopic scale. The self-assembled nanostructures can be finely tuned to a variety of morphologies by altering the molecular parameters of the block copolymers or the environment (Figure 3.1). These self-assembled block copolymer systems have already been found to be appropriate for several applications in nanotechnology, including detergents, paints, electronics, cosmetics, lubricants, tissue engineering, and drug delivery, as determined by the morphology of the structures [1–3].

In an attempt to create synthetic structures that somehow approach natural assemblies in terms of their complexity, functionality, and performance,

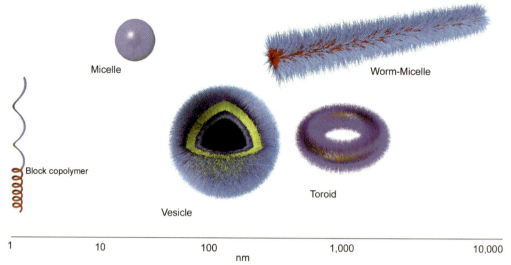

Figure 3.1 Block copolymer nanoassemblies. Several structures ranging from 10 nm to 10000 nm can be formed using block copolymers as the building units.

nanostructures have been designed to respond in a controlled manner to external stimuli. Stimulus-responsive nanostructures, which are also referred to as "smart," "intelligent", or "environmentally sensitive" nanostructures, are systems that exhibit sharp changes in response to physical stimuli such as heat, ultrasound, and light, or to chemical stimuli such as pH, ions in solution, and chemical substances. These responses depend on the stimulus applied, and they may comprise modifications in the properties of the assemblies such as shape, volume, and permeation rates.

3.2
Smart Nanoassemblies for Drug and Gene Delivery

An "ideal" drug carrier should emulate a system that is capable of residing *in vivo* for long periods of time, targeting particular cell types, incorporating a large set of therapeutic agents, and releasing those molecules in the appropriate environment. Micelles, rods, or polymersomes have been used as carriers of therapeutic agents to target specific sites in the body [3, 4]. Moreover, several micellar delivery systems for anticancer drugs are currently undergoing clinical trials [5]. Rods, threads, and fibers have also been shown to be useful as support media for cell growth *in vitro*, and have been used to direct nerve regrowth *in vivo* or as the basis of more complex mechanical systems [6–8].

Polymeric self-assembled nanostructures have a great potential to reach pathological sites while avoiding biological barriers in the human, due to their nanoscopic size (typically between 10 and 100 nm). It has been determined that the so-called enhanced permeability and retention (EPR) effect plays a major role in tissue targeting to, for example, tumor tissues [9, 10]. The EPR effect is attributed to the angiogenic tumor vasculature, which has a higher permeability compared to normal tissues due to its discontinuous endothelium, and to the somewhat undeveloped lymphatic drainage in tumors. These characteristics lead to the extravasation of colloidal particles in tumor and other inflamed tissues, and their subsequent retention there. Moreover, the polymeric nanostructures must circulate in the bloodstream for a sufficient time in order to achieve this targeting. Thus, the surface properties of polymeric micelles represent important factors as they determine the biological fate of the material. The reduction of nonspecific recognition and uptake by the reticuloendothelial system (RES) is critical for the prolonged circulation of the carrier. Colloidal carriers bearing hydrophilic polymers on the surface of particles have been shown to reduce the opsonization and subsequent uptake by the RES cells of the liver, spleen, and bone marrow [11–13].

The development of smart nanoassemblies that dynamically morph their properties due to sensitivity to chemical or physical stimuli, magnifies to an even greater extent the significance of block copolymer nanostructures in biological applications. In this way, smart polymeric nanostructures can respond either to pathological or physiological endogenous stimuli present in the body, or to externally applied stimuli such as temperature, light, or ultrasound. One sophisticated and rational approach is a conversion of the core-forming segment of polymeric micelles from a hydrophobic to a more hydrophilic state under a stimulus, which causes an eventual dissolution of the micelles. The therapeutic agent should be stably associated with the hydrophobic core, and release of the drug would be expected to occur along with destabilization of the nanostructure. Irreversible changes in the structure of the nanoassemblies can be induced by cleavage of the covalent bonds, whereas reversible changes can result from changes in hydrogen-bonding capability, in the solubility-temperature of polymers, the protonation, the pH-sensitivity of polymers, and from changes in the redox potential.

In this chapter we will introduce smart nanoassemblies as applied to drug and gene delivery fields, organized in groups depending on their stimuli sensitivity.

3.3
Endogenous Triggers

3.3.1
pH-Sensitive Nanoassemblies

3.3.1.1 Drug Delivery
The mildly acidic pH of solid tumors and inflammatory tissues [14] (pH ~6.8), and the pH of the endosomal and lysosomal compartments of cells [15] (pH 5–6), provides a potential trigger for the release of systemically administered drugs from a pH-sensitive carrier, as blood and normal tissues have a pH of 7.4. Moreover, as macromolecular carriers enter the cells via endocytosis and are localized in

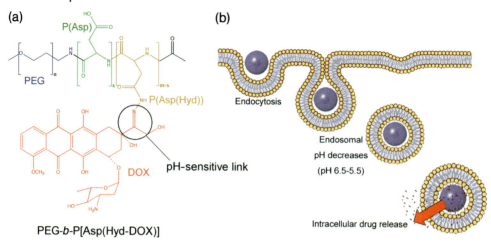

Figure 3.2 PEG-P(ASP(Hyd))/DOX conjugate forming pH-sensitive micelles. At endosomal pH, the micelles release DOX molecules [17].

the endosomes or in the lysosomes [16], the pH represents a very useful stimulus for the design of drug carriers that release their contents only within the intracellular environment. Conversely, drug-loaded nanoassemblies that are stable at low pH (i.e., in the stomach) and degrade at near-physiological pH (i.e., in the intestines) are very attractive for controlled drug release following their oral administration.

pH-sensitive nanoassemblies can be constructed by the use of an acid-labile bond between the drug and the carrier polymer. As drug release is dependent on the chemical or enzymatic hydrolysis of the bond between the drug and the polymer backbone, drug release from these nanoconstructs is usually slower than that of physically loaded drugs. Recently, pH-sensitive doxorubicin (DOX)-loaded polymeric micelles were prepared by chemically conjugating DOX to the side chain of poly(ethylene glycol)-b-poly(aspartic acid) (PEG-b-P(Asp)) copolymers via an acid-labile hydrazone bond (Figure 3.2) [17]. These micelles specifically released DOX at endosomal pH conditions (pH 5.0), whereas DOX was retained in the micelle core at physiological pH. These pH-sensitive DOX micelles were very effective against subcutaneous murine colon adenocarcinoma 26 *in vivo*, while their toxicity proved to be negligible due to minimal drug leakage [18].

Gillies *et al.* reported the construction of PEG-b-P(Asp) functionalized with trimethoxybenzylidene acetals as acid-labile linkages [19]. The cyclic benzylidene acetals and the copolymer backbone increased the hydrophobicity of the core due to the aromatic rings, while the acetal groups hid the polarity of the diol. Although these micelles were relatively stable at physiological pH, hydrolysis of the acetal bonds occurred at pH 5. In addition, the generation of diols increased the hydrophilicity of the polymers. During this process, disassembly of the micelles occurred as the hydrophobic dye was released.

Protonation of the block copolymer backbone can also trigger destabilization of the micelles. Block copolymers with such characteristics normally contain L-histidine [20, 21], pyridine [22], and tertiary amine groups in their hydrophobic segments [23]. In these systems, polymeric micelles are formed at a pH above the pK_a of the protonatable group, and therefore the hydrophobic segment essentially is uncharged. As the pH falls below the pK_a, however, ionization of the polymer causes increased hydrophilicity and electrostatic repulsions of the polymers, leading to destabilization of the micelles. In this way, PEG-b-poly(L-histidine) (PEG-b-P(His)) was used to prepare pH-sensitive polymeric micelles incorporating DOX [20]. The prepared micelles showed an accelerated release of drug as the pH was decreased, with ionization of the P(His) block forming the micelle core determining the pH-dependent critical micelle concentration (CMC) and stability of the system. Moreover, control of the transition pH is possible by combining different block copolymers. In light of these findings, Lee et al. prepared mixed micelles from PEG-b-P(His) and PEG-b-poly(lactic acid) (PEG-b-PLA) [21]. The PEG-b-P(His) micelles destabilized at physiological pH, whereas the mixed polymeric micelles of PEG-b-P(His) and PEG-b-PLA showed improved micellar stability at pH 7.4 and dissociated over a pH range of 6.0 to 7.2, depending on the proportion of PEG-b-PLA present. Similar pH-sensitive mixed micelles were prepared from biotin-P(His)-b-PEG-b-PLLA (poly(L-lactic acid)) and PEG-b-P(His). At pH >7, the P(His) attached to the biotin was mostly deionized and became hydrophobic, thus interacting with the micellar PLA core. However, as the pH was slowly decreased the P(His) segments became progressively ionized and extended outwards through the polyethylene (PEG) brush surrounding the core, thus exposing the biotin moieties for ligand–receptor interactions. At pH values <6.5, protonation of P(His) in the PEG-b-P(His) block copolymer contained in the core caused the induction of micellar dissociation.

The above-described pH-sensitive nanoassemblies release their contents after dissociation of the micelles. However, a different type of pH-sensitive nanoassembly was designed to release the encapsulated contents after aggregation or collapse of the nanoassemblies. As an example, Leroux et al. prepared random copolymers of N-isopropylacrylamide (NIPAAm) and methacrylic acid (MAA) substituted with alkyl chains at either the terminal chain ends, or distributed randomly over the copolymer chain to induce micelle formation [24]. When chloroaluminum phthalocyanine (AlClPc), a widely used photosensitizer for the photodynamic treatment of cancer, was incorporated into these micelles [25], the addition of 5 mol% MAA to the copolymers caused the hydrophobic core to distort following neutralization of the MAA as the pH fell below 5.7–5.8 at 37 °C. This phenomenon was thought to cause the release of the entrapped photosensitizer and to alter the intracellular localization of the drug in a favorable way, making it more photoactive.

Smart polymeric micelles may also represent a promising approach for the oral delivery of hydrophobic drug molecules. Sant et al. developed pH-sensitive micelles composed of block copolymers of PEG as the hydrophilic block and poly(alkyl acrylate-co-methacrylic acid) [PEG-b-(PAA-co-MAA)] [26]. Due to the presence of pendant carboxylic groups on the MAA segments in the core, the copolymers self-assembled at pH <4.7, whereas, above this value the micelles dissociated owing to ionization of the COOH moieties. The pH at which micellization occurred was decreased with a reduction in the length of the hydrophobic block. Three poorly water-soluble drugs, namely indomethacin, fenofibrate, and progesterone, were successfully loaded into these micelles. It was also possible to trigger drug release in a pH-dependent manner by changing the pH of the release medium from 1.2 to 7.2; this clearly demonstrated the potential of pH-responsive polymeric micelles for targeting drugs to the intestine following oral administration.

3.3.1.2 Gene Delivery

Gene therapy refers to the potential use of nucleic acids, irrespective of whether this involves plasmid DNA, antisense oligonucleotides or siRNA, to modulate the expression of genes in cells for therapeutic purposes. Among the highlights of gene therapy are:

- The replacement of a deficient gene in a genetically inherited disease, with a normal copy restoring the production of a functional protein.
- The correction of genetic defects beyond inherited disorders, as modulation of the regulation of gene expression is involved in numerous acquired diseases.
- The integration of functions in cells that are not originally present, and which could serve a therapeutic purpose.

Gene delivery may be divided into two main categories, depending on the vectors used for nucleic acid transfer, namely *viral* and *nonviral*. The first vectors to be developed were based on the use of viruses or pseudoviral particles. Viral vectors may pose serious problems in terms of immunogenicity, toxicity and potential oncogenicity, thus risking their use as therapeutic drugs [27]. Nonviral gene delivery involves the use of cationic lipids and cationic polymers to deliver the genes. Synthetic self-assembled gene vectors based on cationic polymers, termed polyplexes, are considerably safer and easier to produce. Polyplexes have also been progressively ameliorated as gene vectors, and specific DNA delivery to several tissues has been achieved *in vivo* by using either systemic or localized delivery [28]. Cationic polymers mask the negative charge of the plasmid DNA (pDNA) and package it into small particles, thus protecting the pDNA from both enzymatic and hydrolytic degradation [29–31]. Polycations with a relatively low pK_a value, such as poly(ethylenimine) (PEI), present a high transfection activity, most likely because they buffer the endosomal acidification and produce an increase in the ion osmotic pressure in endosomes. This is followed by protonation of the amines, which leads in turn to a disruption of the endosomal membranes and release of the endosomal contents into the cytoplasm. This whole process is referred to as the "proton-sponge effect" (Figure 3.3) [32]. Such polycations require a high ratio of cationic amino groups in the polycations to phosphate anions in the DNA (N/P ratio), in order to form a polyplex which has a high stability and efficient transfection activity. Moreover, although free PEI contributes to the increased gene expression, it also produces a considerable increase in the toxicity of the carrier [33].

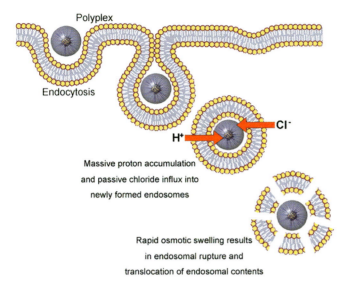

Figure 3.3 The "proton sponge effect." Polyplexes containing weak base components buffer the endosomal acidification and produce an increase in the ion osmotic pressure in endosomes; this leads to disruption of the endosomal membranes. Finally, the endosomal contents are delivered into the cytoplasm.

Conversely, polycations with a pK_a >9.0, such as poly(L-lysine) (P(Lys)), form stable polyplexes even at a relatively lower N/P ratios [34]. The introduction of buffering units into a polycation with high pK_a value improves the transfection efficiency based on the proton-sponge effect, but decreases the stability of the complex due to a lower affinity towards DNA [35]. The buffering capacity of these units is also lessened due to protonation of the polycations following their complexation with DNA. Thus, as the efficiencies of the polyplexes are still too low for clinical use, the next crucial point of gene delivery will be to construct virus-like polyplexes using smart polymer conjugates. By using that approach, the creation of effective gene vectors for clinical applications should resolve the problems of poor stability and high toxicity of the current polyplexes, and also provide the buffering capacity to enhance transfection, without an excess of free polymers.

It is essential that the smart gene nanostructures recognize the biological signals and undergo designed structural changes which match the different steps of gene delivery. In this respect, Oishi et al. reported the creation of hepatocyte-targeted polyion complex (PIC) micelles with a pH-sensitive PEG shell as a smart delivery system for antisense oligodeoxynucleotides (ODNs) [36]. These PIC micelles were prepared from P(Lys) and a lactosylated PEG-ODN conjugate (Lac-PEG-ODN), which had an acid-labile linkage (β-thiopropionate) between the PEG and ODN segments. The lactose-PIC micelles achieved an elevated antisense effect against luciferase gene expression in human hepatoma (HuH-7) cells, this being significantly higher than that produced by either ODN or Lac-PEG–ODN alone, and also the lactose-free PIC micelle, most likely due to an asialoglycoprotein receptor-mediated endocytosis process. In addition, a significant decrease in the antisense effect was observed for a lactosylated PIC micelle without the acid-labile linkage. This suggested that the pH-sensitive release of the active antisense ODN molecules into the cytoplasm is a key event in the antisense effect of this micelle. Conversely, the use of a polycation with low pK_a, such as branched PEI instead of the P(Lys), to prepare the PIC micelle led to a decrease in the antisense effect, most likely due to a buffer effect of the branched PEI in the endosomal compartment that prevented cleavage of the acid-labile linkage in the conjugate.

Another approach for developing a smart gene delivery system consists of an A–B–C-type triblock copolymer using a biocompatible fragment in the A-fragment, a polycation with low pK_a value and buffering effect as the B-fragment, and a polycation with high pK_a to condense the DNA as the C-fragment. Fukushima et al. showed that PEG-b-poly[(3-morpholinopropyl) aspartamide] (PMPA; pK_a = 6.2)-b-P(Lys) (PEG-b-PMPA-b-P(Lys)) formed smart PIC micelles where the P(Lys) backbone condensed DNA and the uncomplexed PMPA backbone covered the P(Lys)/DNA polyplex core (Figure 3.4) [37]. These PIC micelles had a diameter of 88.7 nm, a zeta potential of 7.3 mV, and exhibited much higher transfection efficiency against HuH-7 cells than did micelles prepared from PEG-b-PLL (poly(lactic acid)) or the combination of PEG-b-PLL and PEG-b-PMPA. The improvement in transfection efficiency of these three-layered polyplex micelles can be related to the buffering capacity of the PMPA segment in the polyplex micelle. Nevertheless, positively charged polyplexes might potentially induce cytotoxicity and form aggregates with the plasma proteins present in the biological media, thus restricting their *in vivo* applicability. In order to overcome this problem, two strategies have been followed:

- The first approach consists of using a block copolymer of a PEG segment and a cationic polyaspartamide segment carrying an ethylenediamine unit at the side chain (PEG-b-P[Asp(DET)]) (Figure 3.5) [38]. This block copolymer led to the formation of stable polyplexes with a core of tightly packed pDNA. Ethylenediamine undergoes a clear, two-step protonation with a characteristic *gauche–anti* conformational transition providing an effective buffering function in the acidic endosomal compartment. Thus, after endocytosis of these polyplexes, the ethylenediamine unit in the block copolymer is expected to facilitate the efficient translocation of the micelle towards the cytoplasm due to the proton-sponge effect. The PIC micelles prepared from PEG-b-P[Asp(DET)] accomplished an appreciably high gene transfection efficacy and a remarkably low cytotoxicity against several cell lines, including primary osteoblasts. Moreover, these polyplexes showed a high efficiency *in vivo* for the treatment of vascular lesions, with markedly reduced cytotoxicity and thrombogenicity [39].

- A second approach was proposed by Lee et al. using polymeric micelles of PEG-b-poly[(N′-citraconyl-2-aminoethyl)aspartamide] (PEG-b-P(Asp(EDA-Cit))) [40]. This block copolymer has the ability to switch the charge from anionic to cationic at the endosomal pH due to degradation of the citraconic amide side chain at pH 5.5. This rapid charge-conversion can cause the PIC micelles to

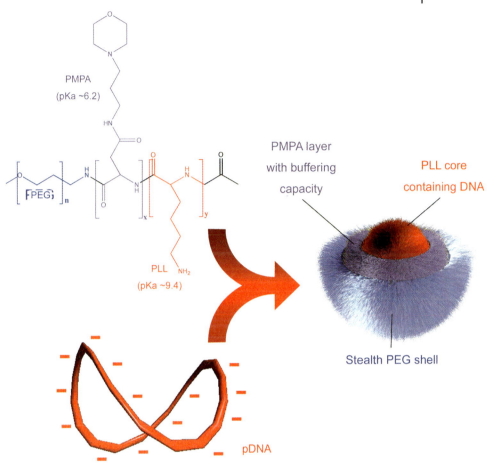

Figure 3.4 Chemical structure of the PEG-PMPA-PLL triblock copolymer, and schematic illustration of the three-layered polyplex micelles with spatially regulated structure [37].

promptly release the loaded protein in response to the endosomal pH. Consequently, this pH-sensitive charge-conversion polymer has shown great promise for the design of gene carriers that become cationic at the early endosomal stage, yet still have the ability to achieve endosomal escape due to the proton-sponge effect.

3.3.2
Oxidation- and Reduction-Sensitive Polymeric Nanoassemblies

Redox-sensitive nanostructures from block copolymers can result in a change of the assembly morphology and the selective release of encapsulated drugs when an electric current is applied externally. Moreover, the redox-triggering can occur

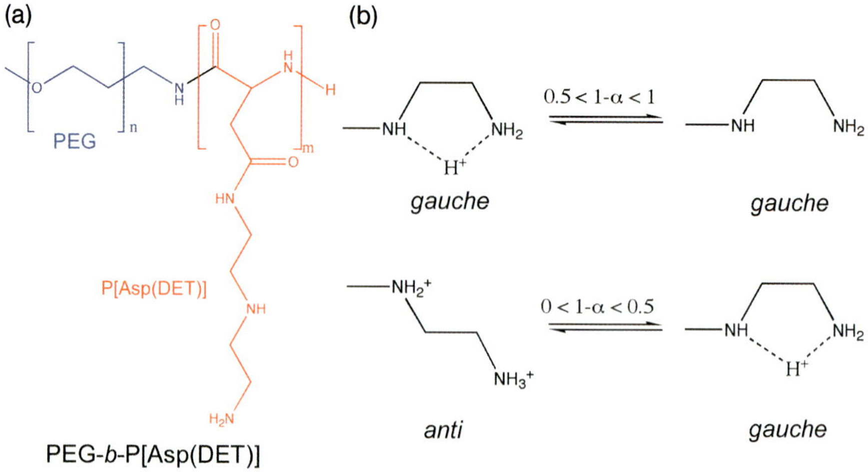

Figure 3.5 (a) PEG-b-P[Asp(DET)] copolymer bearing an ethylenediamine unit at the side chain leads to the formation of stable polyplexes with smart buffering properties; (b) Ethylenediamine presents a two-step protonation with a unique *gauche–anti* conformational transition providing an effective buffering function at the endosome. The proton-free form may take both *gauche* and *anti* conformation [38].

at inflammation sites and solid tumors, since those pathologies present activated macrophages that release oxygen-reactive species. Hubbell et al. synthesized amphiphilic A–B–A block copolymers which consisted of the hydrophobic poly(propylene sulfide) and PEG (PEG-b-PPS-b-PEG), which formed polymeric vesicles in water [41]. Following exposure of these vesicles to oxidative agents, the thioethers in the PPS block were oxidized to poly(propylene sulfoxide) and eventually to poly(propylene sulfone), leading in turn to hydrophilization of the originally hydrophobic block. As a result, the vesicles became destabilized. This oxidative conversion was also accomplished by incorporating glucose oxidase (GOx) into the vesicles [42]. After incubation in 0.1 M glucose solution, the GOx-containing polymersomes were disassembled due to the oxidation of glucose by GOx to produce H_2O_2. Another possibility would be to take advantage of the redox tunability of metal-containing compounds. In this way, redox-active micelles were prepared from amphiphilic block copolymers bearing a hydrophobic ferrocenylalkyl moiety (FPEG) [43]. Oxidation of the ferrocenyl moiety caused the micelles to be disrupted into unimers. However, when loaded with perylene, these redox-active FPEG micelles released the drug in a controlled manner by applying a selective and electrochemical oxidation of the FPEG.

Another type of reduction-sensitive nanostructures is represented by polymeric assemblies containing disulfide bonds. Polyplex micelles with a disulfide-crosslinked core efficiently release the loaded pDNA in response to reductive intracellular conditions; that is, at 50- to 1000-fold higher glutathione concentrations than are encountered in the extracellular environment (Figure 3.6) [44–47].

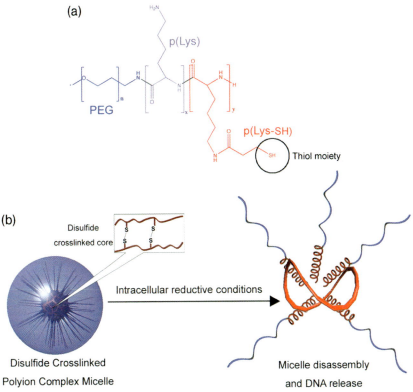

Figure 3.6 Disulfide crosslinking is used to stabilize PIC micelles. (a) Molecular structure of the PEG-P(L-Lysine-SH) block copolymer; (b) Under intracellular reductive conditions, the disulfide bonds are cleaved and the PIC micelles release their contents.

The intracellular glutathione reductively cleaves the disulfide links, which leads to a destabilization of the system. This type of micelle also achieved sufficient tolerability against destabilization by anionically charged biocomponents, and induced an efficient transfection in the cell. As a result, the transfection efficiencies achieved *in vivo* were relatively high [47].

3.3.3
Other Endogenous Triggers

Several other internal triggers, including enzymes or peptides, can be used to control the structure and properties of smart block copolymer nanoassemblies. The construction of smart polyplexes with two types of gene-transfection activation has been reported [48]. These systems utilize cationic polymers which are responsive to cyclic AMP-dependent protein kinase A (PKA) or to caspase-3, PAK, and

PAC, respectively. The PAK polymer incorporates a substrate for PKA, ARRASLG, while the PAC polymer has a substrate sequence for caspase-3, DEVD, and a cationic oligolysine, KKKKKK. These polymers formed stable complexes with DNA. However, the PKA or caspase-3 signal breaks up the PAK–DNA or PAC–DNA complexes, respectively, releasing the DNA and activating the gene transfection activity.

3.4
External Stimuli

3.4.1
Temperature

As local heating can also be exploited to destabilize smart nanoassemblies, several thermosensitive copolymers that use this approach are currently undergoing investigation for biomedical applications. The thermosensitive nanoassemblies are characterized by a lower critical solution temperature (LCST), below which water is bound to the thermosensitive polymer block so as to prevent both intrapolymer and interpolymer interactions, thus rendering the nanoassembly water-soluble. Whilst the formation of hydrogen bonds between the thermosensitive polymer and water lowers the free energy of mixing, the ordered molecular orientations of water on the polymer lead to negative entropy changes and positive contributions to free energy. The nonpolar hydrophobic groups in the thermosensitive polymer are incompatible with water, and facilitate an ordered clustering of the surrounding water molecules so as to decrease the entropy. Increasing the temperature of the system causes the water cluster to become destabilized to compensate the thermal energy. Above the LCST, water is released from the polymer chain (Figure 3.7a). The positive entropic contribution then grows and dominates the heat term, which causes the monophase system to be come unbalanced and leads to phase separation due to polymer association. To date, the most extensively investigated thermosensitive polymer is poly(N-isopropylacrylamide) (PNIPAAm; Figure 3.8a); this molecule has a sharp LCST in water at approximately 32 °C (slightly lower than body temperature), which makes it extremely attractive for the design of thermosensitive drug delivery systems [49]. The LCST of a thermosensitive polymer can also be modulated by copolymerizing it with hydrophobic comonomers to reduce the LCST, or with hydrophilic comonomers to increase the LCST. PNIPAAm copolymers may be used either as a hydrophilic segment or as a hydrophobic segment of polymeric micelles, switching the dispersivity depending on the LCST (Figure 3.7b). By using this approach, Okano *et al.* were able to prepare DOX-loaded polymeric micelles of PNIPAAm-b-poly(butyl methacrylate) (PNIPAAm-b-PBMA) and PNIPAAm-b-poly(styrene) (PNIPAAm-b-PS) [50]. Below the LCST of PNIPAAm, this system demonstrated a core–shell micellar structure, but on heating above the LCST the DOX was rapidly released from the PNIPAAm-b-PBMA micelles as a result of structural distortion of the relatively flexible PBMA core, caused by collapse of the PNIPAAm shell. In con-

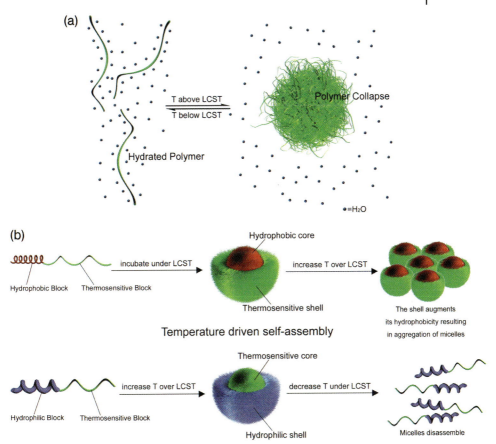

Figure 3.7 (a) Thermosensitive polymers can undergo reversible transformation from hydrated to hydrophobic/collapse state upon heating; (b) Several strategies can be used with thermosensitive polymers, such as using them as the micelle shell below the LCST or the micelle core over the LCST.

trast, the PNIPAAm-*b*-PS micelles did not show any enhanced DOX release when the temperature was increased above the LCST, mainly because the rigid PS core was insensitive to collapse of the PNIPAAm. As PNIPAAm exists in its precipitated form at body temperature, this system proved not to be suitable for *in vivo* application without modification. Nevertheless, the copolymerization of NIPAAm with the hydrophilic dimethylacrylamide (DMAAm) resulted in a random copolymer (P(NIPAAm-*co*-DMAAm)) with a LCST slightly above body temperature (40 °C) [51]. The release of DOX from P(NIPAAm-*co*-DMAAm)-*b*-PLA micelles was very slow at 37 °C, but showed a sudden increase at 42.5 °C, which suggested that this system might have potential benefit for hyperthermic treatments.

A system in which PNIPAAm was utilized as the core-forming block was reported by Feijen *et al.* [52]. Here, the PEG-*b*-PNIPAAm block copolymer was

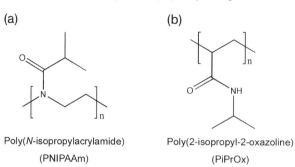

Poly(N-isopropylacrylamide) Poly(2-isopropyl-2-oxazoline)
(PNIPAAm) (PiPrOx)

Figure 3.8 Two examples of thermosensitive polymers having a LCST. (a) Poly(N-isopropylacrylamide) (PNIPAAm); (b) Poly(2-isopropyl-2-oxazoline) (PiPrOx). Both polymers have been used in the construction of nanoassemblies of block copolymers.

water-soluble below the LCST of PNIPAAm, but above this temperature it formed polymeric micelles with a collapsed PNIPAAm core and a PEG outer shell. The temperature at which micelles are formed is known as the critical micelle temperature (CMT). The heating rate is a critical parameter for the size of PEG-b-PNIPAAm micelle, as a higher heating rate causes rapid dehydration of the thermosensitive segments. Any subsequent collapse of these segments precedes the aggregation between polymers and, as a result, micelles with a well-defined core–shell structure are formed.

Poly(2-isopropyl-2-oxazoline) (PiPrOx) is another promising thermosensitive polymer (Figure 3.8b). This material possess an isopropyl group in the side 2-position and, as PNIPAAm, the aqueous solutions of PiPrOx have a LCST at near-physiological conditions [53, 54]. Park et al. prepared novel thermosensitive PIC micelles in an aqueous medium via the complexation of a pair of oppositely charged block copolymers containing the thermosensitive PiPrOx segments, PiPrOx-b-P(Lys) and PiPrOx-b-P(Asp) [55]. These PIC micelles had a constant cloud-point temperature of approximately 32 °C under physiological conditions, regardless of their concentration. Since the LCST of PiProx can be modulated by copolymerization [56], these PiPrOx-PIC micelles have high potential as a size-regulated, temperature-responsive nanocontainers for loading charged compounds.

The main drawback of thermosensitive drug delivery systems is that the thermal treatment required for the controlled destabilization of the micelles and subsequent drug release is not always feasible in clinical practice. However, this issue can be overcome using secondary external triggers. Sershen et al. developed a photothermally modulated hydrogel using NIPAAm-b-acrylamide in combination with photoactive gold nanoshells [57]. The nanoshells strongly absorbed near-infrared (NIR) irradiation (1064 nm) and converted it to heat, resulting in a collapse of the hydrogel. As an example, laser irradiation of this system led to the controlled release of methylene blue and proteins.

3.4.2
Light

The attractive feature of light-responsive polymeric assemblies is that the drug release can be induced at a specific time and site of light exposure, with ultraviolet (UV), visible (VIS) or NIR light being applied as the trigger. Since NIR light shows a deeper tissue penetration and minimal damage to healthy cells, its use is of particular interest for biomedical applications [58]. The structural changes of the nanoassemblies induced by light can be either irreversible or reversible; for example, cleavage of the side chains of the block copolymers induced an irreversible destabilization of the structures. The micelles prepared from PEG-b-polymethacrylate copolymers bearing a photolabile pyrene chromophore moiety in the side chain (PPy) demonstrated photosensitivity [59]. Likewise, pyrenyl methyl esters were cleaved following UV irradiation, thus transforming the hydrophobic polymethacrylate segment to hydrophilic poly(methacrylic acid) (PMA). The same group also developed a PEG-b-poly(2-nitrobenzyl methacrylate) system [60], where the cleavage of 2-nitrobenzyl moieties occurred by photolysis either via one-photon UV (365 nm) or two-photon NIR (700 nm) excitation. The formation of carboxylic acid after irradiation shifted the hydrophilic/hydrophobic balance and resulted in either break-up of the micelles or in swelling of the micelle core when it was crosslinked with a diamine.

The exposure of photoactive groups to light may also generate reversible structural changes due to deviations in the hydrophilic–hydrophobic balance. The molecular units showing photochemical-induced transitions include azobenzenes (Figure 3.9a), cinnamoyl derivatives (Figure 3.9b), and spirobenzopyran (Figure 3.9c).

A reversible isomerization between the *trans* (E) and *cis* (Z) geometric isomers of azobenzenes can be induced by light or heat (Figure 3.9a). The *cis* isomer is thermodynamically less stable, and isomerizes to the *trans* form due to thermal energy; the illumination can reduce this conversion time to minutes. Moreover, the photoisomerization wavelengths can be tuned by modulating the substituent groups at the chromophores. In that way, azo-functionalized hyperbranched polyesters [61] and polypeptides [62] showed photoinduced transformations. Photosensitive polymeric rods were formed with helical polypeptides by the hydrophobic interactions between the azo groups in a planar, apolar, *trans* configuration [63]. The photoisomerization of these azo-moieties to the skewed, polar, *cis* configuration, inhibited the interactions between azo-groups, leading to a disintegration of the nanoassemblies. Moreover, azobenzene–polymethacrylate-b-poly(acrylic acid) copolymers self-assembled in dioxane–water mixtures into micelles and vesicles [64], although the nanoassemblies broke up after UV irradiation. Subsequent illumination with visible light led to all of the nanoassemblies being reformed due to the *cis*-to-*trans* isomerization of the azobenzene, thus indicating the reversibility of the system.

Cinnamate undergoes either *trans*-to-*cis* photoisomerization, thus generating cinnamate residues with an increased hydrophilicity, or photodimerization after

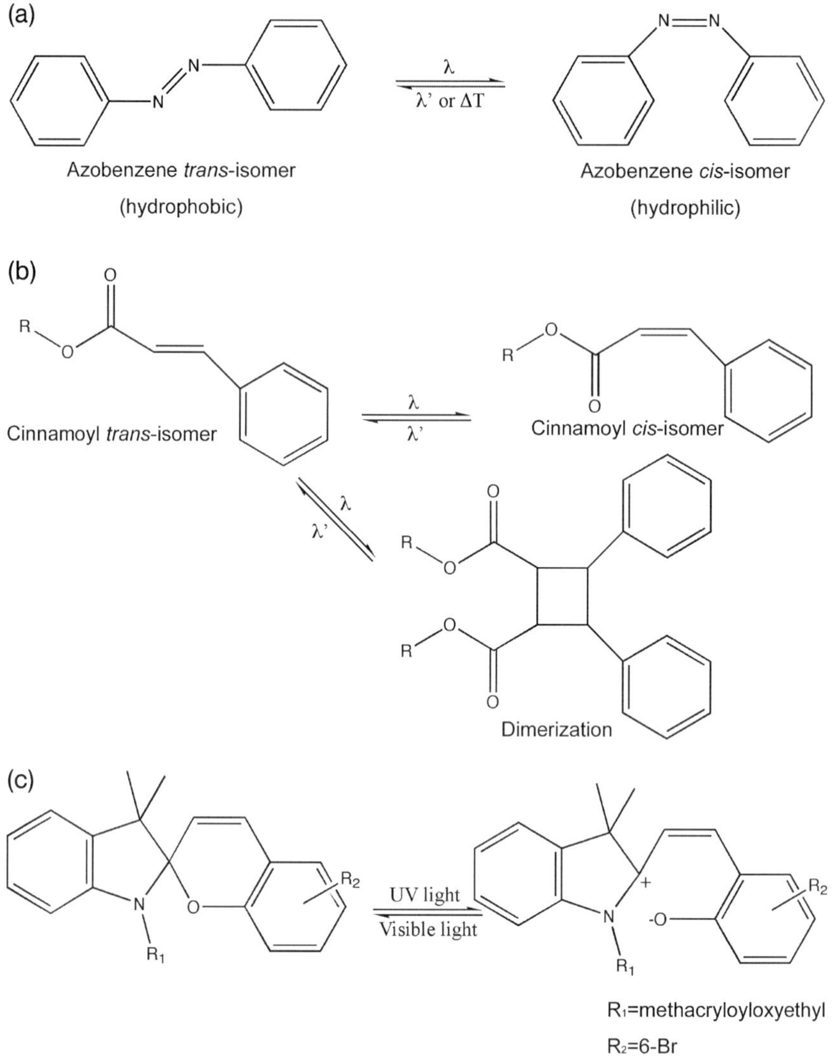

Figure 3.9 Light-sensitive groups normally used to modify block copolymers in order to obtain light-sensitive nanostructures. (a) Azobenzenes undergo isomerization by light and heat between the *trans* and *cis* geometric isomers; (b) Cinnamate undergoes *trans*-to-*cis* photoisomerization, producing residues with higher hydrophilicity or dimers; (c) Spirobenzopyran undergoes reversible photoisomerization into a zwitterionic merocyanine.

irradiation with UV light (Figure 3.9b). Accordingly, the partial modification of poly(HPMA) by cinnamate (9 mol%) produced both temperature- and light-sensitive polymers [65]. UV irradiation of the polymer in aqueous solution resulted in a 6 °C increase in the LCST due to *trans*-to-*cis* isomerization. However, at a high cinnamate content the polymer was photocrosslinked due to the photodimerization. The parameters that affect the LCST in this system are the polymer concentration, the amount of cinnamate moieties per polymer chain, and the extent of isomerization. Moreover, photocrosslinked cinnamoyl microcapsules were used to encapsulate cyclodextrin, and subsequently release it by UV illumination as a result of photocleavage and microcapsular break-up [66].

Block copolymers bearing merocyanine were self-assembled due to the attractive electrostatic dipole–dipole interaction of the zwitterionic moieties. The zwitterionic merocyanine is the photoisomerization product from spirobenzopyran (SBP) after UV irradiation (Figure 3.9c). For example, HPMA copolymers bearing various amounts of spirobenzopyran moieties aggregated into large clusters after UV exposure in water due to zwitterionic formation after 20 min illumination [67]. After incubation under visible light, however, the rapid reversal of the metastable zwitterionic form to the neutral form caused the nanostructures to be dissociated. As the neutral form of this copolymer had a low solubility at high ionic strength, this photoreversible cluster behavior was of completely opposite nature in 1 M NaCl.

3.4.3
Ultrasound

Ultrasound has been used successfully as a noninvasive trigger for the *in vitro* and *in vivo* release of drugs from PEG-*b*-poly(propylene oxide) (PPO)-*b*-PEG (poloxamer) micelles (Figure 3.10) [68–70]. Ultrasound not only causes the release of drug from these micelles but also enhances the intracellular uptake of both the released and encapsulated drug. Moreover, ultrasound can penetrate deeply into the interior of the body, and can also be focused and precisely controlled. The application of ultrasound in the frequency range of 20 to 90 kHz to pluronic micelles loaded with DOX or ruboxyl (a paramagnetic anthracyclin, an analogue of Rubomicin; Rb)

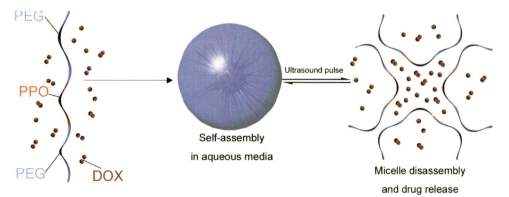

Figure 3.10 Poloxamers (Pluronics®) have been used for the preparation of DOX-loaded polymeric micelles [68–70]. Upon ultrasound application, the micelles release the drug, whereas eliminating the ultrasound leads to reformation of the micelle.

caused release of the drug. The highest rate of release was observed at 20 kHz, but fell with increasing ultrasonic frequency, despite much higher power densities. The release of DOX was also higher than that of Rb, due to a stronger interaction and a deeper incorporation of Rb into the micelle cores. When using pulsed ultrasound, there is a rapid release and re-encapsulation behavior that takes within half a second for each pulse. The released drug is quickly re-encapsulated between ultrasound pulses; this suggests that, on leaving the sonicated volume, the non-extravasated and noninternalized drug would circulate in the encapsulated form, thus preventing any unwanted drug interactions with normal tissues. The *in vivo* enhancement of cellular drug uptake for this system was attributed to an increase in drug release from the micelles, which implied that ultrasound could be focused on a localized tumor and an anti-cancer agent released from the micelles and delivered directly to the malignant tissues. Unfortunately, ultrasound may also promote permeabilization of the cell membrane, followed by an increase in cellular drug incorporation.

3.5
Future Perspectives

Several approaches to the development of smart nanoassemblies have been discussed in this chapter, with a combination of stimuli-sensitive processes being used to control the properties of the nanostructures and provide a precise customization of drug delivery behavior *in vivo*.

Moreover, external stimuli-responsive nanoassemblies – the fate of which can be followed *in vivo* (e.g., smart drug delivery systems loaded with magnetic resonace imaging (MRI) contrast agents [71], or fluorescent dyes for NIR imaging [72]) – will facilitate the use of an external trigger and enhance the therapeutic effects of smart drug carriers.

The incorporation of molecules that target specific cellular signals on the outer surfaces of smart nanoconstructs [21, 36, 73–76], or the construction of nanoassemblies with copolymers that have specific interaction with cells [77], is essential for designing carrier systems with specific cellular recognition. Such recognition can be precisely tuned by constructing end-functionalized block copolymers, the self-assembly of which will lead to the formation of nanostructures with pilot molecules on their exterior. In this way, a specific drug delivery to a target tissue and specific activation of that delivered drug within the targeted cell, may enhance the efficacy and minimize any adverse side effects during drug targeting. The combination of a specific cellular uptake of nanoassemblies with intracellular drug release might also permit the accurate management of drug distribution, leading to an enhanced or innovative therapeutic activity. Although several of the pilot molecule-directed block-copolymer assemblies reported here have still to be optimized, in some cases the activities achieved are elevated to a significant extent, thus highlighting the huge potential of cell-targeted, smart nanoassemblies.

References

1 Lazzari, M. and López-Quintela, M.A. (2003) *Adv. Mater.*, **15**, 1583–1594.
2 Hamley, I.W. (2003) *Angew. Chem. Int. Ed. Engl.*, **42**, 1692–1712.
3 Kataoka, K., Harada, A. and Nagasaki, Y. (2001) *Adv. Drug Deliv. Rev.*, **47**, 113–131.
4 Allen, C., Maysinger, D. and Eisenberg, A. (1999) *Colloids Surf. B Biointerfaces*, **16**, 3–27.
5 Matsumura, Y. (2007) *J. Drug Target.*, **15**, 507–517.
6 Martina, M. and Hutmacher, D.W. (2007) *Polym. Int.*, **56**, 145–157.
7 Hasirci, V., Vrana, E., Zorlutuna, P., Ndreu, A., Yilgor, P., Basmanav, F.B., and Aydin, E. (2006) *J. Biomater. Sci. Polym. Ed.*, **17**, 1241–1268.
8 Jagur-Grodzinski, J. (2006) *Polym. Adv. Tech.*, **17**, 395–418.
9 Matsumura, Y. and Maeda, H. (1986) *Cancer Res.*, **46**, 6387–6392.
10 Maeda, H., Wu, J., Sawa, T., Matsumura, Y. and Hori, K. (2000) *J. Control. Release*, **65**, 271–284.
11 Stolnik, S., Illum, L. and Davis, S.S. (1995) *Adv. Drug Deliv. Rev.*, **16**, 195–214.
12 Kwon, G., Suwa, S., Yokoyama, M., Okano, T., Sakurai, Y. and Kataoka, K. (1994) *J. Control. Release*, **29**, 17–23.
13 Avgoustakis, K., Beletsi, A., Panagi, Z., Klepetsanis, P., Livaniou, E., Evangelatos, G. and Ithakissios, D.S. (2003) *Int. J. Pharm.*, **259**, 115–127.
14 Engin, K., Leeper, D.B., Cater, J.R., Thistlethwaite, A.J., Tupchong, L. and McFarlane, J.D. (1995) *Int. J. Hyperthermia*, **11**, 211–216.
15 Murphy, R.F., Powers, S. and Cantor, C.R. (1984) *J. Cell Biol.*, **98**, 1757–1762.
16 Savic, R., Luo, L., Eisenberg, A. and Maysinger, D. (2003) *Science*, **300**, 615–618.
17 Bae, Y., Fukushima, S., Harada, A. and Kataoka, K. (2003) *Angew. Chem. Int. Ed. Engl.*, **42**, 4640–4643.
18 Bae, Y., Nishiyama, N., Fukushima, S., Koyama, H., Yasuhori, M. and Kataoka, K. (2005) *Bioconj. Chem.*, **16**, 122–130.
19 Gillies, E.R. and Frechet, J.M. (2003) *Chem. Commun.*, **14**, 1640–1641.
20 Lee, E.S., Shin, H.J., Na, K. and Bae, Y.H. (2003) *J. Control. Release*, **90**, 363–374.
21 Lee, E.S., Na, K. and Bae, Y.H. (2003) *J. Control. Release*, **91**, 103–113.
22 Martin, T.J., Prochazka, K., Munk, P. and Webber, S.E. (1996) *Macromolecules*, **29**, 6071–6073.
23 Tang, Y., Liu, S.Y., Armes, S.P. and Billingham, N.C. (2003) *Biomacromolecules*, **4**, 1636–1645.
24 Leroux, J., Roux, E., Le Garrec, D., Hong, K. and Drummond, D.C. (2001) *J. Control. Release*, **72**, 71–84.
25 Taillefer, J., Jones, M.C., Brasseur, N., van Lier, J.E. and Leroux, J.C. (2000) *J. Pharm. Sci.*, **89**, 52–62.
26 Sant, V.P., Smith, D. and Leroux, J.-C. (2004) *J. Control. Release*, **97**, 301–312.
27 Merdan, T., Kopecek, J. and Kissel, T. (2002) *Adv. Drug Deliv. Rev.*, **54**, 715–758.
28 Wagner, E. (2004) *Pharm. Res.*, **21**, 8–14.
29 De Smedt, S.C., Demeester, J. and Hennink, W.E. (2000) *Pharm. Res.*, **17**, 113–126.
30 Brown, M.D., Schatzlein, A.G. and Uchegbu, I.F. (2001) *Int. J. Pharm.*, **229**, 1–21.
31 Han, S., Mahato, R.I., Sung, Y.K. and Kim, S.W. (2000) *Mol. Ther.*, **2**, 302–317.
32 Behr, J.P. (1997) *Chimia*, **51**, 34–36.
33 Boeckle, S. (2004) *J. Gene Med.*, **6**, 1102–1111.
34 Zauner, W. (1998) *Adv. Drug Deliv. Rev.*, **30**, 97–113.
35 Putnam, D. (2001) *Proc. Natl Acad. Sci. USA*, **98**, 1200–1205.
36 Oishi, M., Nagatsugi, F., Sasaki, S., Nagasaki, Y. and Kataoka, K. (2005) *Chembiochem*, **6**, 718–725.
37 Fukushima, S., Miyata, K., Nishiyama, N., Kanayama, N., Yamasaki, Y. and Kataoka, K. (2005) *J. Am. Chem. Soc.*, **127**, 2810–2811.
38 Kanayama, N., Fukushima, S., Nishiyama, N., Itaka, K., Jang, W.-D., Miyata, K. and Kataoka, K. (2006) *ChemMedChem*, **1**, 439–444.
39 Akagi, D., Oba, M., Koyama, H., Nishiyama, N., Fukushima, S., Miyata, T., Nagawa, H. and Kataoka, K. (2007) *Gene Ther.*, **14**, 1029–1038.

40 Lee, Y., Fukushima, S., Bae, Y., Hiki, S., Ishii, T. and Kataoka, K. (2007) *J. Am. Chem. Soc.*, **129**, 5362–5363.
41 Napoli, A., Valentini, M., Tirelli, N., Muller, M. and Hubbell, J.A. (2004) *Nat. Mater.*, **3**, 183–189.
42 Napoli, A., Boerakker, M.J., Tirelli, N., Nolte, R.J., Sommerdijk, N.A. and Hubbell, J.A. (2004) *Langmuir*, **20**, 3487–3491.
43 Takeoka, Y., Aoki, T., Sanui, K., Ogata, N., Yokoyama, M., Okano, T., Sakurai, Y. and Watanabe, M. (1995) *J. Control. Release*, **33**, 79–87.
44 Kakizawa, Y., Harada, A. and Kataoka, K. (1999) *J. Am. Chem. Soc.*, **121**, 11247–11248.
45 Kakizawa, Y., Harada, A. and Kataoka, K. (2001) *Biomacromolecules*, **2**, 491–497.
46 Miyata, K., Kakizawa, Y., Nishiyama, N., Harada, A., Yamasaki, Y., Koyama, H. and Kataoka, K. (2004) *J. Am. Chem. Soc.*, **126**, 2355–2361.
47 Miyata, K., Kakizawa, Y., Nishiyama, N., Yamasaki, Y., Watanabe, T., Kohara, M. and Kataoka, K. (2005) *J. Control. Release*, **109**, 15–23.
48 Katayama, Y., Fujii, K., Ito, E., Sakakihara, S., Sonoda, T., Murata, M. and Maeda, M. (2002) *Biomacromolecules*, **3**, 905–909.
49 Schild, H.G. (1992) *Prog. Polym. Sci.*, **17**, 163–249.
50 Chung, J.E., Yokoyama, M. and Okano, T. (2000) *J. Control. Release*, **65**, 93–103.
51 Kohori, F., Sakai, K., Aoyagi, T., Yokoyama, M., Yamato, M., Sakurai, Y. and Okano, T. (1999) *Colloids Surf. B Biointerface*, **16**, 195–205.
52 Topp, M.D.C., Dijkstra, P.J., Talsma, H. and Feijen, J. (1997) *Macromolecules*, **30**, 8518–8520.
53 Uyama, H. and Kobayashi, S. (1992) *Chem. Lett.*, 1643–1646.
54 Park, J.-S., Akiyama, Y., Winnik, F.M. and Kataoka, K. (2004) *Macromolecules*, **37**, 6786–6792.
55 Park, J.-S., Akiyama, Y., Yamasaki, Y. and Kataoka, K. (2007) *Langmuir*, **23**, 138–146.
56 Park, J.-S. and Kataoka, K. (2007) *Macromolecules*, **40**, 3599–3609.
57 Sershen, S.R., Westcott, S.L., Halas, N.J. and West, J.L. (2000) *J. Biomed. Mater. Res.*, **5**, 293–298.
58 Weissleder, R. (2001) *Nat. Biotechnol.*, **19**, 327–331.
59 Jiang, J. and Tong, Z. (2005) *J. Am. Chem. Soc.*, **127**, 8290–8291.
60 Jiang, J., Tong, X., Morris, D. and Zhao, Y. (2006) *Macromolecules*, **39**, 4633–4640.
61 Wang, G. and Wang, X. (2002) *Polym. Bull.*, **49**, 1–8.
62 Pieroni, O., Fissi, A. and Popova, G. (1998) *Prog. Polym. Sci.*, **23**, 81–123.
63 Minoura, N., Higuchi, M. and Kinoshita, T. (1997) *Mater. Sci. Eng. C. Biomim. Mater. Sens. Syst.*, **4**, 249–254.
64 Wang, G., Tong, X. and Zhao, Y. (2004) *Macromolecules*, **37**, 8911–8917.
65 Laschewsky, A. and Rekai, E.D. (2000) *Macromol. Rapid Commun.*, **21**, 937–940.
66 Yuan, Z., Fischer, K. and Schärtl, W. (2005) *Langmuir*, **21**, 9374–9380.
67 Konak, C., Rathi, R.C., Kopeckova, P. and Kopecek, J. (1997) *Macromolecules*, **30**, 5553–5556.
68 Pitt, W.G., Husseini, G.A. and Staples, B.J. (2004) *Expert Opin. Drug Deliv.*, **1**, 37–56.
69 Husseini, G.A., Myrup, G.D., Pitt, W.G., Christensen, D.A. and Rapoport, N.Y. (2000) *J. Control. Release*, **69**, 43–52.
70 Marin, A., Muniruzzaman, M. and Rapoport, N. (2001) *J. Control. Release*, **10**, 69–81.
71 Nakamura, E., Makino, K., Okano, T., Yamamoto, T. and Yokoyama, M. (2006) *J. Control. Release*, **114**, 325–333.
72 Ghoroghchian, P.P., Frail, P.R., Susumu, K., Blessington, D., Brannan, A.K., Bates, F.S., Chance, B., Hammer, D.A. and Therien, M.J. (2005) *Proc. Natl Acad. Sci. USA*, **102**, 2922–2927.
73 Yamamoto, Y., Nagasaki, N., Kato, M. and Kataoka, K. (1999) *Colloids Surf. B Biointerfaces*, **16**, 135–146.
74 Nagasaki, Y., Yasugi, K., Yamamoto, Y., Harada, A. and Kataoka, K. (2001) *Biomacromolecules*, **2**, 1067–1070.
75 Torchilin, V., Lukyanov, A.N., Gao, Z. and Papahadjopoulos-Sternberg, B. (2003) *Proc. Natl Acad. Sci. USA*, **100**, 6039–6044.
76 Bae, Y., Nishiyama, N. and Kataoka, K. (2007) *Bioconj. Chem*, **18**, 1131–1139.
77 Holowka, E.P., Sun, V.Z., Kamei, D.T. and Deming, T.J. (2007) *Nat. Mater.*, **6**, 52–57.

4
A Comprehensive Approach to the Alignment and Ordering of Block Copolymer Morphologies

Massimo Lazzari and Claudio De Rosa

4.1
Introduction

4.1.1
Motivation

Self-assembly is an enormously powerful concept in modern materials science, which was first associated with the use of synthetic strategies for the preparation of nanostructures only about 15 years ago [1–3]. Since then, a large variety of carefully designed building blocks have been proposed and employed for working "from atoms up" with the aim of fabricating two-dimensional (2-D) and three-dimensional (3-D) structures. In particular, within the past decade, interest in a potentially ideal nanoscale tool has been growing exponentially, namely phase-separated block copolymers (BCs) [4–17].

Immiscibility among the BC constituents is common, and phase separation results in the series of morphologies, for example, lamellar, gyroid, hexagonal and body-centered cubic for diblock copolymers [4, 7, 9], the size and shape of which may be conveniently tuned by changing the molecular weights and compositions of the BCs [5, 18–20]. Part of the enormous potential for nanomaterial fabrication of these self-organized patterns of chemically distinct domains that have periodicity in the mesoscale – that is, between a few tens and hundreds of nanometers – has already been demonstrated [21–40]. However, in our opinion their development into practical routes suitable for industrial applications will probably only be fully exploited when a few key limitations have been efficiently overcome. As an example, writing and replication processes in microelectronics require spatial and orientational control of patterns which, in the case of BCs, entails the solution of the nontrivial problem of large-area ordering and precise orientation of domains.

Regular structures can be generated by controlling the typical disadvantages of spontaneous phase separation. First, similar to polycrystalline materials, the self-assembly of BCs in bulk is prone to form grains with a high level of local order but a very short persistence length, especially in the case of cylinders and lamellae

Advanced Nanomaterials. Edited by Kurt E. Geckeler and Hiroyuki Nishide
Copyright © 2010 WILEY-VCH Verlag GmbH & Co. KGaA, Weinheim
ISBN: 978-3-527-31794-3

which, on a larger scale, correspond to bulk materials with isotropic properties. Although annealing permits a consistent annihilation of the corresponding grain boundary defects, in addition to other defects such as dislocation, disclination, terracing and asymmetry defects [41–47], the formation of defect-free structures points to more efficient strategies based on the application of external forces or spatial constraints. In the same way, long-range oriented patterns, for example, cylindrical or lamellae domains parallel to the substrate, with all of the axes aligned in a single direction rather than randomly in the parallel plane, may only be induced through a careful engineering of surface effects or by well-oriented fields.

It could be stated that for many technological applications such an appealing spontaneous organization has to be directed by some form of templating (in this context considered in its broader sense), where the BC components could not only interact with each other but also take advantage of external controlling interactions.

Numerous methods to induce and control BC domain orientation have been explored, particularly for substrate-supported films to create perfectly periodic 2-D patterns, ranging from the first demonstration of the effectiveness of the application of electric fields [48] to the recent combination of both active and passive forces to obtain laterally ordered arrays of cylindrical domains with areal densities of up to 1.5 terabits per cm^2 [39], through the elegant application of chemically nanopatterned templates [31]. Most methods proposed to date will be critically presented in this chapter, and special attention will be given to the combination of techniques to yield 3-D control.

4.1.2
Organization of the Chapter

The scope of this chapter is to provide an overview of the variety of methods that can induce long-range ordered BC morphologies, focusing on the most promising areas of ongoing research. The main strategies applied so far may be classified into three different approaches:

1. Control of orientation by applying external fields, such as electric, magnetic, thermal, mechanical and solvent evaporation. Depending on the field imposed, thin films and/or samples in bulk may be oriented.

2. Induction of large-area ordering by facilitating the self-assembly, generally of thin films, on templates either topographically or chemically nanopatterned.

3. Modulation of substrate and surface interactions as a result of:
    a) Preferential interaction of one block with the surface.
    b) Neutralization of attractions to the substrate or to the surface.
    c) Epitaxial crystallization of domains onto a crystalline substrate.
    d) Directional eutectic crystallization of a BC solvent.
    e) Graphoepitaxy.
    f) 2-D geometric confinements.

Before highlighting recent advances in such control strategies, a preliminary section specifically for all researchers who are new to this field will offer practical information on film preparation, and on the methods used in practice to achieve thermodynamic equilibrium morphologies. Because one of the aims of this book is to offer solutions to experimental difficulties and research needs, the decision was taken to highlight the practical aspects of phase separation and orientation with respect to the theoretical concepts.

4.2
How to Help Phase Separation

Except for some specific instances – for example, direct use as photonic crystals or in the case of amphiphilic BC for drug and genetic delivery – applications of BCs require the preparation of thin films with thicknesses ranging from a few tens of nanometers (in some cases down to thicknesses of less than the corresponding equilibrium period of the BC) to several micrometers. Films with low surface roughness may be produced by spin-coating or dip-coating from relatively dilute solutions, that is, approximately 1–5% by weight, onto solid substrates with uniform flatness. The thickness and the uniformity of the film surface mainly depend on the concentration of the solution, the volatility of the solvent, and the specific instrumental speed – that is, the spin speed or withdrawal speed, respectively. Silicon wafers, eventually coated with metals, semiconductors, carbon or polymers are often utilized as the support. During dip-coating processes, and also for films prepared from direct casting, the solvent may evaporate slowly, thus allowing a stable organization of macromolecules close to thermodynamic equilibrium. In contrast, in the case of spun-cast films the solvent is driven off so quickly that nonequilibrium structures could be observed. Moreover, the concentration gradient due to fast solvent evaporation can in fact have a significant effect on domain orientation, as discussed in Section 4.

In thin films, the self-assembled BC morphologies are influenced not only by molecular weights, polydispersity and composition, but also by other variables such as the selectivity of the solvent for one block, surface–interfacial interactions, and the interplay between structure periodicity and film thickness, which can cause significant deviations from the predicted phases in the bulk state. An elegant demonstration and a didactic example of the influence of surface effects in thin films has been reported by Krausch and coworkers [49]. Their experimental data for a triblock copolymer of ABA type are compared with those of the simulated film in Figure 4.1. Several more studies have demonstrated the importance of confinement effects, and will be treated in detail later.

Independently of the casting techniques, even for a film prepared taking all necessary precautions, it is not possible to obtain a perfectly ordered morphology over a large area. In fact, an optimization of the nanostructures is almost indispensable and can be carried out by different annealing processes, with the double objective of obtaining the equilibrium morphology and eliminating the

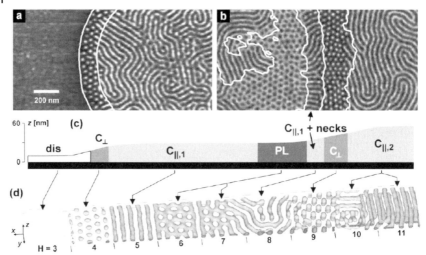

Figure 4.1 (a,b) Tapping-mode scanning force microscopy phase images of thin films of a cylinder-forming polystyrene-block-polybutadiene-block-polystyrene (PS-b-PB-b-PS) after solvent annealing, showing a sequence of phases as a function of the height profile (c), which is in good agreement with the simulation of an $A_3B_{12}A_3$ copolymer film (d). Owing to preferential wetting, the silicon substrate is covered with a homogeneous 10 nm-thick PB layer which, with increasing thickness from left to right, is followed by isolated domains of PS, parallel-oriented PS cylinders, perforated lamella, parallel-oriented PS cylinders again, and finally two layers of parallel-oriented cylinders. Reproduced with permission from Ref. [49]; © 2002, The American Physical Society.

defects. Annealing is usually performed before any further manipulation, even though some approaches include the alignment of domains within the orientation process.

In principle, *thermal annealing* is the simplest option and the most commonly used method. This consists of controlled heating at a temperature above the glass transition temperature (T_g) of the constituent blocks, preferably in an inert atmosphere or under vacuum, for a specific time. For polymers with high molecular weights and complex architectures, the high degree of chain entanglements and the difficult diffusion of one polymer block through the domains of the other blocks pose large kinetic barriers to equilibrium. In such cases, and for partially crystalline BCs with high fusion temperatures, the annealing window for the relevant conditions of temperature and time that are theoretically necessary and those of the order–disorder transition or polymer decomposition, may be insufficient to reach equilibrium. On the other hand, the degradation of one or more blocks has been used to stabilize the morphologies for further applications, through decomposition, partial oxidation or crosslinking [23, 26, 27, 50–53], or even to induce a hierarchical transition of morphologies [54].

Alternative annealing approaches entail an increase in chain mobility by the addition of some type of plasticizer, either transient, such as a solvent, or low-molecular-weight homopolymers [55, 56], which also swell each domain. Exposing thin films to vapors of a good solvent potentially allows long-range equilibrium morphologies to be produced without any thermal treatment. However, as genuinely neutral solvents are rare, the use of solvents selective for one block may induce structures far removed from the equilibrium situation [49, 57–62], which remain frozen after solvent removal and are sometimes difficult to obtain by other methods [60]. As a conjecture, a fully selective solvent should lead to a micellar-like microdomain phase. Once more, Krausch and coworkers were the first to demonstrate for a polystyrene-*block*-poly(2-vinylpyridine)-*block*-poly(*t*-butyl acrylate) (PS-*b*-P2VP-*b*-P*t*BA) triblock copolymer, that the choice of annealing solvent strongly influenced the types of metastable structures observed, and also investigated the time development of the microdomain structure [60]. More recently, the morphology evolution has been followed in other BCs, for example in polystyrene-*block*-poly(methyl methacrylate) (PS-*b*-PMMA) thin films with different thicknesses exposed to PMMA-selective solvent vapors (Figure 4.2) [63, 64].

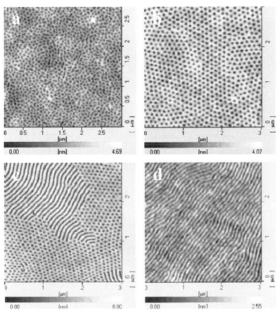

Figure 4.2 Atomic force microscopy topographic images of PS-*b*-PMMA (M_n PS = 133 000, M_n PMMA = 130 000) thin films cast onto an Si substrate with an SiO$_x$ surface layer exposed to chloroform vapor. (a) After 40 h, the initially featureless structure shows a disordered microstructure that evolves, at 60 h (b), into a hexagonally packed nanocylinder structure. (c) Further treatment, 80 h, promotes the evolution after 100 h (d) to a mixed morphology, which completely develops into stripes having the repeat spacing (L_0), of 90 nm. For longer annealing times, the film returns to a flat surface. It has also been shown that only when the film thickness is less than $1/2L_0$ can the packed nanocylinder structures form. Reproduced with permission from Ref. [63]; © 2004, The American Chemical Society.

Other representative examples in controlling the orientation and lateral ordering of microstructures are poly(ethylene oxide) (PEO)-containing block copolymers. A simple solvent annealing of cylinder-forming PS-*b*-PEO and PS-*b*-PMMA-*b*-PEO BC thin films can lead to a perpendicular orientation with a high degree of lateral ordering [65, 66]. In these systems, the relative humidity of the vapor during solvent annealing was shown to play a fundamental role in achieving such order [67], while the addition of small amounts of an alkali halide or metal salts to cylinder-forming PS-*b*-PEO, where the salt complexes with the PEO block, was found to significantly enhance the long-range positional order [68]. In a similar way, the incorporation of diverse PEO additives into PS-*b*-PMMA diblock copolymer thin films gave a vertical alignment of the PMMA cylinder, due mainly to an interaction with the water vapor under the high humidity, solvent-swollen processing conditions, as a direct consequence of the miscibility of the PEO with PMMA block [69, 70].

Finally, an in-depth understanding of the formation of either equilibrium structures or kinetically trapped morphologies during solvent annealing requires not only an accurate choice of solvent but also the proper control of experimental parameters, such as vapor pressure, treatment time, solvent extraction rate and, in case, relative humidity. In practice, this knowledge facilitates the preparation of reproducible nanopatterns in thin films, which also, in the case of metastable assemblies, show a long-term stability, at least for T_gs of all the components that are well above room temperature.

Another plasticizing agent that presents a modest equilibrium sorption and can be easily removed is supercritical CO_2. Although its rapid diffusion in most polymers allows an equally rapid equilibrium distribution in thin films, CO_2 annealing has so far been limited to a surprisingly low number of BCs, essentially PS-*b*-PMMAs, having molecular weights up to $300\,000\,\text{g}\,\text{mol}^{-1}$ [71, 72].

4.3
Orientation by External Fields

Since the first macroscopic alignment of cylindrical domains of an industrial triblock copolymer (a PS-*b*-PB-*b*-PS; Kraton 102) by extrusion carried out by the Keller group during the early 1970s [73, 74], many more mechanical flow fields have been proposed to control BC alignment. Following the Keller belief that BC microstructuring and orientation were exciting research subjects, but with very limited applicability (E. Pedemonte, personal communication), and after the shear flow experiments on the same copolymer by Skolios and coworkers [75–77], during the 1990s research interests returned strongly to alignment by relatively weak external fields. Another reason for this was the greater availability of BCs with different architectures and chemical compositions [5, 18–20]. Particularly in recent years, most investigations moved from bulk materials to thin films, due to their nanotechnological potential, with a special focus on the use of electric fields. Active external fields will be discussed later in the chapter, while those that are passive, such as with interfacial interactions, are considered in the next sections.

4.3.1
Mechanical Flow Fields

Extrusion [73, 74], compression [78–82], flows involving oscillatory shearing [75–77, 82–91] and other steady shearing [92–98], and techniques that combine different flow fields [99–104] have been successfully applied to induce alignment in BCs, either in a microphase-separated molten state or, to a lesser extent, in gels of solvent-swollen microdomains [105–112], with the objective of forming single crystal structures. Large-amplitude oscillatory shear (LAOS), which was first proposed at the end of the 1980s to orient commercial triblock copolymers [82, 83], is the most widely employed technique, also due to the easy characterization of the shear field with respect to other methods. As an example, common parallel plate-type rheometers enable the modulation of valuable experimental variables to be carried out, such as shear rate, frequency, strain amplitude, and temperature.

Early studies on the effect of mechanical flow fields on the orientation of BC domains have already been summarized in many excellent publications and books, for example, by Fasolka and Mayes [9], Hamley [4, 113], the group working with Thomas [12, 114] and others [84, 115, 116], and so need not be reviewed here. Di-, tri-, and multiblock copolymers in bulk with different microstructures, such as lamellar [74, 78, 79, 84–91, 96, 98], cylindrical [80–82, 94, 95, 97, 101, 103], spherical [92, 93, 97] and gyroid [102], can be preferentially oriented parallel or, where applicable, perpendicular or mixed-perpendicular to the flow direction. Most experimental observations have been supported and even anticipated by theory, with several good theoretical works being published in recent years [117–126]. Further developments concerned the extension of the shearing techniques to thin films. Albalak and Thomas [99, 100] proposed a novel casting method, termed *roll-casting*, in which a BC solution is subjected to a flow between two counter-rotating rolls (Figure 4.3a). The solution is compressed and, as a result of solvent evaporation and shearing, the microphase separates into a film with a high degree of orientation and close to single-crystal characteristics. Only cylinder morphology from triblock commercial polymers has been oriented in this way [101, 102], with a film thickness intrinsically limited by the geometry of the device to hundreds of nanometers. Thinner films of sphere- and cylinder-forming polystyrene-*b*-poly(ethylene propylene) (PS-*b*-PEP) have been shear-aligned by simple flowing techniques [93, 94], and extended to arbitrarily large areas. Films of 30–50 nm spread onto a silicon substrate are covered by an elastomer pad, which is then slowly pulled forward by a constant force (Figure 4.3b). The steady shear must be applied at temperatures between the T_g and the temperature of order–disorder transition, T_{ODT}, of the copolymer, thus providing an essentially infinite orientational order in all directions without the typical limitations of sheared bulk samples, related to the presence of multigrains.

(a)

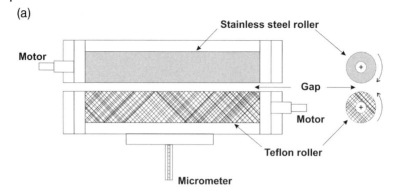

(b)

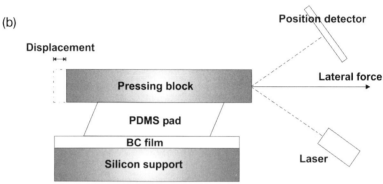

Figure 4.3 Schematic of apparatus used to induce alignment in BC thin films. (a) In the roll-caster (upper view and section), two independent rollers – one of stainless steel and the other of Teflon – are driven at the same angular velocity by two independent motors and separated by a micrometer-controlled gap; (b) In the shear-alignment set-up, the polydimethylsiloxane (PDMS) pad is pressed with a constant weight against the film; the support is heated at the annealing temperature while the displacement of the pressing block is continuously monitored with a laser–mirror–optical sensor assembly.

Unconventional methods, which to some extent are based on shearing and a stretching field, such as electrospinning [127–130] and orientation by spin coating [131, 132], have been proposed in the past few years, but still need to be investigated in depth. *Electrospinning* in particular [133, 134] seems to offer great potential for the fabrication of nanofibers with internal oriented structuring from cylinder-forming copolymers with at least one partially crystalline block.

4.3.2
Electric and Magnetic Fields

Following the pioneering studies of Amundson *et al.* on lamellar diblock copolymer thin films [135–137], static electric fields have been widely used in copolymer melts to macroscopically orient lamellar or cylindrical morphologies [48, 138–146], with most of the investigations being focused on PS-*b*-PMMA. Although morphologies parallel to the substrate could be obtained using an in-plane field, a uniaxial orientation along the field and perpendicular to the substrate is preferably induced between the electrodes compressing the film. As initially proposed by Russell and coworkers, a convenient procedure consists of the preparation of films between two Kapton sheets coated on one side with aluminum, with the film thickness controlled through Kapton spacers. Kapton is a commercially available polyimide with excellent mechanical properties that allows further manipulation after alignment of the domains [23], and also facilitates sample characterization, such as for microtoming. In order to avoid electric shorting, one of the electrodes must be placed in contact with the polymer, while the second is inverted with the Kapton side facing the polymer.

The general statement that the orientation of BC microdomains is possible if the applied field is high enough for a given difference in the dielectric constant between blocks, can be expressed by Equation 4.1 [147, 148]:

$$E_c = \Delta\gamma^{1/2} \frac{2(\varepsilon_A + \varepsilon_B)^{1/2}}{\varepsilon_A - \varepsilon_B} t^{-1/2} \quad (4.1)$$

where E_c is the critical electric field strength, ε_A and ε_B are the dielectric constants of blocks, and t is the film thickness, which also takes into account the surface interactions through the difference between the interfacial energies of each block with the substrate, $\Delta\gamma$. In the absence of preferential interactions with the substrate – that is, $\Delta\gamma = 0$ – the domains do not need any external field to orient normal to the substrate (see Section 4.5.1). From a practical point of view, Equation 4.1 permits the evaluation of the critical parameter for a given set of the different parameters, eventually imposed by the same experimental design. The thickness of the region to be oriented is limited to a few millimeters by the electric field strengths, in the range of approximately 1 to 100 V μm^{-1}, considered as the upper limit that prevents dielectric breakdown.

In addition, large-scale topographic structures could be generated by electrohydrodynamic (EHD) film instability, a technique in which a liquid dielectric interface is destabilized by a perpendicular electric field [149]. The interplay between EHD structure formation and the structural control over BC microphases allowed lithographically controlled micrometer-sized features to be obtained, where the cylindrical or lamellar domains are oriented parallel to the patterned structure axis; that is, normal to the substrate [144, 150].

Theoretical studies have been conducted to investigate various aspects of the dynamics of mesophase formation and orientation, mainly in diblock copolymer melts [147, 148, 151–154]; here, special mention should be made of the capacitor analogy proposed by Pereira and Williams for symmetric BCs [147]. The appeal of this model actually lies in its simplicity which, at a more qualitative level, also permits justification of the alignment of lamellae perpendicular to the electrodes on the basis of preferential travel of the electric field lines along easy paths, though each and every block has the highest polarizability. In the same way, parallel orientation is unfavorable, as the electric field lines would be forced to cross both blocks evenly. Tsori *et al.* have recently proposed that, in these systems, alignment may also occur by other strong orienting forces, such as by the presence of a conductivity or mobility contrast [155], postulating an important role for the ions remaining in the polymers after synthesis in the reorientation mechanism [153].

The alignment pathways in thin films of BCs with various morphologies have been recently followed using *in situ* small-angle X-ray scattering (SAXS) [140, 143] and scanning force microscopy (SFM) [145]. The application of an electric field (perpendicular to the plane of the film) to the disordered melt of an asymmetric BC induced the progressive growth of microphase-separated structures oriented parallel to the field only at temperatures below the order–disorder transition temperature. This occurs up to a stationary, equilibrium state, and also for a copolymer already having ordered domains parallel to the substrate, either a cylindrical PS-*b*-PMMA [140] or a lamellar polystyrene–polyisoprene (PS-PI) system [143], alignment parallel to the field could be achieved within minutes. An analysis of the scattering patterns suggested that, during the reorientation, the initial large grains are broken up into smaller pieces by the amplification of interfacial fluctuations, which are then able to rotate in the field. The final state of the symmetric BCs of PS and PI consists of many small grains with a high degree of orientation of the lamellae parallel to the electric field, but with a random orientation in the perpendicular plan. This observation also finds some theoretical support in the prediction by Tsori [148, 153, 154]. The limits imposed by the film thickness and the interfacial energy effects, as predicted in Equation 4.1, have been faced experimentally by the Russell group for symmetric and asymmetric BC thin films between substrates with modulated interfacial interactions [142, 156, 157]. In the presence of a preferential interaction of one block with the substrate (see Section 4.5.1) such effects are dominant for film thicknesses up to approximately 10 L_0, regardless of the electric field applied normal to the surface (see, e.g., Figure 4.4a). For thicker films, the effects dissipate with distance from the surface and the domains orient in the direction of the electric field, at least in the interior of the film (Figure 4.4b), with a distance of propagation inversely proportional to the strength of the interfacial interactions [157].

All of these melt-based procedures suffer severe limitations because of the melt viscosity itself, which is directly dependent on the molecular weight and architecture of BCs, and the thickness of the region to be oriented. In addition, the achievement of high degrees of orientation eventually requires high temperatures and electric field strengths close to the decomposition conditions. An alternative approach to circumvent such restrictions consists in the addition of a neutral solvent with the effect of increasing chain mobility, as already discussed above for thermal annealing, thus facilitating large-scale domain alignment [158–160]. Concentrated diblock (PS-*b*-PI) [158] and triblock [polystyrene-*block*-poly(hydroxyethyl methacrylate)-*block*-poly(methyl methacrylate), PS-*b*-PHEMA-*b*-PMMA] [159] copolymer solutions have been aligned under a dc electric field during solvent evaporation, leading to highly anisotropic bulk samples (Figure 4.5). The orientation kinetics and mechanisms of alignment – that is, grain orientation and nuclea-

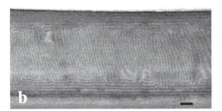

Figure 4.4 Cross-sectional transmission electron microscopy images of ≈300 nm (a) and ≈700 nm (b) symmetric PS-*b*-PMMA films between PDMS-coated Kapton electrodes annealed at 170 °C under ≈40 V μm^{-1} electric field for 6 and 16 h, respectively. The scale bar represents 100 nm. Reproduced with permission from Ref. [142]; © 2004, The American Chemical Society.

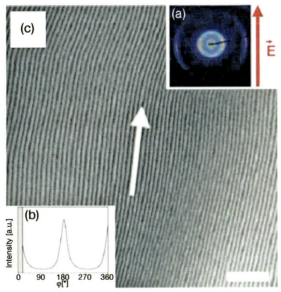

Figure 4.5 (a) 2-D small-angle X-ray scattering pattern; (b) Azimuthal intensity distribution for a first-order reflection; (c) TEM image of a bulk sample prepared from a 40 wt% solution in toluene of symmetric PS-*b*-PI (M_n = 80 000). The arrows indicate the direction of the electric field. Scale bar = 400 nm. Reproduced with permission from Ref. [158]; © 2003, The American Chemical Society.

tion and growth of domains – have been corroborated by simulations [158] based on dynamic density functional theory (DFT) [161]. In summary, several groups have demonstrated that the application of unidirectional electric fields, coupled with an appropriate choice of materials and experimental conditions, allow an effective fabrication of patterns with a high degree of orientation and high aspect ratio, suitable for nanotechnological applications. However, the orientation of domains is azimuthally degenerate and a further improvement to achieve morphologies with orientation controlled in the three dimensions logically requires the application of a second, orthogonal, external field. The first sequential combination toward long-range, 3-D ordered thin films was shown to control the orientation of lamellar microdomains through the application of an elongational flow field to obtain an in-plane orientation in ≈30 μm-thick films, and an electric field normal to the surface [162]. The concurrent application of the orthogonal field (e.g., a mechanical shearing) is expected to reduce the presence of defects and grain boundaries in an even more effective fashion.

A further alignment approach is available for materials that exhibit anisotropic susceptibility due to an anisotropic molecular structure. Magnetic field-induced orientation has been achieved for liquid crystalline (LC) diblock copolymers with a dielectric diamagnetic isotropy, possibly through the magnetic alignment of LC mesogens [163–165], and also for BCs with a crystallizable block through an accurate control of the crystallization process [166]. The latter investigation is particularly likely to open a new fruitful research field and, at the same time, pave the way to an alternative method of relatively wide applicability, which only has the prerequisite of a high nucleation density during crystallization. Magnetic fields also offer the ability to apply very high fields without the risks of electric fields, associated with the danger and limit of electric breakdown. In addition, the orientation of PEO-based diblock copolymers with a poly(methacrylate) containing an azobenzene moiety in the side chain as a hydrophobic LC segment was produced by using a LASER beam, that induced the alignment of photoresponsive azobenzenes and transferred the molecular ordering at supramolecular level [167]. Macroscopic parallel array of nanocylinders can be obtained over an arbitrary area, in principle even on curved surfaces.

4.3.3
Solvent Evaporation and Thermal Gradient

The observation of lamellar and cylindrical microdomains in thin films perpendicular to the surface as a result of solvent evaporation was first reported by Turturro and coworkers [168], and subsequently investigated in more detail by Kim and Libera for a similar triblock copolymer [169, 170]. On the basis of these and other studies on either spun-cast or solution-cast films from solutions in a good solvent for all the blocks, a reasonable mechanism of orientation could be proposed (Figure 4.6). In the first moments after deposition, the T_g of the swollen film is still well below room temperature, thus allowing free chain mobility. With the decrease in the solvent concentration, the BC undergoes a transition from the disordered to the ordered state and, as in thin films the diffusion of the solvent produces a gradient of concentration, the ordering front rapidly propagates from the air surface to the substrate. The consequent decrease of T_g below room temperature, for at least one block, locks in the structures which, due to the high

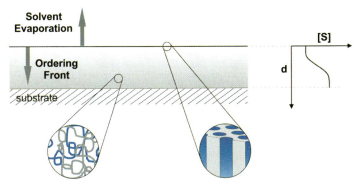

Figure 4.6 Schematic of the solvent evaporation in a diblock copolymer thin film. The diffusion produces a gradient in the concentration of the solvent, [S], as a function of depth, d, which induces an ordering front from the film surface to the substrate.

directionality of the solvent gradient, are highly oriented normal to the surface. This behavior has been reported so far for films of less than one-half micron thick, for example, on PS-PB systems [168–170], PS-*b*-PEO [65, 171, 172], polystyrene-*block*-polyferrocenyldimethylsilane (PS-*b*-PFS) [173], and PS-b-P4VP [174], whereas the applicability of these conditions can be extended to any BCs having the T_g of one block above room temperature. Commonly, the short-range order of cylindrical microdomains of as-spun films may be significantly enhanced by solvent annealing. In the case of PS-*b*-PEO thin films, a better control over domain orientation was obtained by resorcinol complexation of the PEO cylinder-forming block [175]. This is possibly due to a lower degree of crystallinity of the PEO/resorcinol complexes with respect to the pure PEO, that has the effect of enhancing the mobility of polymers during film formation, and therefore also the orientation of domains parallel to the solvent gradient.

After the excellent results obtained by the Hashimoto group [176, 177], the use of temperature gradient apparatus to produce highly oriented single crystals was also reported for thin films [178, 179]. As an example, square-millimeter well-oriented samples of a cylinder-forming BC were obtained, with a density of disclinations estimated at $4 \times 10^{-4} \mu m^{-2}$ (corresponding to an average disclination separation of 50 μm) [178].

4.4
Templated Self-Assembly on Nanopatterned Surfaces

The combination of top-down strategies to fabricate patterns that direct the bottom-up organization of organic or inorganic building blocks is a strategy often used in the micrometer and, to a minor extent, in the nanometer regime. Also, in the case of BCs, long-range order and orientation may be induced if self-assembly is forced to occur into/onto a guide, either topographically or chemically patterned, or in other 2-D or 3-D confinements. In this section, only the effect of surfaces with features having periodicity commensurable with those of the BC morphologies is considered. The study of the control of the registration and orientation of domains by topographic confinements (in the form of surface relief or groove structures) with a length scale much larger than the BC lattice parameters had a somehow different theoretical and historical development, and is better known as *graphoepitaxy*. It will therefore be discussed later as it follows such different nomenclature.

Rockford and coworkers decorated miscuit silicon wafers with gold to fabricate striped and essentially flat surface of nonpolar gold and polar silicon oxide, with periodicity ranging from 40 to 70 nm [180, 181]. Films of symmetric PS-*b*-PMMA solution cast onto such substrates showed, after thermal annealing, lamellar orientation normal to the plane and parallel to the striping, with the greatest ordering for patterns commensurable with the bulk lamellar period of the BC. A mismatch in the length scale of 10% and 25% in thick and ultrathin films, respectively, appeared to be sufficient to lose control over orientation. A novel integrated fabrication strategy that makes use of advanced lithographic techniques has been tuned, by Nealey and coworkers, to produce perfect periodic domain ordering [182, 183]. Although Nealey actually discussed epitaxial self-assembly, we prefer to reserve that term for a different methodology to control spatial and orientation order, as introduced by De Rosa *et al.* [184] (for a definition of epitaxy, see Section 4.5.2). As schematically reported in Figure 4.7, in the first essential step of this procedure a self-assembled monolayer (SAM) is precisely patterned throughout the photoresist using extreme ultraviolet interferometric lithography [185, 186]. Following the conversion of the topographic pattern to a chemical pattern and the photoresist removal, a symmetric PS-*b*-PMMA is spin-coated onto the chemical pattern and, as the modified regions present polar groups that preferentially wet the PMMA block, the self-assembly results in lamellae oriented perpendicular to the substrate. High-quality, defect-free BC films require pretty much the same period of the chemical pattern and the BC, with commensurability within just a few percent. For values within approximately 10%, the morphology can still be surface-directed but it is not perfect (Figure 4.8). An extension of this method, through the use of ternary blends of diblock copolymers and the corresponding homopolymers, showed that is possible to pattern chemical templates with periodicities substantially deviating from those of the BC [31]. The redistribution of homopolymer macromolecules during thermal annealing facilitates defect-free assembly of the blend as a whole, to form 50–80 nm periodic arrays in addition to pattern structures with different bend geometries. More recently, the validity of this approach was proven by other groups, which demonstrated that PS-*b*-PMMA can self-assemble in a well-aligned, long-range ordered hexagonal lattice commensurate with chemically prepatterned templates prepared by electron beam lithography (EBL) [187] or into a laterally

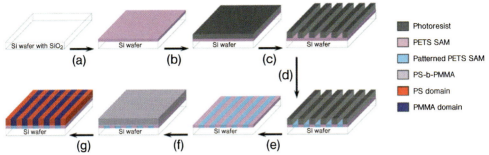

Figure 4.7 Schematic representation of the fabrication process of chemically nanopatterned surfaces that template the self-assembly of symmetric PS-*b*-PMMA (M_n = 104 000). (a) A self-assembled monolayer of phenylethyltrichlorosilane is deposited on a silicon wafer; (b) A photoresist is then spin-coated and patterned with alternating lines and spaces by ultraviolet interferometric lithography (c); (d) The topographic pattern is converted into a chemical pattern by irradiation with soft X-rays in the presence of oxygen; (e) After the photoresist removal, a toluene solution of PS-*b*-PMMA is spin-coated onto the patterned SAM (f). (g) Thermal annealing facilitates surface-directed self-assembly. Reproduced with permission from Ref. [183]; © 2003, Nature Publishing Group.

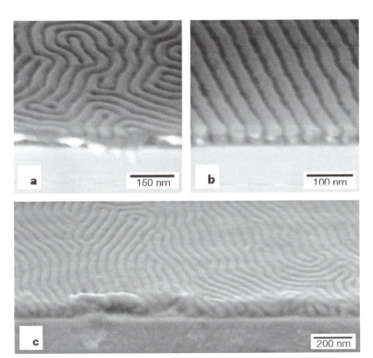

Figure 4.8 Cross-sectional images of symmetric PS-*b*-PMMA (M_n = 104 000; L_0 = 48 nm) films on unpatterned surfaces (a) and chemically nanopatterned surfaces with 47.5 nm periodicity (b), prepared as reported in Figure 4.7. When the surface pattern is greater than L_0 (c), lamellae exhibit an imperfect ordering. Reproduced with permission from Ref. [183]; © 2003, Nature Publishing Group.

stacked one-dimensional (1-D) lamellar assembly along surfaces chemically stripe patterned by conventional lithography [188]. A further optimization of blend composition, polymer chemistry and lithographic techniques will make it possible to fabricate nonregular shaped structures of different types, which implicitly afford the production of nanodevices requiring patterns more complex than simple periodic arrays.

4.5
Epitaxy and Surface Interactions

Control of the orientation of microdomains in the microstructure of BCs can be obtained through a bias field induced by surface interactions. It is well known that a certain type of substrate strongly influences the structure of polymers. A trivial example is the action of nucleating agents on the crystallization, which has often been attributed to epitaxy of the growing polymer crystal. The interactions between the surfaces of a substrate or of confining walls and the molecules of amorphous or semicrystalline block copolymers, determine the orientation of domains on the substrate. Different types of interaction can be established depending on the nature of the surface and of the BC.

Specific interactions, such as epitaxy and directional crystallization, are involved in semicrystalline or amorphous BCs and crystalline substrates. It will be shown in the following sections that epitaxy and/or graphoepitaxy can be used in combination with the directional crystallization providing powerful methods for creating regular surface patterns in BC thin films.

4.5.1
Preferential Wetting and Homogeneous Surface Interactions

The microstructures in thin films block copolymers generally depend on additional variables such as thickness of the film and two types of surface interactions – that is, interactions of the BC with the superstrate (often air) and with the substrate. In particular, the behavior of BC thin films depends primarily on the interfacial interactions and commensurability of the period of the BC with the film thickness. The chemical modification of BCs and various surface-patterning techniques (e.g., soft lithography and holography) allow a better control of the microdomain structures of BC thin films. The simplest interaction of a BC film deposited on a substrate is the preferential wetting of one block at an interface so as to minimize the interfacial and surface energies. As a consequence, a parallel orientation of microdomains, lamellae and cylinders is often induced at the interface, and this orientation tends to propagate throughout the entire film [9, 189–200]. Quantization of the film thickness occurs in thin films with dimensions of the order of only a few microdomain repeats, when the thickness is incommensurate with the natural period of the block copolymer. For lamellar-forming block copolymers, discrete integer or half-integer values of the repeat period occur, leading to the formation of terraces (i.e., islands and holes) at the polymer/air interface [193–196].

The microstructure can be altered by variation of the film thickness on the substrate and preferential interactions of blocks with the substrate [59, 198, 199, 201–203]. Symmetric boundary conditions are established when one of the blocks interacts preferentially with both the substrate and the air surface [193], whereas asymmetric conditions pertain when one block is preferentially wetted by the substrate and the other block by the superstrate [197].

The control of orientation of the microdomains can also be achieved by confining a BC between two surfaces; that is, adding a superstrate to a BC film supported on a substrate [204–207]. Strong or weak interactions of BCs with the surfaces can be created by coating the surface walls with a homopolymer or a random copolymer, respectively, containing the same chemical species as the confined BC [206]. In the case of a neutral surface, for example, by using a random copolymer, the lamellar microdomains rearrange themselves so that the direction of periodicity is parallel to the substrate [206, 208–210]. Moreover, decreasing the confined film thickness – that is, creating a large incompatibility strain of the natural domain period of the BC and the film thickness – induces a heterogeneous in-plane structure where both parallel and perpendicular lamellae are located near the confining substrate [207]. Various theoretical studies have predicted the structural behavior of BC thin films in a confined geometry [211–220] and are basically consistent with experimental results.

Great influences of both film thickness and surface interaction on the orientations of the microdomains have also been found for BCs forming cylinder [190, 221–229] and sphere [230, 231] morphologies. A parallel-to-vertical transition of cylindrical PB microdomains was observed, depending on film thickness [190, 223, 224, 228]. Moreover, strong preferential interactions between one block and the substrate in the case of confined and supported block copolymer thin film, induce a transition from cylindrical microdomains to a layered structure near the substrate surface [221, 222, 224]. It has also been shown that in cylinder-forming PS-*b*-PB BC thin films, the PS cylinders transformed into a perforated interlayer, penetrated by PB channels that connect the two outermost PB surface layers, for film thicknesses that are significantly less than their respective unperturbed chain dimensions [224]. For the reverse structure – that is, PB cylinders in a polystyrene matrix – the cylinders transformed into spheres as the film thickness decreased, then to hemispheres, and finally to a bilayer of surface-segregated PB covering a PS-rich interlayer [224].

The alignment of microdomains in both substrate-supported and -confined films has also been achieved by tuning the specific interaction between the BC and the substrate through modification of the surface [27, 139, 210, 232, 238]. An example is the application of a random copolymer brush anchored on the substrate, which allows the orientation of the domains to be changed from parallel to perpendicular by altering the composition of the random copolymer [238]. A new approach

128 | 4 Alignment and Ordering of Block Copolymer Morphologies

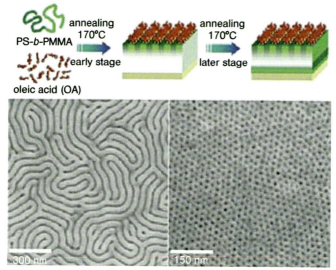

Figure 4.9 Schematic on the surfactant-assisted orientation of PS-*b*-PMMA thin film. Reproduced with permission from Ref. [239]; © 2008, Wiley-VCH.

developed by Char and coworkers is based in the addition of a low-molecular-weight surfactant, oleic acid, that has the effect to mediate the self-assembly of the BC thin film (Figure 4.9) [239]. At the early stage of the thermal annealing, oleic acid (whose acid groups have a preferential affinity with the hydrophobic part of the PMMA block) segregates to the upper part of the film, and preferentially in the PMMA domains. The energetically neutral conditions at the surface rapidly induce the perpendicular orientation of microdomains into the inner film.

4.5.2
Epitaxy

Epitaxy is defined as the oriented growth of a crystal on the surface of a crystal of another substance (the substrate). The growth of the crystals occurs in one or more strictly defined crystallographic orientations defined by the crystal lattice of the crystalline substrate [240–242]. The resulting mutual orientation is due to a 2-D or, less frequently, a 1-D structural analogy, with the lattice matching in the plane of contact of the two species [240]. The term epitaxy, literally meaning "on surface arrangement," was introduced in the early theory of organized crystal growth based on structural matching [240–242]. Discrepancy between atomic or molecular spacings is measured by the quantity $100(d - d_0)/d_0$, where d and d_0 are the lattice periodicities of the adsorbed phase and the substrate, respectively. In general 10–15% discrepancies are considered as an upper limit for epitaxy to occur in polymers [242].

Inorganic substrates were first used for the epitaxial crystallization of polymers [243–257]. Successive studies have demonstrated the epitaxial crystallizations of polyethylene (PE) and linear polyesters onto crystals of organic substrates, such as condensed aromatic hydrocarbons (naphthalene, anthracene, phenanthrene, etc.), linear polyphenyls, and aromatic carboxylic acids [258–262]. In the case of PE, a unique orientation of the crystals grown on all the substrates is observed, with different contact planes depending on the substrate [258–260].

An example of an epitaxial relationship between PE crystals and an organic substrate is shown in Figure 4.10, with the substrate constituted by a crystal of benzoic acid (BA) [259] (monoclinic structure with $a = 5.52$ Å, $b = 5.14$ Å, $c = 21.9$ Å and $\beta = 97°$, with a melting temperature of 123 °C). A clear match between the PE interchain distance (the b-axis of PE equal to 4.95 Å) and the b-axis periodicity of benzoic acid crystal (5.14 Å), and between the c-axis periodicity of PE (2.5 Å) and the a-axis of benzoic acid crystal (5.52 Å), produces the crystallization of PE onto preformed crystals of benzoic acid with lamellae standing edge-on, that is, normal to the surface of benzoic acid crystal (Figure 4.10). The PE polymer chains lie flat on the substrate surface with their chain axis parallel to the substrate surface and parallel to the a-axis of benzoic acid crystal, and the b-axis of PE parallel to the b-axis of the benzoic acid crystal [259]. The (100) plane of PE is in contact with the (001) exposed face of benzoic acid.

In order to apply epitaxial control, BCs need to have at least one crystalline block, which can interact with a crystalline surface. The morphology of semicrystalline block copolymers is the result of the interaction between microphase separation of the component blocks and crystallization of the crystallizable block. The process pathway of structure formation is of primary importance, in the sense that the forming structures set off those emerging to the described geometry into which the new structure must evolve. The final morphology is therefore path-dependent; different microdomain structures are obtained if the crystallization occurs from a homogeneous melt (in this case the crystallization drives the

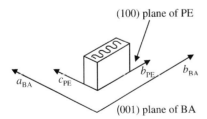

$(100)_{PE} \parallel (001)_{BA}$ $b_{PE} \parallel b_{BA}$ and $c_{PE} \parallel a_{BA}$

Figure 4.10 PE lamella oriented edge-on on the (001) face of a BA crystal substrate after epitaxial crystallization. The (100) plane of PE is in contact with the (001) plane of BA, and b- and c-axes of PE are parallel to b- and a-axes of BA, respectively [259].

microphase separation), or it occurs from an already microphase-separated heterogeneous melt (in this case microphase separation precedes crystallization and provides a microstructure within which crystallization takes place.) In the latter case, crystallization is confined to occur within the pre-existing microseparated domains; crystallization confined in preformed lamellar, spherical or within and outside cylindrical microdomains has been observed [12, 78, 80]. Different situations occur when the crystallization is controlled by epitaxy. Although the versatility of epitaxy in the case of crystallization of polymers has clearly been shown, the method of using organic substrates to control the crystallization and the morphology of semicrystalline BCs was introduced only in 2000, when De Rosa and coworkers [184–263] first used epitaxy to control the molecular and microdomain orientation of PE-b-PEP-b-PE and PS-b-PE semicrystalline BC thin films. These authors used BA [184, 263, 264] or anthracene (AN) [264, 265] and obtained precise control of the molecular orientation of the crystalline block and subsequent overall long-range order of the BC microdomains.

The selected area electron diffraction pattern and the transmission electron microscopy (TEM) bright-field image of a thin film of PE-b-PEP-b-PE epitaxially crystallized onto a BA crystal are shown in Figure 4.11a and b, respectively. The diffraction pattern essentially presents only the 0kl reflections of PE, and therefore corresponds to the b^*c^* section of the reciprocal lattice of PE (Figure 4.11a). This indicates that the chain axis of the crystalline PE lies flat on the substrate surface and is oriented parallel to the a-axis of the BA crystals, as in the case of the PE homopolymer (Figure 4.10) [184]. The (100) plane of PE is in contact with the (001) plane of BA; therefore, the crystalline PE lamellae stands edge-on on the substrate surface, with the b- and c-axes of PE oriented parallel to the b- and a-axes of BA, respectively. The bright-field TEM image (Figure 4.11b) indicates that epitaxy has produced, instead of a spherulitic structure, a highly aligned lamellar structure with long, thin crystalline PE lamellae with a thickness of 10–15 nm, evidenced as dark regions, oriented along the $[010]_{PE}\|[010]_{BA}$ direction. The dark-field image created by using the strongest 110 reflection (see Figure 4.11c) reveals the same parallel array of crystalline edge-on PE lamellae oriented along the b-axis of BA crystals. This can be seen as bright regions due to the lamellae all being arranged in Bragg diffraction conditions determined by the 2-D epitaxy [184]. A scheme of the orientation of the PE-b-PEP-b-PE obtained on BA crystals is shown in Figure 4.12 [184]. The biaxial matching of the BA and PE lattices creates a highly ordered lamellar BC microdomain state. The widths of the crystalline PE lamellae are highly uniform, the PE crystals and the intervening noncrystalline PEP are all parallel, and the orientations of both the c- and b-axes of PE crystals over many micron-sized regions are very high [184].

4.5.3
Directional Crystallization

In polymer–diluent binary mixtures, if both the polymer and solvent are crystallizable above room temperature, then a eutectic-like behavior can be observed. In

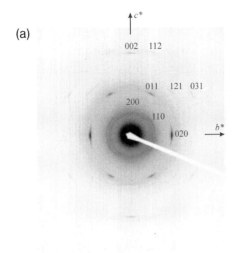

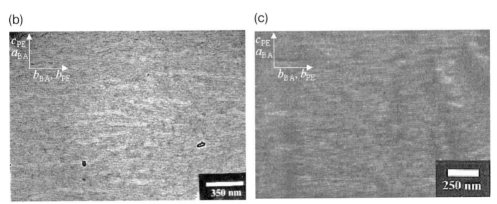

Figure 4.11 (a) Selected area electron diffraction pattern, (b) TEM bright-field image, and (c) TEM (110) dark-field image of a thin film of PE-*b*-PEP-*b*-PE epitaxially crystallized onto a BA crystal. In the TEM bright-field image, the dark regions correspond to the denser crystalline PE phase, which form long lamellae standing edge-on on the substrate surface and preferentially oriented with the *b*-axis of PE parallel to the *b*-axis of the BA (panel b). In the dark-field image, the bright regions correspond to the crystalline PE lamellae in the Bragg condition (panel c). Reproduced with permission from Ref. [184]; © 2000, The American Chemical Society.

fact, the melting temperatures of both the polymer and solvent are depressed up to the eutectic composition [266–271].

The presence of solvent in a noncrystalline BC depresses the order–disorder temperature of the BC, depending on the solvent quality. Several theories predict the phase behavior of BC–organic solvent mixtures [272–274]. Therefore, it can be expected that mixtures of BCs with crystallizable solvents exhibit eutectic behavior. In theory, the two liquidus lines of the binary mixture of a block copolymer and

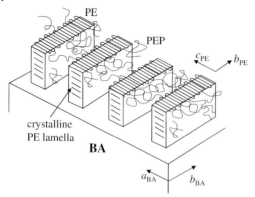

Figure 4.12 Schematic model of the crystalline and amorphous microdomains in the PE-b-PEP-b-PE triblock copolymer epitaxially crystallized on the BA crystal. The epitaxy shows the relative orientation of crystalline PE lamellae on the BA crystal: $(100)_{PE}||(001)_{BA}$ and $c_{PE}||a_{BA}$, $b_{PE}||b_{BA}$. Reproduced with permission from Ref. [184]; © 2000, The American Chemical Society.

an organic diluent—that is, the freezing point depression curve of the organic diluent and the order–disorder temperature depression curve of the BC—can meet each other at a certain composition and induce the characteristic eutectic transformation of a liquid to two solids: the crystalline solvent and the phase-separated solid block copolymer. Since the order–disorder transition in a block copolymer is a weak first-order transition, the phase-separated block copolymer at the invariant point can have some nonzero diluent content, as seen in the mixtures of block copolymer and solvents [274]. If this were the case, the phase separation at the invariant point would be from homogeneous liquid into a pure solid (crystalline solvent) and another solid (solvent-swollen BC microdomains). Park et al. [275] used this hypothesis of eutectic formation to globally organize the eutectic mixture for the first time and induce the alignment of microdomains in the solid BC. These authors used organic diluents, such as BA and AN, and amorphous asymmetric PS-b-PI and symmetric PS-b-PMMA diblock copolymers [275]. In this process, the first requirement is that the diluent must dissolve the BC above the melting temperature of the solvent. Second, the diluent must have a tendency to crystallize directionally into large, plate-like crystals. When the organic diluent, which initially was a solvent for the BC, directionally crystallizes along its fast growth direction under a temperature gradient, the BC undergoes microphase separation, due to a rapid decrease in the solvent concentration. At the same time, the orientation of the microdomains nucleated from the eutectic transforming solution is determined by the fast directional growth of the organic diluent. In the case of BA and AN, the fast growth directions are both b-axis, and consequently the intermaterial dividing surface of the microdomains of the BC tends to orient along this direction, which also corresponds to the temperature gradient direction [275]. A polarized optical microscope image of directionally crystallized BA crystal is shown in Figure 4.13a [275]. The BA crystal is elongated along the fast growth b-axis and presents

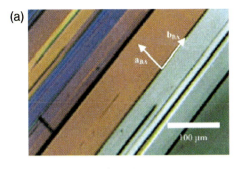

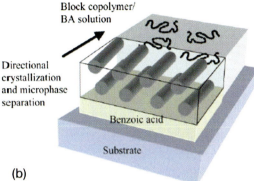

Figure 4.13 (a) Polarized optical microscopy image of directionally crystallized BA crystals. The large, flat and elongated BA crystals are aligned with the b-axis parallel to the growth front direction. Reproduced with permission from Ref. [275]; © 2001, The American Chemical Society; (b) Schematic model of the microstructure formation of a cylinder forming BC under directional crystallization of the BA from homogeneous solution. Reproduced with permission from Ref. [12]; © 2003, Elsevier.

a flat (001) surface. The different thicknesses of the BA crystals lead to the different colors under the microscope. A schematic model of cylindrical microdomains well aligned from the directional eutectic transformation of the homogeneous BC solution is shown in Figure 4.13b [12]. This process occurs within a few seconds (the growth velocity of the BA crystal is nearly $1\,cm\,s^{-1}$) and the microstructure is formed and then kinetically trapped.

Examples of the ordered microstructures obtained by this process are shown in Figure 4.14 for symmetric PS-b-PMMA (Figure 4.14a) and asymmetric PS-b-PI (Figure 4.14b,c) diblock copolymers prepared by directional solidification with BA crystals [275]. In the case of PS-b-PMMA, the darker regions correspond to the RuO_4-stained PS microdomains (Figure 4.14a). Edge-on parallel alternating lamellae of PS and PMMA are well aligned along the fast growth direction of the BA crystals (the b-axis), as shown in the schematic model of Figure 4.14d [275]. The well-aligned parallel lamellae extend over regions larger than $50\,\mu m^2$. The fast Fourier transform (FFT) power spectrum in the inset of Figure 4.14a shows a spotlike first reflection located on the meridian, indicating the nearly

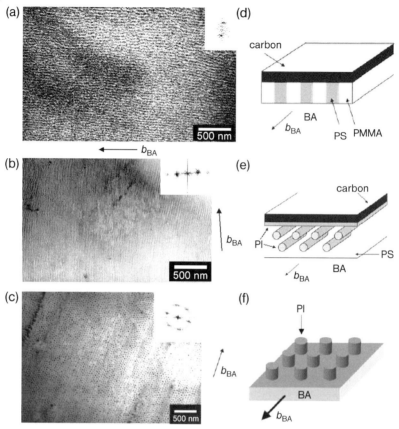

Figure 4.14 TEM bright-field images of thin films directionally solidified with BA of the symmetric PS-b-PMMA (20-nm thick) stained with RuO$_4$ (a), the asymmetric PS-b- PI stained with OsO$_4$ with a thickness of 50 nm (b) and 20 nm (c). The dark regions correspond to the stained PS lamellae (a), parallel PI cylinders (b) and vertically oriented PI cylinders (c) well aligned along the fast growth direction of the BA crystals (b-axis). The insets show the FFT power spectrum of the TEM micrographs. Schematic models of the microstructures of PS-b-PMMA (d), PS-b-PI (e) and ultrathin film of PS/PI (f) processed with BA. Reproduced with permission from Ref. [275]; © 2001, The American Chemical Society; and with permission from Ref. [12]; © 2003, Elsevier.

single-crystal-like microdomain structure. The fast directional microstructure formation during the phase separation with a thin film thickness approximately less than a half-lamellar period avoids preferential wetting of one of the blocks on the substrate, leading to the oriented lamellae microdomain structure where the interface of the microdomains is parallel to the normal of the substrate surface [275]. The structure is kinetically driven and subsequently vitrified at room temperature. Importantly, for regions thicker than approximately a half-lamellar period, the perpendicular lamellae orientation switches to an in-plane parallel one, and large planar regions are produced. The bright-field TEM image of a film of the asymmetric PS-b-PI with a thickness of approximately 50 nm, directionally solidified with BA (Figure 4.14b) shows OsO$_4$-stained darker PI cylinders, lying in-plane and well oriented along the crystallographic *b*-axis of the BA, as reported in the scheme of Figure 4.14e [275]. The ordered parallel cylinder structure also extends over regions larger than 50 μm^2. The average diameter of the PI cylindrical microdomains is approximately 20 nm, while the average distance between the cylinders is 40–50 nm. In both of these examples, the initial homogeneous solution confined between the glass substrates transforms due to the imposed directional solidification of large crystals of BA having (001) surfaces coexisting with a thin liquid layer near the eutectic composition. Dropping the temperature further then causes this layer also to directionally solidify by thickening the preexisting BA crystal with the simultaneous formation of a thin, metastable vertically oriented *lamellar* microdomain film (Figure 4.13b) [275]. In the case of the PS-*b*-PMMA, the vertical lamellar structure is vitrified due to the high glass transition temperatures of both blocks. In the case of PS-*b*-PI with a low volume fraction of PI block, however, it has been hypothesized that the vertical lamellar microstructure is transformed into in-plane cylindrical microstructure due to the interfacial instability of thin lamellae, film thickness, and preferential wetting of the PI block [275]. Similar ordered microstructures have been obtained employing other organic crystallizable solvents as AN which has a melting temperature of 216 °C, about 100 °C higher than that of BA [275].

This process has also been used in combination with thickness effects to achieve vertically aligned cylinders employing ultrathin films [275]. Cylindrical microdomains, indeed, orient vertically for relatively thin films due to incommensurability effects, which is similar to what was observed for lamellar diblock copolymers [276]. A bright-field TEM image of a thinner film (of approximately 20 nm thickness) of PS-*b*-PI, prepared from a more dilute solution, directionally solidified with BA and stained with OsO$_4$, is shown in Figure 4.14c [275]. The dark OsO$_4$-stained vertically aligned cylindrical PI microdomains are oriented into rows along the *b*-axis of the BA crystal and packed in an approximate hexagonal lattice, as shown in the schematic model of the microstructure reported in Figure 4.14f [275]. The aligned vertical cylinders extend over regions larger than 50 μm^2. The FFT power spectrum in the inset of Figure 4.14c shows spotlike first reflections with sixfold symmetry, indicating approximate hexagonal packing of the cylindrical PI microdomains.

Various di- and tri-block copolymers with different architecture and properties such as rubbery–glassy, glassy–glassy, amorphous–crystalline, amorphous–liquid crystalline, and ABC-type terpolymers have been successfully organized with this process [275].

4.5.4
Graphoepitaxy and Other Confining Geometries

Graphoepitaxy is a process whereby an artificial surface topography of a crystalline or amorphous substrate influences and controls the orientation of the crystal growth in thin films [277–280]. Graphoepitaxy has been used (often in combination with epitaxy) to obtain high orientation of polymeric crystals onto substrates constituted by films of other polymers [281–285]. For instance, a variety of polymers, including PE, nylons, polyesters, and liquid crystalline polymers, have been grown in highly oriented form on the oriented surface of poly(tetrafluoroethylene) (PTFE) crystals [281]. In the case of PE, the orientation of the crystals is such that the *c*-axis lies in the plane of the film and the *bc* plane of the unit cell is in contact with the substrate surface [285]. Different orientations, for instance, a fiber-like alignment, have been observed for other polymers [285]. This suggests that the alignment of materials on PTFE films can occur by graphoepitaxial or epitaxial mechanisms, depending on the materials. In some instances both the lattice structure and the surface topography of the PTFE films promote the alignment, so that the two mechanisms can operate in conjunction [285].

Graphoepitaxy was utilized by Segalman and coworkers [286, 287] to control the orientation of BC microdomains. The procedure is an example of the combined top-down and bottom-up approaches to patterning. Segalman's group used topographically alternating mesa and well patterns fabricated by conventional photolithography and chemical etching techniques to align spherical P2VP microdomains of a PS-*b*-P2VP diblock copolymer. A large area single crystal of the P2VP spheres was obtained on the mesas and wells [286].

Nanostructures with long-range order were also developed using graphoepitaxy in combination with BC lithography [288, 289]. Cheng *et al.* employed a diblock copolymer of PS-*b*-PFS, in which the organometallic PFS block provides an excellent (10 : 1) etching contrast in an oxygen plasma [287]. A topographically patterned silica substrate was fabricated by interference lithography, after which the substrate was used for a templating BC self-assembly. Monolayer films of the BC were deposited by spin-casting and annealed onto the patterned silica substrate. The film was then oxidized by etching in an oxygen plasma. Scanning electron microscopy (SEM) images of annealed and etched PS-*b*-PFS films, reported in Figure 4.15, revealed ordered arrays of PFS spheres in grooves of different widths [288]. A thin PFS–PS brush layer is present at the groove edge and bottom where the PFS block wets the silica substrate. This surface-induced thin layer was shown to drive the ordering of the PS–PFS block copolymer microdomains, and resulted in PFS spherical microdomains parallel to the groove edges. In the microstructure with grooves having a width of 240 nm (Figure 4.15c), and comparable to the typical block copolymer grain size, a near-perfect alignment of PFS spherical microdomains was achieved. The microdomain patterns were transferred onto the underlying silica substrate using a reactive ion etcher (RIE) process with a CHF_3 plasma, so as to create well-ordered arrays of silica posts [288]. The oxidized PFS domains are perfectly ordered by the guided self-assembly of the BC in the prepatterned substrate, and can potentially be useful for fabricating 2-D photonic crystal waveguide structures. More recently, other investigations have confirmed the general applicability of this approach employing spherical, cylindrical or lamellar diblock copolymers to yield 2-D patterns into different lithographic features [35, 36, 38, 290, 291]. However, as the pre-pattern size should be determined by the grain size of the block copolymers, only patterns with a small periodicity are applicable. To overcome this problem, graphoepitaxy should be combined with

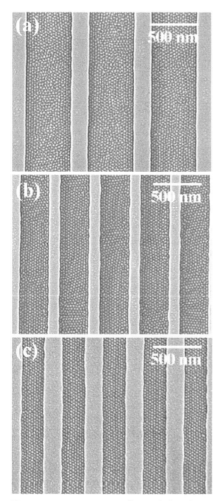

Figure 4.15 SEM images of annealed and oxygen plasma-treated PS-b- PFS films on silica gratings with (a) 500 nm-wide grooves, (b) 320 nm-wide grooves, and (c) 240 nm-wide grooves. Reproduced with permission from Ref. [288]; © 2002, The American Institute of Physics.

methods that allow the grain size of the BCs to be increased, which can then in turn be controlled with the micron size patterns available, for example, from soft lithography [12]. As discussed in the next sections, the combination of graphoepitaxy or epitaxy with directional crystallization provides possible routes to achieve this result.

Unlike graphoepitaxy, which can be considered to approximate a 1-D confinement of a thin film between (often parallel) solid walls, the 2-D confinement of BCs has been predicted to induce more complicated effects on phase-separation and morphology orientation [292–294]. Recent investigations have confirmed for various PS-based diblock copolymers that, in the case of a nonplanar geometry of confinements, such as in cylindrical pores, the confining of dimensions incommensurable with the copolymer period and the imposed curvature leads to the production of unusual morphologies [295–297]. The self-assembly of asymmetric diblock copolymers introduced as a melt into nanoporous alumina templates occurs with an alignment of the cylinders along the pore axis, favored by the preferential wetting of the pore with the majority block [296]. Furthermore, for symmetric BCs, concentric lamellae oriented parallel to the long axes of the pores have been observed, with an overall number of lamellae and forbidden segregations that are directly dependent on the pore diameter and the preferential wetting of one block, respectively [296, 297].

4.5.5
Combination of Directional Crystallization and Graphoepitaxy

The idea of combining graphoepitaxy from a topographical pattern fabricated by conventional photolithography procedures and the directional crystallization of a solvent was first introduced by Park et al. [298]. The thin film of the BC was confined between crystals of BA, obtained from a directional eutectic crystallization of the homogeneous BC solution, and the topographic substrate pattern. The confinement induced a thickness variation of the BC film, and resulted in the two different orientations of the microdomains [298]. The patterned substrate was produced via standard lithographic techniques, and consisted of 30 nm-high, 2 μm × 2 μm mesas arranged in a square array with a 4 μm spacing, as shown in the schematic diagram of Figure 4.16a. The square-shaped mesas are made by selective etching of a thermally grown silicon oxide on a 10 cm wafer. A PS-b-PI diblock copolymer was directionally solidified with benzoic acid on the patterned substrate [298]. A low-magnification, bright-field TEM micrograph of the OsO_4-stained BC film is shown in Figure 4.16b. The prepatterned substrate structure produces thickness variations in the directionally solidified BC thin films; the films are thinner on the mesas and thicker in the plateau regions. As described in Section 4.5.3 (see Figure 4.14b,c), the directional eutectic solidification produces an orientation of PI cylinders along the b-axis of the benzoic acid crystal or normal to the substrate, depending on the thickness [275, 298].

The BC film directionally solidified on the patterned substrate was then subjected to O_2-RIE to selectively remove the cylindrical PI microdomains [298]. A higher-magnification atomic force microscopy (AFM) image after O_2-RIE, shown in Figure 4.17a, reveals the double orientation of the cylindrical PI microdomains. In the thicker film regions (ca. 50 nm thick), the cylindrical PI microdomains are well aligned along the b-axis of the BA crystal, with the cylinder axis parallel to the substrate and to the walls of the mesas. In the thinner film regions (ca. 20 nm thick), the PI cylinders are hexagonally packed with their cylinder axes perpendicular to the substrate and with the a_1 hexagonal lattice direction parallel to b-axis of BA; that is, along the [10] direction of the square substrate pattern [298]. The SEM

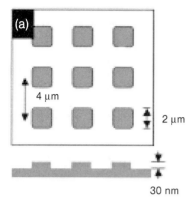
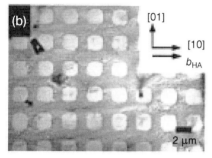

Figure 4.16 (a) Schematic of the topographically patterned silicon oxide substrate; (b) Bright-field TEM image of a thin film of PS/PI block copolymer, directionally solidified with BA on the prepatterned substrate. The low-magnification image depicts a typical region larger than 1000 μm² where the cylindrical PI microdomains are aligned via directional solidification. The replicated micron-scale pattern structure is seen due to the different film thicknesses in the two types of region. The inset shows the basis vectors of the cylindrical PI microdomain lattice and the fast growth direction of the BA crystal (b axis). Reproduced with permission from Ref. [298]; © 2001, The American Institute of Physics.

image of Figure 4.17b shows that the microdomains successively transform from parallel to vertical when they pass from the plateau into the mesa regions [298]. The in-plane PI cylindrical microdomains aligned along the b-axis direction of the BA crystal in the plateau regions transform abruptly into hexagonally packed, vertically oriented cylinders. The cylindrical hole structures on the mesa areas after O_2-RIE are also shown in Figure 4.17c, in height-contrast AFM. The vertically ordered PI cylindrical domains, which appear dark, on the mesa regions are essentially empty. A scheme of this process of directional solidification of the BC, combined with topographically mediated substrate patterning, is shown in Figure 4.17d. The BC films confined between the top BA crystal and the bottom prepatterned substrate undergo thickness variation (h_1 and h_2), leading to two different microdomain orientations. Control over the position and orientation of cylindrical

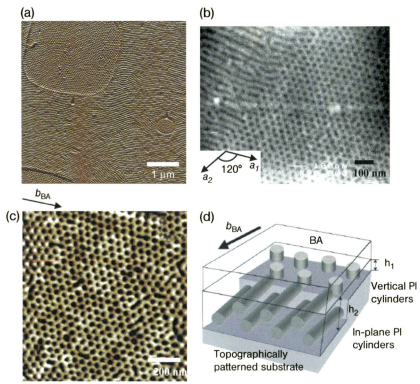

Figure 4.17 (a,c) AFM and (b) SEM images of a thin film of asymmetric PS-b-PI, directionally solidified with BA on the prepatterned substrate and subsequently etched by O_2-RIE. The tapping mode AFM image (panel a) shows that the cylindrical PI microdomains, with two different orientations with respect to the substrate, are well aligned along the fast growth direction of the BA crystals, as indicated by the arrow. The square-shaped mesa regions exhibit vertically oriented, hexagonally packed PI cylinders. The thicker matrix regions show the in-plane PI cylinders. The SEM image reveals that aligned PI cylinders transform their in-plane to vertical orientation with respect to the substrate. The two hexagonal lattice directions of the vertically ordered PI cylinders are shown in inset (b). The height mode AFM image on a mesa region (c) indicates that vertically ordered cylindrical PI microdomains are selectively removed by O_2-RIE, which appear dark. Reproduced with permission from Ref. [298]; © 2001, The American Institute of Physics; (d) Scheme of the orientation of microdomains of PS-b-PI between the top BA crystal and the bottom prepatterned substrate. Reproduced with permission from Ref. [12]; © 2003, Elsevier.

domains on the patterned substrate provides nanolithographic templates for micromagnetics and microphotonics [298].

4.5.6
Combination of Epitaxy and Directional Crystallization

The method of combination of directional crystallization and epitaxy was introduced by De Rosa *et al.* [263]. The process is based on the directional solidification of the eutectic solution of a semicrystalline BC in a crystallizable organic solvent and subsequent epitaxial crystallization of the crystalline block onto the organic crystalline substrate [263]. The organic substrate (e.g., BA or AN) at a temperature higher than its melting temperature is a solvent for the BC, and becomes a substrate, when it crystallizes at lower temperatures, for the epitaxial crystallization of the BC [263–265, 299].

Examples of the ordered microstructures obtained with this method are provided by PS-*b*-PEP-*b*-PE and PS-*b*-PE semicrystalline BCs containing crystallizable PE blocks directionally solidified with BA or AN [263–265].

The electron diffraction pattern of PS-*b*-PEP-*b*-PE triblock copolymer film directionally solidified and epitaxially crystallized with BA is shown in Figure 4.18a [264]. As in the case of the melt-compatible PE-*b*-PEP-*b*-PE (see Figure 4.11a) [184], the pattern of Figure 4.18a essentially presents only the 0*kl* reflections of PE, indicating orientation of the PE crystals similar to that found for the PE-*b*-PEP-*b*-PE triblock copolymer [184], and for the homopolymer, as shown in Figures 4.10 and 4.12. The crystalline PE lamellae are oriented edge-on on the substrate surface with the (100) plane of PE in contact with the (001) plane of BA. The *b*- and *c*-axes of PE are parallel to the *b*- and *a*-axes of benzoic acid, respectively.

The bright-field TEM image of the unstained film directionally solidified and epitaxially crystallized onto BA, reported in Figure 4.18b [264], shows the highly oriented lamellar structure with long thin crystalline lamellae oriented along the $[010]_{PE}||[010]_{BA}$ direction. The bright-field image of the same sample after staining with RuO_4, as shown in Figure 4.18c, indicates that the PS blocks, corresponding to the darker-stained regions, are organized in parallel cylinders the axes of which are generally oriented along the elongated direction $[010]_{PE}||[010]_{BA}$ of the crystalline lamellae [264].

As the order–disorder temperature of PS-*b*-PEP-*b*-PE (>250 °C) is much higher than the crystallization temperature of the PE block (almost 65 °C) [300, 301], the directional crystallization of BA induces the orientation of the cylindrical PS microdomains along the fast growth direction of the BA crystal (*b*-axis) [264] (as discussed in Section 4.5.3 for amorphous BCs) [275]. Subsequently, epitaxial crystallization of the crystalline PE block onto the BA crystalline substrate induces molecular orientation of the PE chains, leading to oriented crystalline PE lamellae in the presence of pre-existing ordered PS cylinders [264]. A schematic model (Figure 4.18d) displays the final microstructure generated by a combination of the directional crystallization and epitaxy. The process induces alignment of the PS cylinders and PE lamellae along the same direction, resulting in a multilayered ordering of the PS cylinders [264].

A different microstructure of the same PS-*b*-PEP-*b*-PE is obtained when AN is used as the crystallizable solvent [265], due to a different epitaxial relationship between PE and AN crystals [258]. The selected area diffraction (SAD) pattern of the PS-*b*-PEP-*b*-PE film epitaxially crystallized onto AN is shown in Figure 4.19a [265]. The pattern exhibits 0*kl* reflections of two sets of PE crystals having two different orientations [265], corresponding to the b^*c^* sections of the PE reciprocal lattice symmetrically rotated by an angle of almost 70° with respect to the a^*-axis of the AN. This angle of 70° corresponds to the angle made by the [1 1 0] and

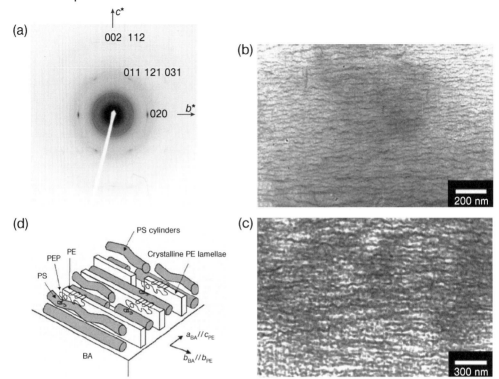

Figure 4.18 Selected area diffraction pattern (a), TEM bright-field image of unstained (b) and stained with RuO$_4$ (c) thin film of PS-b-PEP-b-PE triblock copolymer directionally solidified and epitaxially crystallized onto BA. The dark regions in the TEM bright-field image of the unstained film (panel b) correspond to the crystalline PE phase, which form long lamellae oriented edge-on, and aligned with the b-axis parallel to the b-axis of BA. The dark regions in the TEM bright-field image of the film stained with RuO$_4$ (panel c) correspond to the PS phase, which form cylinders aligned parallel to the crystalline lamellae; (d) Schematic model of the microstructure in thin films of the PS-b-PEP-b-PE terpolymer directionally solidified and epitaxially crystallized onto BA. The directional solidification induces in-plane cylindrical PS microdomains aligned along the fast growth direction of BA crystals. The following epitaxial crystallization of PE blocks onto the BA crystals induces the formation of long, thin crystalline PE lamellae in between the PS cylinders. The PE crystals have their (100) planes in contact with the (001) plane of the BA crystal with $a_{BA}\|c_{PE}$ and $b_{BA}\|b_{PE}$. Reproduced with permission from Ref. [264]; © 2003, Wiley-VCH.

[1 $\bar{1}$ 0] directions in the ab-plane of the AN crystal (a_{AN} = 8.65 Å; b_{AN} = 6.04 Å; c_{AN} = 11.18 Å; β = 124.7°; melting temperature of 216 °C). The c-axis of PE is therefore oriented parallel to two equivalent [1 1 0] and [1 $\bar{1}$ 0] directions of the AN crystal, and two sets of PE lamellae oriented edge-on on the AN crystals are produced, with the (100) plane of PE in contact with the (001) plane of AN (Figure 4.19b) [265]. The epitaxial relationship between PE and AN crystals can be readily explained in terms of matching between the b-axis of PE and the inter-row spacing

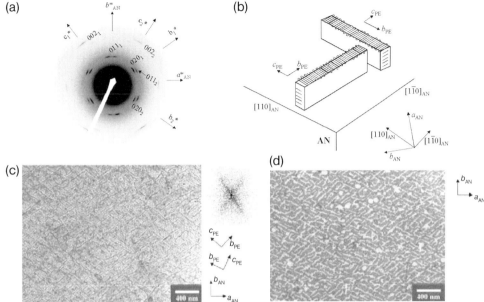

Figure 4.19 (a) Selected area electron diffraction pattern of a thin film of PS-b-PEP-b-PE block terpolymer epitaxially crystallized onto AN; (b) Scheme showing the two orientations of edge-on PE lamellae onto the (001) exposed face of the AN crystals. The (100) plane of PE is normal to the electron beam and parallel to the (001) face of the AN crystals. The c-axes of the two sets of PE crystals are parallel to the [1 0 9] and [1⁻10] direction of the AN crystal; (c) TEM bright-field image of an unstained thin film of PS-b-PEP-b-PE epitaxially crystallized on AN. The dark regions correspond to the crystalline PE phase, which forms cross-oriented PE lamellae standing edge-on; (d) TEM bright-field image of the film of PS-b-PEP-b-PE epitaxially crystallized on AN and stained with RuO_4. The dark regions correspond to the PS phase, which forms a double-oriented structure similar to that of the PE microdomains. Reproduced with permission from Ref. [265]; © 2001, Wiley-VCH.

of the (110) plane of AN and a second matching of 2c of PE with the Bragg distance of the (110) plane (d_{110}) of AN [258]. Similar lattice matching occurs between PE and other aromatic substances [258].

The bright-field TEM image of the unstained film of the PS-b-PEP-b-PE epitaxially crystallized onto AN (Figure 4.19c) provides a direct visualization of the crystalline PE lamellae that presents a cross-oriented texture. The lamellae having a thickness of 10–20 nm are oriented along the two different [1 1 0] and [1 1̄ 0] directions of the AN crystals [265]. A bright-field image of a different region of the same sample after staining with RuO_4 (Figure 4.19d) indicates that the PS blocks (the darker regions) are organized into a similar, double-oriented pattern [265]. The directional solidification of the mixture of the terpolymer and AN develops the orientation of the cylindrical PS microdomains along the fast growth direction of the AN crystals (b axis direction) before crystallization of the PE. The epitaxial crystallization induces a high orientation of the molecular chains of the crystalline phase, resulting in PE lamellae standing edge-on on the substrate surface, oriented in two directions due to the crystallographic degeneracy. This structure in turn induces reorganization and reorientation of the pre-existing PS cylinders [265].

This thin-film process suggests the possibility of creating new 2-D microdomain textures via self-assembled semicrystalline BCs using a combination of one or more crystallizable blocks and various crystalline organic substrates [265].

The asymmetric PS-b-PE diblock copolymer is an interesting example of crystallization of PE confined in PE cylinders formed by self-assembly in the melt [263, 264]. The electron diffraction pattern of an unstained film of PS-b-PE film epitaxially crystallized onto BA is shown in Figure 4.20a. The pattern is similar to those obtained for the PS-b-PEP-b-PE (Figure 4.18a) [264] and PE-b-PEP-b-PE (Figure 4.11a) [184] epitaxially crystallized onto BA and presents only the 0kl reflections of PE. This indicates that the PE lamellae are oriented edge-on on the (001) surface of BA, with the (100) plane of PE in contact with the (001) plane of BA and the b- and c-axes of PE parallel to the b- and a-axes of BA, respectively, as shown in Figures 4.10 and 4.12. A bright-field image of a thin film of PS-b-PE directionally solidified and epitaxially crystallized onto BA and stained with RuO_4, is shown in Figure 4.20b [263]. The presence of a very well-ordered array of light-unstained PE cylinders in the dark-stained PS matrix is apparent [263]. The cylinders are vertically aligned and packed on a hexagonal lattice, having an average lattice constant of almost 40 nm. The order extends to large distances (>20 µm).

The PE crystals have been visualized with dark-field imaging, employing the strong 110 reflection of PE (Figure 4.20c). Small rectangular PE crystals are observed in the dark-field image of Figure 4.20c (the light regions), well aligned along the b-axis direction of the BA crystals and packed on a pseudo-hexagonal lattice, the size and orientation of which are the same as seen in the bright-field image of Figure 4.20b [263]. The PE crystals are 7 nm thick and 20 nm long, with their longest dimension parallel to the [1 1 0]* reciprocal lattice direction [263]. The size and spacing of PE microdomains observed in bright- and dark-field images show that there is precisely one crystalline PE lamella centered in each cylinder, with the b-axis of the lamella oriented parallel to the b-axis of the BA, confirming the epitaxy of PE on BA demonstrated by the electron diffraction pattern (Figure 4.12a). A scheme of the nanostructure obtained in PS/PE thin film is shown in Figure 4.20d [263, 264].

The same PE-b-PS was also processed with AN to manipulate both molecular and microstructure orientation [264]. The electron diffraction pattern of an unstained film is shown in Figure 4.21a. As in the case of Figure 4.19a for the PS-b-PEP-b-PE triblock copolymer, the pattern exhibits the 0kl reflections of two sets of PE crystals in two different orientations, indicating that PE lamellae stand edge-on on the substrate surface with the c-axes of PE oriented parallel to the [1 1 0] or [1 $\bar{1}$ 0] directions (Figure 4.19b) [264]. A RuO_4-stained bright-field TEM image of the PS-b-PE diblock copolymer directionally solidified and epitaxially crystallized onto AN is shown in Figure 4.21b. Also, in this case the light-unstained

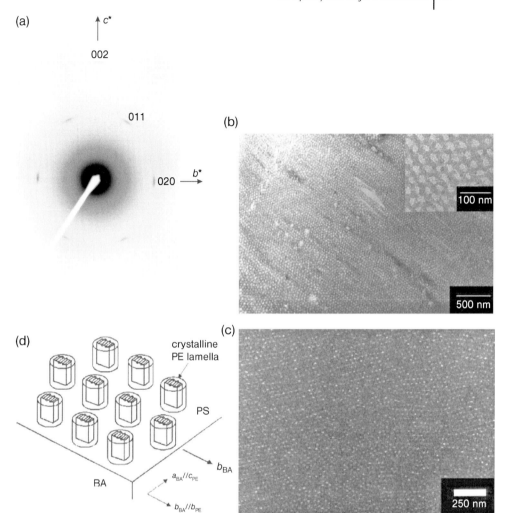

Figure 4.20 Selected area electron diffraction pattern (a), TEM bright-field image (b) and TEM dark-field image (c) of a thin film of PS-b-PE directionally solidified and epitaxially crystallized onto BA. In the TEM bright-field image, the film of PS-b- PE was stained with RuO$_4$; therefore, the darker regions correspond to the stained PS matrix, while the light regions correspond to the crystalline PE microdomains, which form hexagonally packed cylinders oriented perpendicular to the substrate surface. Inset, magnified region of (panel b) showing the noncircular shape of the PS-PE interface. The dark-field image (panel c) of the unstained film was created using the (110) diffraction reflection. Small rectangular PE crystals are observed, well aligned along the b-axis direction of the BA crystals and packed on a pseudohexagonal lattice, the size and orientation of which is the same as seen in panel (b). Reproduced with permission from Ref. [263]; © 2000, Nature Publishing Group; (d) Schematic models of the microstructures in thin films of PS-b-PE directionally solidified and epitaxially crystallized onto BA. Vertically oriented PE cylinders are packed on a hexagonal lattice and contain one crystalline PE lamella oriented edge-on on the substrate with the b-axis parallel to the b-axis of the BA crystal. Reproduced with permission from Ref. [264]; © 2003, Wiley-VCH.

(a)

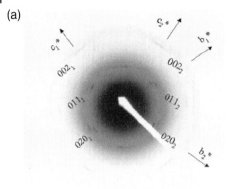

(b)

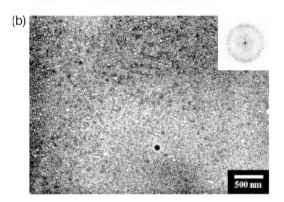

(c)

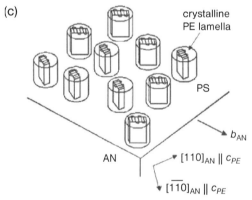

Figure 4.21 Selected area electron diffraction pattern (a) and TEM bright-field image (b) of a thin film of PS-*b*-PE directionally solidified and epitaxially crystallized onto AN. In the TEM bright-field image the film of PS-*b*-PE was stained with RuO_4; therefore, the darker regions correspond to the stained PS matrix, while the light regions correspond to the crystalline PE microdomains, which form hexagonally packed cylinders oriented perpendicular to the substrate surface; (c) Schematic model of the microstructure in thin films of PS-*b*-PE, directionally solidified and epitaxially crystallized onto AN. Vertically oriented PE cylinders are packed on a hexagonal lattice and contain one crystalline PE lamella oriented edge-on on the substrate with two different orientations, in which the *c*-axes of PE lamellae are parallel to the [1 1 0] or [1 $\bar{1}$ 0] directions of AN crystals. Reproduced with permission from Ref. [264]; © 2003, Wiley-VCH.

PE cylinders are vertically aligned but the order in the lateral packing is lower than that observed for the same BC crystallized onto BA (Figure 4.20b) [264]. The lattice has, indeed, many defects, resulting in very small grain size with short-range order. This is probably because the shape anisotropy of AN crystals in the growth process is lower than that of BA crystals and the epitaxial crystallization of the PE block on AN, which causes the double texture of the crystalline PE lamellae, may disturb the lateral ordering of the PE cylinders [264]. A scheme of the microstructure obtained is shown in Figure 4.21c. The PE cylinders are vertically aligned and contain one PE crystalline lamella oriented along two equivalent crossed directions [264].

It has been suggested by De Rosa *et al.* [263] that the mechanism of the pattern formation of PS-*b*-PE on BA or AN is based on the combination of epitaxy and directional crystallization, and is basically similar to that for PS-*b*-PEP-*b*-PE (Figure 4.18) [263, 264]. The directional crystallization of the organic crystal (BA or AN) produces the formation of the globally oriented PE cylindrical microdomains due to the polymer first undergoing the order–disorder transition. However, in this instance the subsequent epitaxial crystallization of the PE block influences the final microstructure, where vertically oriented cylindrical PE microdomains [263] (Figure 4.20d) instead of in-plane PS cylinders are formed (Figure 4.18d) [264]. The morphological evolution of the system is depicted schematically in Figure 4.22. The initial homogeneous solution confined between the glass substrates (Figure 4.22a) transforms, due to the imposed directional eutectic solidification, into large crystals of BA having (001) surfaces coexisting with a thin liquid layer near the eutectic composition (Figure 4.22b). Decreasing the temperature also causes this layer to solidify directionally, by thickening the pre-existing BA crystal, and the formation of a thin, metastable vertically oriented lamellar microdomain film (Figure 4.22c). The PE microdomains are aligned along the fast growth direction of substrate crystals, as shown in Figure 4.13b. Many sequences of events are possible, depending on the affinity of the respective blocks for the substrate surfaces, the polymer film thickness, the microdomain period, and the type of subsequent epitaxy and orientation of chains in the crystallizable block with respect to the crystalline substrate. In the case of the PS-*b*-PE/BA system, the structure transforms due to the instability of the flat interface at this composition and from the in-plane PE *c*-axis orientation induced by the epitaxy, so that vertical cylinders readily form (Figure 4.22d). The depression of the T_g of PS by plasticization from the BA allows the intermediate structure to reorganize when the crystallization of PE occurs, even at 60 °C. The domains evolve from an aligned precursor state (Figure 4.22d), into the final structure with vertical cylinders containing one PE crystalline lamella (Figure 4.22e) [263].

In the case of PS-*b*-PEP-*b*-PE, due to the presence of the PEP mid-block and the appropriate condition for the epitaxy of the PE block, in-plane crystalline PE lamellae were produced without hindering the pre-existing PS cylinders (see Figure 4.18d) [264]. However, in the case of PS-*b*-PE, the fact that the epitaxial crystallization of the PE block must take place inside the cylindrical microdomains and the

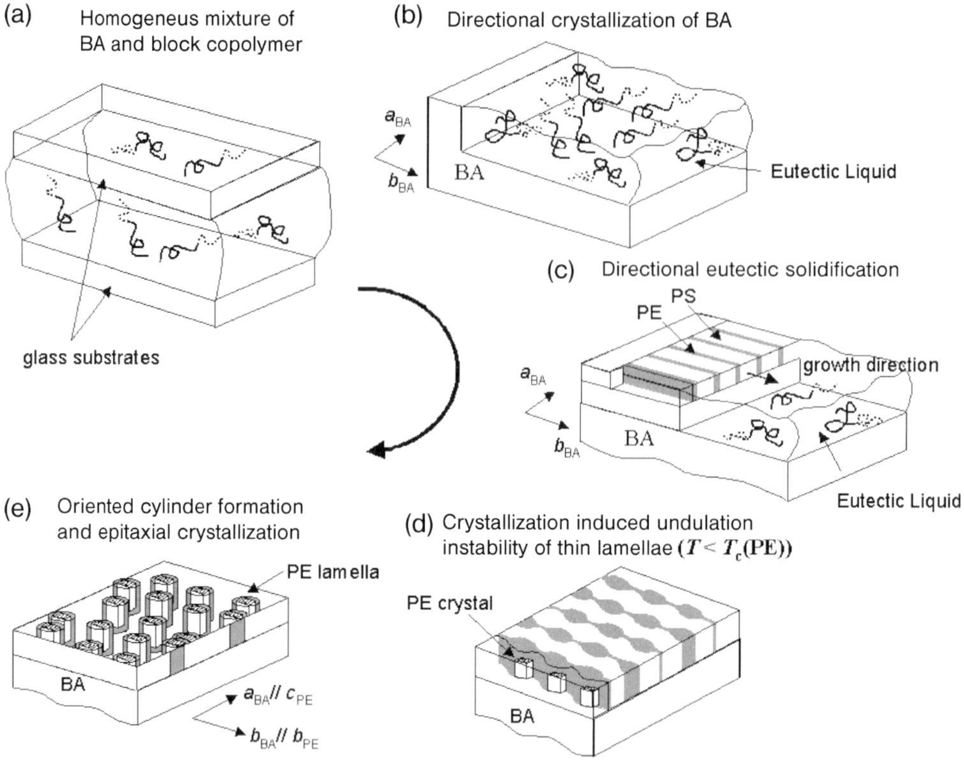

Figure 4.22 Structural evolution during the directional eutectic solidification and epitaxial crystallization of the BC from the crystallizable solvent. (a) Homogeneous solution of PS-b-PE in BA between two glass substrates; (b) Directional solidification forms crystals of BA coexisting with a liquid layer of more concentrated polymer; (c) Second directional solidification, showing the eutectic liquid layer transforming into a BA crystal (which grows on the pre-eutectic BA crystal) and an ordered lamellar BC; (d) Because of the highly asymmetric composition of the BC and the epitaxial crystallization of the PE in contact with the BA substrate, the flat interfaces of vertically oriented lamellae are unstable, and spontaneously deform in order to achieve a more preferred interfacial curvature and allow epitaxial growth of PE; (e) The layers transform into an array of vertically oriented, pseudohexagonally packed semicrystalline PE cylinders. A single, chain-folded PE lamella is formed in each cylinder. The PE crystals have their (100) planes contacting the (001) plane of the BA crystal with $a_{BA}||c_{PE}$ and $b_{BA}||b_{PE}$. Reproduced with permission from Ref. [263]; © 2000, Nature Publishing Group.

c-axis of the PE block must contact with the crystalline BA or AN surface, make the two controlling forces of directional crystallization and epitaxy determine the final microstructure orientation in a more cooperative manner. Indirect evidence of the contribution of the PE epitaxy to the vertically ordered microstructure is given by the results reported by Park et al. [275], who introduced the method of directional crystallization (see Section 4.5.3) using a PS-b-PI that has a volume fraction and molecular weight similar to those of the PS-b-PE. The authors found in-plane PI cylinder orientation with the same experimental conditions due to the absence of the epitaxy (see Figure 4.14b) [275]. Plasticization of the PS matrix from BA or AN enables the subsequent epitaxial crystallization to transform the microstructure into vertical cylinders [263, 264].

4.6
Summary and Outlook

The importance of developing methods which allow control over the alignment of domains arises from the opportunities that they provide for the exploitation of BCs in nanotechnological applications, such as in nanolithography or for the fabrication of photonic materials and other optical and electronic devices. In this chapter we have attempted to describe most of techniques proposed essentially during the past decade to induce long-range ordered and preferentially oriented morphologies, considering as seminal work the partial alignment of cylindrical domains in BC melts obtained during the early 1970s by the application of a mechanical field [73].

Macroscopically oriented thin films have been achieved by methods of general applicability, including roll-casting, shearing, and through the use of unidirectional electric fields. Less-direct approaches have established a control of surface interactions or entail a fine-tuning of surface topography and geometric confinements, eventually through lithographically patterned substrates. Moreover, when BCs fulfill specific structural or molecular requirements, other strategies are available: (i) semicrystalline BCs or copolymers having the T_g of one block above room temperature may undergo morphology alignment by epitaxy or solvent evaporation, respectively; (ii) BCs soluble in a crystallizable solvent are suitable for alignment by directional crystallization, eventually in combination with epitaxy or graphoepitaxy. Other approaches, essentially proposed for bulk BC systems, achieve only moderate orientation with multigrain structures, for example, through the application of different mechanical flow fields, or appear to be too complicated to assume nimble developments (e.g., for orientation by temperature gradients).

Finally, a scenario could be foreseen in which only a few of the techniques proposed so far will allow the development of processes for the preparation of defect-free nanostructured materials suitable for technological applications. This is despite the fact that most of them are of remarkable interest as model studies which, in the best case, could assist researchers working in this field and on the development of other complementary technologies, or they may simply be intended for academic exercises. In particular, there is a considerable interest in the combination of different methods of alignment for the production of "truly" defect free 3-D nanostructures, with the goal of overcoming the azimuthal degeneration of domains, intrinsic for unidirectional interactions. The best potential is logically offered by the simultaneous application or orthogonal directing interactions. However, in our opinion, it is worth devoting additional effort to further explore the integration of ultraviolet interferometric lithography and, more generally, the implementation of the combination of top-down and bottom-up approaches.

Acknowledgments

M.L. and C.D.R. gratefully acknowledge financial support from the Ministerio de Ciencia e Innovación (MAT2008-06503 and the program Ramón y Cajal) and MIUR (Prin 2007), respectively. M.L. has also benefited from useful discussions with Prof. M. A. Lopez-Quintela, University of Santiago, Spain.

References

1. Seto, C.T. and Whitesides, G.M. (1990) *J. Am. Chem. Soc.*, **112**, 6409.
2. Manka, J.S. and Lawrence, D.S. (1990) *J. Am. Chem. Soc.*, **112**, 2440.
3. Koert, U., Harding, M.M. and Lehn, J.-M. (1990) *Nature (London)*, **346**, 339.
4. Hamley, I.W. (1998) *The Physics of Block Copolymers*, Oxford University Press, Oxford.
5. Hadjichristidis, N., Pispas, S. and Floudas, G. (2003) *Block Copolymers. Synthetic Strategies, Physical Properties and Applications*, John Wiley & Sons, Inc., New York.
6. Lazzari, M., Liu, G. and Lecommandoux, S. (eds) (2006) *Block Copolymers in Nanoscience*, Wiley-VCH Verlag GmbH, Weinheim.
7. Bates, F.S. and Fredrickson, G.H. (1999) *Phys. Today*, **52**, 32.
8. Fasolka, M.J. and Mayes, A.M. (2001) *Annu. Rev. Mater. Res.*, **31**, 323.
9. Krausch, G. and Magerle, R. (2002) *Adv. Mater.*, **14**, 1579.
10. Ryu, D.Y., Jeong, U., Kim, J.K. and Russell, T.P. (2002) *Nat. Mater.*, **1**, 114.
11. Hamley, I.W. (2003) *Angew. Chem., Int. Ed.*, **42**, 1692.
12. Park, C., Yoon, J. and Thomas, E.L. (2003) *Polymer*, **44**, 6725.
13. Lazzari, M. and Lopez-Quintela, M.A. (2003) *Adv. Mater.*, **15**, 1583.
14. Segalman, R.A. (2005) *Mater. Sci. Eng. R Rep.*, **48**, 191.
15. Ruzette, A.-V. and Leibler, L. (2005) *Nat. Mater.*, **4**, 19.
16. Cheng, J.Y., Ross, C.A., Smith, H.I. and Thomas, E.L. (2006) *Adv. Mater.*, **18**, 2505.
17. Lazzari, M., Rodríguez-Abreu, C., Rivas, J. and Lopez-Quintela, M.A. (2006) *J. Nanosci. Nanotechnol.*, **6**, 892.
18. Hatada, K., Kitayama, T. and Vogl, O. (eds) (1997) *Macromolecular Design of Polymeric Materials*, Marcel Dekker, New York.
19. Hadjichristidis, N., Pitzikalis, M., Pispas, S. and Iatrou, H. (2001) *Chem. Rev.*, **101**, 3747.
20. Matyjaszewski, K. and Davis, T.P. (eds) (2002) *Handbook of Radical Polymerization*, John Wiley & Sons, Inc., New York.
21. Park, M., Harrison, C., Chaikin, P.M., Register, R.A. and Adamson, D.H. (1997) *Science*, **276**, 1401.
22. Li, R.R., Dapkus, P.D., Thompson, M.E., Jeong, W.G., Harrison, C., Chaikin, P.M., Register, R.A. and Adamson, D.H. (2000) *Appl. Phys. Lett.*, **76**, 1689.
23. Thurn-Albrecht, T., Schotter, J., Kästle, G.A., Emley, N., Shibauchi, T., Krusin-Elbaum, L., Guarini, K., Black, C.T., Tuominen, M.T. and Russell, T.P. (2000) *Science*, **290**, 2126.
24. Park, M., Chaikin, P.M., Register, R.A. and Adamson, D.H. (2001) *Appl. Phys. Lett.*, **79**, 257.
25. Cheng, J.Y., Ross, C.A., Chan, V.Z.-H., Thomas, E.L., Lammertink, R.G.H. and Vancso, G.J. (2001) *Adv. Mater.*, **13**, 1174.

26 Guarini, K.W., Black, C.T., Milkove, K.R. and Sandstrom, R.L. (2001) *J. Vac. Sci. Technol. B*, **19**, 2784.

27 Kim, H.-C., Jia, X., Stafford, C.M., Kim, D.H., McCarthy, T.J., Tuominen, M.T., Hawcker, C.J. and Russell, T.P. (2001) *Adv. Mater.*, **13**, 795.

28 Naito, K., Hieda, H., Sakurai, M., Kamata, Y. and Asakawa, K. (2002) *IEEE Trans. Magn.*, **38**, 1949.

29 Cheng, J.Y., Meyes, A.M. and Ross, C.A. (2004) *Nat. Mater.*, **3**, 823.

30 Yoon, J., Lee, W. and Thomas, E.L. (2005) *MRS Bull.*, **30**, 721.

31 Stokykovich, M.P., Müller, M., Kim, S.O., Solak, H.H., Edwards, E.W., de Pablo, J.J. and Nealey, P.F. (2005) *Science*, **308**, 1442.

32 Han, E., In, I., Park, S.-M., La, Y.-H., Wang, Y., Nealey, P.F. and Gopalan, P. (2007) *Adv. Mater.*, **19**, 4448.

33 Chen, F., Akasaka, S., Inoue, T., Takenaka, M., Hasegawa, H. and Yoshida, H. (2007) *Macromol. Rapid Commun.*, **28**, 2137.

34 Lazzari, M., Scalarone, D., Hoppe, C.E., Vazquez-Vazquez, C. and Lòpez-Quintela, M.A. (2007) *Chem. Mater.*, **19**, 5818.

35 Bita, I., Yang, J.K.W., Jung, Y.S., Ross, C.A., Thomas, E.L. and Berggren, K.K. (2008) *Science*, **321**, 939.

36 Bosworth, J.K., Paik, M.Y., Schwartz, E.L., Huang, J.Q., Ko, A.W., Smilgies, D.-M., Black, C.T. and Ober, C.K. (2008) *ACS Nano*, **2**, 1396.

37 Kim, T.H., Hwang, J., Hwang, W.S., Huh, J., Kim, H.-C., Kim, S.H., Hong, J.M., Thomas, E.L. and Park, C. (2008) *Adv. Mater.*, **20**, 522.

38 Yamaguchi, T. and Yamaguchi, H. (2008) *Adv. Mater.*, **20**, 1684.

39 Park, S., Lee, D.Y., Xu, J., Kim, B., Hong, S.W., Jeong, U., Xu, T. and Russell, T.P. (2009) *Science*, **323**, 1030.

40 Liu, X. and Stamm, M. (2009) *Nanoscale Res. Lett.*, **4**, 459.

41 Thomas, E.L., Anderson, D.M., Henkee, C.S. and Hoffman, D. (1988) *Nature (London)*, **334**, 598.

42 Gido, S.P. and Thomas, E.L. (1994) *Macromolecules*, **27**, 6137.

43 Gido, S.P. and Thomas, E.L. (1997) *Macromolecules*, **30**, 3739.

44 Hahm, J., Lopes, W.A., Jaeger, H.M. and Sibener, S.J. (1998) *J. Chem. Phys.*, **109**, 10111.

45 Goldacker, T., Abetz, V., Stadler, R., Erukhimovich, I. and Leibler, L. (1999) *Nature (London)*, **398**, 138.

46 Hahm, J. and Sibener, S.J. (2001) *J. Chem. Phys.*, **114**, 4730.

47 Tsarkova, L., Knoll, A., Krausch, G. and Magerle, R. (2006) *Macromolecules*, **39**, 3608.

48 Morkved, T.L., Lu, M., Urbas, A.M., Ehrichs, E.E., Jaeger, H.M., Mansky, P. and Russell, T.P. (1996) *Science*, **273**, 931.

49 Knoll, A., Horvat, A., Lyakhova, K.S., Krausch, G., Sevink, G.J.A., Zvelindovsky, A.V. and Magerle, R. (2002) *Phys. Rev. Lett.*, **89**, 035501.

50 Hashimoto, T., Tsutsumi, K. and Funaki, Y. (1997) *Langmuir*, **13**, 6869.

51 Liu, G., Ding, J., Guo, A., Hertfort, M. and Bazett-Jones, D. (1997) *Macromolecules*, **30**, 1851.

52 Liu, G., Ding, J., Hashimoto, T., Kimishima, K., Winnik, F.M. and Nigam, S. (1999) *Chem. Mater.*, **11**, 2233.

53 Leiston-Belanger, J.M., Russell, T.P., Drockenmuller, E. and Hawker, C.J. (2005) *Macromolecules*, **38**, 7676.

54 La, Y.-H., Edwards, E.W., Park, S.-M. and Nealey, P.F. (2005) *Nano Lett.*, **5**, 1379.

55 Jeong, U., Ryu, D.Y., Kho, D.H., Kim, J.K., Goldbach, J.T., Kim, D.H. and Russell, T.P. (2004) *Adv. Mater.*, **16**, 533.

56 Ahn, D.U. and Sancaktar, E. (2008) *Soft Matter*, **4**, 1454.

57 Meiners, J.-C., Quintel-Ritzi, A., Mlinek, J., Elbs, H. and Krausch, G. (1997) *Macromolecules*, **30**, 4945.

58 Kim, G. and Libera, M. (1998) *Macromolecules*, **31**, 2569.

59 Fasolka, M.J., Banerjee, P., Mayes, A.M., Pickett, G. and Balazs, A.C. (2000) *Macromolecules*, **33**, 5702.

60 Elbs, H., Drummer, C., Abetz, V. and Krausch, G. (2002) *Macromolecules*, **35**, 5570.

61 Buck, E. and Fuhrmann, J. (2001) *Macromolecules*, **34**, 2172.

62 Peng, J., Xuan, Y., Wang, H.F., Li, B.Y. and Han, Y.C. (2005) *Polymer*, **46**, 5767.

63 Xuan, Y., Peng, J., Cui, L., Wang, H., Li, B. and Han, Y. (2004) *Macromolecules*, **37**, 7301.

64 Chen, Y., Huang, H., Hu, Z. and He, T. (2004) *Langmuir*, **20**, 3805.

65 Kim, S.H., Misner, M.J., Xu, T., Kimura, M. and Russell, T.P. (2004) *Adv. Mater.*, **16**, 226.

66 Bang, J., Kim, S.H., Drokenmuller, E., Misner, M.J., Russell, T.P. and Hawker, C.J. (2006) *J. Am. Chem. Soc.*, **128**, 7622.

67 Bang, J., Kim, S.H., Stein, G.E., Russell, T.P., Li, X., Wang, J., Kramer, E.J. and Hawker, C.J. (2007) *Macromolecules*, **40**, 7019.

68 Kim, S.H., Misner, M.J., Yang, L., Gang, O., Ocko, B.M. and Russell, T.P. (2006) *Macromolecules*, **39**, 8473.

69 Park, S.C., Kim, B.J., Hawker, C.J., Kramer, E.J., Bang, J. and Ha, J.S. (2007) *Macromolecules*, **40**, 8119.

70 Park, S.C., Jung, H., Fukukawa, K., Campos, L.M., Lee, K., Shin, K., Hawker, C.J., Ha, J.S. and Bang, J. (2008) *J. Polym. Sci. Part A: Polym. Chem.*, **46**, 8041.

71 RamachandraRao, V.S., Gupta, R.R., Russell, T.P. and Watkins, J.J. (2001) *Macromolecules*, **34**, 7923.

72 Vogt, B.D., RamachandraRao, V.S., Gupta, R.R., Lavery, K.A., Francis, T.J., Russell, T.P. and Watkins, J.J. (2003) *Macromolecules*, **36**, 4029.

73 Keller, A., Pedemonte, E. and Willmouth, F.M. (1970) *Nature (London)*, **225**, 538.

74 Folkes, M.J., Keller, A. and Scalisi, F.P. (1973) *Colloid Polym. Sci.*, **251**, 1.

75 Skoulios, A. (1977) *J. Polym. Sci. Polym. Symp.*, **58**, 369.

76 Hadziioannou, G., Mathis, A. and Skoulios, A. (1979) *Colloid Polym. Sci.*, **257**, 15.

77 Hadziioannou, G., Mathis, A. and Skoulios, A. (1979) *Colloid Polym. Sci.*, **257**, 136.

78 Kofinas, P. and Cohen, R.E. (1995) *Macromolecules*, **28**, 336.

79 Drzal, P.L., Barnes, J.D. and Kofinas, P. (2001) *Polymer*, **42**, 5633.

80 (a) Quiram, D.J., Register, R.A., Marchand, G.R. and Adamson, D.H. (1998) *Macromolecules*, **31**, 4891.
(b) Quiram, D.J., Register, R.A. and Marchand, G.R. (1997) *Macromolecules*, **30**, 4551.
(c) Quiram, D.J., Register, R.A., Marchand, G.R. and Ryan, A.J. (1997) *Macromolecules*, **30**, 8338.

81 van Asselen, O.L.J., van Casteren, I.A., Goossens, J.G.P. and Meijer, H.E.H. (2004) *Macromol. Symp.*, **205**, 85.

82 Morrison, F.A. and Winter, H.H. (1989) *Macromolecules*, **22**, 3533.

83 Morrison, F.A., Winter, H.H., Gronski, W. and Barnes, J.D. (1990) *Macromolecules*, **23**, 7200.

84 Wiesner, U. (1997) *Macromol. Chem. Phys.*, **198**, 3319.

85 Pinheiro Scott, B. and Winey, K.I. (1998) *Macromolecules*, **31**, 4447.

86 Leist, H., Maring, D., Thurn-Albrecht, T. and Wiesner, U. (1999) *J. Chem. Phys.*, **110**, 8225.

87 Hermel, T.J., Wu, L.F., Hahn, S.F., Lodge, T.P. and Bates, F.S. (2002) *Macromolecules*, **35**, 4685.

88 Stangler, S. and Abetz, V. (2003) *Rheol. Acta*, **42**, 569.

89 Wu, L., Lodge, T.P. and Bates, F.S. (2004) *Macromolecules*, **37**, 8184.

90 Oelschlaeger, C., Gutmann, J.S., Wolkenhauer, M., Spiess, H.-W., Knoll, K. and Wilhelm, M. (2007) *Macromol. Chem. Phys.*, **208**, 1719.

91 Mendoza, C., Pietsch, T., Gindy, N. and Fahmi, A. (2008) *Adv. Mater.*, **20**, 1179.

92 Sebastian, J.M., Graessley, W.W. and Register, R.A. (2002) *J. Rheol.*, **46**, 863.

93 Angelescu, D.E., Waller, J.H., Register, R.A. and Chaikin, P.M. (2005) *Adv. Mater.*, **17**, 1878.

94 Angelescu, D.E., Waller, J.H., Adamson, D.H., Deshpande, P., Chou, S.Y., Register, R.A. and Chaikin, P.M. (2004) *Adv. Mater.*, **16**, 1736.

95 Luo, K.F. and Yang, Y.L. (2004) *Polymer*, **45**, 6745.

96 Lisal, M. and Brennan, J.K. (2007) *Langmuir*, **23**, 4809.

97 Register, R.A., Angelescu, D.E., Pelletier, V., Asakawa, K., Wu, M.W., Adamson, D.H. and Chaikin, P.M. (2007) *J. Photopolym. Sci. Technol.*, **20**, 493.

98 Hong, Y.-R., Adamson, D.H., Chaikin, P.M. and Register, R.A. (2009) *Soft Matter*, **5**, 1687.
99 Albalak, R.J. and Thomas, E.L. (1993) *J. Polym. Sci.: Part B, Polym. Phys.*, **31**, 37.
100 Albalak, R.J. and Thomas, E.L. (1994) *J. Polym. Sci.: Part B, Polym. Phys.*, **32**, 341.
101 Honeker, C.C., Thomas, E.L., Albalak, R.J., Hajduk, D.A., Gruner, S.M. and Capel, M.C. (2000) *Macromolecules*, **33**, 9395.
102 Dair, B.J., Avgeropoulos, A. and Hadjichristidis, N. (2000) *Polymer*, **41**, 6231.
103 Villar, M.A., Rueda, D.R., Ania, F. and Thomas, E.L. (2002) *Polymer*, **43**, 5139.
104 Kwon, Y.K., Ko, Y.S. and Okamoto, M. (2008) *Polymer*, **49**, 2334.
105 Mortensen, K., Brown, W. and Norden, B. (1992) *Phys. Rev. Lett.*, **68**, 2340.
106 Hamley, I.W., Koppi, K.A., Rosedale, J.H., Bates, F.S., Almdal, K. and Mortensen, K. (1993) *Macromolecules*, **26**, 5959.
107 Hamley, I.W., Pople, J.A., Fairclough, J.P.A., Ryan, A.J., Booth, C. and Yang, Y.W. (1998) *Macromolecules*, **31**, 3906.
108 Daniel, C., Hamley, I.W., Mingvanish, W. and Booth, C. (2000) *Macromolecules*, **33**, 2163.
109 Hamley, I.W. (2000) *Curr. Opin. Colloid Interface Sci.*, **5**, 342.
110 Hamley, I.W. (2001) *Phyl. Trans. R. Soc.*, **359**, 1017.
111 Mortensen, K., Theunissen, E., Kleppinger, R., Almdal, K. and Reynaers, H. (2002) *Macromolecules*, **35**, 7773.
112 Castelletto, V., Hamley, I.W., Crothers, M., Attwood, D., Yang, Z. and Booth, C. (2004) *Macromol. Sci. Phys. B*, **43**, 13.
113 Hamley, I.W. (2001) *J. Phys.: Condens. Matter*, **13**, R643.
114 Honeker, C.C. and Thomas, E.L. (1996) *Chem. Mater.*, **8**, 1702.
115 Fredikson, G.H. and Bates, F.S. (1996) *Annu. Rev. Mater. Sci.*, **26**, 501.
116 Watanabe, H. (1997) *Acta Polym.*, **48**, 215.
117 Morozov, A.N., Zvelindovsky, A.V. and Fraaije, J.G.E. (2001) *Phys. Rev. E*, **64**, 051803.
118 Corberi, F., Gonnella, G. and Lamura, A. (2002) *Phys. Rev. E*, **66**, 016114.
119 Zvelindovsky, A.V.M. and Sevink, G.J.A. (2003) *Europhys. Lett.*, **62**, 370.
120 Fraser, B., Denniston, C. and Muser, M.H. (2005) *J. Polym. Sci. Pol. Phys.*, **43**, 970.
121 Zvelindovsky, A.V. and Sevink, G.J.A. (2005) *J. Chem. Phys.*, **123**, 074903.
122 Chen, P.L. (2005) *Phys. Rev. E*, **71**, 061503.
123 Rychkov, I. (2005) *Macromol. Theory Simul.*, **14**, 207.
124 Huang, Z.F. and Vinals, J. (2006) *Phys. Rev. E*, **73**, 060501.
125 Daud, A., Morais, F.M., Morgado, W.A.M. and Martins, S. (2007) *Physica A*, **374**, 517.
126 Kindt, P. and Briels, W.J. (2008) *J. Phys. Chem.*, **128**, 124901.
127 Fong, H. and Reneker, D.H. (1999) *J. Polym. Sci. Polym. Phys.*, **37**, 3488.
128 Ruotsalainen, T., Turku, J., Heikkilä, P., Ruokolainen, J., Nykänen, A., Laitinen, T., Torkkeli, M., Serimaa, R., Harlin, ten Brinke, G., Harlin, A. and Ikkala, O. (2005) *Adv. Mater.*, **17**, 1048.
129 Ma, M.L., Hill, R.M., Lowery, J.L., Fridrikh, S.V. and Rutledge, G.C. (2005) *Langmuir*, **21**, 5549.
130 Kalra, V., Mendez, S., Lee, J.H., Nguyen, H., Marquez, M. and Joo, Y.L. (2006) *Adv. Mater.*, **18**, 3299.
131 Hahm, J. and Sibener, S.J. (2000) *Langmuir*, **16**, 4766.
132 Li, X., Han, Y.C. and An, L.J. (2002) *Langmuir*, **18**, 5293.
133 Li, D. and Xia, Y. (2004) *Adv. Mater.*, **16**, 1151.
134 Dzenis, Y. (2004) *Science*, **304**, 1917.
135 Amundson, K., Helfand, E., Davis, D.D., Quan, X., Patel, S. and Smith, S.D. (1991) *Macromolecules*, **24**, 6546.
136 Amundson, K., Helfand, E., Davis, D.D., Quan, X. and Smith, S.D. (1993) *Macromolecules*, **26**, 2698.
137 Amundson, K., Helfand, E., Davis, D.D., Quan, X., Hudson, S.D. and Smith, S.D. (1994) *Macromolecules*, **27**, 6559.
138 Onuki, A. and Fukuda, J. (1995) *Macromolecules*, **28**, 8788.
139 Thurn-Albrecth, T., Steiner, R., DeRouchey, J., Stafford, C.M., Huang, E., Bal, M., Tuominen, M., Hawker, C.J.

and Russell, T.P. (2000) *Adv. Mater.*, **12**, 787.
140 Thurn-Albrecth, T., DeRouchey, J., Russell, T.P. and Kolb, R. (2002) *Macromolecules*, **35**, 8106.
141 Elhadj, S., Woody, J.W., Niu, V.S. and Saraf, R.F. (2003) *Appl. Phys. Lett.*, **82**, 872.
142 Xu, T., Zhu, Y., Gido, S.P. and Russell, T.P. (2004) *Macromolecules*, **37**, 2625.
143 DeRouchey, J., Thurn-Albrecth, T., Russell, T.P. and Kolb, R. (2004) *Macromolecules*, **37**, 2538.
144 Xiang, H., Lin, Y., Russell, T.P. and Kolb, R. (2004) *Macromolecules*, **37**, 5358.
145 Olszowka, V., Hund, M., Kuntermann, V., Scherdel, S., Tsarkova, L., Boker, A. and Krausch, G. (2006) *Soft Matter*, **2**, 1089.
146 Wang, J.-Y., Chen, W. and Russell, T.P. (2008) *Macromolecules*, **41**, 7227.
147 Pereira, G.G. and Williams, D.R.M. (1999) *Macromolecules*, **32**, 8115.
148 Tsori, Y. and Andelman, D. (2002) *Macromolecules*, **35**, 5161.
149 Schaeffer, E., Thurn-Albrecth, T., Russell, T.P. and Steiner, U. (2000) *Nature (London)*, **403**, 874.
150 Voicu, N.E., Ludwigs, S. and Steiner, U. (2008) *Adv. Mater.*, **20**, 3022.
151 Ashok, B., Muthukumar, M. and Russell, T.P. (2001) *J. Chem. Phys.*, **115**, 1559.
152 Kyrylyuk, A.V., Zvelindovsky, A.V., Sevink, G.J.A. and Fraaje, J.G.E.M. (2002) *Macromolecules*, **35**, 1473.
153 Tsori, Y., Tournilhac, F. and Leibler, L. (2003) *Macromolecules*, **36**, 5873.
154 Tsori, Y., Tournilhac, F., Andelman, D. and Leibler, L. (2003) *Phys. Rev. Lett.*, **90**, 145504.
155 Dürr, O., Dietrich, W., Maas, P. and Nitzan, A. (2002) *J. Phys. Chem.*, **106**, 6149.
156 Thurn-Albrecth, T., DeRouchey, J., Russell, T.P. and Jaeger, H.M. (2000) *Macromolecules*, **33**, 3250.
157 Xu, T., Hawker, C.J. and Russell, T.P. (2003) *Macromolecules*, **36**, 6178.
158 Böker, A., Elbs, H., Hänsen, H., Knoll, A., Ludwigs, S., Zettl, H., Zvelindovski, A.V., Sevink, G.J.A., Urban, V., Abetz, V., Müller, A.H.E. and Krausch, G. (2003) *Macromolecules*, **36**, 8078.
159 Böker, A., Knoll, A., Elbs, H., Abetz, V., Müller, A.H.E. and Krausch, G. (2002) *Macromolecules*, **35**, 1319.
160 Boker, A., Schmidt, K., Knoll, A., Zettl, H., Hansel, H., Urban, V., Abetz, V. and Krausch, G. (2006) *Polymer*, **47**, 849.
161 Zvelindovski, A.V. and Sevink, G.J.A. (2003) *Phys. Rev. Lett.*, **90**, 049601.
162 Xu, T., Goldbach, J.T. and Russell, T.P. (2003) *Macromolecules*, **36**, 7296.
163 Osuji, C., Ferreira, P.J., Mao, G., Ober, C.K., Vander Sande, J.B. and Thomas, E.L. (2004) *Macromolecules*, **37**, 9903.
164 Tomikawa, N., Lu, Z.B., Itoh, T., Imrie, C.T., Adachi, M., Tokita, M. and Watanabe, J. (2005) *Jpn. J. Appl. Phys. 2*, **44**, L711.
165 Tao, Y., Zohar, H., Olsen, B.D. and Segalman, R.A. (2007) *Nano Lett.*, **7**, 2742.
166 Grigorova, T., Pispas, S., Hadjichristidis, N. and Thurn-Albrecht, T. (2005) *Macromolecules*, **38**, 7430.
167 Yu, H., Iyoda, T. and Ikeda, T. (2006) *J. Am. Chem. Soc.*, **128**, 11010.
168 Turturro, A., Gattiglia, E., Vacca, P. and Viola, G.T. (1995) *Polymer*, **21**, 3987.
169 Kim, G. and Libera, M. (1998) *Macromolecules*, **31**, 2569.
170 Kim, G. and Libera, M. (1998) *Macromolecules*, **31**, 2670.
171 Kimura, M., Mister, M.J., Xu, T., Kim, S.H. and Russell, T.P. (2003) *Langmuir*, **19**, 9910.
172 Lin, Z., Kim, D.H., Wu, X., Boosahda, L., Stone, D., LaRose, L. and Russell, T.P. (2002) *Adv. Mater.*, **14**, 1373.
173 Temple, K., Kulbaba, K., Power-Billard, K.N., Manners, I., Leach, K.A., Xu, T., Russell, T.P. and Hawker, C.J. (2003) *Adv. Mater.*, **15**, 297.
174 Park, S., Wang, J.-Y., Kim, B., Xu, J. and Russell, T.P. (2008) *ACS Nano*, **2**, 766.
175 Scalarone, D., Tata, J., Lazzari, M. and Chiantore, O. (2009) *Eur. Polym. J.*, **45**, 2520.
176 Hashimoto, T., Bodycomb, J., Funaki, Y. and Kimishima, K. (1999) *Macromolecules*, **32**, 952.
177 Hashimoto, T., Bodycomb, J., Funaki, Y. and Kimishima, K. (1999) *Macromolecules*, **32**, 2075.

178 Angelescu, D.E., Waller, J.H., Adamson, D.H., Register, R.A. and Chaikin, P.M. (2007) *Adv. Mater.*, **19**, 2687.
179 Mita, K., Takenaka, M., Hasegawa, H. and Hashimoto, T. (2008) *Macromolecules*, **41**, 8789.
180 Rockford, L., Liu, Y., Mansky, P., Russell, T.P., Yoon, M. and Mochrie, S.G.J. (1999) *Phys. Rev. Lett.*, **82**, 2602.
181 Rockford, L., Mochrie, S.G.J. and Russell, T.P. (2001) *Macromolecules*, **34**, 1487.
182 Yang, X.M., Peters, R.D., Nealey, P.F., Solak, H.H. and Cerrina, F. (2000) *Macromolecules*, **33**, 9575.
183 Kim, S.O., Solak, H.H., Stoykovich, M.P., Ferrier, N.J., de Pablo, J.J. and Nealey, P.F. (2003) *Nature (London)*, **424**, 411.
184 De Rosa, C., Park, C., Lotz, B., Fetters, L.J., Wittmann, J.C. and Thomas, E.L. (2000) *Macromolecules*, **33**, 4871.
185 Yang, X.M., Peters, R.D., Kim, T.K. and Nealey, P.F. (1999) *J. Vac. Sci. Technol. B*, **17**, 3203.
186 Solak, H.H., David, C., Gobrecht, J., Golovkina, V., Cerrina, F., Kim, S.O. and Nealey, P.F. (2003) *Microelectron. Eng.*, **56**, 67–68.
187 Tada, Y., Hakasaka, S., Yoshida, H., Hasegawa, H., Dobisz, E., Kercher, D. and Takenaka, M. (2008) *Macromolecules*, **41**, 9267.
188 Shin, D.O., Kim, B.H., Kang, J.-H., Jcong, S. J., Park, S.H., Lee, Y. H. and Kim, S.O. (2009) *Macromolecules*, **42**, 1189.
189 Matsen, M.W. (1998) *Curr. Opin. Colloid Interface Sci.*, **3**, 40.
190 Henkee, C.S., Thomas, E.L. and Fetters, L.J. (1988) *J. Mater. Sci.*, **23**, 1685.
191 Coulon, G., Deline, V.R., Russell, T.P. and Green, P.F. (1989) *Macromolecules*, **22**, 2581.
192 Anastasiadis, S.H., Russell, T.P., Satija, S.K. and Majkrzak, C.F. (1989) *Phys. Rev. Lett.*, **62**, 1852.
193 Russell, T.P., Coulon, G., Deline, V.R. and Miller, D.C. (1989) *Macromolecules*, **22**, 4600.
194 Anastasiadis, S.H., Russell, T.P., Satija, S.K. and Majkrzak, C.F. (1990) *J. Chem. Phys.*, **92**, 5677.
195 Russell, T.P., Menelle, A., Anastasiadis, S.H., Satija, S.K. and Majkrzak, C.F. (1991) *Macromolecules*, **24**, 6269.
196 Collin, B., Chatenay, D., Coulon, G., Ausserre, D. and Gallot, Y. (1992) *Macromolecules*, **25**, 1621.
197 Coulon, G., Dailant, J., Collin, B., Benattar, J.J. and Gallot, Y. (1993) *Macromolecules*, **26**, 1582.
198 Mayes, A.M., Russell, T.P., Bassereau, P., Baker, S.M. and Smith, G.S. (1994) *Macromolecules*, **27**, 749.
199 Carvalho, B.L. and Thomas, E.L. (1994) *Phys. Rev. Lett.*, **73**, 3321.
200 Joly, S., Ausserre, D., Brotons, G. and Gallot, Y. (2002) *Eur. Phys. J. E*, **8**, 355.
201 Walton, D.G., Kellogg, G.J., Mayes, A.M., Lambooy, P. and Russell, T.P. (1994) *Macromolecules*, **27**, 6225.
202 Smith, A.P., Douglas, J.F., Meredith, J.C., Amis, E.J. and Karim, A. (2001) *Phys. Rev. Lett.*, **87**, 015503.
203 Smith, A.P., Douglas, J.F., Meredith, J.C., Amis, E.J. and Karim, A. (2001) *J. Polym. Sci. Part B Polym. Phys.*, **39**, 2141.
204 Lambooy, P., Russell, T.P., Kellogg, G.J., Mayes, A.M., Gallagher, P.D. and Satija, S.K. (1994) *Phys. Rev. Lett.*, **72**, 2899.
205 Koneripalli, N., Singh, M., Levicky, R., Bates, F.S., Gallagher, P.D. and Satija, S.K. (1995) *Macromolecules*, **28**, 2897.
206 Kellogg, G.J., Walton, D.G., Mayes, A.M., Lambooy, P., Russell, T.P., Gallagher, P.D. and Satija, S.K. (1996) *Phys. Rev. Lett.*, **76**, 2503.
207 Koneripalli, N., Levicky, R., Bates, F.S., Ankner, J., Kaiser, H. and Satija, S.K. (1996) *Langmuir*, **12**, 6681.
208 In, I., La, Y.H., Park, S.M., Nealey, P.F. and Gopalan, P. (2006) *Langmuir*, **22**, 7855.
209 Ji, S., Liu, C.-C., Son, J.G., Gotrik, K., Craig, G.S.W., Gopalan, P., Himpsel, F.J., Char, K. and Nealey, P.F. (2008) *Macromolecules*, **41**, 9098.
210 Kim, S.H., Misner, M.J. and Russell, T.P. (2008) *Adv. Mater.*, **20**, 4851.
211 Turner, M.S. (1992) *Phys. Rev. Lett.*, **69**, 1788.
212 Shull, K.R. (1992) *Macromolecules*, **25**, 2122.
213 Pickett, G.R., Witten, T.A. and Nagel, S.R. (1993) *Macromolecules*, **26**, 3194.

214 Kikuchi, M. and Binder, K. (1994) *J. Chem. Phys.*, **101**, 3367.
215 Brown, G. and Chakrabarti, A. (1995) *J. Chem. Phys.*, **102**, 1440.
216 Pickett, G.T. and Balazs, A.C. (1997) *Macromolecules*, **30**, 3097.
217 Matsen, M.W. (1997) *J. Chem. Phys.*, **106**, 7781.
218 Tang, W.H. and Witten, T.A. (1998) *Macromolecules*, **31**, 3130.
219 Geisinger, T., Muller, M. and Binder, K. (1999) *J. Chem. Phys.*, **111**, 5251.
220 Frischknecht, A.L., Curro, J.G. and Frink, L.J.D. (2002) *J. Chem. Phys.*, **117**, 10398.
221 Liu, Y., Zhao, W., Zheng, X., King, A., Singh, A., Rafailovich, M.H., Sokolov, J., Dai, K.H., Kramer, E.J., Schwarz, S.A., Gebizlioglu, O. and Sinha, S.K. (1994) *Macromolecules*, **27**, 4000.
222 Turner, M.S., Rubinstein, M. and Marques, C.M. (1994) *Macromolecules*, **27**, 4986.
223 van Dijk, M.A. and van den Berg, R. (1995) *Macromolecules*, **28**, 6773.
224 Radzilowski, L.H., Carvalho, B.L. and Thomas, E.L. (1996) *J. Polym. Sci. B Polym. Phys.*, **34**, 3081.
225 Kim, H.C. and Russell, T.P. (2001) *J. Polym. Sci. B Polym. Phys.*, **39**, 663.
226 Suh, K.Y., Kim, Y.S. and Lee, H.H. (1998) *J. Chem. Phys.*, **108**, 1253.
227 Huinink, H.P., Brokken-Zijp, J.C.M., van Dijk, M.A. and Sevink, G.J.A. (2000) *J. Chem. Phys.*, **112**, 2452.
228 Konrad, M., Knoll, A., Krausch, G. and Magerle, R. (2000) *Macromolecules*, **33**, 5518.
229 Knoll, A., Magerle, R. and Krausch, G. (2004) *J. Chem. Phys.*, **120**, 1105.
230 Szamel, G. and Muller, M. (2003) *J. Chem. Phys.*, **118**, 905.
231 Yokoyama, H., Mates, T.E. and Kramer, E.J. (2000) *Macromolecules*, **33**, 1888.
232 Mansky, P., Russell, T.P., Hawker, C.J., Pitsikalis, M. and Mays, J. (1997) *Macromolecules*, **30**, 6810.
233 Mansky, P., Liu, Y., Huang, E., Russell, T.P. and Hawker, C. (1997) *Science*, **275**, 1458.
234 Huang, E., Russell, T.P., Harrison, C., Chaikin, P.M., Register, R.A., Hawker, C.J. and Mays, J. (1998) *Macromolecules*, **31**, 7641.
235 Huang, E., Rockford, L., Russell, T.P. and Hawker, C.J. (1998) *Nature (London)*, **395**, 757.
236 Huang, E., Pruzinsky, S., Russell, T.P., Mays, J. and Hawker, C.J. (1999) *Macromolecules*, **32**, 5299.
237 Huang, E., Mansky, P., Russell, T.P., Harrison, C., Chaikin, P.M., Register, R.A., Hawker, C.J. and Mays, J. (2000) *Macromolecules*, **33**, 80.
238 Harrison, C., Chaikin, P.M., Huse, D.A., Register, R.A., Adamson, D.H., Daniel, A., Huang, E., Mansky, P., Russell, T.P., Hawker, C.J., Egolf, D.A., Melnikov, I.V. and Bodenschatz, E. (2000) *Macromolecules*, **33**, 857.
239 Son, J.G., Bulliard, X., Kang, H., Nealey, P.F. and Char, K. (2008) *Adv. Mater.*, **20**, 3643.
240 Royer, L. (1928) *Bull. Soc. Fr. Mineral. Crystallogr.*, **51**, 7.
241 van deer Mere, J.H. (1949) *Discuss. Faraday Soc.*, **5**, 206.
242 Swei, G.S., Lando, J.B., Rickert, S.E. and Mauritz, K.A. (1986) *Encyclopedia Polym. Sci. Eng.*, **6**, 209.
243 Willems, J. (1955) *Naturwissenschaften*, **42**, 176.
244 Willems, J. and Willems, L. (1957) *Experientia*, **13**, 465.
245 Willems, J. (1958) *Discuss. Faraday Soc.*, **25**, 111.
246 Fischer, E.W. (1958) *Discuss. Faraday Soc.*, **25**, 204.
247 Wellinghoff, S.H., Rybnikar, F. and Baer, E. (1974) *J. Macromol. Sci. (Phys.)*, **B10**, 1.
248 Koutsky, J.A., Walton, A.C. and Baer, E. (1967) *J. Polym. Sci. Polym. Lett. Ed.*, **5**, 177.
249 Koutsky, J.A., Walton, A.C. and Baer, E. (1967) *J. Polym. Sci. Polym. Lett. Ed.*, **5**, 185.
250 Ashida, M., Uedn, Y. and Watanabe, T. (1978) *J. Polym. Sci. Polym. Phys. Ed.*, **16**, 179.
251 Rickert, S.E. and Baer, E. (1978) *J. Mater. Sci. Lett.*, **13**, 451.
252 Lovinger, A.J. (1980) *Polym. Prep. Am. Chem. Soc.*, **21**, 253.
253 Martinez-Salazzas, J., Barham, P.J. and Keller, A. (1984) *J. Polym. Sci. Polym. Phys. Ed.*, **22**, 1085.

254 Wittmann, J.C. and Lotz, B. (1986) *J. Mater. Sci.*, **21**, 659.
255 Tuinstra, F. and Baer, E. (1970) *J. Polym. Sci. Polym. Lett. Ed.*, **8**, 861.
256 Lovinger, A.J. (1983) *J. Polym. Sci. Polym. Phys. Ed.*, **21**, 97.
257 Hobbs, S.Y. (1971) *Nature Phys. Sci.*, **12**, 234.
258 Wittmann, J.C. and Lotz, B. (1981) *J. Polym. Sci. Polym. Phys. Ed.*, **19**, 1837.
259 Wittmann, J.C., Hodge, A.M. and Lotz, B. (1983) *J. Polym. Sci. Polym. Phys. Ed.*, **21**, 2495.
260 Wittmann, J.C. and Lotz, B. (1989) *Polymer*, **30**, 27.
261 Wittmann, J.C. and Lotz, B. (1990) *Prog. Polym. Sci.*, **15**, 909.
262 Kopp, S., Wittmann, J.C. and Lotz, B. (1995) *Makromol. Chem. Macromol. Symp.*, **98**, 917.
263 De Rosa, C., Park, C., Lotz, B. and Thomas, E.L. (2000) *Nature (London)*, **405**, 433.
264 Park, C., De Rosa, C., Lotz, B., Fetters, L.J. and Thomas, E.L. (2003) *Macromol. Chem. Phys.*, **204**, 1514.
265 Park, C., De Rosa, C., Lotz, B., Fetters, L.J. and Thomas, E.L. (2001) *Adv. Mater.*, **13**, 724.
266 Smith, P. and Pennings, A.J. (1974) *Polymer*, **15**, 413.
267 Smith, P. and Pennings, A.J. (1976) *J. Mater. Sci.*, **11**, 1450.
268 Hodge, A.M., Kiss, G., Lotz, B. and Wittmann, J.C. (1982) *Polymer*, **23**, 985.
269 Wittmann, J.C. and St. John Manley, R. (1977) *J. Polym. Sci. Polym. Phys. Ed.*, **15**, 1089.
270 Wittmann, J.C. and St. John Manley, R. (1977) *J. Polym. Sci. Polym. Phys. Ed.*, **15**, 2277.
271 Dorset, D.L., Hanlon, J. and Karet, G. (1989) *Macromolecules*, **22**, 2169.
272 Fredrickson, G.H. and Leibler, L. (1989) *Macromolecules*, **22**, 1238.
273 Lodge, T.P., Pan, C., Jin, X., Liu, Z., Zhao, J., Maurer, W.W. and Bates, F.S. (1995) *J. Polym. Sci. Polym. Phys.*, **33**, 2289.
274 Hamley, I.W., Fairclough, J.P.A., Ryan, A.J., Ryu, C.Y., Lodge, T.P., Gleeson, A.J. and Pedersen, J.S. (1998) *Macromolecules*, **31**, 1188.
275 Park, C., De Rosa, C. and Thomas, E.L. (2001) *Macromolecules*, **34**, 2602.
276 Van Dijk, M.A. and van den Berg, R. (1995) *Macromolecules*, **28**, 6773.
277 Smith, H.I. and Flanders, D.C. (1978) *Appl. Phys. Lett.*, **32**, 349.
278 Smith, H.I., Geis, M.W., Thompson, C.V. and Atwater, H.A. (1983) *J. Cryst. Growth*, **63**, 527.
279 Kobayashi, T. and Takagi, K. (1984) *Appl. Phys. Lett.*, **45**, 44.
280 Flanders, D.C., Shaver, D.C. and Smith, H.I. (1978) *Appl. Phys. Lett.*, **32**, 597.
281 Wittmann, J.C. and Smith, P. (1991) *Nature (London)*, **352**, 414.
282 Hansma, H., Motamedi, F., Smith, P., Hansma, P. and Wittmann, J.C. (1992) *Polym. Commun.*, **33**, 647.
283 Dietz, P., Hansma, P.K., Ihn, K.J., Motamedi, F. and Smith, P. (1993) *J. Mater. Sci.*, **28**, 1372.
284 Fenwick, D., Ihn, K.J., Motamedi, F., Wittmann, J.C. and Smith, P. (1993) *J. Appl. Polym. Sci.*, **50**, 1151.
285 Fenwick, D., Smith, P. and Wittmann, J.C. (1996) *J. Mater. Sci.*, **31**, 128.
286 Segalman, R.A., Yokoyama, H. and Kramer, E.J. (2001) *Adv. Mater.*, **13**, 1152.
287 Segalman, R.A., Hexemer, A., Hayward, R.C. and Kramer, E.J. (2003) *Macromolecules*, **36**, 3272.
288 Cheng, J.Y., Ross, C.A., Thomas, E.L., Smith, H.I. and Vancso, G.J. (2002) *Appl. Phys. Lett.*, **81**, 3657.
289 Cheng, J.Y., Ross, C.A., Thomas, E.L., Smith, H.I. and Vancso, G.J. (2003) *Adv. Mater.*, **15**, 1599.
290 Cheng, J.Y., Zhang, F., Chuang, V.P., Mayes, A.M. and Ross, C.A. (2006) *Nano Lett.*, **6**, 2099.
291 Ruiz, R., Ruiz, N., Zhang, Y., Sandstrom, R.L. and Black, C.T. (2007) *Adv. Mater.*, **19**, 2157.
292 He, X.-H., Song, M., Liang, H.-J. and Pan, C.-Y. (2001) *J. Chem. Phys.*, **114**, 10510.
293 Sevink, G.J.A., Zvelindovsky, A.V., Fraaije, J.G.E.M. and Huinink, H.P. (2001) *J. Chem. Phys.*, **115**, 8226.
294 Bosse, A.W., Garcia-Cervera, C.J. and Fredrickson, G.H. (2007) *Macromolecules*, **40**, 9570.

295 Shin, K., Xiang, H., Moon, S.I., Kim, T., McCarthy, T.J. and Russell, T.P. (2004) *Science*, **306**, 76.

296 Xiang, H., Shin, K., Kim, T., Moon, S.I., McCarthy, T.J. and Russell, T.P. (2004) *Macromolecules*, **37**, 5660.

297 Sun, Y., Steinhartm, M., Zschech, D., Adhikari, R., Michler, G.H. and Gösele, U. (2005) *Macromol. Rapid Commun.*, **26**, 369.

298 Park, C., Cheng, J.Y., De Rosa, C., Fasolka, M.J., Mayes, A.M., Ross, C.A. and Thomas, E.L. (2001) *Appl. Phys. Lett.*, **79**, 848.

299 Thomas, E.L., De Rosa, C., Park, C., Fasolka, M., Lotz, B., Mayes, A.M. and Yoon, J. (2005) US Patent No. 6, 893, 705 (MIT).

300 Park, C., Simmons, S., Fetters, L.J., Hsiao, B., Yeh, F. and Thomas, E.L. (2000) *Polymer*, **41**, 2971.

301 Park, C., De Rosa, C., Fetters, L.J. and Thomas, E.L. (2000) *Macromolecules*, **33**, 7931.

5
Helical Polymer-Based Supramolecular Films
Akihiro Ohira, Michiya Fujiki, and Masashi Kunitake

5.1
Introduction

Nature's elegant bottom-up fabrication of hierarchical superhelical assemblies over a wide range of scales has inspired the development of functional advanced materials from efficient approaches [1].

In natural systems, hierarchical helical architectures, such as the double helix of DNA, the triple helix of collagen, and the α-helical coiled-coils of myosin and keratin on the molecular level, are ubiquitous [2]. They also occur frequently in artificial systems, and it is well known that amphiphilic compounds, such as phospholipids, amino acids, nucleic acid derivatives, and carbohydrate compounds, fabricate a variety of helical superstructures including tubes, disks, bundles, fibers, and cholesteric liquid crystals [1, 3]. Until recently, extensive studies have been conducted into the helical supramolecular assemblies of organic molecular materials [1, 3–5] and, indeed, recent progress in the areas of polymer and supramolecular sciences has led us to construct highly ordered and well-defined helical architectures for synthetic helical polymers [6]. Akagi *et al.* reported the synthesis of hierarchical helical polyacetylene fiber under an asymmetric reaction field consisting of a chiral nematic (N^*) liquid crystal [7]. Nolte *et al.* demonstrated that the formation of various helical superstructures of amphiphilic block copolymer containing a polystyrene tail and a charged helical polyisocyanide head group derived from isocyano-L-alanine-L-alanine and isocyano-L-alanine-L-histidine [8]. Yashima *et al.* succeeded in the construction of both two-dimensional (2-D) and three-dimensional (3-D) smectic self-assemblies composed of helical polyisocyanides [9]. The same group also created a one-dimensional (1-D) helical supramolecular architecture of double-stranded helical polyacetylenes, which can form 2-D, self-assembled structures on the substrate [10]. Today, the control of such highly ordered and hierarchical helical assemblies of synthetic helical polymers is a major challenge in polymer and supramolecular sciences, especially from the point of view of developing the materials sciences. Generally, the driving forces in the formation of helical assembled structures are varied, including a chiral interaction and a polar interaction between solute–solvent and solute–solute, such as hydrogen bonding, electrostatic interactions, coordination bonding, and π-π stacking. If a simple system were to organize a hierarchical helical structure only by weak intermolecular interactions such as van der Waals interactions, it might be helpful to understand the chiral interactions between the molecules. In this chapter, we will review several helical polymer-based supramolecular assemblies in order to provide a better understanding of the driving forces in the formation of these helical assemblies. First, we describe helical polymer-based 1-D and 2-D architectures from the perspective of the construction of supramolecular organization with helical polymers. Next, we describe helical polymer-based functional films, focusing on the functionalities, such as switch, memory, transfer and amplification, that are derived from the intrinsic chirality of helical polymers. Two different types of well-known helical polymer were chosen. A poly(γ-L-glutamate)-derivative polypeptide and a synthetic helical polysilane polymer were selected because these are well studied and possess ideal helical conformation, thus allowing us to investigate both the intramolecular and intermolecular interactions. Studies of the crystal structure of poly(γ-L-glutamate) derivatives have focused primarily on the packing structure of the α-helices. Kaufman *et al.* initially reported that long *n*-alkyl side chains of polymers based on acrylic and methacrylic acids can pack into paraffin-like crystallites [11]. Watanabe *et al.* discovered the thermotropic liquid crystal nature of these polymers and investigated the molecular packing and thermotropic behaviors for a series of poly(γ-L-glutamate)s with *n*-alkyl side chains, ranging from amyl to octadecyl [12].

Polysilane can be prepared by an easy synthetic route, and exhibits unique electronic and optical properties that are attributed to the delocalized σ electrons along the Si–Si main chain [13]. Much research has been conducted on the behavior of polysilane in solution in order to elucidate polymer conformation [14]. Consequentially, it has been found that the optical and electronic properties of polysilane may change easily with the side chain structure, solvent, or temperature. Optically active polysilanes with various chiral substituents in the side chains adopt helical conformations with a preferential screw-sense [14]. Furthermore, it has been found that helical polysilanes exhibit many interesting characteristics both in solution and in solid states [15]. The relationship between the global conformation of a polymer chain and the physical properties in dilute solution has already been established from spectroscopic evidence in helical polysilanes [16].

Research into the supramolecular assembly based on a helical polymer is expected to contribute significantly to developments within the life sciences and materials sciences. In the field of life sciences, it is important to clarify how the existence of the helical configuration is related to human vital activity. In the field of materials sciences, a helical conformation is used for functional materials, for example in electronic and/or ionic conductors, using the directivity with a well-defined helical conformation, molecular sensing and sensor systems using the regularity of the helical conformation. Helical and superhelical assemblies are ubiquitous in nature, as exemplified in biological polymers such as DNA and structural proteins, wherein the information stored in the molecular building blocks is utilized for supramolecular organization into higher-order helical architectures at different hierarchical levels, from nanoscopic to macroscopic lengths. Inspired by this elegant spontaneous fabrication into higher-order helical architectures at different hierarchical levels in nature, the self-organization of synthetic organic molecules has been widely exploited to generate novel functional supramolecular architectures. However, in order to achieve this system, it is necessary to design a programmed molecule which can self-organize spontaneously into higher-ordered supramolecular structures with controlled morphology, dimension, and size. As the helical polymer possesses a programmed and well-organized structure, a supramolecular assembly based on a helical polymer would be promising for controlling the hierarchy in multi-scale so as to create a natural system.

5.2
Helical Polymer-Based 1-D and 2-D Architectures

Today, the precise control of topology at the single polymer chain level (1-D), especially for helical polymers, is a "hot issue," because understanding the 1-D change of topology is the ultimate goal for the development of polymer-based nanomaterials, and will take advantage of the solubility, flexibility, stability, and electronic properties of these materials [17]. The 2-D, self-organized microscopic and/or mesoscopic structures composed of helical polymers are also of considerable interest both for advancing the fundamental understanding of structural properties, such as helicity and rigidity, and for developing potential applications, including liquid crystals and molecular sensing and sensory systems at the monolayer level [18]. In addition to the structural properties, understanding the interactions that operate between the molecules is also crucial for self-organization. The results of this research will be useful for the construction of hierarchical 3-D bulk phases.

In general, the supramolecular 1-D and 2-D architectures composed of helical polymers can be visualized by using scanning probe microscopy (SPM) techniques, such as scanning tunnel microscopy (STM) and atomic force microscopy (AFM). These not only provide the opportunity for direct observations on atomic flat substrates with submolecular resolution, but also reveal the structural information of individual polymer chains, whereas most spectroscopic characterizations have provided averaged data for multiple molecules or aggregates. Observations of biopolymers and synthetic polymers [19–22] in the isolated or aggregated state using SPM (mainly AFM) have been conducted in order to analyze molecular characteristics such as chain length, end-to-end length, polymer diameters, and chain rigidity [23, 24]. In fact, AFM has proved very advantageous for observing helical characteristics, such as helicity, helical pitch, and helix reversal [9].

In this section, as examples of 1-D structures, such as rods and circles, we describe the changes in the 1-D architectures of isolated helical polysilanes with different chain lengths on the surfaces. Additionally, we describe a mesoscopic 2-D superhelical assembly of polysilane formed by homochiral intermolecular interactions. The observation of the associated superstructure of the synthetic helical polymers was significant, especially for a wide range of scales, and also provided a better understanding of the mechanism of formation of the highly ordered and hierarchical architectures. We also present details of the formation of 2-D epitaxial adlayers of poly(γ-L-glutamate)s as examples of 2-D helical, polymer-assembled structures. It should be noted here that not only the observation of both 1-D and 2-D architectures but also the direct visualization of intramolecular and intermolecular interactions existing between helical polymers were clearly demonstrated by AFM, with molecular-level resolution.

5.2.1
Formation of Various 1-D Architectures of Helical Polysilanes on Surfaces

The topology of an isolated helical polysilane chain bearing a fluoroalkyl side group (Scheme 5.1, polymer **1**) changes from a rod-like to a circular structure on a mica surface, with the main chain length determined by means of AFM [25]. Polymer **1** is an ideal model molecule with different lengths; it is a 1-D, semi-flexible semiconductor that has exhibited unique properties that have contributed to intramolecular CF/Si interactions [26].

5.2.1.1 Direct Visualization of 1-D Rod, Semi-Circle and Circle Structures by AFM

Figure 5.1 exemplifies typical AFM images of **1** on mica surfaces. In these images, three different forms of topology – namely, rod, semi-circle, and circle structures – coexisted and can be clearly distinguished. Figures 5.1c, d, and e show AFM images of rod, semi-circle, and circle structures of isolated **1**, respectively, in a relatively narrow area. The average height of both the rod and circle structures was approximately 0.86 nm and was almost uniform, which corresponds to the real height of adsorbates.

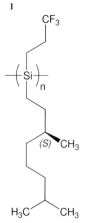

Scheme 5.1 Chemical structure of fluoroalkyl-containing helical polysilane **1**.

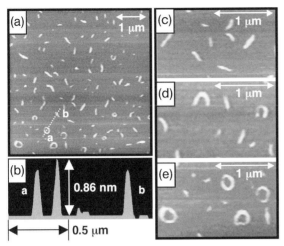

Figure 5.1 Typical atomic force microscopy (AFM) images of isolated polysilane **1** chains on mica surfaces in (a) large area, (b) cross-sectional profile of line a–b in (a), (c) rod-like structure, (d) rod-like and semi-circle structures, and (e) rod-like, semi-circle, and circle structures.

All of the observed rod, semi-circle, and circle structures were not aggregated forms but isolated single polymer chains. The distribution of the average number of contour lengths (L_n) is shown in Figure 5.2.

In the region of shorter lengths in Figure 5.2 (region I), only rod-like structures were observed (see Figure 5.1c), and this was essentially consistent with the results that **1** possesses a stable, semi-rigid, helical conformation by the intramolecular CF/Si interaction in the solution system [26]. However, it should be emphasized here that the chain topology eventually transformed from rod to circle structures as the main chain length increased on the surfaces. In particular, semi-circle structures were observed in the very narrow region of medium length (400~500 nm) in Figure 5.2 (region II), thereby showing that the topological switching between the rod and circle structures is attributable to the discontinuous transition phenomenon. The distribution of the end-to-end distance also exhibited a discrete decrease as contour length increased, as shown in Figure 5.3.

Figure 5.4 shows a high-resolution AFM image of the higher-order helical conformation in a circle structure. The observed helical pitch in the AFM images was determined to be approximately 44.2 ± 4.2 nm, although it was much longer than that expected from the 7_3 helical structure model (0.45 nm). This might be attributed not only to a tip broadening effect but also to an intramolecular CF/Si interaction, leading to a long-period macromolecular right-handed helicity [27]. Except for the limitation of the AFM spatial resolution, and because such helical conformations cannot be seen within the rod and semi-circle structures, significant long

164 | 5 Helical Polymer-Based Supramolecular Films

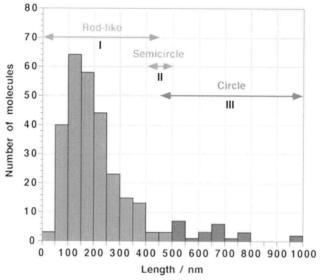

Figure 5.2 Contour length distribution of **1** estimated from AFM images. Regions I, II, and III indicate the existence of rod-like, semi-circle, and circle structures, respectively.

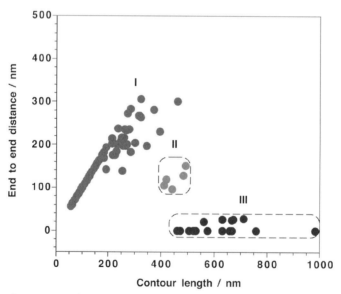

Figure 5.3 Distribution of the end-to-end distance of **1** obtained from AFM images as a function of contour length. The numbers I, II, and III correspond to the rod, semi-circle, and circle structures, respectively.

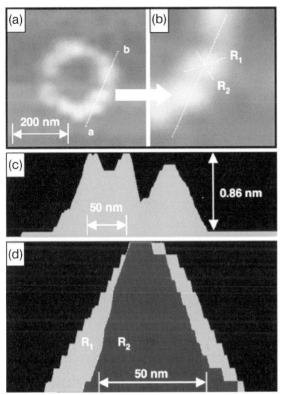

Figure 5.4 High-resolution AFM image of **1**. (a) Circle structure; (b) Highly zoomed image of (a); (c) Cross-section profile of line a–b in (a); (d) Cross-sectional profiles of segments R_1 and R_2. The greater width of R_1 in the profile indicates that the polymer chain possesses a right-handed helical structure. Each bright spot and its cross-sectional profile clearly indicate that the polymer chain has a helical structure. From the cross-sectional profile, it can be seen that the local conformation of R_2 is flat on the bottom, but R_1 is bulged on the top [33].

chains might be required to configure the long-period helicity. In several AFM images and their cross-section profiles, the height and width were frequently different at the right and left or up and down points in the circle structure (e.g., Figure 5.1b).

5.2.1.2 Driving Force for the Formation of 1-D Architectures

The chain dynamics were realized on the surfaces, and not in the solution system. Although these structures cannot be formed in the solution system due to thermal fluctuation, the mechanism of topology switching of single polymer chains is different from that of the associated forms. The dominant driving force of topology switching would originate from a degree of freedom, as the topology of a single polymer chain on the surface depends heavily on the chain length.

The proposed structural models regarding the circle structure are shown in Figure 5.5. Imperfect open-circle structures with a narrow end-to-end gap were frequently observed, as illustrated in Figure 5.5 (open circle A). In the case of B, it is hypothesized that the intramolecular CF/Si interaction allows it to form and stabilize a perfect closed-circle structure, because the fluorine atoms on the side chains may attach to the Si atoms on the main chain, close to the end groups. Besides this effect, a Si–O–Si bond may be formed between the end-termini via hydrolysis of the SiH end group by water adsorbed onto the mica surfaces [28]. In the case of circle C, the end-termini of the polymer chain are crossing each other, while in circle D the end-termini are not crossing but are parallel to each other. These structural models can be the precursors of toroidal structures.

As expected, much longer polymer chains frequently formed toroid-like structures, as shown in Figure 5.6. These circle and toroid-like architectures may be

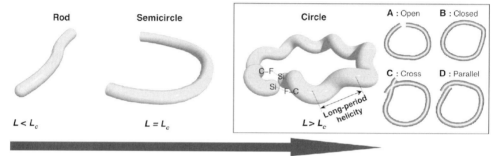

Figure 5.5 Schematic representation of morphology switching of single polysilane chain 1 with chain length dependence based on AFM observations. Here, L_c indicates the critical length of the semi-circle structure as an intermediate state between the rod and circle structures. In the circle structure, (a), (b), (c), and (d) indicate the proposed models (open, closed, cross, and parallel, respectively).

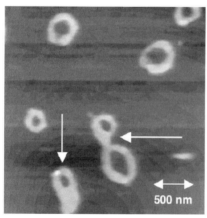

Figure 5.6 Typical AFM image of toroid-like structures adsorbed onto mica surfaces.

useful in device applications because a nanoscaled circle architecture composed of organic/inorganic materials has been expected to provide a superior device performance due to the lack of defects in the end-terminus part, which strongly affects energy loss [29].

5.2.2
Formation of Mesoscopic 2-D Hierarchical Superhelical Assemblies

Optically active polysilanes, such as poly[(S)-3,7-dimethyloctyl-3-methylbutylsilane] (**2**) and poly[(R)-3,7-dimethyloctyl-3-methylbutylsilane] (**3**), which are able to undergo the helix–helix transition [30] at −20 °C in iso-octane (as shown in Scheme 5.2) were used in this study. Notably, we present here both the construction of mesoscopic highly ordered and hierarchical–superhelical assemblies based on only weak homochiral intermolecular interactions.

5.2.2.1 Direct Visualization of a Single Polymer Chain

Figure 5.7 shows the typical AFM images of single polymer chains of polymers **2**, **3**, and racemic **4**, deposited onto the freshly cleaved mica surfaces and their section analysis. The samples were prepared by casting a dilute iso-octane solution of polymer sample (approximate range of 5~10 µg ml^{-1}), after which the strands

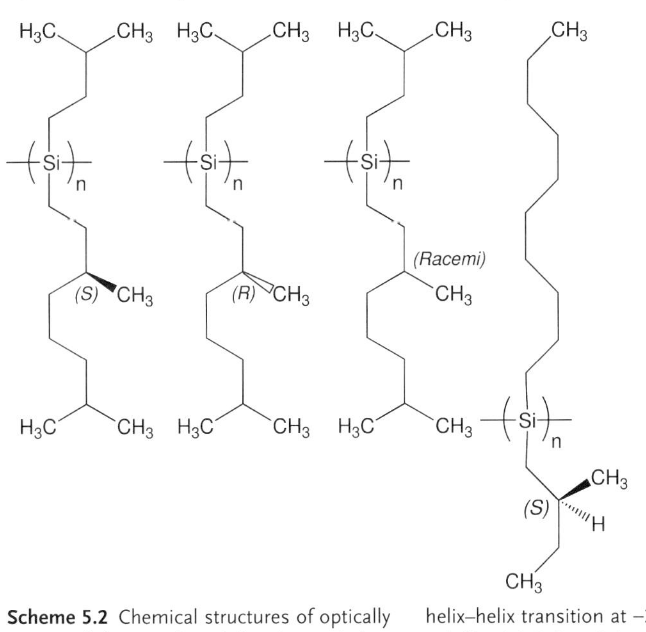

Scheme 5.2 Chemical structures of optically active poly{(S)-3,7-dimethyloctyl-3-methyl butylsilane} **2**, poly{(R)-3,7-dimethyloctyl-3-methyl butylsilane} **3**, which can undergo a helix–helix transition at −20 °C in iso-octane, poly{(rac)-3,7-dimethyl-octyl-3-methylbutylsilane} **4**, and poly{n-decyl-(S)-2-methylbutylsilane} **5**.

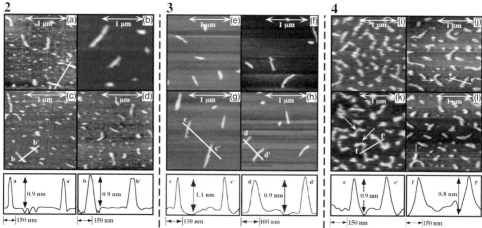

Figure 5.7 Typical AFM images of single polymer chains of **2**, **3**, and **4**, and their cross-sectional profiles [lines a–a' in (c) and b–b' in (d) of **2**, lines c–c' in (g) and d–d' in (h) of **3**, and lines e–e' in (j) and f–f' in (k) of **4**] onto the mica surfaces. **2**: $M_w = 2.9 \times 10^5$, $M_w/M_n = 3.43$. **3**: $M_w = 2.6 \times 10^5$, $M_w/M_n = 3.53$. **4**: $M_w = 3.2 \times 10^5$, $M_w/M_n = 5.39$. The concentrations of casting solution were ca. $7\,\mu g\,ml^{-1}$ for **2**, ca. $5\,\mu g\,ml^{-1}$ for **3**, and ca. $10\,\mu g\,ml^{-1}$ for **4**.

corresponding to the single polymer chains were randomly and separately adsorbed onto the mica surface. Both, the **2** and **3** polymer chains typically possessed an isolated and stretched conformation. Although the information on the helical conformation, such as the helicity and helical pitch, could not be obtained due to a lack of resolution, these AFM images demonstrate the rod-like conformation, which is expected from the empirical relationship between the molar extinction coefficient ε, the viscosity index α (**2**: 1.47, **3**: 1.32 in toluene at 70 °C), and the relatively long persistence length (~60 nm) [31]. A different polysilane, poly{n-decyl-(S)-2-methylbutylsilane} (**5**), which is a stiff polymer with a persistence length of 70 nm, was observed by means of AFM as a rod-like chain onto the sapphire surfaces [23b]. Therefore, these results were consistent with the previous report.

However, slightly bent chains were frequently observed when the polymer chains of the racemic polymer **4** (M_w: 5.2×10^5, M_w/M_n: 5.39), which is a random copolymer system composed of equal amounts of R-comonomer and S-comonomer, were compared to those of **2** and **3** (marked by arrows in Figure 5.7j, k, and l). Additionally, the molar absorption coefficient ε was lower than that of **2** and **3** (2.8×10^4 (Si repeat unit)$^{-1}$) in iso-octane at 20 °C, and the α value, which indicates the rigidity of the polymer chain, was also relatively lower than that of **2** and **3** (1.15 in tetrahydrofuran at 40 °C), as shown in Figure 5.8. In addition to the empirical relationship between ε and the viscosity index of polysilanes [14b], the present AFM observations suggest that the topology of **4** is more flexible than that of **2** and **3**.

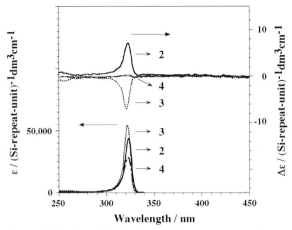

Figure 5.8 Circular dichroism and UV absorption spectra in homogeneous iso-octane solutions of **2** (at 20 °C, solid line), **3** (at −5 °C, dotted line) and **4** (at 20 °C, broken line). Either the ε or $\Delta\varepsilon$ value was normalized (Si repeat unit)$^{-1}$.

5.2.2.2 Formation of Superhelical Assemblies by Homochiral Intermolecular Interactions

The polymer chain, which is much longer than the persistence length of **2** (M_w: 3.9 × 10^5, M_w/M_n: 3.02), formed a unique aggregated structure. Figure 5.9 shows the typical AFM images of the superhelical assemblies of **2**. The highly oriented assembled polymer chains were observed on the mica surfaces (Figure 5.9a), which might be attributed to a dewetting property and evaporation process of iso-octane on the mica surfaces. In the case of a fast evaporation process at 80 °C, no uniform superhelical assemblies were found.

Figures 5.9b and c show the assembled polymer chains and cross-section analysis in the narrow area. The lines a–b and c–d in Figure 5.9b indicate the height and width, respectively, of the branching points in the main assembled polymer chain. The height of two different polymer chains splitting from the branching points was approximately 17 nm at the lower part and 39 nm at the higher part. The lines e–h in Figure 5.9c indicate the height and width of the main assembled polymer chain. The height of the main assembled polymer chain, corresponding to approximately 70 nm, is higher than the total of the branching chain (17 + 39 nm), and shows that the main assembled coil is formed by a twisting of branches. These results reveal that the assembled polymer chain is composed of several twined polymer chains.

Figure 5.10a shows the high-resolution AFM image of coiled-coil assemblies of polymer **2**. Right-handed superhelical polymer chains, which possess a long helical pitch, were clearly observed. In the case of the coiled-coil forms of polymer **3** (M_w: 7.9 × 10^5, M_w/M_n: 4.78), superhelical polymer assemblies were also observed, which have the opposite handedness (left-handed) as polymer **2**, as shown in

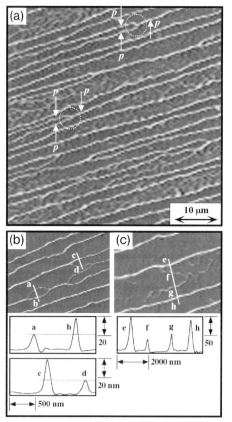

Figure 5.9 AFM images and cross-sectional profiles [lines a–b and c–d in (b) and e–f–g–h in (c)] of highly ordered superhelical assemblies of **2** onto the mica surfaces. $M_w = 3.9 \times 10^5$, $M_w/M_n = 3.02$. The concentration of the casting solution was ca. 500 µg ml^{-1}.

Figure 5.10b. It should be noted that each helical (*S* or *R*) polysilane was configured in its corresponding unidirectional superhelical structure. It has been suggested that the **2** and **3** chains possess the *P* (right-handed) and *M* (left-handed) helical screw-senses, respectively. In our case, large-scaled, right-handed and left-handed superhelical fibers are composed of right-handed **2** and left-handed **3**, respectively [32]. In Figure 5.9a, here again, it was found that the branched chains (branching points are denoted by a dotted circle), which are components of the main assembled chain, also possessed a right-handed helical structure (marked by arrows). Each branched chain and its small aggregate maintained the same helicity, leading to the formation of uniform and hierarchical superhelical assemblies with the same helicity as the components. It has been reported that, in both natural and artificial systems, heretofore, the helicities of the components were

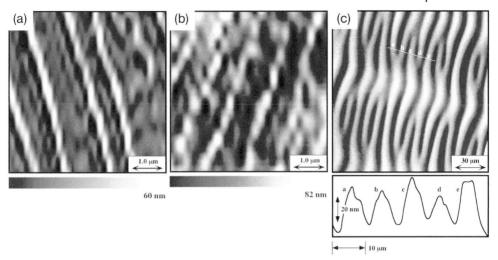

Figure 5.10 AFM images of superhelical structures of **2**, **3**, and **4** polymer chains onto mica surfaces. (a) **2** ($M_w = 3.9 \times 10^5$, $M_w/M_n = 3.02$); (b) **3** ($M_w = 7.9 \times 10^5$, $M_w/M_n = 4.78$); (c) **4** ($M_w = 3.2 \times 10^5$, $M_w/M_n = 5.39$).

Table 5.1 Numerical data of assembled fibers of **2, 3,** and **4** polymer on the mica surfaces (see Scheme 5.2).

Polymer	$M_w \times 10^5$	M_w/M_n	Height of aggregate (nm)	Helical pitch (nm)	Screw-sense of fibers
2	3.9	3.02	65 ± 12	646 ± 82	Right-handed
3	7.9	4.78	49 ± 10	636 ± 77	Left-handed
4	3.2	5.39	51 ± 9.5	–	–

different from that of the hierarchical superstructure that eventually formed; for example, the right-handed superhelix of collagen is composed of three left-handed helices [2a, 5k, 33]. Given that the same helicity was maintained during the process of association without reference to the state of the aggregate (number of polymer chains), homochiral intermolecular interactions would be a dominant driving force for the formation of uniform and hierarchical superhelical assemblies. The data reporting on the height, helical pitch and handedness of the coiled-coil chains are listed in Table 5.1.

A comparison of the helical fibers constructed from **2** and **3** showed, surprisingly, almost the same value for the heights and helical pitches, but a difference in the handedness. A similar phenomenon has been reported in a biological system, where Larson and coworkers observed the left-handed superhelical structure composed of RecA and double-stranded DNA by means of AFM [33]. In that case, the pitches of the superhelices of the complexes (RecA and DNA filament) were very similar to each other, regardless of the number of component filaments. Moreover, from observations of the heights of the helical fibers of **2** and **3**, it is believed that the formation of the highly ordered structure in the large scale is achieved by the assembly of several polymer chains. This observation is an example of mesoscopic large-scaled, highly ordered superhelical assemblies that possess both of the helical senses constructed by a synthetic helical polymer. Most of the synthetic helical polymers have a polar functional group in the main or side chain, and therefore an associated structure is formed by the combination of various strong intermolecular interactions, such as hydrogen bonding, electrostatic interactions, coordination bonding, and π-π stacking. In the present case, only weak intermolecular interactions can account for the polymer association, because the polysilanes **2** and **3** are composed of nonpolar groups with side chains. It is noteworthy that no assembled superhelical structures were observed with **4**. As for the morphology of the observed structure of **4**, although large-scaled stripe patterns were formed on the substrate, the structures possessed no helicity. As discussed above, there was no major difference between **2** or **3** and **4** in the conformation of a single polymer chain. These results suggest that the homochirality of the single helical polymer chain rather than the chain conformation structure contributes to the formation of the large-scaled, highly ordered superhelical assembly [34]. Similar phenomena have been observed in the 2-D crystallization of small chiral organic molecules on the surfaces. For example, homochiral enantioselective crystallization from racemic mixture has been reported [34a]. Kühnle and coworkers have shown that cysteine molecules formed homochiral pairs (D-D and L-L) from racemic mixture using STM. Chiral transfer from single molecules into self-assembled monolayers on the metal surfaces has also reported by Fasel *et al.* [34b]; here, enantiopure *P* and *M* chiral single molecules on the surfaces formed the enantiopure self-assembled monolayers with clockwise and anticlockwise formats, respectively. In addition to the enantiopure self-assembly, it has been reported that the enhancement of chiral interactions in two dimensions was observed in the enantioseparation of chiral molecules, using STM [34c]. These results also suggested that homochiral intermolecular interaction would be an important factor to form the 2-D chiral surfaces.

5.2.3
Formation of 2-D Crystallization of Poly(γ-L-Glutamates) on Surfaces

Poly(γ-L-glutamate) (**6**) is a polypeptide that is well known for forming a stable α-helical conformation, even when the substituted side chains are varied, as shown in Scheme 5.3. Polymers of **6** have longer alkyl side chains and can form 2-D self-organized arrays on highly oriented pyrolytic graphite (HOPG) [20c].

In the following section, the formation of 2-D epitaxial arrays on surfaces and a comparison of structures between 2-D epitaxial arrays and 3-D bulk phases will be described, based on the results of AFM and wide-angle X-ray diffraction (XRD) studies.

6A: n=14
6B: n=18 M_w: 37,914
6C: n=18 M_w: 56,059
6D: n=22

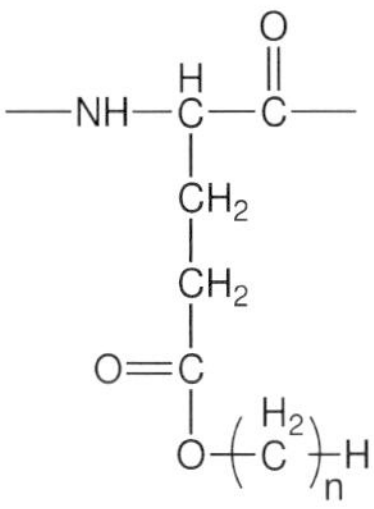

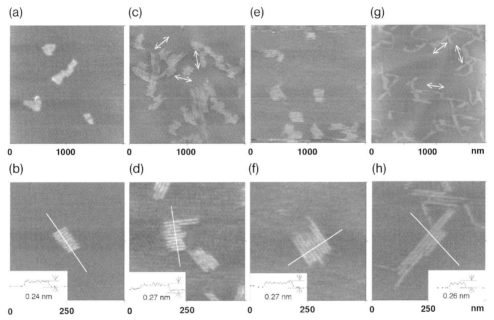

Scheme 5.3 Chemical structure of poly(γ-L-glutamate) derivatives **6**, where *n* is defined as the number of methylene units in the alkyl group.

Figure 5.11 Typical AFM images and cross-sectional profiles (insets) of polymers **6** adsorbed onto HOPG substrate. (a, b) **6A**; (c, d) **6B**; (e, f) **6C**; (g, h) **6D**. The Z-scale of the images is constant at 4 nm. Cross-sectional profiles of the island structures are shown in the insets of the images. The arrows indicated the running directions of the rods in the 2-D array, and the directions corresponded to $\langle \bar{1}2\bar{1}0 \rangle$ directions of the graphite basal plane. In whole images, the entire surface is uniformly covered by bright islands, representing the polymer adsorbates, and the dark portion surrounding islands is bare graphite surface.

5.2.3.1 Direct Visualization of 2-D Self-Organized Array by AFM

Figure 5.11 shows a collection of high-resolution island images for the **6** series, which all revealed a similar island structure. The island-like arrays consisted of rods, which possessed an entirely straight conformation in a parallel arrangement. A single rod, which was separated from the islands, was not found at all. It is important to emphasize here that the corrugations of the islands were constant for each of the members within the **6** series; this suggests that the alkyl chains on **6** were flat and contacted graphite in the same manner. It also indicates that the polymers adsorbed on the HOPG were not in an aggregated form, but rather in a monolayer form. The stripped patterns in the arrays were triangular in arrangement, with the running stripes directly rotated 60° from each other (see Figure 5.11c and g, where the marked arrows indicate the running direction of the stripes in the islands). This clearly indicated that the formation of the island structures is by epitaxial adsorption.

5.2.3.2 Orientation in 2-D Self-Organized Array

The orientation of the adsorbed structure was estimated by the relationship between the alignment of the alkyl chain and the graphite substrate. Figure 5.12 shows the schematic representation of the structural model proposed for **6**. In the array model, the intervals of the rods correspond to intermolecular distances between flat-oriented adjacent polymers. Extended alkyl side chains with the all-*trans* conformation are divided to both sides of the helical main chain, and align perpendicular to the main chain. The driving force for the formation of the array is predominantly an epitaxial interaction between the alkyl side chains and the graphite surface. The rods of the **6** polymers run in the $\langle \bar{1}2\bar{1}0 \rangle$ direction. When the rod model was superimposed on a graphite lattice, the alkyl side chains, which were angled at 90° to the direction of the rods, were aligned in the $\langle 10\bar{1}0 \rangle$ direction. It is well known that alkyl chains adsorb epitaxially onto HOPG, and that the alkyl chains are aligned in the $\langle 10\bar{1}0 \rangle$ direction [35], clearly showing the accuracy of the proposed model and indicating that the formation of the array is predominantly due to epitaxial adsorption of the side groups.

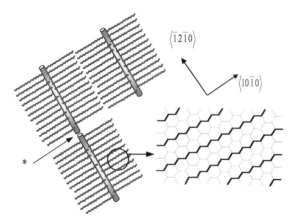

Figure 5.12 Schematic representation of the proposed structural models for the 2-D array, displaying the rod structure and the epitaxial alignment of alkyl side chains on the HOPG substrate of **6**.

5.2.3.3 Intermolecular Weak van der Waals Interactions in 2-D Self-Organized Arrays

Although the interactions between the alkanes and graphite are stronger than the intermolecular interactions, the lateral intermolecular interactions are important in the formation of an ordered 2-D structure [36]. In addition to the epitaxial interaction between the alkyl side chains and the graphite surface, the formation of an island structure composed of rods suggests the existence of the attractive intermolecular interactions, including van der Waals interactions, between the alkyl chains of adjacent rods. Alkyl chains have also been reported to align "side by side" with parallel or perpendicular conformations at the surface [36, 37]. In the case of the **6** series, the interval between the alkyl chains is very close to that reported for a flat conformation. From the model, it was expected that the length of the rods was attributable to the length of each helical polymer. However, the observed lengths of the strings were regularly several times longer than expected, and often overestimated even by gel permeation chromatography (GPC). For example, the observed lengths of the rods for **6B** (M_w = 37 900; M_w/M_n = 1.1; expected length =15 nm) and **6C** (56 100; 1.2; 22 nm) were 27–44 and 17–98 nm, respectively, indicating that each rod consisted of several polymers to form "head-to-head" or "head-to-tail" structures, shown by the arrow in Figure 5.11. The inter-polymer "side-by-side" arrangement is due to the intermolecular interactions between the alkyl chains of adjacent rods. Although the average degree of polymerization for **6D** was almost the same as that of **6C** (Figure 5.1e), the rods of **6D** (Figure 5.1g) were longer than those of other **6**s. This observation was due not to differences in the molecular masses of the polymers, but rather to the enhancement of intermolecular interactions with increasing alkyl chain length. More interestingly, the rods seem to have a tendency to align with the edges of rods. The islands in Figure 5.11b, d and f are typical examples, with relatively ordered rod ends. This suggests that attractive lateral intermolecular interactions exist between rods at the surface.

5.2.3.4 Comparison of Structures between a 2-D Self-Organized Array and 3-D Bulk Phase

The interval distances between the parallel rods depend on the length of the alkyl side chains of the polymers. Figure 5.13 shows plots of the intervals between the parallel rods against the number of alkyl side chains. The intervals increase linearly with the length of the alkyl side chains, and from the slope of the linear plot the length per methylene unit is approximately 2.2 Å; this corresponds to increments of an extended alkyl chain with all-*trans* conformations, and indicates that the alkyl side chains are perpendicular to the helical main chain. The extrapolated value of the interval to $n = 1$, which should correspond to the interval for poly(γ-methyl-L-glutamates) (PMLG, $n = 1$), was 19.9 Å. This value is appropriate for the diameter of PMLG, and validates the model subsequently proposed. In Figure 5.13, the results of the bulk crystal analysis were plotted with those of the 2-D array. The slope for the PG*n* bulk crystal, in which the side chains were crystallized, was approximately 1.3 Å/CH_2 (a methylene unit), which is almost half the

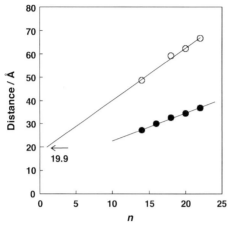

Figure 5.13 Variations in the interval distances (○) between neighboring rods in the 2-D array and the corresponding distances (●) between the center of **6** in the bulk crystal, which was revealed through (100) reflections.

value of that for the 2-D crystal. This difference is due to the interdigitating structure of the bulk crystal.

The 2-D array of **6** on HOPG forms under submonolayer coverage conditions. All results, including the correlation between the interval of rods and alkyl chain lengths, clearly proved the proposed model for the epitaxial adsorption of **6**. Furthermore, epitaxial helical polymer–substrate and polymer–polymer interactions were also found which can lead to the formation of controllable architectures. In particular, the existence of some of the intermolecular interactions between the end-termini of polymers was quite interesting and important for the design of 1-D supramolecular architectures.

5.2.4
Summary of Helical Polymer-Based 1-D and 2-D Architectures

In this section, attention will be focused on the formation of 1-D and 2-D self-organized structures composed of helical polymers. From Section 5.2.1, the proper adsorbate–substrate interaction, which should be neither too weak nor too strong, enables the detection of the topology switching and 2-D self-organization. The control of chain length leads to the construction of various 1-D helical architectures onto surfaces; these have the rigid characteristics of helical polymers, which are unique and advantageous for constructing various hierarchical systems from 1-D to 3-D. In Section 5.2.2, the rigidity of a single polymer chain and the mesoscopic hierarchical helical superstructures in the polysilane aggregates were observed, using AFM. The dominant driving force for the formation of right- and left-handed superhelical structures is based on homochiral weak intermolecular interactions.

Furthermore, it should be noted that a helical polymer chain that is much longer than the persistence length could easily form the entangled structure, consequently leading to the formation of mesoscopic highly ordered superhelical structures. In Section 5.2.3, in addition to adsorbate–substrate interactions, the interactions of lateral (alkyl side chain) and longitudinal (end-termini of main chain) directions were shown to also be required for 2-D self-organization.

It should be emphasized here that both the polymer–substrate and polymer–polymer interactions can lead to the formation of controllable architectures. Furthermore, these interactions are not exceptionally strong, but they are sufficient to form well-organized structures.

In the following section, as an example of functional 3-D architecture, attention will be focused on a unique optical activity based on a helical conformation, which has potential use for functional materials in the solid state.

5.3
Helical Polymer-Based Functional Films

Previously, we have described the formation of 1-D and 2-D architectures composed of helical polymers. These well-defined helical, polymer-based supramolecular architectures can be designed and controlled, except for some limitations, and therefore expanding these to the life sciences and/or materials sciences applications based on optical activity remains a major challenge [1b]. As noted above, synthetic helical polymers exhibit many unique phenomena, including chiroptical memory, amplification [38], formation of thermotropic cholesteric liquid crystals [12, 39], molecular chirality recognition [40], and helix–helix (*PM*) transition [30, 32, 41–49]. The *PM* transition phenomenon, which involves the reversible switching between the *P*-(plus, right-handed) and *M*-(minus, left-handed) screw-sense segments along the helical backbone, is especially promising for chiroptical materials [30, 32, 38, 40–49]. The *PM* transition, driven by external stimuli such as temperature [30, 32, 45, 49], photoradiation [46] and additives [40j], is currently understood to be one of the general characteristics of helical polymers.

Molecule-based chiroptical properties, such as memory and the helicity switch using the *PM* transition phenomenon in helical polymers, will be useful for data storage, optical devices, chromatographic chiral separation and liquid crystals for display [50]. In addition to memory and the helicity switch using the *PM* transition phenomenon, chirality transfer and/or amplification in polymer systems have also been studied extensively in aggregates, while the complexation of achiral polymers with chiral additives has been investigated in solution systems [40j, 51, 52].

In this section, attention is focused on the fabrication of helical polymer-based solid films as supramolecular assemblies that possess functionality based on chiroptical properties, such as switch [30, 53], memory [30], transfer, and amplification [54]. A focus on these chiroptical properties, either in bulk thin film or in a polymer matrix, is required from a practical viewpoint [48, 55]. In the present system, a temperature control is used as the "trigger" to realize the memory, switch, and chirality transfer/amplification. Those chiroptical properties can be detected by using circular polarized light (CPL), which is able to distinguish between right- or left-handed polarity (helicity), thus highlighting their potential application as circular polarization recording/erasing systems.

5.3.1
Chiroptical Memory and Switch in Helical Polysilane Films

The *PM* transition phenomenon in helical polymer films is primarily applied for molecular-based chiroptical memory and switches. In general, there are two types of memorizing system which function by means of a *PM* transition phenomenon in the solid state. One system is in the Re-Writable (RW) mode, while the other is in the Write-Once Read-Many (WORM) mode. In the RW mode, the right- or left-handed helicity can be erased when data are renewed. In contrast, in the WORM mode the right- or left-handed helicity is immobilized once for the storage of data, and these are then read (detected) using CPL.

The basic requirement for a switching system is the bistable state of a molecule, which can be assigned as either the chiroptical inversion "−1 and +1" switch through the helix–helix (*PM*) transition or the on-off "0 and +1" switch through the helix–coil transition. Normally, these chiroptical properties can be characterized by using circular dichroism (CD). The conformational transition temperature (T_c), such as for the helix–helix or helix–coil transition, is the critical factor when designing the chiroptical memory and switch in the solid state. Since below T_c the segmental motion of the polymer chain is frozen, polymers having relatively low T_c values may represent good candidates for practical application.

Polysilanes possessing relevant enantiopure chiral side chains will be successful candidates for chiroptical memory and switches at the molecular level. Furthermore, the T_c of polysilanes in the solid state can easily be designed by adjusting the molecular mass.

In this section, we present some examples of easy, versatile approaches for chiroptical memory with RW and WORM modes, and chiroptical switches based on the inversion "−1" and "+1" and on-off "0" and "+1" switch in solid films. Certain polysilanes, including poly{(S)-3,7-dimethyloctyl-3-methylpentylsilane} (2) and poly{(S)-3,7-dimethyloctyl-*n*-propylsilane} (7), were used, each of which can undergo helix–helix or helix–coil transitions at a certain temperature in dilute solution states by controlling their molecular mass and/or thermal modulation (Scheme 5.4).

5.3.1.1 Memory with Re-Writable Mode and Inversion "−1" and "+1" Switch

Polymers **2** and **3** exhibited a strong molecular mass dependence on the phase transition temperature T_c (estimated using differential scanning calorimetry; DSC) due to the *PM* transition, as shown in Figure 5.14. The enthalpy change (ΔH) estimated from the heating was only approximately 0.05 kcal mol^{-1} per repeat unit [15a]. Given that the helix–helix transition of polysilane is caused by the order–disorder transition of side chain groups in solution, the entropy gain based on the

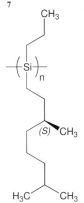

Scheme 5.4 Chemical structures of optically active poly{(S)-3,7-dimethyloctyl-n-propylsilane} **7**, which can undergo a helix–coil transition at 80°C ($\Delta\varepsilon/\varepsilon = 0$ at 80°C) in iso-octane.

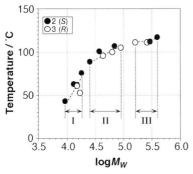

Figure 5.14 Change of transition temperature (T_c) estimated from DSC thermograms as a logarithmic function of molecular mass. Filled and open circles indicate **2** and **3**, respectively. Regions I, II, and III indicate reversible, irreversible, and no change of CD signals for the cast films in the heating and cooling process, respectively.

order–disorder transition of side chains could easily overcome the tiny amount of enthalpy, thus paving the way for the induction of a helix–coil transition, even in the solid state.

For the cast film of **2** onto quartz substrate comprising the low-molecular-mass fraction (M_w: 1.3×10^4, M_w/M_n: 1.16, corresponding to region I in Figure 5.14), a typical bisignate CD signal based on exciton coupling was clearly observed, and seen reversibly to switch between almost mirror-imaged CD spectra in the heating–cooling cycles (see Figure 5.15).

In this case, T_c was approximately 47°C, which was estimated from the temperature dependence of Kuhn's dissymmetry ratio ($g_{solid} = \Delta OD/OD$), where OD is the optical density of the UV absorbance at λ_{max} and $\Delta OD/OD$ is defined as $2(OD_L - OD_R)/(OD_L + OD_R)$ of the CD intensity at the extremum (λ_{ext}) of the first Cotton band, as shown in Figure 5.16. The transition temperature estimated from CD almost corresponded to that estimated by DSC.

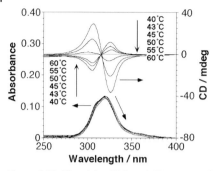

Figure 5.15 Ultraviolet (UV) and CD spectra of the solid film of **2** in the heating process.

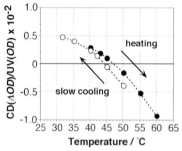

Figure 5.16 Temperature dependence of relative Kuhn's dissymmetry ratios $\Delta OD/OD$ for the cast film of **2** in the heating (●) and cooling (○) processes. $M_w = 1.3 \times 10^4$, $M_w/M_n = 1.16$. The heating rate was 20 °C min^{-1}, and the cooling rate 2 °C min^{-1}.

In this case, the phase transition temperature in the film state seemed to be approximately 60 °C higher than the *PM* transition temperature in iso-octane. A higher transition energy might be required to induce the phase transition in the solid state, and therefore the phase transition would originate from a macroscopic geometric change, such as a twisted superhelical structure, among the polymeric backbones based on exciton coupling in the films [1b, 56]. The reversible switch occurs between the *laevo* and *dextro* helical polymer shapes, which is followed by changes in the helical supramolecular geometry, as observed in the CD spectrum and illustrated in Figure 5.17. The molecular mass control alone is sufficient to change the transition temperature of **2** in region I of Figure 5.14. Moreover, the chiroptical property of the inversion "−1" and "+1" switch can make it possible to obtain chiroptical memory with RW mode.

Chiroptical memory with RW mode can be achieved by controlling the cooling rate of the film with a low molecular mass. Figure 5.18 shows the multiple thermal cycle responses of the CD intensities at the extrema for the cast film of **2**. Cycling

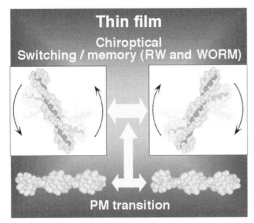

Figure 5.17 Schematic representation of helical supramolecular geometry switch on the basis of the helix–helix transition of **2**.

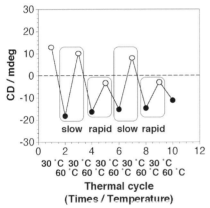

Figure 5.18 Thermal cycle responses of extrema of CD intensities for the solid film of **3**. Thermal cycles were conducted by heating to 60 °C (●) and followed by slow (○) or rapid (shaded circle) cooling to 30 °C. Rapid cooling was achieved by dipping the film into the ice water. The heating rate was 20 °C min^{-1}, and the slow cooling rate 2 °C min^{-1}. $M_w = 1.3 \times 10^4$, $M_w/M_n = 1.16$.

was conducted between by heating above the T_c and by slow cooling or rapid quenching below the T_c. In the slow cooling process, the sign of the CD signal changed, indicating a chiroptical inversion "−1" and "+1" switch. On the other hand, a chiroptical memory with RW mode occurred during the rapid quenching from above the T_c. However, the sign of the CD signal did not change during the rapid quenching. This memory effect was resettable by heating to above the T_c. The chiroptical inversion "−1" and "+1" switch and chiroptical memory with RW mode are feasible by controlling both the cooling conditions in the solid film and the molecular mass of **2**.

5.3.1.2 Memory with Write-Once Read-Many (WORM) Mode

Another memory state, the WORM mode, can be achieved by controlling the molecular mass only, although in this case the management of cooling conditions is not required. With regards to the cast films in region II in Figure 5.14, the phase transition was seen only on heating, and the state above the T_c remained unchanged during the cooling process. This irreversible transition behavior can be confirmed by the g_{solid}-value (defined as $\Delta OD/OD$), as shown in Figure 5.19. This irreversible change in the CD signal indicates the nonerasable memory as the WORM mode. It should be noted that the transition temperature of the WORM mode in region II of Figure 5.14 is also adjustable by controlling the molecular mass.

The dynamic chiroptical properties of polysilanes in the solid state, such as inversion switching and memory with RW and WORM modes, can be managed by a combination of molecular mass control and thermal modulation.

5.3.1.3 On-Off "0" and "+1" Switch Based on Helix–Coil Transition

As noted above, the basic requirement for a switching system is the bistable state of a molecule, which can be rapidly interconverted by an external stimulus, even in the solid state. In addition to the achievement of a chiroptical inversion "−1 and +1" switch in polymer **2** and **3** solid films by means of molecular mass control and thermal modulation, an on-off "0 and +1" switch, based on changes in the magnitude of the helicity in polymer **7** film, can also be achieved by thermal modulation. The DSC thermogram of **7** displayed a second-order transition at −47 °C and a first-order transition at −8 °C corresponding to the glass and helix–coil transitions, respectively, as shown in Figure 5.20. The heat flow (ΔH is ~0.004 kcal mol^{-1} per repeat unit) for the helix–coil transition in **7** was relatively small, with only weak van der Waals interactions likely to permit the helix–coil transition, even in the solid state (this is the same as the previously discussed case in Section 5.3.1.1).

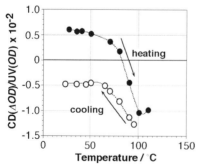

Figure 5.19 Temperature dependence of relative Kuhn's dissymmetry ratios $\Delta OD/OD$ for the cast film of **2** in the heating (●) and cooling (○) processes. $M_w = 2.5 \times 10^4$, $M_w/M_n = 1.25$.

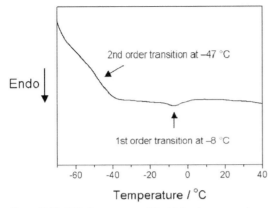

Figure 5.20 DSC thermogram of **7** obtained on second heating with a heating rate of 10 °C min^{-1}.

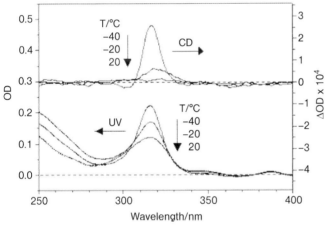

Figure 5.21 Ultraviolet (UV) and CD spectra of the solid film of **7**. $M_w = 330\,000$, $M_w/M_n = 1.87$.

Figure 5.21 shows the UV and CD spectra of the **7** film cast from iso-octane onto a quartz substrate. Although there was no detectable CD absorption at 20 °C, **7** showed a strong Cotton effect at −40 °C, indicating a coil-to-helix transition of **7** in the solid state with decreasing temperature. This transition can be well correlated with DSC results. The positive Cotton effect could result from a single polymer chain because of the similarity between the UV and CD absorption profiles. Furthermore, no bisignate CD signal attributable to exciton coupling was observed, even at −60 °C, which is completely different from the cases of the previous Sections 5.3.1.1 and 5.3.1.2. Although **7** showed a strong Cotton effect at low temperatures both in solution and in the solid state, the global conformations of polymer chains in both states may be significantly different. Due to the excluded volume of other chains and the chain entanglements, the entire conformation of the polymer chains in the solid state would be difficult to change with decreasing temperature, even above their glass transition temperature (T_g). These results clearly indicate that the macroscopic geometric changes in the inversion "−1" and "+1" switch and the on-off "0" and "+1" switch are different.

The helix–coil transition in the thin solid film can be explained by some parameters, such as *fwhm*, λ_{max} and $\Delta OD/OD$, as shown in Figure 5.22. The *fwhm* of the UV absorption band became narrower when the temperature decreased from 20 to −40 °C and remained constant below −40 °C (around T_g), indicating that the homogeneity of the polymer backbone was induced due to the decreasing degree of freedom in the Si–Si bond rotations. The λ_{max} shifted slightly to a longer wavelength with decreasing temperature from 20 to −20 °C, and shifted to a shorter wavelength below −20 °C. It is likely that the chiroptical property in the solid state would be induced through the movement and/or disappearance of the helical reversal within the localized segments, which is in stark contrast to that of the solution state.

Figure 5.23 shows the multiple thermal cycle responses of $\Delta OD/OD$ in the range from −20 to 20 °C. The intensity changed during heating and cooling, indicating a switching property. The helix–coil transition in the solid state is highly reversible, can be applicable to the on-off switching system, and is assigned as "0 and +1".

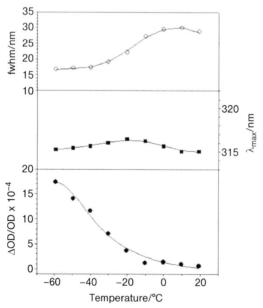

Figure 5.22 Temperature dependence of relative Kuhn's dissymmetry ratios $\Delta OD/OD$, λ_{max}, and *fwhm* of the solid film of 7. $M_w = 330\,000$, $M_w/M_n = 1.87$.

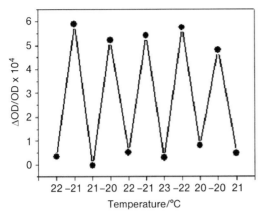

Figure 5.23 Thermal cycle responses of relative Kuhn's dissymmetry ratios $\Delta OD/OD$ of the solid film of **7**. Thermal cycles were conducted by heating to 20 °C, followed by cooling to −20 °C.

5.3.2
Chiroptical Transfer and Amplification in Binary Helical Polysilane Films

In addition to controlling the chiroptical properties such as memory and switch, chiroptical transfer and amplification have also been achieved based on the concept of the "sergeants and soldiers" principle [57] in binary helical polysilane films.

Optically active poly{n-decyl-(S)-2-methylbutylsilane} (as "sergeants", **5**: $M_w = 2.11 \times 10^5$, $M_w/M_n = 1.77$) with an almost 7_3 helix and optically inactive poly(n-decyl-3-methylbutyl-silane) (as "soldiers", **8**: $M_w = 1.11 \times 10^5$, $M_w/M_n = 1.75$) and poly(n-decyl-i-butylsilane) (as "soldiers", **9**: $M_w = 3.36 \times 10^4$, $M_w/M_n = 1.17$) were used for this study, as shown in Scheme 5.5. Polymers **8** and **9** have equal populations of right and left 7_3 helices; however, the rigidities of **5**, **8**, and **9** are different. The polymers **5** and **9**, with β-branching methyl groups in the side chains, were classified into rigid rod-like polymers with a persistent length of about 70 nm, while **8** belonged to a semi-rigid polymer with a persistent length of only 6 nm [31].

The CD spectra of a grafted **5**-spin-coated **8** bilayer system prepared on quartz substrate were measured before and after annealing, as shown in Figure 5.24. The CD spectra of both the grafted **5**-spin-coated **8** bilayer system and a grafted **5** system alone showed a bisignate CD signal with a positive Cotton band at about 309 nm and a negative Cotton band at about 324 nm, as shown in Figure 5.24. For the grafted **5**-spin-coated **8** bilayer system, both positive- and negative-Cotton CD intensity were greatly increased after thermal annealing. The enhancement in the CD signal intensities was a result of a chiroptical transfer to spin-coated **8** either from the chiral side chain of grafted **5** or the polymer backbone itself, and it was drastically amplified after annealing. On the other hand, no significant changes

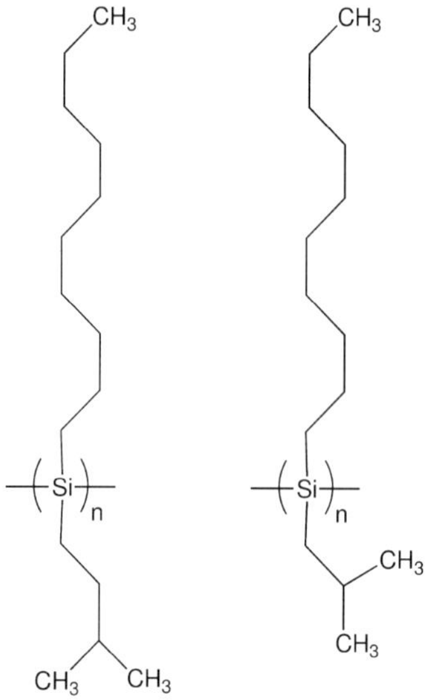

Scheme 5.5 Chemical structures of optically inactive poly(n-decyl-3-methylbutylsilane) (as "soldiers", **8**) and poly(n-decyl-i-butylsilane) (as "soldiers", **9**).

of CD signal intensities were observed for only immobilized grafted **5** after annealing. In Figure 5.24, a decrease of the UV absorbance of the grafted **5**-spin-coated **8** bilayer system was observed after the annealing treatment. This result implies that most of the polymer chains lie down before annealing in the quartz substrate plane, as the film was prepared by a spin-coating technique. The thermal annealing treatment may impact some of the polymer chain segments of semi-rigid spin-coated **8** in the perpendicular orientation and/or tilt of the substrate plane, leading to the decrease of the apparent UV absorbance at 321 nm, as illustrated in Figure 5.25 [58].

Figure 5.26 shows the change in the relative Kuhn's dissymmetry ratios ($g_{solid} = \Delta OD/OD$), after (g_{AA}) and before (g_{BA}) annealing with the varied film thickness. As was evident, the ratios increased with the decrease in the thickness of **8**. The magnitude of the g_{AA}/g_{BA} value was found to be significantly higher in grafted **5** (filled squares) than in spin-coated **5** (filled circles), presumably due to the partial penetration of spin-coated **5** polymer chains into the surface during the spin-coating process of **8**, leading to the similar solvent solubility behaviors of **5** and **8**. A tiny amount of immobilized, optically active **5** could induce and effectively amplify the optical activity in the optically inactive **8** layer by thermal treatment. Weak van der Waals interactions at the surface between these two polymers might be responsible for the transfer and amplification of the optical activity in **8**. The positional segmental movement of polymer chains in **8** (melting point 40 °C, as estimated by DSC) would easily occur, which would be responsible for the helical

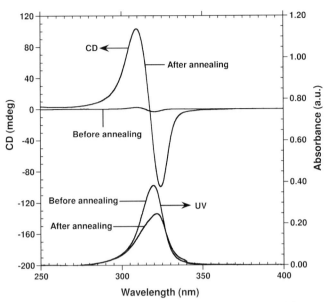

Figure 5.24 Circular dichroism (CD) and UV spectra of grafted **5**-spin-coated **8** bilayer system onto quartz substrate before and after annealing. The initial UV absorbance of **8** was 0.04, and total UV absorbance of grafted **5**-spin-coated **8** bilayer system was 0.39. In this case, the spectra do not significantly affect the spectral intensity with rotation of the solid films at different angles.

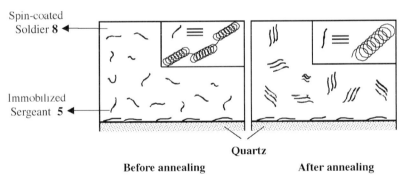

Figure 5.25 Schematic representation of thermo-driven chiroptical transfer and amplification in the "soldier" optically inactive polysilane **8** from the "sergeant" optically active helical polysilane **5** on quartz.

transfer from **5** to **8** in the solid film. This could be regarded as a "sergeants and soldiers" system in the solid state. In the case of the **5**-spin-coated **9** bilayer system, the optical activity transfer phenomenon was not seen, due to the higher transition energy barrier (no transition was observed by DSC) originating from the rigid rod-like conformation by the help of β-branched side chain.

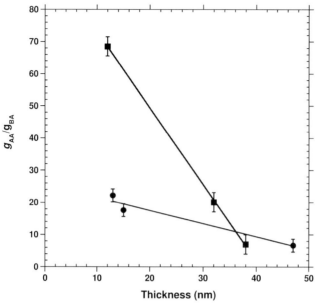

Figure 5.26 Relative Kuhn's dissymmetry ratios $\Delta OD/OD$ of a grafted **8**-spin-coated **9** bilayer system (■) and spin-coated **8**-spin-coated **9** (●) on a quartz surface before (g_{BA}) and after (g_{AA}) annealing, as a function of the film thickness.

5.3.3
Summary of Helical Polymer-Based Functional Films

In this section, attention is focused on the fabrication of helical polymer-based solid films as supramolecular assemblies that possess functionality based on chiroptical properties, such as switch, memory, transfer, and amplification. In Section 5.3.1, simple molecular mass control and thermal modulation were shown to be capable of realizing various chiroptical properties, such as the inversion "−1" and "+1" switch and memory with RW and WORM modes. Furthermore, different types of polysilane were used to construct on-off "0" and "+1" switching. For practical use of those systems, without the threat of polymer degradation by UV light for example, a laser-assisted heat-mode RW system which is based on a dye dispersed in the switchable polysilanes may be feasible, provided that the applied wavelengths of the dye and laser are much longer than those of the polysilanes. A high-power, circularly polarized laser-diode would be effective to directly heat the dye, leading to an indirect heating of the polysilane film above the T_c. Controlling the cooling rate of the film would enable the switching or memorizing of the chiroptical state above the T_c. In order to access data from the polysilane film, a controlled modulation of the laser to the dye, including the power, pulse irradiation time, and periods, would be effective to erase with a heat mode and to read with a photon mode.

In Section 5.3.2, the optical activity of rigid rod-like polysilane was discussed, both in terms of grafted and spin-coated films, with a preferential helical sense, which was transferred and greatly amplified to an optically inactive semi-rigid, helical spin-coated polysilane layer at the solid surface by thermal annealing. Selection of the correct rigidity of the optically inactive helical polysilane was important for the emergence of a helical polymer command surface of the "sergeants and soldiers" type, a chiroptical amplification system in the solid film states.

It should be emphasized here that the weak van der Waals interactions between polymers are key to the assembly of various chiroptical systems that possess switch, memory, transfer, and amplification properties. Optically active polysilane is unique in natural and synthetic helical polymers, and allows for the possibility of creating a wide variety of supramolecular architectures with a range of chiroptical properties.

Acknowledgments

The authors thank Professors M. M. Green (Polytechnic University), J. Michl (University of Colorado), R. West (University of Wisconsin), K. Tamao (Kyoto University), G. Basu (Bose Institute), R. G. Jones (University of Kent), T. M. Swager (Massachusetts Institute of Technology), Y. Kawakami (JAIST), A. Teramoto (Ritsumeikan University), T. Sato (Osaka University), and E. Yashima (Nagoya University) for collaborations and fruitful discussions on the consideration of optical properties of polysilanes in solution and in the solid state. They also thank Professors J. Watanabe and S. Kawauchi of the Tokyo Institute of Technology for collaborations and discussions on the structural properties of poly(γ-L-glutamates). Thanks are also extended to T. Hagihara, and F. Asanoma for technical assistance and discussion, and to Doctors T. Imase, S.-Y. Kim, A. Saxena, K. Okoshi, M. Naito, G. Kwak, M. Ishikawa, T. Kawabe, G.-Q. Guo, and Y.-G. Yang for fruitful discussions.

A.O., M.F., and M.K. are supported by grants from CREST-JST, "Syntheses of the Cooperative Hyper Helical Polymers and Understanding of Structure-Property-Functionality Relationship" and "Nanoelectronic-Device Fabrication Based on the Fine Molecular Design." A.O. is also supported by grants from the Ministry of Economy, Trade and Industry, Japan. M.F. is supported in part by grants from the Ministry of Education, Science, Sports, and Culture of Japan and by a Grant-in-Aid for Scientific Research "Experimental Test of Parity Nonconservation at Helical Polymer Level" (No. 16655046) and "Design, Synthesis, Novel Functionality of Nanocircle and Nanorod Conjugating Macromolecules" (No. 16205017).

References

1. (a) Lehn, J.-M. (1995) *Supramolecular Chemistry*, VCH Verlag GmbH, Weinheim.
(b) Green, M.M., Nolte, R.J.M. and Meijer, E.W. (2003) *Materials Chirality: Topics in Stereochemistry*, John Wiley & Sons, Inc., New York.
2. (a) Pauling, L. and Corey, R.B. (1953) *Nature*, **4341**, 59.
(b) Eyre, D.R. (1980) *Science*, **207**, 1315.
(c) Dickerson, R.E., Drew, H.R., Conner, B.N., Wing, R.M., Fratini, A.V. and Kopka, M.L. (1982) *Science*, **216**, 475–485.
(d) Watanabe, J. and Ono, H. (1986) *Macromolecules*, **19**, 1079.
3. (a) Sommerdijk, N.A.J.M., Lambermon, M.H.L., Feiters, M.C., Nolte, R.J.M. and Zwanenburg, B. (1997) *Chem. Commun.*, 1423.
(b) Nakashima, N., Asakuma, S. and Kunitake, T. (1985) *J. Am. Chem. Soc.*, **107**, 509.
(c) Sommerdijk, N.A.J.M., Lambermon, M.H.L., Feiters, M.C., Nolte, R.J.M. and Zwanenburg, B. (1997) *Chem. Commun.*, 455.
(d) Yanagawa, H., Ogawa, Y., Furuta, H. and Tsuno, K. (1989) *J. Am. Chem. Soc.*, **114**, 3414.
(e) Fuhrhop, J.H. and Helfrich, W. (1993) *Chem. Rev.*, **93**, 1565, and references therein.
4. Rai, R., Saxena, A., Ohira, A. and Fujiki, M. (2005) *Langmuir*, **21**, 3957.
5. (a) Kunitake, T. (1992) *Angew. Chem. Int. Ed. Engl.*, **31**, 709.
(b) Fuhrhop, J.H. and Helfrich, W. (1993) *Chem. Rev.*, **93**, 1565.
(c) Schunr, J.M. (1993) *Science*, **262**, 1669.
(d) Engelkamp, H., Middlebeek, S. and Nolte, R.J.M. (1999) *Science*, **284**, 785.
(e) Prins, L.J., Huskens, J., de Jong, F., Timmerman, P. and Reinhoudt, D.N. (1999) *Nature*, **398**, 498.
(f) Hirschberg, J.H.K.K., Brunsveld, L., Vekemans, J.A.J.M. and Meijer, E.W. (2000) *Nature*, **407**, 167.
(g) Volker, B., Huc, I., Khoury, R.G., Krische, M.J. and Lehn, J.-M. (2000) *Nature*, **407**, 720.
(h) Brunsveld, L., Meijer, E.W., Prince, R.B. and Moore, J.S. (2001) *J. Am. Chem. Soc.*, **123**, 7978.
(i) Koga, T., Matsuoka, M. and Higashi, N. (2005) *J. Am. Chem. Soc.*, **127**, 17596.
(j) Lohr, A., Lysetska, M. and Würthner, F. (2006) *Angew. Chem. Int. Ed. Engl.*, **44**, 5071.
(k) Ajayaghosh, A., Varghese, R., Jacob, S. and Vijayakumar, C. (2006) *Angew. Chem. Int. Ed. Engl.*, **45**, 1141.
6. (a) Ikeda, Y., Jmshidi, K., Tsuji, H. and Hyon, S.-H. (1987) *Macromolecules*, **20**, 906.
(b) Brizzolara, D., Cantow, H.-J., Diederichs, K., Keller, E. and Domb, A.J. (1996) *Macromolecules*, **29**, 191.
7. Akagi, K., Piao, G., Kaneko, S., Sakamaki, K., Shirakawa, H. and Kyotani, M. (1998) *Science*, **282**, 1683.
8. Cornelissen, J.J.L.M., Fischer, M., Sommerdijk, N.A.J.M. and Nolte, R.J.M. (1998) *Science*, **280**, 1427.
9. (a) Maeda, K., Takeyama, Y., Sakajiri, K. and Yashima, E. (2004) *J. Am. Chem. Soc.*, **126**, 16284.
(b) Kajitani, T., Okoshi, K., Sakurai, S., Kumaki, J. and Yashima, E. (2006) *J. Am. Chem. Soc.*, **128**, 708.
10. Yashima, E., Maeda, K. and Furusho, Y. (2008) *Acc. Chem. Res.*, **41**, 1166 and references therein.
11. Kaufman, H.S., Sacher, S., Alfrey, T. and Frakuchen, I. (1948) *J. Am. Chem. Soc.*, **70**, 3147.
12. (a) Watanabe, J., Fukuda, Y., Gehani, R. and Uematsu, I. (1984) *Macromolecules*, **17**, 1004.
(b) Watanabe, J., Ono, H., Uematsu, I. and Abe, A. (1985) *Macromolecules*, **18**, 2141.
(c) Watanabe, J. and Ono, H. (1986) *Macromolecules*, **19**, 1079.
(d) Watanabe, J., Goto, M. and Nagase, T. (1987) *Macromolecules*, **20**, 298.
(e) Watanabe, J. and Nagase, T. (1987) *Polym. J.*, **19**, 781.
(f) Watanabe, J. and Nagase, T. (1987) *Macromolecules*, **21**, 171.
13. (a) Miller, R.D. and Michl, J. (1989) *Chem. Rev.*, **89**, 1359.

(b) Michl, J. and West, R. (2000) In *Silicon-Containing Polymers: The Science and Technology of Their Synthesis and Applications* (eds R.G. Jones, W. Ando and J. Chojnowski), Kluwer, Dordrecht, p. 499.

14 (a) Fujiki, M. (1994) *J. Am. Chem. Soc.*, **116**, 6017.
(b) Fujiki, M. (1996) *J. Am. Chem. Soc.*, **118**, 7424.
(c) Fujiki, M. (2001) *Macromol. Rapid Commun.*, **22**, 539.
(d) Fujiki, M.J. (2003) *Organomet. Chem.*, **685**, 15.

15 (a) Yuan, C.-H. and West, R. (1994) *Macromolecules*, **27**, 629.
(b) Frey, H., Möller, M., Turetskii, A., Lotz, B. and Matyjaszewski, K. (1995) *Macromolecules*, **28**, 5498.
(c) Terao, K., Terao, Y., Teramoto, A., Nakamura, N., Terakawa, I., Sato, T. and Fujiki, M. (2001) *Macromolecules*, **34**, 2682.
(d) Terao, K., Terao, Y., Teramoto, A., Nakamura, N., Fujiki, M. and Sato, T. (2001) *Macromolecules*, **34**, 4519.
(e) Teramoto, A., Terao, K., Terao, Y., Nakamura, N., Sato, T. and Fujiki, M. (2001) *J. Am. Chem. Soc.*, **123**, 12303.
(f) Natsume, T., Wu, L., Sato, T., Terao, K., Teramoto, A. and Fujiki, M. (2001) *Macromolecules*, **34**, 7899.

16 Sato, T., Terao, K., Teramoto, A. and Fujiki, M. (2003) *Polymer*, **44**, 5477.

17 (a) Sheiko, S.S. and Möller, M. (2001) *Chem. Rev.*, **101**, 4099.
(b) Brunsveld, L., Folmer, B.J.B., Meijer, E.W. and Sijbesma, R.P. (2001) *Chem. Rev.*, **101**, 4071.
(c) Ungar, G. and Zeng, X.-B. (2001) *Chem. Rev.*, **101**, 4157.

18 (a) Satrijo, A. and Swager, T.M. (2005) *Macromolecules*, **38**, 4054.
(b) Satrijo, A., Meskers, S.C.J. and Swager, T.M. (2006) *J. Am. Chem. Soc.*, **128**, 9030.
(c) Thomas, S.W. III, Joly, G.D. and Swager, T.M. (2007) *Chem. Rev.*, **107**, 1339.
(d) Maeda, K., Morioka, K. and Yashima, E. (2007) *Macromolecules*, **40**, 1349.

19 (a) Kanno, T., Tanaka, H., Miyoshi, N. and Kawai, T. (2000) *Jpn. J. Appl. Phys.*, **39**, L269–L270.
(b) Hamai, C., Tanaka, H. and Kawai, T. (2000) *J. Phys. Chem.*, **104**, 9894–9897.
(c) Hansma, H.G. (2001) *Annu. Rev. Phys. Chem.*, **52**, 71–92.

20 (a) Ohnishi, S., Hara, M., Furuno, T. and Sasabe, H. (1996) *Jpn. J. Appl. Phys.*, **35**, 6233.
(b) Takeda, S., Ptak, A., Nakamura, C., Miyake, J., Kageshima, M., Jarvis, S.P. and Tokumoto, H. (2001) *Chem. Pharm. Bull.*, **49**, 1512.
(c) Imase, T., Ohira, A., Okoshi, K., Sano, N., Kawauchi, S., Watanabe, J. and Kunitake, M. (2003) *Macromolecules*, **36**, 1865–1869.

21 (a) Nakajima, K., Ikehara, T. and Nishi, T. (1996) *Carbohydr. Polym.*, **30**, 77.
(b) Kibry, A.R., Gunning, A.P. and Morris, V.J. (1996) *Biopolymers*, **38**, 355.
(c) Balnois, E., Stoll, S., Wilkinson, K.J., Buffle, J., Rinaudo, M. and Milas, M. (2000) *Macromolecules*, **33**, 7440.
(d) Camesano, T.A. and Wilkinson, K.J. (2001) *Biomacromolecules*, **2**, 1184.

22 (a) Callroll, D.L., Czerw, R., Teklead, D. and Smith, D.W. Jr. (2000) *Langmuir*, **16**, 3574.
(b) Hasegawa, T., Matsuura, K., Ariga, K. and Kobayashi, K. (2000) *Macromolecules*, **33**, 2772.
(c) Cornelissen, J.J.L.M., Donners, J.J.J.M., de Gelder, R., Graswinckel, W.S., Metselaar, G.A., Rowan, A.E., Sommerdijk, N.A.J.M. and Nolte, R.J.M. (2001) *Science*, **293**, 676.
(d) Nishimura, T., Takatani, K., Sakurai, S., Maeda, K. and Yashima, E. (2002) *Angew. Chem. Int. Ed. Engl.*, **41**, 3602.
(e) Samori, P., Ecker, C., Gössl, I., de Witte, P.A.J., Cornelissen, J.J.L.M., Metselaar, G.A., Otten, M.B.J., Rowan, A.E., Nolte, R.J.M. and Rabe, J.P. (2002) *Macromolecules*, **35**, 5290.
(f) Li, B.S., Cheuk, K.K.L., Yang, D., Lam, J.W.Y., Wan, L.J., Bai, C. and Tang, B.Z. (2003) *Macromolecules*, **36**, 5447.
(g) Kiriy, N., Jähne, E., Adler, H.-J., Schneider, M., Kiriy, A., Gorodyska, G., Minko, S., Jehnichen, D., Simon, P.,

Fokin, A.A. and Stamm, M. (2003) *Nano Lett.*, **3**, 707.
(h) Sakurai, S., Kuroyanagi, K., Morino, K., Kunitake, M. and Yashima, E. (2003) *Macromolecules*, **36**, 9670.

23 (a) Kumaki, J., Nishikawa, Y. and Hashimoto, T. (1996) *J. Am. Chem. Soc.*, **118**, 3321.
(b) Ebihara, K., Koshihara, S., Yoshimoto, M., Maeda, T., Ohnishi, T., Koinuma, H. and Fujiki, M. (1997) *Jpn. J. Appl. Phys.*, **36**, L1211.

24 (a) Shinohara, K., Yasuda, S., Kato, G., Fujita, M. and Shigekawa, H. (2001) *J. Am. Chem. Soc.*, **123**, 3619.
(b) Sakurai, S., Okoshi, K., Kumaki, J. and Yashima, E. (2006) *Angew. Chem. Int. Ed. Engl.*, **45**, 1245.

25 Ohira, A., Kim, S.-Y., Fujiki, M., Kawakami, Y., Naito, M., Kwak, G. and Saxena, A. (2006) *Chem. Commun.*, 2705.

26 Kim, S.-Y., Saxena, A., Kwak, G., Fujiki, M. and Kawakami, Y. (2004) *Chem. Commun.*, 538.

27 Sakurai, S., Ohira, A., Suzuki, Y., Fujito, R., Nishimura, T., Kunitake, M. and Yashima, E. (2004) *J. Polym. Sci. A*, **42**, 4621.

28 Saxena, A., Okoshi, K., Fujiki, M., Naito, M., Guo, G., Hagihara, T. and Ishikawa, M. (2004) *Macromolecules*, **36**, 367.

29 (a) Donley, C.L., Zaumseil, J., Andreasen, J.W., Nielsen, M.M., Sirringhaus, H., Friend, R.H. and Kim, J.-S. (2005) *J. Am. Chem. Soc.*, **127**, 12890.
(b) Mano, T., Kuroda, T., Sanguinetti, S., Ochiai, T., Tateno, T., Kim, J., Noda, T., Kawabe, M., Sakoda, K., Kido, G. and Koguchi, N. (2005) *Nano Lett.*, **5**, 425.

30 (a) Fujiki, M. (2000) *J. Am. Chem. Soc.*, **122**, 3336.
(b) Fujiki, M., Koe, J.R., Montonaga, M., Nakashima, H., Terao, K. and Teramoto, A. (2001) *J. Am. Chem. Soc.*, **123**, 6253; Addition/Correction, (2001), *J. Am. Chem. Soc.*, **123** (35), 8644.

31 Fujiki, M., Koe, J.R., Terao, K., Sato, T., Teramoto, A. and Watanabe, J. (2003) *Polym. J.*, **35**, 297.

32 (a) Ohira, A., Okoshi, K., Fujiki, M., Kunitake, M., Naito, M. and Hagihara, T. (2004) *Adv. Mater.*, **16**, 1645.
(b) Ohira, A., Kunitake, M., Fujiki, M., Naito, M. and Saxena, A. (2004) *Chem. Mater.*, **16**, 3919.

33 Shi, W.-X. and Larson, R.G. (2006) *Nano Lett.*, **6**, 144.

34 (a) Kühnle, A., Linderoth, T.R., Hammer, B. and Besenbacher, A. (2002) *Nature*, **415**, 891. Recently, two-dimensional crystallization of small chiral organic molecules on the surfaces has attracted much attention in the field of "surface chirality". For example, homochiral enantioselective crystallization from racemic mixture has been reported. Kühnle and coworkers have demonstrated that cysteine molecules formed homochiral pairs (D-D and L-L) from racemic mixture by using scanning tunneling microscopy (STM).
(b) Fasel, R., Parschau, M. and Ernst, K.-H. (2003) *Angew. Chem. Int. Ed. Engl.*, **42**, 5178. Chiral transfer from single molecules into self-assembled monolayers on the metal surfaces has been reported by Fasel and coworkers. In that case, enantiopure *P* and *M* chiral single molecules on the surfaces formed the enantiopure self-assembled monolayers with clockwise and anticlockwise, respectively.
(c) Fasel, R., Parschau, M. and Ernst, K.-H. (2006) *Nature*, **439**, 449. In addition to the enantiopure self-assembly, these authors have more recently reported that the enhancement of chiral interactions in two dimensions was observed in the enantioseparation of chiral molecules by STM study These results also suggested that homochiral intermolecular interaction is an important factor to form the 2-D chiral surfaces.

35 (a) Percec, P., Ahn, C.-H., Ungar, G., Yeardley, D.J.P., Moeller, M. and Sheiko, S.S. (1998) *Nature*, **391**, 161–164.
(b) Percec, V., Holerca, M.N., Magonov, S.N., Yeardley, D.J.P., Ungar, G., Duan, H. and Hudson, S.D. (2001) *Biomacromolecules*, **2**, 706–728.

36 Yin, S., Wang, C., Qiu, X., Xu, B. and Bai, C. (2001) *Surf. Interface Anal.*, **32**, 248.

37 Rabe, J.P. and Buchholz, S. (1991) *Science*, **253**, 424.

38 (a) Mayer, S., Maxein, G. and Zentel, R. (1998) *Macromolecules*, **31**, 8522.

(b) Langeveld-Voss, B.M.W., Janssen, R.A.J. and Meijer, E.W. (2000) *J. Mol. Struct.*, **521**, 285.
(c) Nagai, K., Maeda, K., Takeyama, Y., Sakajiri, K. and Yashima, E. (2005) *Macromolecules*, **38**, 5444.
(d) Ousaka, N., Inai, Y. and Kuroda, R. (2008) *J. Am. Chem. Soc.*, **130**, 12266.

39 Okoshi, K., Kamee, H., Suzaki, G., Tokita, M., Fujiki, M. and Watanabe, J. (2002) *Macromolecules*, **35**, 4556.

40 (a) Yashima, E., Matsushima, T. and Okamoto, Y. (1995) *J. Am. Chem. Soc.*, **117**, 11596.
(b) Yashima, E., Nimura, T., Matsushima, T. and Okamoto, Y. (1996) *J. Am. Chem. Soc.*, **118**, 9800.
(c) Yashima, E., Matsushima, T. and Okamoto, Y. (1997) *J. Am. Chem. Soc.*, **119**, 6345.
(d) Yashima, E., Maeda, K. and Okamoto, Y. (1998) *J. Am. Chem. Soc.*, **120**, 8895.
(e) Yashima, E., Maeda, K. and Okamoto, Y. (1999) *Nature*, **399**, 449.
(f) Nakako, H., Mayahara, Y., Nomura, R., Tabata, M. and Masuda, T. (2000) *Macromolecules*, **33**, 3978.
(g) Nakako, H., Nomura, R. and Masuda, T. (2001) *Macromolecules*, **34**, 1496.
(h) Maeda, K., Goto, H. and Yashima, E. (2001) *Macromolecules*, **34**, 1160.
(i) Onouchi, H., Maeda, K. and Yashima, E. (2001) *J. Am. Chem. Soc.*, **123**, 7441.
(j) Yashima, E., Maeda, K. and Sato, O. (2001) *J. Am. Chem. Soc.*, **123**, 8159.
(k) Kumaki, J., Kawauchi, T., Ute, K., Kitayama, T. and Yashima, E. (2008) *J. Am. Chem. Soc.*, **130**, 6373.
(l) Nishino, T. and Umezawa, Y (2008) *Anal. Chem.*, **80**, 6968.

41 (a) Pohl, F.M. and Jovin, T.M. (1972) *J. Mol. Biol.*, **67**, 375.
(b) Pohl, F.M. (1976) *Nature*, **260**, 365.
(c) Pohl, F.M., Thomae, R. and DiCapua, E. (1982) *Nature*, **300**, 545.

42 Toriumi, H., Saso, N., Yasumoto, Y., Sasaki, S. and Uematsu, I. (1979) *Polym. J.*, **11**, 977.

43 (a) Maeda, K. and Okamoto, Y. (1998) *Macromolecules*, **31**, 5164.
(b) Maeda, K. and Okamoto, Y. (1999) *Macromolecules*, **32**, 974.

44 (a) Cheon, K.S., Selinger, J.V. and Green, M.M. (2000) *Angew. Chem. Int. Ed. Engl.*, **39**, 1482.
(b) Tang, K., Green, M.M., Cheon, K.S., Selinger, J.V. and Garetz, B.A. (2003) *J. Am. Chem. Soc.*, **125**, 7313.

45 Schenning, A.P.H.J., Fransen, M. and Meijer, E.W. (2002) *Macromol. Rapid Commun.*, **23**, 265.

46 Maxein, G. and Zentel, R. (1995) *Macromolecules*, **28**, 8438.

47 Nakashima, H., Fujiki, M., Koe, J.R. and Motonaga, M. (2001) *J. Am. Chem. Soc.*, **123**, 1963.

48 Watanabe, J., Okamoto, S., Satoh, K., Sakajiri, K., Furuya, H. and Abe, A. (1996) *Macromolecules*, **29**, 7084.

49 (a) Koe, J.R., Fujiki, M., Nakashima, H. and Motonaga, M. (2000) *Chem. Commun.*, 389.
(b) Fujiki, M. (2001) *Macromol. Rapid Commun.*, **22**, 539.
(c) Fujiki, M., Tang, H.-Z., Motonaga, M., Torimitsu, K., Koe, J.R., Watanabe, J., Sato, T. and Teramoto, A. (2002) *Silicon Chem.*, **1**, 67.

50 Feringa, B.L., van Delden, R.A., Koumura, N. and Geertsema, E.M. (2000) *Chem. Rev.*, **100**, 1789.

51 (a) Okamoto, Y. and Nakano, T. (2001) *Chem. Rev.*, **101**, 4013.
(b) Cornelissen, J.J.L.M., Rowan, A.E., Nolte, R.J.M. and Sommerdijk, N.A.J.M. (2001) *Chem. Rev.*, **101**, 4039.

52 (a) Aoki, T., Kaneko, T., Maruyama, N., Sumi, A., Takahashi, M., Sato, T. and Teraguchi, M. (2003) *J. Am. Chem. Soc.*, **125**, 6346.
(b) Nakano, T., Nakagawa, O., Tsuji, M., Tanikawa, M., Yade, T. and Okamoto, Y. (2004) *Chem. Commun.*, 144.

53 Kim, S.-Y., Fujiki, M., Ohira, A., Kwak, G. and Kawakami, Y. (2004) *Macromolecules*, **37**, 4321.

54 Saxena, A., Guo, G., Fujiki, M., Yang, Y., Ohira, A., Okoshi, K. and Naito, M. (2004) *Macromolecules*, **37**, 3081.

55 (a) Muellers, B.T., Park, J.-W., Brookhart, M.S. and Green, M.M. (2001) *Macromolecules*, **34**, 527.
(b) Mayer, S. and Zentel, R. (2000)

Macromol. Rapid Commun., **21**, 927.
(c) Catellani, M., Luzzati, S., Bertini, F., Bolognesi, A., Lebon, F., Longhi, G., Abbate, S., Famulari, A. and Meille, S.V. (2002) *Chem. Mater.*, **14**, 4819.
(d) Roux, C. and Leclere, M. (1992) *Macromolecules*, **25**, 2141.

56 Nakanishi, K. and Berova, N. (1994) in *Circular Dichroism: Principles and Applications*, Chapters 5 and 13 (eds K. Nakanishi, N. Berova and R.W. Woody), John Wiley & Sons, Inc., New York.

57 (a) Green, M.M., Reidy, M.P., Johnson, R.D., Darling, G., O'Leary, D.J. and Willson, G. (1989) *J. Am. Chem. Soc.*, **111**, 6452.
(b) Green, M.M., Park, J.-W., Sato, T., Teramoto, A., Lifson, S., Selinger, R.L.B. and Selinger, J.V. (1999) *Angew. Chem. Int. Ed. Engl.*, **38**, 3138.

58 (a) Fukuda, K., Seki, T. and Ichimura, K. (2002) *Macromolecules*, **35**, 2177.
(b) Ichimura, K. (2000) *Chem. Rev.*, **100**, 1847.

6
Synthesis of Inorganic Nanotubes
C.N.R. Rao and Achutharao Govindaraj

6.1
Introduction

Zero-dimensional nanoparticles and one-dimensional (1D) nanowires and nanotubes are important classes of nanomaterials [1, 2]. The first family of nanotubes is that of carbon nanotubes described by Iijima [3]. Nanotubes are, however, no longer confined to carbon but encompass a variety of inorganic materials [4, 5], and peptides [6]. In this article, our concern is with nanotubes of inorganic materials excluding carbon.

The early examples of inorganic nanotubes synthesized in the laboratory are those of molybdenum and tungsten sulfides by Tenne and coworkers [7]. These layered sulfides form fullerene-type structures and hence also nanotubes. Several methods to prepare nanotubes of Mo and W sulfides and of the analogous selenides have been reported in the last few years [1, 2]. The synthesis of BN nanotubes has also received considerable attention because of the similarity of the structure of BN to graphite. In the last few years, nanotubes of several inorganic materials including binary oxides, nitrides, halides as well as metals and non-metallic elemental materials have been prepared and characterized [1, 2, 4, 5]. Besides nanotubes of binary compounds, those of complex materials such as perovskite titanates and spinels have also been reported. Composites that involve nanotubes, nanowires, and nanoparticles are also known. In this article, we shall provide a status report on the synthesis of inorganic nanotubes. In doing so, we shall cover the synthetic strategies employed for the different classes of inorganic nanotubes. In view of the vast literature that has emerged in the last 2–3 years, we were unable to cite all the papers in this area and have restricted ourselves to representative ones. We apologize for any oversight or error in judgement.

Advanced Nanomaterials. Edited by Kurt E. Geckeler and Hiroyuki Nishide
Copyright © 2010 WILEY-VCH Verlag GmbH & Co. KGaA, Weinheim
ISBN: 978-3-527-31794-3

6.2
General Synthetic Strategies

Several strategies have been employed for the synthesis of inorganic nanotubes. In the case of molybdenum and tungsten sulfides and such layered chalcogenides, decomposition of precursor compounds such as the trisulfides (e.g., MoS_3 or WS_3) and ammonium thiometallate or selenometallate has been successful. An important method, used particularly in the case of oxide nanotubes, is the hydrothermal and solvothermal route, carried out in the presence of surfactants or other additives in certain instances. Electric arcing and laser ablation have been used to synthesize nanotubes of BN and other materials. Sol–gel chemistry is useful for the synthesis of nanotubes, especially of oxides. Chemical vapor deposition (CVD) is commonly used for the synthesis of some of the nanotubes.

A popular method of synthesis in the last few years has employed templates. The templates can be porous membranes of alumina or polycarbonate. The pores are used to deposit the relevant materials or these precursors, followed by annealing and removal of the template. Deposition of the material in the porous channels is carried out by the sol–gel method or by an electrochemical procedure. Electrochemical anodization is commonly used for the synthesis of nanotubes of TiO_2, ZnO, and such oxides. The porous membrane method has emerged to be a general means of preparing a large variety of inorganic nanotubes and nanowires. Carbon nanotubes, surfactants, polymer gels, and liquid crystals have all been used as templates, wherein the precursor material is covered over the templates, followed by annealing and removal (burning or dissolution) of the template. In what follows, we shall discuss the synthesis of various inorganic nanomaterials where we will indicate the method and give the most essential aspects of the procedure. In order for the reader to obtain greater details, we have provided a large list of references.

6.3
Nanotubes of Metals and other Elemental Materials

Synthesis of gold nanotubes was reported by Martin and co-workers [8, 9] in the 1990s. They prepared Au tubules with lengths of up to a few micrometers and diameters of a few hundred nanometers by electrochemically depositing gold into the pores of a microporous alumina membrane (AM). To obtain gold tubules, initially the AM pore walls were chemically derivatized by attaching a molecular anchor such as a cyanosilane, so that the electrodeposited metal preferentially deposits on the pore wall, which leads to tubule formation.

Electrodeposition in membrane pores is an important method for the synthesis of metal nanotubes. Thus, Au nanotube arrays have been prepared by direct electrodeposition in the nanochannels of alumina templates. The nanochannel alumina templates with pore diameters of about 105 nm and 45 nm were used to synthesize nanotubes with average outer diameters respectively of 105 nm and

45 nm and a wall thickness of 15 nm. The lengths went up to several micrometers. The alumina templates are readily removed by treatment with NaOH. The nanotube arrays so obtained have a well-controlled microstructure and are polycrystalline with a face-centered cubic structure [10]. Au nanotubes have also been prepared by electroless deposition in the pores of track-etched polycarbonate membranes that contain 10 μm thick and 220 nm diameter pores [11]. The inner surfaces of the polycarbonate membranes were first sensitized with a Sn^{2+} salt and then activated by forming a layer of Ag, before depositing Au for a period of 2 h. The gold nanotubes were cleaned with 25% HNO_3 solution for 15 h. Hydrophobic or hydrophilic self-assembled monolayers on gold nanotubes can be formed by rinsing the samples in ethanol for 20 min, followed by immersion in a solution of ethanol that contains $HS(CH_2)_{15}CH_3$ or $HS(CH_2)_{15}COOH$. The Au nanotubules are polycrystalline and have lengths of up to 6 μm and inner diameters of ~1 nm. By controlling the Au deposition time, Au nanotubules of effective inside diameters of molecular dimensions (<1 nm) can be prepared. They found electroless deposition allows for more uniform gold deposition in a short duration of time. Since the electroless plating method used to deposit Au nanotubes in polymeric templates does not work in AMs, Martin and coworkers [12] have developed a modified electroless plating strategy that can be used to deposit high-quality Au nanotubes within the pores of alumina templates.

Three-dimensional (3D) Au nanotube arrays with smooth as well as nanoporous walls have been obtained by using anodic alumina and conducting polyaniline nanorod templates [13]. In this procedure, polyaniline nanorods were predeposited electrochemically in the interior of a porous alumina membrane, which was then used as a template for the formation of vertically aligned Au nanotubes. For the synthesis of Au nanotube arrays with nanoporous walls, gold/silver alloy nanowalls were electrodeposited from cyanide solutions that contain gold/silver ions (mole ratio, $Au^+/Ag^+ = 1:3$). The nanoporous walls were generated by de-alloying (selective dissolution of the less noble metal) the gold/silver alloy shells with concentrated nitric acid, which also dissolves the polyaniline nanorods. The porous architecture is formed because of an intrinsic dynamic pattern formation process, in which the more noble metal (Au) atoms tend to aggregate into two-dimensional clusters through a phase separation process at the solid–acid interface. The length, average inner diameter, and wall thickness of the Au nanotubes with smooth walls was ~4 μm, ~196 nm, and 62 (±18 nm), respectively. The Au nanotubes with smooth nanoporous walls (nanopore diameter of ~8 nm) had similar physical dimensions as the nanotubes.

Gold nanotubes embedded within the pores of the polycarbonate template membranes were subjected to reactive ion etching (RIE) using an oxygen plasma to selectively etch approximately 2.3 mm of the polycarbonate, leaving the polycrystalline gold nanotubes intact. Figure 6.1 shows field-emission scanning electron microscopy (FESEM) images of the top surface of a template membrane after electroless deposition of gold followed by RIE [14]. Single crystalline and bamboo-like Au nanotube arrays growing in the [111] direction, with a diameter of 100–150 nm and a length of 10 μm, and standing perpendicular to the Ti metal foil

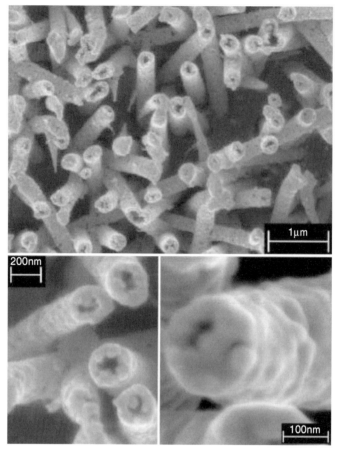

Figure 6.1 Field-emission SEM images of gold nanotubes at different magnifications. Micrographs were taken on the etched side of the membrane. Reproduced with permission from [14]. Copyright 2004, Wiley-VCH.

substrates are obtained by using a radiation track-etched hydrophilic polycarbonate membrane [15]. By using water-dissolvable Na_2SO_4 nanowires as templates, Au nanoparticle tubes have been obtained by the self-assembly of Au nanoparticles [16]. The as-synthesized Au nanoparticle tubes were further calcined at 300 °C (5 °C min^{-1}) for 30 min to transform them into polycrystalline Au nanotubes. At this temperature, the Au nanoparticles (3–5 nm diameter) melt and form nanotubes of ~1 μm length and ~80 nm diameter. Using an electroless deposition procedure, continuous, polycrystalline Au nanotubes with controllable shape, size (tens of nanometer in diameter), shell thickness (~5 nm), and length (up to 5 μm) can be grown by using Co nanoparticles as sacrificial templates [17]. Here, the alignment of Co nanoparticles into a 1D structure is induced by manipulation of the magnetic field. In this reaction, Au^{3+} is reduced to Au^0 by the Co^0 nanoparticles as given by the following reaction:

$$3Co^0 + 2Au^{3+} \rightarrow 3Co^{2+} + 2Au^0 \tag{6.1}$$

Goethite (FeOOH) nanorods have been used as templates to grow Au nanotubes with a length of a few hundred nanometers and an aspect ratio between 3 and 4 [18]. The uniform growth of gold nanoshells on goethite nanorods was achieved by SiO_2-mediated assembly/attachment of Au nanoparticles/seeds on these rods, followed by a one-step seeded growth by the catalyzed reduction of $HAuCl_4$ using formaldehyde. The successful attachment of small Au seeds on goethite nanorods requires modification of the nanorod surface, which was carried out by depositing a thin layer of silica using tetraethoxysilane (TEOS), followed by silanization using (3-aminopropyl)trimethoxysilane (APS). Gold nanoparticles (~3 nm in diameter) are prepared by the reduction with tetrakishydroxymethylphosphonium chloride (THPC), which were used for the self-assembly of Au seeds on the surface of goethite rods. The growth of Au shells on the goethite surface was accomplished by the selective reduction of $HAuCl_4$ with a weak reducing agent such as formaldehyde, catalyzed by the Au seeds. In a typical synthesis, 100 mL of the dispersion of goethite was added to 48 g of poly(vinyl pyrrolidone) (PVP) and the mixture stirred for 24 h before transferring to ethanol (100 mL). Twelve millilitres of the goethite–PVP dispersion was diluted with 200 mL of ethanol that contained 20 mL of ammonium hydroxide and 0.75 mL of TEOS, under mechanical stirring. The goethite nanorods coated with silica were washed several times with ethanol. Silanization was carried out by adding 0.6 mL of pure APS to 10 mL of the goethite@silica dispersion under magnetic stirring, followed by additional washing with ethanol to remove excess APS. For the assembly of Au nanoparticles on the nanorod surface, 5 mL of the solution of silanized goethite particles was centrifuged and re-dispersed in 5 mL of the Au colloid solution in an ultrasound bath for 1 to 2 min. This step was repeated until no further adsorption of gold on the surface of the particles was observed. Further growth of the Au coating on the nanorods was carried out as follows. A mixture of 0.425 mL of 0.01 M $HAuCl_4$ and 10 mL of 1.8×10^{-3} M K_2CO_3 was aged in the dark for a day so that Au^{III} ions were reduced to Au^I ions. This solution (5 mL) was mixed with 0.01–0.03 mL of the Au-covered goethite solution and 0.01 mL of formaldehyde solution. The thickness and surface roughness of the obtained shells could be adjusted by simply varying the concentration ratio between the seeds (modified goethite rods) and the growth reagents ($HAuCl_4$ and formaldehyde).

Platinum and PtPd alloy nanotubes with polycrystalline walls and a face-centered cubic structure (50 nm in diameter, 5–20 μm long, and 4–7 nm wall thickness) can be synthesized by the galvanic replacement reaction of Ag nanowires [19]. In this procedure, Ag nanowires are refluxed for 10 min with platinum acetate in aqueous solution. When an aqueous platinum acetate solution is mixed with a dispersion of Ag nanowires, the galvanic replacement reaction generates a tubular sheath whose morphology is complementary to that of the Ag nanowire (step 1 in

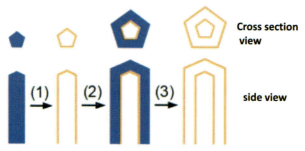

Scheme 6.1 Schematic representation for the fabrication of multiple-walled nanotubes composed of Pt/Ag alloys.

Scheme 6.1). Following the same scheme for multiple-walled nanoshells, coaxial nanotubes with more than two walls can be prepared. Mesoporous Pt nanotubes have been prepared by incorporating lyotropic liquid crystals in the pores of an AM before depositing the metal (Fig. 6.2) [20]. By using a mixed non-ionic–cationic liquid crystalline surfactant (nonaethylene glycol monododecyl ether ($C_{12}EO_9$), polyoxyethylene (20) sorbitan monostearate (Tween 60) and water), polycrystalline Pt nanotubes (with a face-centered cubic structure) with small inner (3–4 nm) and outer (6–7 nm) diameters have been obtained [21]. In a typical synthesis, the liquid crystalline phase that contains hexachloroplatinic acid (H_2PtCl_6), $C_{12}EO_9$, Tween 60, and water at a molar ratio of 1:1:1:60 was treated with hydrazine to cause the reduction of the metal salts confined in the lyotropic mixed crystals of the two different surfactants to yield the metal nanotubes. The resulting solid was separated, and washed with water and ethanol prior to drying in air. The same procedure has been used to prepare nanotubes of other metals such as Pd and Ag.

Hierarchical assemblies of hollow Pd nanostructures have been grown by using Co nanoparticles as self-sacrificial templates [22]. Assemblies of hollow Pd nanostructures (from 80 nm Pd nanoparticles) are obtained with raspberry-like or nanotube-like geometries, by the replacement reaction between H_2PdCl_4 and Co nanoparticles, which occurs rapidly, wherein the reduced Pd atoms nucleate and grow into very small particles, and eventually evolve into a thin shell around the cobalt nanoparticles. The replacement reaction is given by:

$$Co + PdCl_4^{2-} \rightarrow Pd + Co^{2+} + 4C^{-1} \tag{6.2}$$

The polycrystalline 1D Pd nanostructures have a face-centered cubic structure with lengths that run to several micrometers and diameters up to ~60 nm.

Ni and Co microtubules were prepared by Han et al. [23]. by the pyrolysis of composite fibers consisting of a poly(ethylene terephthalate) (PET) core fiber with the electroless-plated metal at the exterior. Ni microtubules prepared by this method were single-crystalline, but the Cu microtubules were polycrystalline. Electrodeposition in the pores of AMs has been carried out by Bao et al. [24]. to obtain ordered arrays of Ni nanotubules. The pore walls were modified by these

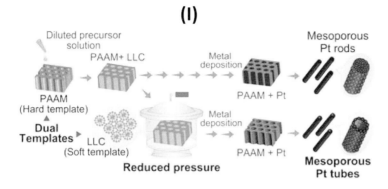

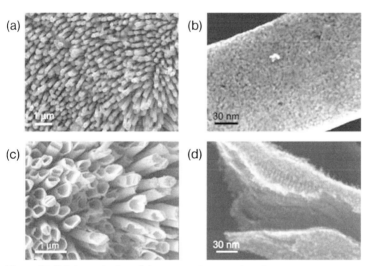

Figure 6.2 I) Schematic view of the preparative procedure for mesoporous Pt nanorods and nanotubes. II) SEM images of: a,b) mesoporous Pt nanorods, and c,d) mesoporous Pt nanotubes. Figures (b) and (d) are highly magnified images of (a) and (c), respectively. Reproduced with permission from [20]. Copyright 2008, RSC.

workers with an organic amine to assist the formation of nanotubes (Fig. 6.3). In the absence of the amine, nickel nanowires were obtained. Nickel, when electrodeposited in the pores, binds preferentially to the pore walls because of its strong affinity towards the amine. The AM is removed by treatment with NaOH. The top-view SEM image in Figure 6.3b shows open-ends of the Ni nanotubules after the removal of the top layer of the AM with NaOH solution. The electron

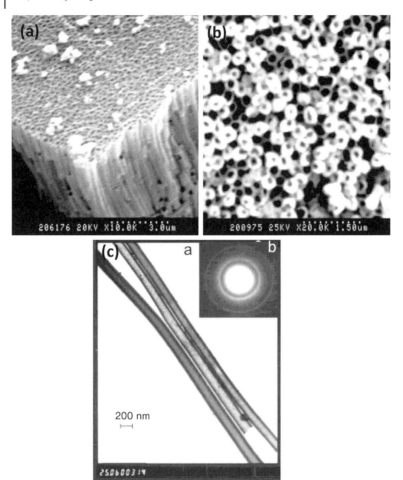

Figure 6.3 SEM images of a) Ni nanotubules after dissolution of the alumina template, b) showing the top-view of an array of Ni nanotubules after partial removal of the template, and c) transmission electron microscopy (TEM) image of the Ni tubules after dissolution of the alumina template. Inset shows electron diffraction pattern of the Ni nanotubules. Reproduced with permission from [24]. Copyright 2001, Wiley-VCH.

diffraction pattern of the Ni nanotubules in the inset of Figure 6.3c shows diffuse rings, which indicates that the Ni nanotubules are polycrystalline and have a face-centered cubic structure. The transmission electron microscopy (TEM) image of the Ni nanotubules in Figure 6.3c shows that the diameter of the Ni nanotubules are uniform with average outer diameter of 160 ± 20 nm. The deposition follows a bottom-up approach. Thus, the growth of nanotubules starts at the Au cathode at the bottom of the pores. The length and the wall thickness of the Ni nanotubules depend on experimental conditions, such as pore-wall modifying agent, and electrodeposition parameters. A low current density appears to be a key factor. For

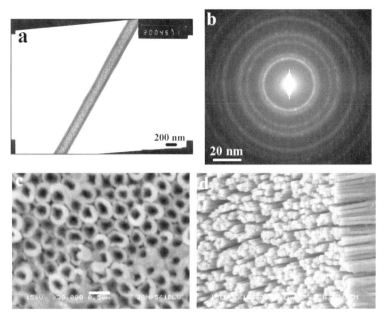

Figure 6.4 a) TEM image and b) SAED pattern of a Ni nanotube after removing the alumina template. c) Top-view and d) side-view SEM images of an ordered array of Ni nanotubes after partial removal of the template. The experimental parameters are: current density, 0.13 mA cm^{-2}, concentration of P123 of 37 g L^{-1}, and deposition time of 48 h. Reproduced with permission from [25]. Copyright 2006, Wiley-VCH.

instance, with a current density of 0.3 mA cm^{-2} and an electrodeposition time of 24 h, Ni nanotubules of 20 μm in length and 30 nm in wall thickness were obtained, while with electrodeposition for 48 h, Ni nanotubules grow up to 35 μm in length and 60 nm in wall thickness. Ordered magnetic Ni nanotubes have been prepared by employing electrodeposition by adding an amphiphilic triblock co-polymer (Pluronic P123) to the electrodeposition solution [25]. By adjusting experimental parameters such as current density and electrodeposition time, the wall thickness and length of the nanotube are controlled (Fig. 6.4). The Ni nanotubes prepared at a current density of 0.13 mA cm^{-2} with an electrodeposition solution that contains 37 g L^{-1} of P123 and a deposition time of 48 h yielded uniform tubes of about 50 nm wall thickness, 60 μm length, and an outer diameter of ~250 nm, which corresponds to the pore diameter of the alumina template. Weak, diffuse rings in the selected-area electron diffraction (SAED) pattern of the nanotubes showed they were polycrystalline.

The preparation and growth mechanism of Ni, Co, and Fe nanotubes have been reported by Cao *et al.* [26]. wherein the nanotubes are actually constructed from non-layered materials of the metal, and the tubular growth is directed by the current in the template-based electrodeposition process (Fig. 6.5). Typically, the length of the as-synthesized tubular structures can reach about 60 μm, which

6 Synthesis of Inorganic Nanotubes

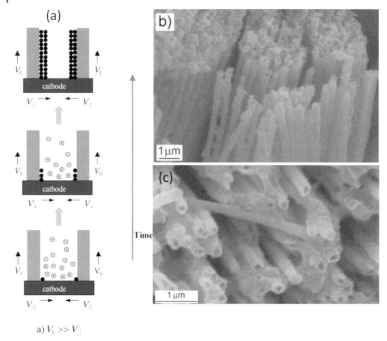

a) $V_\| \gg V_\perp$

Figure 6.5 a) Schematic diagram of the steps in the growth of metal nanotubes by template-based electrochemical deposition method. b,c) SEM images of cobalt and iron nanotube arrays, respectively. Reproduced with permission from [26]. Copyright 2006, Wiley-VCH.

corresponds to the thickness of the AM. Most of them have an outer diameter of 50–100 nm, which corresponds to the pore diameter of the AM, and an inner diameter of about 30–50 nm. The electron diffraction patterns reveal that the nanotubes are single-crystalline and have a body-centered cubic structure for iron, a face-centered cubic structure for nickel, and a hexagonal close packed structure for cobalt. Employing AMs, aligned Fe nanotubes have been grown electrochemically [27]. The wall thickness of the nanotubes could be controlled by changing the deposition parameters. Co nanotubes have also been grown by electrodeposition in an AM [28] Electrodeposition in a rotating electric field produces dense arrays of single crystalline Cu nanotubes [29]. The applied rotating field makes the ions graze the surface of the pores in helical paths and makes the deposition occur selectively in the region near the wall of the nanopores. The wall thickness of the metal nanotubes so obtained are in the range of 15–20 nm and can be controlled by changing the amplitude of the rotating field. The nanotubes had a diameter of ~230 nm with lengths going up to several micrometers.

Electrochemical deposition methods have also been used for the synthesis of nanotubes of many other metals including Zn, Sn, and Ag having polycrystalline structures [30]. In order to fabricate these metal nanotube arrays, a thin layer of

Au was deposited on the bottom surface of the porous AM, for use as a working electrode. The circular step edges of the Au nanorings so formed at the bottom of the AM serve as the preferential sites for the deposition of metal ions. Nanotube arrays of Zn and Sn were then obtained within the porous AM by preferentially plating at circular step edges of the Au nanorings. This was accompanied by the evolution of hydrogen gas, which is critical for nanotube formation. It is well known that circular step edges have the ability to catalyze electron transfer to metal ions in solutions. Since the reductive potential of hydrogen is higher than that of Zn and Sn, evolution of hydrogen gas accompanies metal deposition. Hydrogen gas evolves continuously from the cathode through the central region of the nanopores while the metallic elements become attached to the pore walls, which results in the formation of nanotubes. The nanotubes have open ends on top and are arranged in a well-ordered manner. The outer diameter of the nanotubes is in the range 90–110 nm, which corresponds to the pore diameter of the membrane. The inner diameter of the nanotubes is in the range of 40–60 nm. The length of the nanotubes goes up to several micrometers. The compactness of the nanotubes is quite high, about $1 \times 10^9 \, \text{cm}^{-2}$, which corresponds to the pore density of the membrane.

Metallic indium nanotubes can be prepared by direct thermal evaporation of the metal [31]. A metal source is heated in an Ar atmosphere between 900–1100 °C and the vapor is passed over a catalyst such as a Au-coated silicon substrate to obtain crystalline nanotubes. The length of the nanotubes increases with temperature. The diameter of the head portion (100–300 nm) is approximately three times larger than the tail diameter. The diameter of the tail portion remains uniform throughout the nanostructure. Electrodeposition in AMs has been employed to prepare polycrystalline nanotubes of alloys such as BiSb [32], FeCo [33], FeNi [34], and CoCu [35].

Lyotropic liquid crystal templates along with surfactants are used to synthesize metal–boron nanotubes (M–B (M = Fe, Co, and Ni)) [36]. The non-ionic surfactants used are Tween 40 (polyoxyethylene sorbitan monopalmitate), Tween 60, and Tween 80 (polyoxyethylene sorbitan monooleate), and the anionic surfactant is camphorsulfonic acid. In a typical synthesis, $FeCl_3 \cdot 6H_2O$ was dissolved in water that contained (1S)-(+)-10-camphorsulfonic acid (CSA) and Tween 40 at 60 °C and the mixture cooled to 20 °C. To this a mixture of 4 M $NaBH_4$ and 0.1 M NaOH was added and the mixture kept for 48 h in an inert atmosphere. The resulting solid was collected, washed with distilled water and ethanol, and dried in flowing nitrogen. Non-crystalline Co–B and Ni–B nanotubes were prepared from $CoCl_2 \cdot 6H_2O$ or $NiCl_2 \cdot 6H_2O$ and Tween 60 in place of $FeCl_3 \cdot 6H_2O$ and Tween 40 under similar conditions. This method provides a route for the synthesis of metal–boron nanotubes and can also be extended to other tubular materials of non-crystalline alloys. Fe–B nanotubes prepared using Tween 40 and CSA are several micrometers in length and have inner and outer diameters of around 50–55 and 60–65 nm, respectively. The continuous broad halo rings in the SAED pattern further suggest a non-crystalline nature of the Fe–B nanotubes.

One-dimensional nanostructures of Se and Te were reported some time ago by Gautam et al. [37, 38]. Recently, t-Se nanotubes have been grown hydrothermally in the absence of a surfactant or a polymer [39]. In this procedure, an aqueous solution of sodium selenite (Na_2SeO_3), NaOH, and sodium formate ($NaCHO_2$) is reacted in a hydrothermal bomb at 100 °C for 25 h. Zhang et al. [40]. have reported the fabrication of t-Se nanotubes by a hydrothermal–ultrasonic route. Ma et al. [41]. synthesized t-Se nanotubes in micelles of a non-ionic surfactant, while Li et al. [42]. synthesized them by a sonochemical process. Single-crystalline Te nanotubes have been prepared by a solvothermal method using PVP to modulate the reaction time, the reactants being TeO_2 and ethanolamine [43].

6.4
Metal Chalcogenide Nanotubes

MoS_2 and WS_2 are the first chalcogenides whose fullerene type structures and nanotubes were prepared in the laboratory [44, 45]. The method involved heating of metal oxide nanorods in H_2S. A similar strategy was also employed to prepare the corresponding selenide nanotubes. Recognizing that amorphous MoS_3 and WS_3 are the likely intermediates in the formation of the disulfides, the trisulfides have been directly decomposed in a H_2 atmosphere to obtain the disulfide nanotubes [46]. Diselenide nanotubes were similarly obtained by the decomposition of metal triselenides [47]. The trisulfide route provides a general route for the synthesis of the nanotubes of many metal disulfides such as NbS_2 and HfS_2 that have a single crystalline nature [48, 49]. The decomposition of precursor ammonium salts $(NH_4)_2MX_4$ (X = S, Se; M = Mo, W), is even better, all the products, except the dichalcogenides, being gases [46, 47]. Metal trichalcogenides are intermediates in the decomposition of the ammonium salts as well. These nanotubes have a hexagonal structure with a layer spacing of ~0.6 nm, which corresponds to a d-spacing of (002), with an external diameter of ~25 nm, a wall thickness of ~10 nm, and lengths of up to several micrometers. Employing trisulfides and triselenides as starting materials, nanotubes of TiS_2, HfS_2, NbS_2, $NbSe_2$ and related layered metal chalcogenide nanotubes have been prepared [50]. Recently, MoS_2 and WS_2 have been made by using gas phase reactions using metal chlorides and carbonyls (Fig. 6.6) [51]. Solar ablation can also be used to generate MoS_2 nanotubes [52]. Nanotubes and onions of GaS and GaSe have been generated through laser and thermally induced exfoliation of the bulk powders [53].

CdS nanotubes and related structures have been prepared by the thermal evaporation of CdS powder (Fig. 6.7) [54]. Nanotubes of CdS and CdSe had earlier been prepared by using surfactants as templates [55]. CdS occurs in the hexagonal structure in the nanotubes. The diameter and length are in the ranges of 40–160 nm and 3–4 μm, respectively. The CdSe nanotubes are generally long, with lengths up to 5 mm. The outer diameter of the nanotubes is in the 15–20 nm range while the diameter of the central tubule is in the 10–15 nm range. These nanotubes are polycrystalline and form through the oriental attachment of nanoparticles. Recently, CdS, ZnS, and CuS nanotubes have been made by the hydrogel-assisted route (Fig. 6.8) [56]. The TEM image of CdS in Figure 6.8a, obtained

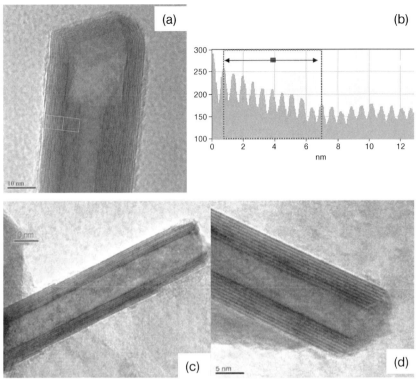

Figure 6.6 a) HRTEM image of the MoS$_2$ nanotube, b) the line profile of the boxed area in (a) gives an interlayer spacing of 6.2 Å. Reproduced with permission from [51a] c,d) TEM images of WS$_2$ nanotubes obtained in the reaction between WCl$_5$ and H$_2$S in the vertical reactor. Reproduced with permission from [51b].

after the removal of the hydrogel template, demonstrates the hollow nature of the nanotubes. The lengths of the nanotubes extend to a few hundred nanometers, while the diameter of the inner tubule is ~2–3 nm, the outer diameter being in the 20–25 nm range. A SAED pattern of a single nanotube is given in the bottom inset of Figure 6.8a. The diffuse rings correspond to the (100) and (110) Bragg planes of hexagonal CdS showing the nanotubes to be generally polycrystalline. Clearly, the tripodal cholamide hydrogel fibers act as templates, on which the CdS particles get deposited, giving rise to the nanotubes. The low magnification TEM image in Figure 6.8b suggests a possible assembly or attachment of the initially formed shorter nanotubes to form linear chains (indicated by arrows in the figure). The hydrogel might be responsible for such an attachment, the hydrogel playing a dual role of being a template to produce hollow nanotubes as well as favoring the attachment or assembly of the nanotubes. The ZnS nanotubes prepared similarly (Figs 6.8c and d) have an inner diameter in the ~4–6 nm range,

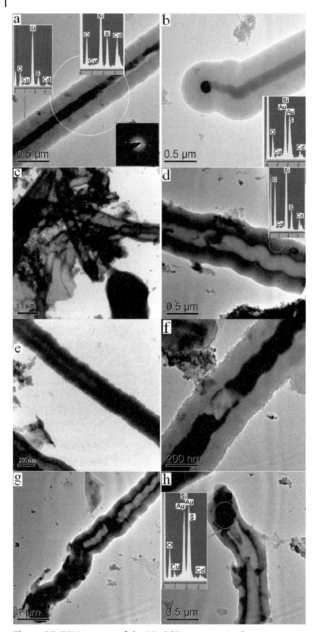

Figure 6.7 TEM images of the 1D CdS nanostructures: a) a core–sheath nanowire with insets that contain the energy dispersive X-ray analysis (EDX) of the core and the sheath and SAED of the core, b) the top of a core–sheath nanowire with an inset of the EDX of the catalyst, c,d) CdS nanotubes, e) tube–wire nanojunction, f) wire–tube–wire nanojunction, g) top part of a nanotube with nanoparticles with the channel, and (h) top part of one nanotube with the catalyst with the inset showing the EDX of the particle. Reproduced with permission from [54]. Copyright 2008, ACS.

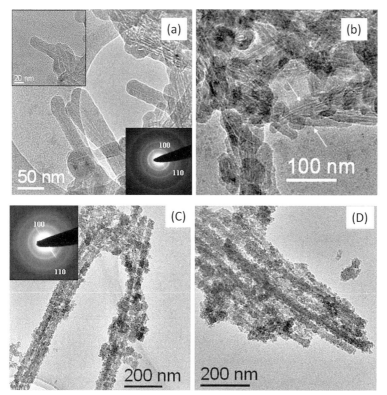

Figure 6.8 a) TEM image of CdS nanotubes obtained after the removal of the hydrogel template. Top inset is a high-magnification image of a single nanotube. Bottom inset is the SAED pattern of the nanotubes. b) TEM image showing a bunch of nanotubes assembled spontaneously, indicated by the arrows. c,d) TEM images showing nanotubes of ZnS obtained using 0.02 mmol of Zn(OAc)$_2$. Reproduced with permission from [56]. Copyright 2006, Elsevier.

with lengths going up to a micrometer. The TEM image shows tiny nanocrystals of ZnS (Fig. 6.8c) making up the walls of the nanotubes. The electron diffraction patterns also reveal the nanotubes to be polycrystalline and of hexagonal structure. The CuS nanotubes prepared by this route have an inner diameter of ~5 nm with lengths extending to a few hundreds of nanometers. The outer diameter is in the 20–30 nm range. The nanotubes are polycrystalline as found from electron diffraction as well as the TEM images. The polycrystalline nanotubes are formed through the oriented attachment of nanoparticles. CuS nanotubes have also been made by the solution reaction of Cu nanowires in ethylene glycol with a suitable sulfur source such as thiourea and thioacetamide at 80 °C [57]. The CuS nanotubes so-produced in large quantities possess a hexagonal structure and have an inner diameter of 30–90 nm, a wall thickness of 20–50 nm, and a length of more than 40 μm. The straight nanotubes are made up of nanoparticles of around 30 nm diameter. This study also shows that sulfur sources such as thiourea and thiacetamide, which release ionic sulfur rather than molecular sulfur at their decomposition temperature, are favorable for the formation of CuS nanotubes, compared with sulfur powder.

CdSe nanotubes have been prepared by a sono-electrochemical method [58], while CuSe nanotubes have been prepared by using trigonal Se nanotubes as templates [59]. The CdSe nanostructures (with a cubic structure) were formed by electrodeposition onto the sonic probe cathode [58]. The deposit could be spheroidal or have a 2D structure. Subsequent sonic shock waves remove the deposit from the probe surface. The sonochemical treatment provides the required energy to roll the nanosheets to form tubular nanoscrolls. After the 2D nanosheets are ejected from the probe, because of the high surface energy of the ends of the nanosheets, the flexible and unstable nanosheets easily roll-up in the sonochemical process. It is well known that collapsing bubbles produced in a liquid solution during sonication can instantaneously generate local spots of high temperature, pressure, and cooling rates. The worm-like morphology of the CdSe prepared by this method involves a tubular structure and the nanotubes (diameter: 80 nm, wall thickness: 10 nm) are entangled with each other. The nanotube with a straight part reveals a round open tip, which is not completely seamed. This indicates that the nanotubes were probably formed through a roll-up process. The CdSe nanotubes show many stacking faults and missing layers in the nanotube walls. The appearance of defects is mainly a result of the high stress and strain present in the nanotubes.

CuSe with a tubular nanostructure is formed by the template mechanism wherein diffusion of Cu atoms occur into t-Se nanotubes [59]. The synthesis process involves the t-Se nanotubes acting as templates and reactants were converted into crystalline nanotubes of CuSe by reacting with Cu nanoparticles freshly produced from an aqueous $CuSO_4$ solution. Apart from CuSe nanotubes, 1D nanocrystallites of Cu_3Se_2, $Cu_{2-x}Se$, and Cu_2Se were also obtained by changing the atom ratio of Cu and Se in the precursors. The tubular nanostructures have the hexagonal structure of CuSe. The wall thickness and diameter of the CuSe nanotubes were around 80 and 300 nm, respectively.

CdTe nanotubes of controlled diameter are prepared by first reacting $CdCl_2$ with thioglycolic acid (TGA) to obtain 1D Cd–TGA nanowires (thickness ~8 nm), the aqueous dispersion of which is then used as a sacrificial template to generate long CdTe nanotubes by reaction with NaHTe [60]. The length of the nanotubes is typically hundreds of micrometers, similar to the initial length of the precursor. The inner and outer diameters of the nanotubes are in the ranges of 12–20 nm and 30–50 nm, respectively. The CdTe nanotubes are polycrystalline and adopt a cubic structure. The average diameter of the nanowires of the 1D Cd-TGA precursors obtained in the presence of poly(acrylic acid) (PAA), could substantially be increased as a function of the amount of PAA. During the formation of CdTe, 1D Cd–TGA precursors are gradually consumed to finally lead to hollow structures, preserving the original shape, size, and morphology of the precursor.

Photochemical decomposition of CSe_2 adsorbed on Ag nanowires yields Ag_2Se nanotubes [61]. The evolution of Ag nanowires to core–shell structures and finally to hollow Ag_2Se nanotubes was studied in detail by TEM analysis. Upon irradiation for 15 min, the TEM image of the nanowires began to show evidence of core–shell nanowires with mean diameters of ~85 nm and ~45 nm cores. The shell grew thicker at the expense of the core with increasing irradiation time, and voids were observed to grow from both ends of the nanowires along the longitudinal axis, ultimately merging to form hollow nanotubes of mean diameters of ~90 nm with ~45 nm voids. The nanotubes are polycrystalline with a structure corresponding to the orthorhombic phase of $R-Ag_2Se$. Trigonal Se nanotubes can be used as templates to prepare Ag_2Se nanotubes [62]. Ag_2Te nanotubes have been generated by the reaction of $AgNO_3$ with sodium tellurate (Na_2TeO_3) in the presence of hydrazine and ammonia by a hydrothermal process in the absence of a template or a surfactant [63]. All these nanotubes are bent and curled, with diameters of 80–250 nm and several tens of micrometers in length. They show the characteristics of tubular structures with open-ended and uncovered hollow interiors. The nanotubes are single-crystalline, and free of dislocations and stacking faults. A structural phase transition of the as-prepared Ag_2Te nanotubes from the low-temperature monoclinic structure (β-Ag_2Te) to the high-temperature face-centered cubic structure (α-Ag_2Te) has been observed.

Bi_2S_3 nanotubes and nanorods are prepared solvothermally at a low temperature of 120 °C, using a mixed solvent (acetone–water, methanol–water, ethanol–water, water, ethylene glycol–water, or glycerol–water) as the reaction medium and urea as the mineralizer [64]. In a typical synthesis, $Bi(NO_3)_3 \cdot 5H_2O$ is dissolved in the mixed solvent and aqueous $Na_2S \cdot 9H_2O$ (S/Bi = 3:1) solution added drop by drop into the solution under vigorous stirring. The mixture of precursors and urea is transferred into a Teflon-lined autoclave and heated solvothermally. The gray-black powder so obtained is washed with distilled water and ethanol several times and dried. A mixture of nanorods and nanotubes is found in the final product synthesized in methanol–water mixtures. The diameter of the nanotubes was more than that of the nanorods (in the 80–100 nm range) with lengths up to a micrometer. The powders synthesized in water are nanotubes, which are polycrystalline with a diameter of 200 nm and a length of about 1 μm. Bi_2S_3 nanotubes synthesized in ethylene glycol–water mixtures are single crystalline, with a diameter in the 200–500 nm range and the lengths up to several micrometers. The inner diameter of the single hollow nanotube is about 100 nm and the walls of the tube are around 100 nm thick. Bi_2S_3 microtubes synthesized in the mixed solvent of glycerol–water have a diameter of about 1 μm and a length of about 7 μm. During the process of solvothermal synthesis, there is a dynamic equilibrium between the Bi_2S_3 solid particles or nuclei and the Bi^{3+} and S^{2-} ions in solution ($2Bi^{3+} + 3S^{2-}$ ⇌ Bi_2S_3). In such an equilibrium, Bi^{3+} and S^{2-} ions tend to dissolve from the small particles into the solution and precipitate onto the surfaces of large particles so that the total energy of the interface between the particles and the solution is decreased. Bi_2S_3 has a layered structure and the weak bonds between the layers give rise to an anisotropic growth of Bi_2S_3 particles during the solvothermal synthesis. At the low reaction temperature, the rate of crystal growth is greater than that of crystal nucleation. The growth of Bi_2S_3 nanosheets is favoured at higher viscosity and surface tension of the reaction medium. The powders synthesized in the mixed solvent of water, ethylene glycol–water, and glycerol–water possess nanosheet structures and the nanosheet nanostructures self-roll to form tube-like structures. The interface energy between bismuth sulfide and the mixed solvent of water, ethylene glycol–water, and glycerol–water with a high surface tension appears to be higher than that between bismuth sulfide and the mixed solvent of acetone–water, ethanol–water, and methanol–water. A solvent with higher viscosity and surface tension favours the formation of tube-like structures and a solvent with lower viscosity and surface tension favours the formation of a rod-like structure during the synthesis. Hence the morphology of the nanostructure depends on the viscosity and surface tension of the mixed solvent used in the solvothermal synthesis. Bi_2S_3 nanotubes have also been prepared by the conventional evaporation method [65], as well as by employing the micelle-template method at 115 °C [66].

Bi_2Se_3 nanotubes can be prepared solvothermally starting with ammonium bismuth citrate and elemental Se in dimethylformamide solution [67]. However, the diameters of the nanotubes are not uniform and vary in the range of 15–150 nm with the wall thickness in the 5–20 nm range. These polycrystalline nanotubes have a rhombohedral structure. Bi_2Te_3 nanotubes have been obtained by the galvanic displacement of nickel nanowires in nitric acid that contains Bi^{3+} and $HTeO_2^+$ ions (Fig. 6.9) [68]. When Ni nanowires are immersed in an nitric acid solution that contains Bi^{3+} and $HTeO_2^+$ ions, the Ni nanowires are galvanically displaced to form Bi_2Te_3, because of the difference in the reduction potentials. The nanotubes have diameters in the 100–130 nm range, a wall thickness of approximately 20 nm, and lengths running to several micrometers. The nanotubes are formed out of highly crystalline rhombohedral Bi_2Te_3 crystals without obvious preferential orientation. The composition of Bi_2Te_3 nanotubes was precisely tuned by adjusting the $[Bi^{3+}]/[HTeO^{2+}]$ ratio. The galvanic displacement reaction can be written as:

$$2Bi^{3+}(aq) + 3HTeO^{2+}(aq) + 9Ni^0(s) + 9H^+(aq) \rightarrow \quad (6.3)$$
$$Bi_2Te_3(s) + 9Ni^{2+}(aq) + 6H_2O(aq)$$

Nanotubular Bi_2Te_3, as well as its alloys with Se and Sb, are obtained by electrodeposition in the nanochannels of alumina templates [69]. Sb_2S_3 nanotubes with thin walls (1.5 nm–2 nm) have been made by the solvothermal reaction of $SbCl_3$ with sulfur in oleylamine solution at a relatively low temperature (175 °C) [70]. The average diameter of the Sb_2S_3 nanotubes (orthorhombic structure) was 10.4 nm, and the length was in the 100 to 300 nm range with an atomic ratio of Sb to S of 1 : 1.51. By changing the concentration of sulfur, the aspect ratios of the initially formed Sb_2S_3 nanoribbons were delicately controlled. Reducing the number of equivalents of sulfur results in a decrease in the length of the nanoribbons with a simultaneous increase in the width. In addition to the wider nanoribbons, a significant amount of nanotubes with 6.7 nm average width were observed.

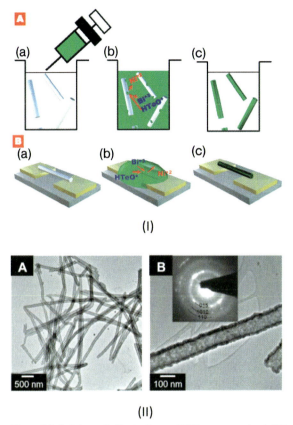

Figure 6.9 I) Schematic illustrations of Bi_2Te_3 nanotube synthesis (A) and individual Bi_2Te_3 nanotube laid across electrodes (B). TEM images and SAED pattern of high aspect ratio Bi_2Te_3 nanotubes (A) synthesized from nickel nanowire (~100 nm in diameter). Tube thickness was approximately 20 nm (B). Reproduced with permission from [68]. Copyright 2007, ACS.

Further reduction in sulfur yields pure small aspect ratio nanotubes. Clearly, the nanotubes are formed by the rolling of the nanoribbons.

Nanotubes of lead chalcogenides are prepared by the reaction of $Pb(NO_3)_2$ with cysteine in ethanolamine solution followed by the subsequent reaction of the nanowire product with the chalcogenide source solution at room temperature [71]. The nanowires are formed by the self-assembly of nanocrystals (10 nm diameter) and the nanowires act as templates for the subsequent formation of lead chalcogenide nanotubes. The nanotubes are polycrystalline, have a typical diameter of about 200 nm and are several micrometers long. The lead chalcogenide nanotubes show face-centered cubic phases with a disorderly aggregation of the nanocrystals. The lead chalcogenide nanocrystals are formed on the surface of the precursor nanowires by an anion-replacement reaction that is the result of the lower solubility of lead chalcogenide in solution. An ion exchange solvothermal reaction between $Na_4P_2S_6$ and $MnCl_2$ gives rise to nanotubes of $Mn_2P_2S_6$ [72]. These nanotubes are single-crystalline (monoclinic) and have uniform outer diameters of 40–50 nm and lengths that range from 110 to 170 nm.

6.5
Metal Oxide Nanotubes

Metal oxide nanotubes have been investigated widely in the last 2–3 years because of the potential applications of some of these materials. They have been prepared by employing several methods including sol–gel chemistry, hydrothermal reactions and use of templates. The template method is used particularly widely in combination with electrodeposition. Besides the popular porous alumina template, carbon nanotubes and other materials are used in the preparation of oxide nanotubes. The template-directed synthesis of oxide nanotubes has been reviewed by Bae et al. [73].

6.5.1
SiO$_2$ Nanotubes

SiO_2 nanotubes are readily prepared by using a variety of templates wherein the templates are coated with the oxide precursor followed by hydrolysis and heat treatment. Besides carbon nanotubes, carbon nanofibres have been used as templates to obtain SiO_2 nanotubes. The same method is also applicable for the synthesis of ZrO_2 and Al_2O_3 nanotubes [74]. Peptide amphiphile (PA) nanofibres have been successfully used as templates. The catalytic activities of PAs that contain lysine, histidine, or glutamic acid have been compared and only the PAs that contain lysine or histidine found to be good as catalytic templates. Depending on the reaction conditions and the size of the PA assembler, the nanotube wall thickness can be varied between 5–9 nm [75]. Self-assembled peptidic lipids that form tubular structures are also used as templates to prepare silica nanotubes [76]. The use of self-assembled structures as templates for preparing silica nanotubes and other nanostructured oxides generally involves the coating of the super-structure with metal alkoxides, sol–gel condensation, and the removal of the templates. This template method has been used to prepare nanotubes of silica and other materials using organogelators including those that are fluorinated [77].

Thermal evaporation of SiO yields SiO_2 nanotubes [78]. This in-situ template-like process (carried out in the presence of GaN and ZnS) in a vertical induction furnace gives amorphous SiO_2 nanotubes wherein both the diameters and lengths can be tuned by changing the reaction temperature. The SiO_2 nanotubes are amorphous, with a diameter of 30 nm and length of several hundred micrometers at 1450 °C, change to a diameter of 100 nm and length of 2–10 micrometers at 1300 °C. At high reaction temperatures, GaN decomposes to give Ga and the newly formed Ga is in the form of small-sized liquid clusters. The Ga clusters are transported by the Ar gas to the low temperature region, where they deposit as liquid droplets on the inner walls of the graphite crucible. The Ga droplets are the favored sites for the adsorption of ZnS vapor generated in the system. On supersaturation, ZnS segregates and gives rise to ZnS nanowires. With an increase in temperature, the decomposition of the SiO results in the formation of Si vapor, which is oxidized by the residual O_2 to SiO_2. This is transported by the Ar gas and is deposited on the surface of the ZnS nanowires, which results in ZnS/SiO_2 hetero-nanowires. Because of the high temperature, the inner ZnS nanowires evaporate, and hollow SiO_2 nanotubes remain. This method can also be used for nanotubes of other inorganic nanomaterials such as ZnS and GaN. CdSe nanocrystals are used as seeds in a one-step thermal evaporation process to prepare long, high density SiO_2 nanotubes [79]. SiO_2 nanotubes generally have smooth surfaces, uniform thickness, and round cross sections of both the interiors and exteriors which are amorphous in nature. The nanotubes are open-ended and have diameters of about 80–110 nm and a length of several hundreds of micrometers. High purity SiO_2 nanotubes are obtained by using track-etched membrane templates that involve O_2 plasma treatment. Oxygen plasma pretreatment ensures pore-filling of the precursor solution and covalent bonding between template and precursor, while pyrolysis of the template–nanostructure composite completely removes organics and produces inorganic nanostructures [80]. These nanotubes are also amorphous and have diameters of 100 nm and lengths of 2–6 micrometers.

Hybrid SiO_2 nanotubes with walls that contain chiral aromatic rings are obtained by using self-assemblies of appropriate amphiphiles [81]. Mesoporous SiO_2 nanotubes with helical channels have been prepared by the self-assembly of surfactants in the presence of chiral molecules [82]. These are formed by the self-assembly of the achiral surfactant sodium dodecyl sulfate (SDS) in the presence of (R)-(+)- and (S)-(−)-2-amino-3-phenylpropan-1-ol (APP) (Fig. 6.10). TEM combined with computer simulations confirm the presence of ordered chiral channels winding around the central axis of the tubes with an inner diameter of 100 nm.

Double-walled silica nanotubes have been obtained by biomimetic synthesis under mild conditions. Pouget et al. [83]. use the peptide lanreotide in which the silica phase and the lanreotide nanotube grow synergistically in a concerted manner by mutually neutralizing their charges (positive on the lanreotide and negative on on the silica) (Fig. 6.11). This requires kinetic coupling of the two chemical processes. The presence of an X-ray diffraction peak at $0.35 Å^{-1}$, characteristic of the β-sheet organization in the lanreotide wall surface, gives a mean diameter of 24.6 nm. The walls of the tubes are separated by 2 nm, which corresponds to the thickness of the lanreotide tube. The walls of the silica tubes are thin (1.4 nm) with lengths up to 3 μm, in agreement with the TEM data shown in Figure 6.11. Calcination at 600 °C converts hybrid organic–inorganic nanostructures into pure silica double-walled nanotubes. A dynamic template mechanism can explain these results. Silica deposition occurs on both sides of the lanreotide molecule, and stops immediately after neutralization of the surface charge. This

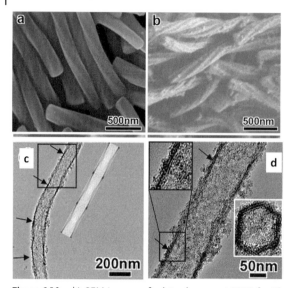

Figure 6.10 a,b) SEM images of calcined silica nanotubes with chiral mesoporous wall structure synthesized with different (R)-(+)-APP/SDS molar ratios of 0 and 0.2, respectively. These materials were synthesized at 30 °C for 6 h and then allowed to age for 1 day at 90 °C. c) Low and d) high magnification TEM images of calcined samples. Reproduced with permission from [82]. Copyright 2007, ACS.

is a well-controlled process and yields a hierarchical structure from the nanometer-scale to the macroscopic level. The double-walled silica tubes form bundles 1–3 μm in length, and bundles of closely packed aligned nanotubes as much as a centimetre long. Shape-coded SiO_2 nanotubes are obtained by using porous alumina along with the sol–gel method [84]. The template synthesis of shape-coded nanotubes begins with the fabrication of a porous alumina film that contains well-defined cylindrical pores with two or more different diameter segments created by multistep anodization of the aluminum substrate. The nanotubes are fabricated with a surface sol–gel method that controls the wall thickness at a single-nanometer level. Amorphous silica nanotubes seeded by copper sulfide nanoparticles have been synthesized by using a supercritical organic solvent [85]. Addition of copper sulfide nanocrystals, monophenylsilane, and small amounts of water and oxygen to supercritical toluene at 500 °C at 10.3 MPa yields silica nanotubes.

6.5.2
TiO_2 Nanotubes

Amongst the nanotubes of various metal oxides, those of TiO_2 have been investigated most widely. TiO_2 nanotubes can be prepared hydrothermally, by anodization of titanium by template-assisted growth as well as seeded growth. Porous alumina templates are specially useful for fabricating dense, uniform, aligned

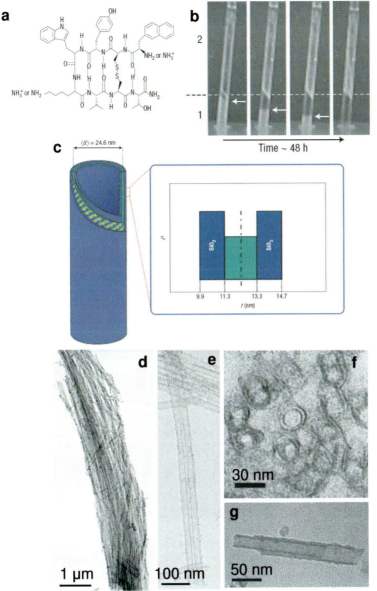

Figure 6.11 Silica mineralization of lanreotide: a) Structure of the lanreotide octapeptide showing the two charged amine sites. b) Time-lapsed pictures of the capillaries during the mineralization process. Initially, a volume of lanreotide gel at 5% (w/w) poured into the bottom of the capillary (region 1) and is covered with the same volume of a 30% (w/w) TEOS/water mixture (region 2) (tube 1). On ageing, the lanreotide turbid gel recedes (white arrows), to yield a transparent region, while long white fibres appear in region 2, starting from the interface (tubes 2 and 3). After 48 h, the lanreotide gel has completely disappeared and only two phases separated by a white ring are observed: the lower clear one containing a dilute lanreotide solution and the upper one filled with mineralized fibres (tube 4). c) Schematic representation of a multi-scale organization of a silica–lanreotide nanotube with an inner and an outer 1.4 nm thick silica shell and a central 2.0 nm thick lanreotide tube as deduced from radial density (ρ) profiles. d) TEM image of double-walled silica nanotube replica obtained after the calcination of silica–lanreotide nanotubes. e) TEM image of a 7 μm long bundle of dried mineralized nanotubes. f) Cross-sectional TEM image of dried fibres, revealing several concentric circles attributable to the double-wall structure. g) TEM image of a fragment of dried nanotube, showing that the internal and external cylinders are independent and free to slide. Reproduced with permission from [83]. Copyright 2007, Nature Publishing Group.

arrays of TiO$_2$ nanotubes on substrates such as glass, silicon, and polymers. Free-standing porous alumina templates have been employed for atomic layer deposition (ALD) of ordered TiO$_2$ nanotube arrays on various substrates (Fig. 6.12) [86]. The diameter and length of the nanotubes, as well as the distance between two neighboring nanotubes, can be controlled by varying the dimensions of the template and the anodization conditions. Typically, hexagonally packed pores with a diameter of ~65 nm and interpore distance of ~110 nm are used. The synthesis of highly ordered arrays of TiO$_2$ nanotubes by potentiostatic anodization of Ti has been reviewed by Grimes [87]. This appears to be an excellent method wherein anodic oxidation is carried out in a dimethyl sulphoxide (DMSO) medium that contains hydrofluoric acid, potassium fluoride, or ammonium fluoride as the electrolyte [88]. Self-aligned, hexagonally close packed TiO$_2$ nanotube arrays, 1000 μm in length with high aspect ratios (~10 000) are obtained by the anodization of titanium. These are polycrystalline in the anatase structure after annealing in oxygen at 280 °C for 1 h. Such nanotubes can be transformed into self-standing membranes [89].

The formation of TiO$_2$ membranes by the anodization of Ti foil in fluorine-containing ethylene glycol has been described [90]. This method yields self-organized, free-standing TiO$_2$ nanotube arrays with ultra-high aspect ratio of the diameter/length (~1500) by simply using solvent-evaporation-induced delamination of the TiO$_2$ barrier layer formed between the TiO$_2$ membrane and Ti foil during anodization. The resulting membrane consists of highly ordered, vertically aligned, one-side open TiO$_2$ nanotube arrays with pore diameter, wall thickness, and length of around 90 nm, 15 nm, and 135 μm, respectively. The as-grown TiO$_2$ nanotubes are amorphous and transform into the anatase structure after annealing at high temperature in air. Aligned TiO$_2$ nanotubes with novel morphologies, such as bamboo-type reinforced nanotubes and 2D nanolace sheets, obtained by an anodization process carried out under alternating-voltage conditions in fluoride-containing electrolytes, have been reported [91]. The experiment was carried out under constant voltage conditions, and after 2 h of anodization at 120 V in an electrolyte that consists of 0.2 mol L^{-1} HF in ethylene glycol, yielded a regular layer of aligned, individual TiO$_2$ nanotubes with thickness of about 10 μm and diameter of 150 nm. If the voltage is lowered to 40 V, the growth of the nanotubes slows down and may even stop. A bamboo-type structure can be grown under certain conditions (i.e., when the voltage is alternated between 120 and 40 V). The spacing between the bamboo rings can be altered by means of changing the time for which the sample is held at 120 V, and spacing is reduced from 200 to 70 nm by reducing the holding time. If anodization takes place for a long time at a low voltage, nanotubular features with a reduced diameter start to grow. This can be exploited to grow a double-layer structure. In this case, branching of the main tube with a diameter of 150 nm into several (typically 2–3) smaller tubes of about 50 nm in diameter occurs. The structures can be transformed to the anatase structure without losing structural integrity by annealing in air at 450 °C. Anodization under constant-voltage conditions leads to an ordered layer that consists of smooth tubes with a defined cylindrical or hexagonal cross section [92, 93]. Fluoride-free aqueous

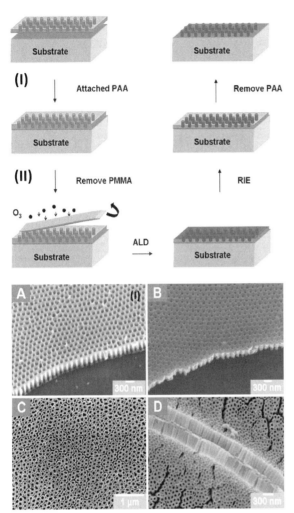

Figure 6.12 I) Schematic of the process to fabricate highly ordered TiO$_2$ nanotube arrays on substrates using ALD on a free-standing porous alumina template. Free-standing AM supported with a poly(methyl methacrylate) (PMMA) layer was first attached onto the substrate. The PMMA was then removed for ALD on the AM template. Substrates were alternatively exposed to TiCl$_4$ and water vapor for ALD at a pressure of 1×10^{-3} Torr. The TiO$_2$ overlayer was etched by RIE, and finally highly ordered ALD TiO$_2$ nanotube arrays were released from the AM template. II) Dense, uniform, highly ordered, and well-aligned ALD TiO$_2$ nanotube arrays on Si and a flexible polyimide film. SEM images of a highly ordered free-standing AM template (~300 nm) attached on Si before (A) and after (B) 150-cycle TiO$_2$ ALD. C) Highly ordered TiO$_2$ nanotube arrays released from the template on the Si substrate. D) Cross-section of well-aligned TiO$_2$ nanotube arrays on a bending polyimide film. Reproduced with permission from [86]. Copyright 2008, ACS.

HCl electrolyte is also used to obtain vertically oriented TiO_2 nanotube arrays [94]. These nanotube arrays, obtained by using a 3 M HCl aqueous electrolyte with an anodization potential of 20 V, have an inner pore diameter of 15 nm, wall thickness of 10 nm, and length up to 600 nm. The nanotubes are polycrystalline and have an anatase structure, with a rutile barrier layer separating the tubes from the underlying metal foil.

Sulfur-doped TiO_2 nanotubular arrays are obtained by potentiostatic anodization of Ti foils followed by the annealing of TiO_2 tubular arrays in a flow of H_2S at 380 °C [95]. Ordered, vertically oriented B-doped TiO_2 nanotube arrays are prepared by forming a nanotube-like TiO_2 film in the anodization process on a Ti sheet followed by CVD treatment with trimethylborate vapor [96]. A double-template-assisted sol–gel method has been used to prepare TiO_2 nanotube arrays with nanopores on their walls [97]. In this method, poly(ethylene glycol) dissolved in a TiO_2 sol is used as a soft template to form nanopores on the walls of TiO_2 nanotube arrays, which were templated from ZnO nanorod hard templates by the dip-coating technique. The microstructure of nanoporous TiO_2 nanotube arrays can change from end-opened to end-closed by increasing the number of dip-coating cycles.

Gas-phase ALD of metal oxides in combination with a micro-contact printing (μ-CP) technique has been used to attain precise atomic-level control over the dimensions (wall thickness) of nanotubes as well as the one-step fabrication of the free-standing oxide nanotubes [98]. In this procedure, octadecyltrichlorosilane (OTS) molecules were transferred onto both sides of the surfaces of the template by the μ-CP technique and self-assembled monolayers (SAMs) formed on the surfaces, thus exposing chemically inert methyl surfaces. The ALD process allows atomic-level control over the thickness of the wall of the nanotubes, and the OTS self-assembled mono-layers function as resistant layers to materials deposition, thus allowing free-standing cylindrical nanotubes to be collected after dissolution of the template without an additional polishing step. This method has been used to obtain high aspect ratio (~300) nanotubes of TiO_2, ZrO_2, and Al_2O_3. Crystalline and homogeneous CeO_2 naoparticles have been used as seeds to grow TiO_2 nanotubes from the hydrolysis of aqueous titanium sulfate solution [99]. Layer-by-layer deposition of a water-soluble titania precursor (titanium(IV) bis(ammonium lactato) dihydroxide, along with the oppositely charged poly(ethylenimine) gives rise to multilayer films. The tubular structure is obtained by depositing inside the cylindrical pores of a polycarbonate membrane followed by calcination [100]. Nanotubular TiO_2 can be obtained by using uncharged or negatively charged L-lysine-based organic gelators as templates [101]. By heating nanotubes of titanic acid (obtained by heating P25-TiO_2 (Degussa) with an NaOH solution at 110 °C) in ammonia, N-doped TiO_2 nanotubes with an anatase structure are obtained [102].

The use of carbon nanotubes to obtain oxide nanotubes is well documented [103, 104]. This method has been used to obtain pure rutile nanotubes [105]. The method involves coating of the carbon nanotubes with amorphous titania through a sol–gel process followed by heating in air to convert titania into anatase. Further heating in nitrogen transforms the anatase predominantly to rutile and finally heating in air again at a higher temperature removes the carbon nanotubes and transforms the remaining anatase to rutile. Brookite-type TiO_2 nanotubes are obtained by heating titanate nanotubes obtained by the reaction of TiO_2 powder with NaOH solution, followed by hydrothermal treatment [106]. Transparent thin films of titanate nanotube arrays with super-hydrophilic characteristics are grown on sapphire substrates by the hydrothermal reaction of sputter-deposited Ti films in an aqueous NaOH solution [107].

6.5.3
ZnO, CdO, and Al_2O_3 Nanotubes

ZnO nanotubes have been prepared by CVD and thermal evaporation, as well as by hydrothermal and solution methods. For example, large-scale ZnO nanotube bundles have been synthesized by a simple wet chemical approach (Fig. 6.13) [108]. In this method, an aqueous solution of $Zn(NO_3)_2$ and hexamethylenetetramine is stirred over long periods and heated at 90 °C. This method yields nanotubes with an inner diameter of ~350 nm and a wall thickness of ~60 nm, and forms radiating

Figure 6.13 Morphological and structural characterization of ZnO nanotube bundles: a,b) Low-magnification FE-SEM images. c) High-magnification FE-SEM image. d) TEM image. The upper left and lower right inset images of (d) are the SAED pattern and the high-resolution transmission electron microscopy (HRTEM) image of a single nanotube, respectively. Reproduced with permission from [108]. Copyright 2007, ACS.

structures. These ZnO nanotubes are single crystalline in nature, having the wurtzite structure, and preferentially grow along the [0001] direction. High aspect ratio nanotubes are obtained by a three-step low temperature process that involves ionic layer absorption, deposition of the ZnO seed layer, followed by hydrothermal annealing of the seed layer and deposition of the 1D ZnO nanostructures [109]. The uniform ZnO nanotubes have a single-crystalline wurtzite structure with lengths that exceed 10 μm and diameters of around 27 nm. By hydrothermal annealing, ZnO nanotubes grown along the <001> direction have been obtained [109]. Nanotube-based paint-brush structures of ZnO have been prepared on Zn foils by solvothermal means by adjusting the pH of the solution [110].

Electrochemical deposition from aqueous solution into porous alumina membranes can be employed to prepare ZnO nanotube arrays [111]. Single-crystalline ZnO nanotube arrays with an hexagonal wurtzite structure have been generated on glass substrates by a two-step solution approach [112]. The method involves the electrodeposition of oriented ZnO nanorods and subsequent coordination-assisted selective dissolution along the *c*-axis to form tubular structures caused by the preferential adsorption of ethylenediamine (EDA) and OH⁻ on different crystal faces. After dissolution in aqueous EDA solution for 10–15 h, the inner/outer wall surfaces of the ZnO nanotubes become smooth with a wall thickness of ~10–30 nm.

Reaction of $Zn(NO_3)_2$ with methenamine in aqueous medium under hydrothermal conditions gives rise to ZnO nanotubes [113]. These nanotubes are hollow with rough surfaces, which indicates a layer-stack structure. The ZnO nanotubes have an hexagonal wurtzite structure with lengths in the range of 1–3 μm and wall thickness in the 50–100 nm range. The cross section is hexagonally faceted, which provides strong evidence that the single nanotube grows along the *c*-axis direction. ZnO nanotube structures have been obtained by the hydrothermal self-assembly of zinc ions at the interface of surfactants, such as cetyltrimethylammonium bromide (CTAB) and the triblock copolymer of poly(ethylene oxide)–poly(propylene oxide)–poly(ethylene oxide) (P123) [114]. These molecules are suitable for the assembly of layered zinc species–surfactant hybrids that can be exfoliated into intermediate single sheets that roll to form tubular ZnO nanostructures because of the heat stress and the crystallization of ZnO sheets. The size of the tubular ZnO is determined by the type of surfactants, the concentration of the surfactant, and the Zn species/surfactant molar ratio. Tubular products can be obtained at a relatively high Zn^{2+}/surfactant molar ratio as the concentrations of CTAB and P123 are higher than their critical micelle concentrations. The hydrophobic side of each surfactant layer is connected through a weak bond, which gives rise to an organic layer, while the hydrophilic side of each surfactant layer adsorbs metal ions or hydrates, to form an inorganic layer. Dehydration and crystallization of amorphous inorganic layers cause the volume to shrink and results in stress in the inorganic layers, and breaks the equilibrium between the organic and the inorganic layers. Crystallography of these ZnO nanotubes is determined by how the nanosheets exfoliate and roll. The large cylindrical ZnO nanotubes have diameters ranging from 200 to 400 nm (wall thickness, 80 nm) with lengths going up to 14 μm. The nanotubes are single-crystalline and possess a wurtzite structure. With an increase in the amount of $Zn(CH_3COO)_2 \cdot 2H_2O$, middle-size ZnO nanotubes with diameters in the range from 40 to 70 nm and lengths up to 550 nm are obtained. The inner diameter of these nanotubes varies from 14 to 30 nm.

ZnO nanotubes with nanometer-scaled holes on the side-walls are obtained by a low-temperature hydrothermal procedure based on a preferential etching strategy [115]. The procedure is as follows. By a two-step heating of a solution mixture of $ZnCl_2$ and ammonia hydrothermally, nanotubes of nearly homogeneous size with ~250 nm diameter, 40 nm wall thickness, and 500 nm length are formed. The first stage involves the transformation of the precursor $Zn(NH_3)_4^{2+}$ to ZnO through hydrothermal decomposition at 95 °C, which leads to the formation of single crystal ZnO nanorods. At the same time, ZnO dissolves as the equilibrium moves to the left. The growth of ZnO rods becomes dominant since the high concentration of $Zn(NH_3)_4^{2+}$ favors the precipitation of ZnO. As the temperature goes down (95 to 75 °C), along with the partial consumption of the $Zn(NH_3)_4^{2+}$, etching of the ZnO nanorods occurs. Formation of porous ZnO nanotubes with nanoholes, which are single crystalline with a wurtzite structure, result from the preferential etching along the c-axis and slow etching along the radial directions. Nanoholes are created on the side-walls of the tubular structure. ZnO nanotubes have been assembled into microsphere superstructures by employing a mixture of poly(ethylene glycol) (PEG), ethanol, and water [116]. Addition of metallic zinc species into the solution mixture leads to aggregation of the PEG polymer coils to Zn^{II}/PEG globules with a diameter of ~500 nm. The globules turn into tube-like-structured ZnO–PEG microsphere assemblies (~2 mm) after ultrasonic pretreatment.

ZnO nanotubes are also produced by the oxidation of Zn nanowires because of the Kirkendall effect. This effect has also been used to produce nanotubes of SiO_2, Co_3O_4, and $ZnCr_2O_4$ as well as of ZnS, CdS, and CdSe (Fig. 6.14) [117]. It is observed that there is a distribution in the diameters of the nanotubes just as in the case of the diameters of the starting nanowires. The Zn nanowires possess a smooth surface with an average diameter of 50 nm and lengths of several tens of micrometers with a zig-zag morphology. The ZnO nanotubes formed from the thermal oxidation of the Zn nanowires have an outer diameter that goes up to 90 nm with lengths that vary from 400 nm to a few micrometers. The ZnO nanotubes have a wall thickness of around 20 nm and an inner diameter close to the diameter of the starting metal nanowires.

CdO nanotubes with a mean diameter of 50 nm have been obtained by the thermal evaporation of Cd powder without using any catalyst or template [118]. The CdO nanotubes are single-crystalline with a cubic structure and have diameters of around 40–65 nm, a wall thickness of around 15 nm, and lengths of over a few tens of micrometers. Some of the nanotubes appear straight in morphology, while others are twisted with some straight parts.

Well-defined uniform Al_2O_3 nanotubes are obtained by the pulse anodization of aluminium in H_2SO_4 solution [119]. Periodic galvanic pulses were employed to achieve mild and hard anodization (HA) conditions, where the pulse duration for

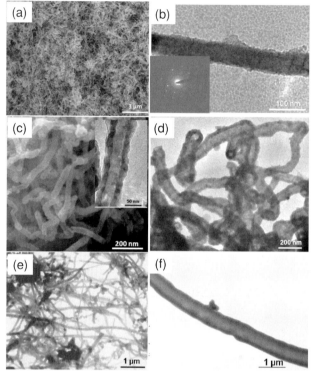

Figure 6.14 Nanowire–nanotube transformation as a result of the Kirkendall effect: a) Low-magnification FESEM images of Zn nanowires. b) TEM image of a Zn nanowire (the inset is the SAED pattern). c) FESEM image of ZnO nanotubes (inset shows the TEM image of a ZnO nanotube). d) TEM image of ZnS nanotubes. e,f) Scanning transmission electron microscope images of Si nanowires and SiO_2 nanotubes, respectively. Reproduced with permission from [117]. Copyright 2008, ACS.

HA determines the length of nanotubes. By properly choosing the pulse parameters, continuous tailoring of the pore structure of the resulting nanoporous anodic alumina (i.e., periodic modulation of pore diameters along the pore axis) is achieved. This also enables weakening of the junction strength between cells, thereby helping in the separation of individual alumina nanotubes from the porous anodic alumina. The average inner and outer diameter of the nanotubes is estimated to be 80 and 95 nm, respectively, with lengths of up to several tens of micrometers. The inner diameter of the alumina nanotubes can be controlled by varying the etching time of the pore walls by using an appropriate etching solution (e.g., H_3PO_4). Aluminium oxide nanotubes have been fabricated by using tris(8-hydroxyquinoline) gallium organic nanowires (GaQ_3) as soft templates to coat alumina using ALD [120]. By dissolving the template in toluene or by heat treatment, alumina nanotubes are prepared, where the alumina shell thickness is controlled by the number of precursor/purge cycles (Fig. 6.15).

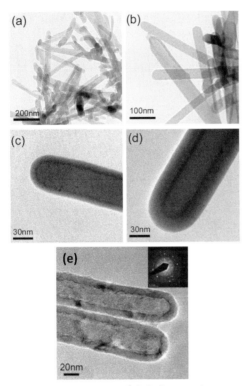

Figure 6.15 TEM images of a) GaQ$_3$ nanowires prepared by thermal evaporation. b,c) GaQ$_3$–Al$_2$O$_3$ core–shell nanowires fabricated by 100 cycles of ALD, and (d) same as (c), after 200 cycles of ALD. e) TEM image of the alumina nanotubes after heat treatment at 900 °C for 1 h. The inset shows the electron diffraction pattern of alumina. Reproduced with permission from [120]. Copyright 2007, ACS.

6.5.4
Nanotubes of Vanadium and Niobium Oxides

Vanadium oxides have received considerable attention because of their structural flexibility and useful catalytic, electrochemical, and other properties. The structure of V$_2$O$_5$ permits the intercalation of various cationic species in the interlamellar space [121]. The cations include alkylammonium ions, which are readily intercalated between the layers under hydrothermal conditions [122]. This system is thus analogous to graphite and layered dichalcogenides. Nesper and coworkers [123, 124] synthesized nanotubules of alkylammonium intercalated VO$_x$ by hydrothermal means. The vanadium alkoxide precursor was hydrolyzed in the presence of hexadecylamine and the hydrolysis product (lamellar structured composite of the surfactant and the vanadium oxide) yielded VO_x nanotubes along with the intercalated amine under hydrothermal conditions. Vanadium in these materials is in the mixed-valent state, and is redox active. The template cannot be removed by calcination as the structural stability is lost above 523 K. It is possible to partially extract the surfactant under mild acidic conditions. Nesper et al. have shown that the alkylamine intercalated in the intertubular space could be exchanged with other alkylamines of varying chain lengths as well as α,ω-diamines [124]. Such mixed valent VO_x nanotubes are also obtained under hydrothermal conditions using V_2O_5 and 3-phenylpropylamine [125]. Most of the VO_x nanotubes obtained by the hydrothermal method are open ended. Very few closed tubes had flat or pointed conical tips. Cross-sectional TEM images of the nanotubular phases show that instead of concentric cylinders (i.e., layers that fold and close within themselves), the tubes are made up of single or double layer scrolls that provide a serpentine-like morphology [124, 126]. Non-symmetric fringe patterns in the tube walls exemplify that most of the nanotubes are not rotationally symmetric and carry depressions and holes in the walls. Diamine-intercalated VO_x nanotubes are multilayer scrolls with narrow cores and thick walls, composed of vanadium oxide layers. Diamine-containing VO_x nanotubes show a smaller number of holes in the wall structure and the tubes are well ordered with uniform distances throughout the tube length [124]. The scroll-like structures of VO_x are not real nanotubes of the type formed by carbon or metal dichalcogenides. Vanadium oxide nanotubes that contain primary monoamines with long alkyl chains have been prepared by employing non-alkoxide vanadium precursors such as $VOCl_3$ and V_2O_5. The amine complexes of the vanadium precursors are then hydrolyzed. Hydrothermal treatment of the precursors gives good yields of VO_x nanotubes that incorporate the amines [127]. In general the inner diameter varies between 15 and 50 nm in the VO_x nanotubes, independent of the precursor. The outer diameters are similar, generally between 50 and 150 nm, with a range of tube lengths. Starting from the alkoxides, the tube length varies from 1 to 15 μm. In the samples obtained starting with $VOCl_3$ and V_2O_5, the average length of the tubes is shorter (1–3 μm). The distribution of the number of layers is similar in all the nanotubes, generally between 6 and 15 layers. The layers are preformed in the lamellar phase and roll up during the autoclave treatment, because the tetragonal structure of the layers gives rise to a more or less square-shaped growth. However, since many scrolls consist of multiple independent layers and the maximum length approaches 15 μm, growth along the tube axis may occur as well. Mn–V oxides have been prepared by mixing V_2O_5 with dodecylamine in the presence of ethanol and water. The amine templates are easily substituted or ion-exchanged with ions like Mn^{2+} in an aqueous alcohol solution to yield Mn–V–O nanotubes [128]. Most of the nanotubes had open ends, while some of them had closed ends, with the side of the tubes wrapped around the end to close it. The Mn^{2+} ions replace the organic cations in the structures and hence are intercalated in between the layers.

Arrays of $V_2O_5 \cdot nH_2O$ nanotubes with diameters of 200 nm and 5 μm length have been fabricated by template-based physical wetting of V_2O_5 sols [129]. Urchine-like

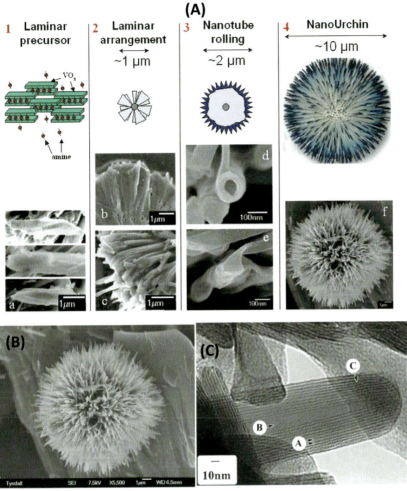

Figure 6.16 Schematic summary of the stages of growth of VO$_x$ nano-urchins. FESEM image of an individual nano-urchin. This fully grown nano-urchin is 12 μm in diameter and is covered in VO$_x$ nanotubes with a volumetric density of ~40 sr^{-1}. HRTEM image of an early stage nanotube. Lattice planes are resolved (A) and have a measured lattice spacing $a = 2.85$ nm. The hollow center (B) extends to the tip of the nanotube. A lattice plane termination dislocation is also observed (C). Reproduced with permission from [130]. Copyright 2006, ACS.

nanostructures (Fig. 6.16) that consist of high-density spherical nanotube radial arrays of vanadium oxide nanocomposites, were successfully synthesized by a simple chemical route using an ethanolic solution of vanadium tri-isopropoxide and alkylamine hexadecylamine for 7 days at 180 °C [130]. M–Nb$_2$O$_5$ nanotube arrays have been prepared starting from H–Nb$_2$O$_5$ nanorods, taking advantage of the phase transformation accompanied by the formation of voids [131].

6.5.5
Nanotubes of other Transition Metal Oxides

A hydrothermal reaction of $KMnO_4$ in HCl gives rise to α-MnO_2 nanotubes [132]. These MnO_2 nanotubes are single-crystalline and possess a tetragonal structure, with an average outer diameter of around 100 nm. The wall thickness is 30 nm and the length goes up to several micrometers. The nanotubes show nearly perfect tetragonal cross sections, consistent with the crystal structure.

By the reaction of $FeCl_3$ with water in the presence of polyisobutylene bis-succinimide (L113B) or the surfactant span80 as well as butanol under solvothermal conditions, Fe_2O_3 nanotubes are produced [133]. Fe_2O_3 nanotubes (rhombohedral structure) prepared from the surfactant (span80 as the template) were tube-like with diameters of 18–29 nm, wall thickness of 3–7 nm, and lengths of 110–360 nm. They have a multi-walled structure, with an interlayer spacing of 2.76 Å. α-Fe_2O_3 nanotubes are also obtained by the deposition of a metal salt solution and NaOH/NH_3 in the pores of templates with the initial formation of an insoluble metal hydroxide precursor, and its subsequent transformation by dehydration and crystallization to metal oxide nanotubular structures [134]. The α-Fe_2O_3 nanotubes so obtained are polycrystalline and have diameters and lengths of ~260 ± 60 nm and 6 ± 3 µm, respectively. These nanotubes possess a rhombohedral structure and consist of small mis-oriented single-crystalline nanocrystalline domains. Ordered Fe_2O_3 nanotube arrays can also be prepared by ALD in an AM [135]. In this method, the thermal decomposition of a homoleptic dinuclear iron(III) *tert*-butoxide complex ($Fe_2(OtBu)_6$) is carried out in the presence of water inside a self-ordered porous anodic AM to yield arrays of nanocrystalline Fe_2O_3 tubes with aspect ratios up to 100. Polycarbonate membranes are used to obtain nanoparticle–nanotube arrays of Fe_2O_3 by employing electrodeposition followed by calcination [136]. The arrays have lengths in the 5–6 µm range and a diameter of ~200 nm, which corresponds closely to the pore dimension and pore diameter, respectively, of the membrane. The open ends of the arrays demonstrate the hollow structure of the product. The nanotubes have a hexagonal structure.

Ferromagnetic γ-Fe_2O_3 nanotubes have been synthesized by a template process with the aid of a high magnetic field [137]. These are polycrystalline, with a cubic spinel structure and lengths of about 30 µm, a wall thickness of 20 nm, and a diameter in the 300–400 nm range. α-Fe_2O_3 is first formed at 500 °C by the thermal decomposition of $Fe(NO_3)_3$ inside the template and α-Fe_2O_3 is then transformed into γ-Fe_2O_3 in the presence of the high magnetic field. Porous alumina templates are also used to obtain Fe_3O_4 nanotubes [138]. The hydrothermal reaction of $Fe(NO_3)_2$ in ethanol at pH 12 gives rise to α-FeOOH nanotubes [139]. The so obtained α-FeOOH nanotubes were ~10 nm in outer diameter and ~6 nm in inner diameter. Electron microscopic images of the nanotubes clearly show the resolved interplanar spacing of about 4.18 Å, which corresponds to the spacing between the (110) planes of orthorhombic α-FeOOH.

Needle-like Co_3O_4 nanotubes have been prepared by employing a one-step self-supported topotactic transformation of nanoneedles of β-$Co(OH)_2$ (Fig. 6.17) [140].

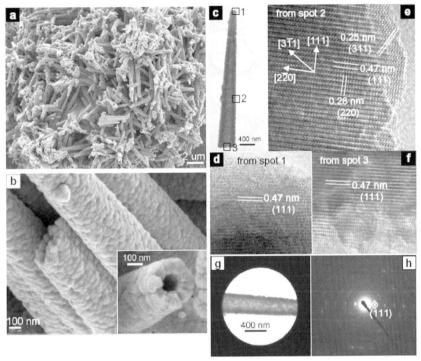

Figure 6.17 a,b) FE-SEM images of needle-like Co_3O_4 nanotubes. The inset in (b) shows a cross-sectional view of a nanotube.
c) Low-magnification TEM image of an individual needlelike Co_3O_4 nanotube.
d–f) HREM images taken from spots 1–3, respectively, of the nanotube shown in (c).
g) TEM image showing a selected area of a Co_3O_4 nanotube for electron diffraction and
h) the corresponding SAED pattern. Reproduced with permission from [140]. Copyright 2008, Wiley-VCH.

The nanotubes can be as long as 10 μm with a variable diameter in the range of 150–400 nm. The nanotubes are cylindrical and constructed from Co_3O_4 building blocks of less than 100 nm. The high-resolution (HR) TEM images in Figures 6.17d–f correspond to spots 1 to 3, respectively, of an individual nanotube (Fig. 6.17c). All show (111) lattice fringes perpendicular to the tube axis with an interplane spacing of 0.47 nm, while the two other sets of lattice fringes seen in Figure 6.17e are (220) and (311) planes, which correspond to interplanar spacings of 0.28 and 0.25 nm, respectively. On the basis of structural analysis, it is found that the tube axis is along the [111] direction with possible small mis-orientations for individual Co_3O_4 nanocrystals. The quasi-single-crystallinity of the cubic Co_3O_4 nanotubes is also confirmed by the SAED pattern (Fig. 6.17h) and the corresponding TEM image in Figure 6.17g. Co_3O_4 nanotubes are also prepared in the pores of AM by CVD, starting with cobaltacetylacetonate [141]. These nanotubes are highly ordered with a uniform diameter in the range of 100–300 nm and lengths of up to tens of micrometers. The nanotubes are composed of cubic polycrystalline

Co_3O_4. Porous AMs have been used to obtain $CoFe_2O_4$ nanotube arrays by employing a sol–gel procedure [142]. These nanotubes are several micrometers in length with a mean outer diameter of 50 nm, which corresponds to the diameter of the alumina pore. The wall thickness is 15 nm.

NiO nanotubes can be fabricated through a MOCVD route using an AM as the template and $Ni(tta)_2$tmeda (Htta = 2-thenoyl-trifluoroacetone, tmeda = tetramethylendiamine) as the Ni source [143]. The nanotubes are polycrystalline and have the bunsenite structure of NiO. The length is around 1 μm, with an outer diameter of ~200 nm with a wall thickness of 20 nm. The nickel–ammine complex ($Ni(NH_3)_x^{2+}$) has also been used as a precursor to produce NiO nanotubes in an AM [144]. These NiO nanotubes are about 60 μm in length, with an outer diameter of about 200 nm and a wall thickness of 60 nm. They are polycrystalline with a face-centered cubic structure.

Synthesis of WO_3 nanotubes has been carried out by various methods, in particular starting from tungstic acid hydrate nanotubes without using any templates [145]. WO_3 nanotubes are obtained on slow calcination at 450 °C. The tungstic acid hydrate nanotubes obtained from the solvothermal reaction have outer diameters of 300–1000 nm and lengths of 2–20 μm. The nanotubes have a nearly rectangular pore of around 250 nm and the surface of the nanotube is not smooth. The polycrystalline nanotubes when calcined at 450 °C for 3 h, retain the open ended tubular structure and the overall dimensions of their precursor nanotubes. The nanotubes have a triclinic structure. The calcined nanotubes are free-standing and show no signs of aggregation. The annealed nanotube walls consist of individual nanoparticles arranged one-dimensionally with numerous self-supported pores, which are formed by the incomplete aggregation of nanoparticles. The sidewalls of the nanotubes are, therefore, porous. Although the diameter of the nanoparticles is approximately 40–80 nm, the nanotubes are 2–20 μm long. HR-TEM images of WO_3 nanotubes show that the lattice fringes of the nanocrystals have a spacing of 0.309 nm, which corresponds to the interplanar distance of the (112) plane of triclinic tungsten trioxide.

6.5.6
Nanotubes of other Binary Oxides

MgO nanotubes can be synthesized by introducing Sn as a catalyst in the CVD process. The nanotubes are formed by the vapor–liquid–solid (VLS) mechanism [146]. MgO nanotubes can also be formed by the thermal evaporation of a mixture of Zn and Mg powders [147]. In_2O_3 nanotubes are prepared by annealing InOOH nanotubes [148], having obtained the latter under solvothermal conditions starting with $InCl_3$ in the presence of a surfactant and formamide in anhydrous ethanol. In_2O_3 tubular nanotubes are also generated by CVD starting with nanoporous structures of InP [149]. The In_2O_3 nanotubes are single crystalline with a cubic structure and have outer diameters of around 200–300 nm and lengths of around 2–6 μm and retain the size and square shape of the pores. Porous In_2O_3 nanotubes can also be prepared by layer-by-layer assembly/deposition on carbon nanotube templates followed by calcination. The layer-by-layer assembly is used to form a polyelectrolyte on the surface of pristine nanotubes, which enhances the adsorption of the metal-complex species on the surface of the carbon nanotube as a result of electrostatic attraction between the charged species [150]. In a typical procedure, the layer-by-layer assembly is formed with a polyelectrolyte such as sodium poly(styrene sulfonate) (PSS) and poly(diallyldimethylammonium chloride) (PDDA) on the surface of pristine carbon nanotubes. A solution mixture of $InCl_3$ and citric acid is added into the solution of the polyelectrolyte-modified carbon nanotubes. A solution of $NaBH_4$ is added into the above-mentioned solution to reduce In^{3+} into indium, which is readily oxidized into In_2O_{3-x} because of the oxygen dissolved in the solution from the surrounding ambient air. The porous In_2O_3 nanotubes are obtained by calcinations (Fig. 6.18). Thus, uniform, porous and polycrystalline In_2O_3 nanotubes with diameters of 30–60 nm can be formed

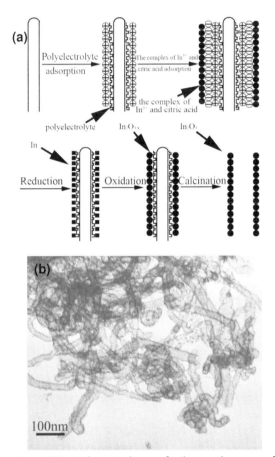

Figure 6.18 a) Schematic diagram for the growth process of In_2O_3 nanotubes. b) TEM image of regular In_2O_3 nanotubes prepared by the calcination of In_2O_3/polyelectrolyte/carbon nanotube nanocomposites at 550 °C in O_2 for 3 h. Reproduced with permission from [150]. Copyright 2007, Wiley-VCH.

using layer-by-layer assembly on carbon nanotube templates, followed by calcination. In_2O_3 nanotubes have a cubic structure and comprize nanoparticles of about 5 nm. The wall thickness is about 9 nm. This technique can also be used for the preparation of nanotubes of other oxides such NiO, SnO_2, Fe_2O_3, and CuO [150].

Y_2O_3 nanotubes are prepared by a non-aqueous electrochemical method involving oxide transfer to Y^{III} precursors [151]. These horn-shaped nanotubes (nanohorns) with a narrow size distribution coexist with small nodular deposits. The horn-shaped structures are hollow. The diameter of these structures tapers from approximately 900 nm at the base to less than 100 nm at the tip. The nanohorns have lengths that range from 3.9 to 16.5 µm. The thickness of the walls of the cylinder is approximately 250 nm, and the inner diameter is approximately 400 nm. These open cap structures have diameters similar to the nanohorn base diameters and are commonly observed at the initial stages of growth. A hydrothermal procedure has also been employed starting with $Y(OH)_3$ nanotubes, followed by calcination of the hydroxide [152].

The one-pot synthesis of SnO_2 nanotubes has been accomplished at room temperature starting from Sn nanorods [153]. The method involves the Kirkendall effect. The SnO_2 nanotubes are polycrystalline, tetragonal with diameters that range from 50 to 60 nm, and lengths from 300 to 500 nm. A 10–20 nm increase in the diameter of SnO_2 nanotubes compared with that of the starting Sn nanorods is seen as a result of the outward flow of Sn during the oxidation. A solution phase synthesis of SnO_2 nanotubes using surfactant-assisted micelles as templates has been reported [154]. In this method, a mixture of $SnCl_4$ with PVP and dimethyl sulfoxide (DMSO) is refluxed in the presence of Na_2S. The nanotubes consist of nanocrystalline particles that have a face-centered cubic structure, a diameter of ~200 nm, and lengths up to a few micrometers. Electrosynthesis of SnO_2 nanotubes employing the track-etched polymer polycarbonate membranes has been carried out [155]. A gold electrode modified with a porous polycarbonate membrane is immersed in an aqueous tin chloride solution. Electrochemistry is employed to control the local pH within the pores and drive the precipitation reaction. Removal of the gold and dissolution of the polymer yields 1D polycrystalline tin oxide particles. The crystallinity of the material is enhanced by annealing at 650 °C. The particles are hollow, with a wall thickness of approximately 10 nm. Nanotubes result from the continuous side-wall precipitation along a reaction front, which are polycrystalline SnO_2 (rutile structure), with a diameter in the 100 nm range and a length between 0.4 and 1.4 µm^2. Nanotubular indium-tin oxide (ITO) has been prepared by the sol–gel process using cellulose paper as a template [156].

ZrO_2 nanotube arrays with diameters of about 130 nm and lengths up to 190 µm are obtained by anodizing zirconium foil in a mixture of formamide and glycerol that contains NH_4F and 3 wt% water [157]. The as-prepared ZrO_2 nanotube arrays are amorphous, with the coexistence of monoclinic and tetragonal phases when annealed from 400 to 600 °C. Monoclinic zirconia nanotubes were obtained at 800 °C with retention of shape. By employing a sol–gel method and AM, yttria-stabilized ZrO_2 nanotubes have been prepared [158]. The length and the diameter of these nanotubes are 50 μm and 200 nm, respectively, in agreement with the dimensions of the template pores. The wall thickness of the nanotubes depends on the impregnation time. The nanotubes after sintering at 800 °C are polycrystalline with a cubic structure.

Formation of CeO_2 nanotubes has been reported by a few workers. CeO_2 nanotubes are produced by a two-step procedure involving precipitation at 100 °C and ageing at 0 °C for 45 days [159]. CeO_{2-x} nanotubes are crystalline, having a cubic fluorite structure. They tend to align the (111) planes parallel to the axis direction. The diameter of the nanotubes ranges from 5 to 30 nm, while the length goes up to several micrometers. The thickness of the wall of the nanotubes is about 5.5 nm. CeO_2 nanotubes are also obtained by annealing layered $Ce(OH)_3$ nanotubes by a simple oxidation–coordination–assisted dissolution process of the $Ce(OH)_3$ nanotubes/nanorods. One-dimensional structures of $Ce(OH)_3$ are synthesized by the hydrothermal treatment of $Ce_2(SO_4) \cdot 9H_2O$ with a 10 M NaOH solution at 130 °C. Starting from $Ce(OH)_3$ nanorods (as well as from narrow cavity nanotubes), it is possible to obtain CeO_2 nanotubes with large cavities by a simple oxidation and dissolution process [160]. These nanotubes with open ends have a cubic structure and display an outer diameter of about 15–25 nm and a length of about 100 nm. Nanoparticles of about 8 nm are attached to the walls and the thickness of the wall is about 5–7 nm. The inner diameter of the nanotubes is about 10–15 nm. In this method, freshly prepared 1D $Ce(OH)_3$ is exposed to air at room temperature for 24 h. The partially oxidized $Ce(OH)_3$ nanotubes with narrow cavities are dispersed in distilled water and subjected to ultrasonication for 2 h after adding 15% H_2O_2. Partial oxidation of the $Ce(OH)_3$ is essential to form the ceria tubular structures. Electrosynthesis of CeO_2 nanotubes using AMs has been reported by employing a non-aqueous electrolyte [161]. ThO_2 nanotubes are reported by the sol–gel method by using a porous alumina template [162].

6.5.7
Nanotubes of Titanates and other Complex Oxides

The hydrothermal method has been employed to synthesize nanotubes of monocrystalline $BaTiO_3$ [163], and tubular $PbTiO_3$ [164]. Pulsed laser ablation also yields $PbTiO_3$ nanotubes within the pores of an AM [165], while sol–gel electrophoretic deposition of an acetic acid-based highly stabilized lead zirconate titanate (PZT) sol in AMs gives rise to PZT nanotubes [166]. The PZT sol with a near-morphotropic phase boundary composition and no polymeric addition was prepared using lead acetate trihydrate, zirconium, and titanium tetra-butoxides. By anodization of a Ti–Zr alloy, $ZrTiO_3$ nanotubes are obtained [167].

Mallouk *et al.* [168]. and Peng *et al.* [169]. have reported the synthesis of niobate-based nanotubes at low temperatures by the exfoliation of acid-exchanged $K_4Nb_6O_{17}$ with tetra(n-butyl)ammonium hydroxide (TBA^+OH^-) in aqueous solution. In the presence of excess acid, around 80% of the potassium ions in $K_4Nb_6O_{17}$ are replaced by protons. The remaining potassium ions appear to reside in the slowly exchanging interlayer galleries that alternate along the stacking axis [168]. The bilamellar colloid initially formed by reacting this material with TBA$^+$OH$^-$ is unstable relative to the formation of unilamellar sheets and tubules. Exfoliation initially produces a bilayer colloid and then transforms (depending on conditions) irreversibly into tubules, unilamellar sheets, or a mixture of the two. The concentration of the colloid and the pH are important factors in controlling the coiling equilibrium. Both H$^+$ and alkali ions help to precipitate the aggregated tubules. The individual tubules are formed by the rolling of sheets of exfoliated K$_{4-x}$H$_x$Nb$_6$O$_{17}$ ($x \approx 3.2$). The tubules are 0.1–1 µm in length, with outer diameters that range from 15 to 30 nm. The colloids and the precipitated tubules are both highly stable.

Potassium hexaniobate nanotubes have also been fabricated from polycrystalline K$_4$Nb$_6$O$_{17}$ at room temperature using the intercalating and exfoliating methods [169]. These are multilayer crystalline nanotubes with interlayer spacings from 0.83 to 3.6 nm, depending on the intercalating molecules such as tetra(n-butyl) ammonium hydroxide (TBA$^+$OH$^-$) and alkylamines (C$_n$H$_{2n+1}$NH$_2$). The number of layers in the wall is in the range of 3 to 8. The outer diameter varies between 20 and 90 nm for the nanotubes obtained with different alkylamines, and the length of the nanotubes ranges from a few hundred nanometers to several micrometers. When a single-layer (–Nb$_6$O$_{17}$–)$_n$ sheet rolls up into a nanotube, C$_n$H$_{2n+1}$NH$_2$ ($n \neq 1$) molecules are already adsorbed on both sides of the sheet and then reside in the interlayer spaces of the nanotube. A model of the spiral structural growth of these nanotubes has been proposed and the tube axis found to be parallel to the [100] direction of the K$_4$Nb$_6$O$_{17}$ crystal. Spiral nanotubes of potassium niobate are obtained by introduction of a polyfluorinated cationic azobenzene derivative, trans-[2-(2, 2, 3, 3, 4, 4, 4-heptafluorobutylamino)ethyl] {2-[4-(4-hexylphenylazo)phenoxy]ethyl} dimethylammonium (abbreviated as C$_3$F$_7$-Azo+), into the layered niobate interlayer by a two-step guest–guest exchange method, with methyl viologen (MV^{2+})-K$_4$Nb$_6$O$_{17}$ as the precursor [170]. When MV^{2+}-intercalated niobate was used as the precursor, the polyfluorinated C$_3$F$_7$-Azo+ results in the quantitative formation of spiral nanotubes from exfoliated nanosheets of the niobate, by rolling along the sandwiched microstructure. Nanotubes of FePO$_4$ have been prepared under solvothermal conditions in the presence of a sodium dodecyl sulfate (SDS) surfactant [171]. Iron phosphate nanotubes have mesoporous walls and diameters of 50–400 nm and lengths of several micrometers. The walls of the nanotubes range from 20 to 40 nm in thickness. The removal of the surfactant by acetate exchange and heat treatment results in amorphous mesoporous nanotubes of FePO$_4$. Mesoporous NiPO$_4$ nanotubes have been prepared in the presence of a cationic surfactant and different bases by the sol–gel method [172]. The solution–liquid–solid (SLS) method has been exploited to obtain tin-filled In(OH)$_3$ nanotubes using liquid droplets of an In–Sn mixture [173].

Nanotubes and how they are formed of other complex oxides reported are: In$_2$Ge$_2$O$_7$ by thermal evaporation [174], InVO$_4$ nanotubes using templates [175], WO$_3$–H$_2$O nanotubes with the aid of intercalated polyaniline [176], chrysotile nanotubes by the hydrothermal method [177], aluminogermanate nanotubes by a simple solution procedure [178], α-FeOOH nanotubes by employing reverse micelles [179], boehmite nanotubes by a hydrothermal procedure [180], and hydroxyapatite nanotubes by a sol–gel procedure in an AM [181].

$La_{0.5}Sr_{0.5}CoO_3$ nanotubes have been prepared using an AM template by the sol–gel method [182], while $ZnAl_2O_4$ spinel nanotubes have been obtained by making use of the Kirkendall effect [183]. $ZrCr_2O_4$ has also been prepared by this method [117]. Nanotubes of $SrAl_2O_4$ [184], as well as $AgIn(WO_4)_2$ [185], are produced under hydrothermal conditions. By controlling the pH, single-walled aluminogermanate nanotubes have been synthesized by the reaction of tetraethylorthogermanate and $AlCl_3$ in NaOH solution [186]. Nickel hexacyanoferrate nanotubes have been fabricated in an AM using the electrokinetic method [187]. Single-crystal calcium sulfate nanotubes can be generated by the reverse micelle method at room temperature [188]. Self-assembled supramolecular $C_{32}H_{70}N_2ZnSO_4$ nanotubes with high thermal stability have been prepared [189]. Bismuth subcarbonate nanotubes are prepared by a simple reflux of bismuth citrate and urea in ethylene glycol [190].

6.6
Pnictide Nanotubes

Several methods to synthesize BN nanotubes including CVD and electrical discharge, as well as templating have been described in the literature [4]. Thin BN tubes of less than 200 nm diameter were first obtained by arc discharge with hollow tungsten electrodes filled with h-BN powder. Following this initial report, a variety of methods have been employed to prepare BN nanotubes. The other methods of synthesis of BN nanotubes include those that are far from equilibrium, such as the electrical arc method [191, 192], arcing between h-BN and Ta rods in a N_2 atmosphere [193], laser ablation of h-BN [194], and continuous laser heating of BN [195]. The last method produces long ropes of BN nanotubes with thin walls. Single-walled BN nanotubes are formed in some cases [191, 196]. Single-walled nanotubes of BN have been deposited on polycrystalline W substrates by using electron-cyclotron resonance nitrogen and electron beam boron sources [197].

Bando and coworkers [198, 199], have carried out extensive work on BN nanotubes. They prepared the nanotubes by the reaction of MgO and B in the presence of ammonia at 1300°C [198]. B_2O_2 (obtained by heating B and MgO at 1300°C) was heated in the presence of NH_3 in a long BN boat, which was placed into a graphite susceptor that was heated from the outside by an RF furnace to obtain BN nanotubes. The BN nanotubes exhibit a 1D nanostructure and have diameters that range from several nanometers to about 70 nm, and lengths that go up to 10 micrometers. They show the presence of mixed hexagonal and rhombohedral BN phases. The nanotubes of diameter around 10 nm exhibit a perfectly cylindrical structure, although edge dislocations are occasionally observed from the rhombohedrally stacking, ordered nanotubes. The concentric tube structures do not contain internal wall closures or internal cap structures. Bando and coworkers [199], have also prepared BN nanotubes by the reaction of MgO, FeO, and B in the presence of NH_3 at 1400°C. The nanotubes are in the pure h-BN phase and have diameters of ~50 nm and lengths of up to several tens of micrometers. The reaction of boric acid or B_2O_3 with N_2 or NH_3 at high temperatures in the presence of carbon or catalytic metal particles has also been employed in the preparation of BN nanotubes [200]. BN nanotubes can be grown directly on substrates at 873 K by a plasma-enhanced laser-deposition technique [201].

B_2O_3-coated multi-walled carbon nanotubes, on reaction with NH_3 at high temperatures, yield BN nanotubes [202]. Bando and coworkers have reviewed various aspects of BN nanotubes and cite methods involving arc-discharge, laser ablation, and the plasma jet method (Fig. 6.19) [203]. Straight BN nanotubes have been grown by heating a mixture of $Mg(BO_2)_2 \cdot H_2O$, NH_4Cl, NaN_3, and Mg powder in an autoclave at 600 °C for 20–60 h. These BN nanotubes had diameters mainly in the range of 30–300 nm and lengths of up to ~5 µm, and a majority of them had at least one closed end [204]. Multi-walled BN nanotubes have been obtained by a reduction–nitridation route wherein the reaction of boron trifluoride etherate and sodium azide was carried out in the presence of Fe–Ni powder in a sealed autoclave at 600 °C for 12 h. The BN nanotubes so obtained have an average outer diameter of 60 nm, an inner diameter of 30 nm, and lengths up to 300 nm [205]. By annealing an Fe film evaporated on boron pellets at 1000 °C in N_2 gives rise to BN nanotubes [206]. Microwave plasma-enhanced CVD has been employed to obtain BN nanotube arrays [207]. In this method, AM are used (for confinement) along with borane and ammonia or nitrogen. BN nanotubes that are 1 mm long are obtained by an optimized ball-milling and annealing method [208]. The annealing tempera-

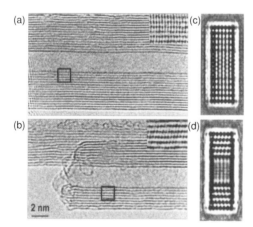

Figure 6.19 a,b) HRTEM images of two zig-zag BN MWNTs. A different stacking order is apparent in the marked areas in (a) and (b), as highlighted in the insets. Hexagonal- (a) and rhombohedral-type (b) stacking (in 12.5° fringe inclinations with respect to the tube axis) are verified by the corresponding computer-simulated HRTEM images (c and d, respectively) for BN MWNTs having the axes strictly parallel to the [10–10] orientations ("zig-zag" NTs). Note a characteristically open tip-end of a BNNT in (b). This feature is frequently seen in BNNTs obtained from high-temperature chemical syntheses. Reproduced with permission from [203b]. Copyright 2007, Wiley-VCH.

ture of 1100 °C is crucial for the growth of the long BN nanotubes because at this temperature there is a fast nitrogen dissolution rate in Fe and the B/N ratio in Fe is 1.

By combining polymer thermolysis and the use of templates, ordered arrays of BN nanotubes have been obtained [209]. In this method, liquid borazine in AM is thermolyzed at high temperatures in a nitrogen atmosphere. Double-walled BN nanotubes are obtained by the reaction of a flowing stream of borazine and ammonia with a nickelocene catalyst vapor in the hot zone of a furnace maintained at 1200 °C [210]. The mechanism of growth of single-walled BN nanotubes by laser vaporization has been examined in detail [211]. It appears that the root-growth model that involves a droplet of B is applicable for this growth. It may be noted that laser vaporization is a unique route for the synthesis of single-walled BN nanotubes.

Single-walled BCN nanotubes are obtained by bias-assisted hot-filament CVD [212]. In this method, a Fe–Mo catalyst supported on MgO powder is used for the decomposition of a mixture of CH_4, B_2H_6, and ethylenediamine. The as-grown SWNTs have clean and smooth surfaces with diameters in the range of 0.8–2.5 nm and lengths up to micrometers. Homogeneous BC_4N nanotube brushes have been obtained recently by the reaction of amorphous carbon nanotubes with boric acid and urea (Fig. 6.20) [213]. The BCN nanotubes have open ends with a diameter of 170 nm, a length of 15 mm, and thickness of around 50 nm. The SAED pattern shows faint rings and a few spots. The X-ray diffraction (XRD) pattern of the BCN nanotube brushes shows broad reflections with d-spacings of 3.43 Å and 2.13 Å, which correspond to the (002) and (100) planes, respectively, similar to the pattern reported for BC_3N. The broad reflections in the XRD pattern and the diffuse rings in the electron diffraction pattern suggest the turbo static nature of the nanotubes.

GaN nanotubes have been grown in AMs by employing MOCVD. The reactants used are trimethylgallium and NH_3 [214]. The diameter of the GaN nanotubes is approximately 200–250 nm and the wall thickness is about 40–50 nm. Nanotubular GaN consists of numerous fine GaN particulates with a size range of 15–30 nm. XRD and TEM analyses indicate that the grains in the GaN nanotubular material is nanocrystalline with a hexagonal structure. By the reaction of $GaCl_3$ with excess NaN_3 in dry benzene, GaN nanotubes with branched tubules have been grown in mesoporous MCM 48 [215]. These GaN nanotubes are polycrystalline with a wurtzite structure and an outer diameter of 50–150 nm. The tube wall thickness is ~10 nm and the lengths go up to several micrometers. Amorphous carbon nanotube brushes, on reaction with $GaCl_3$ followed by reaction with NH_3, give rise to GaN nanotube brushes (Fig. 6.21) [216]. The close packing of 1D GaN nanotube-nanobristles have diameters from 100 to 230 nm and lengths in the 7–10 μm range. The tubes are open at one end and their surface appears to be smooth. The nanotube brushes have a wurtzite structure. The TEM images in Figures 6.21c–e shows a nanotube with an outer diameter of 200 nm and the wall thickness is in the range of 10–15 nm. The HRTEM image of the nanotube wall in Figure 6.21f shows

238 | 6 Synthesis of Inorganic Nanotubes

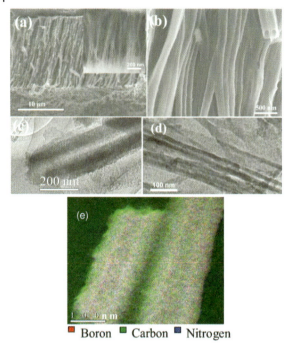

Boron ■ Carbon ■ Nitrogen

Figure 6.20 a) FESEM images of BCN nanotube brushes with the average diameter of a single tube being around 170 nm. The inset shows a FESEM image of BCN nanotube brushes of 40 nm diameter. b) Higher magnification FESEM images of BCN nanotube brushes. TEM images of a BCN nanotube: c) 170 nm diameter and d) 40 nm diameter. e) Elemental mapping of the boron, carbon, and nitrogen of BCN nanotubes obtained from EELS. Reproduced with permission from [213]. Copyright 2008, RSC.

interplanar spacing of 0.244 nm, which indicates the single crystalline nature of the GaN nanotube.

Si_3N_4 nanotubes can be prepared by the sol–gel route in which tetraethoxysilane (TEOS) and phenolic resin are used to prepare a carbonaceous silica xerogel, and ferric nitrate is employed as an additive. The xero-gel is heated at 1300 °C for 10 h in nitrogen [217]. Molybdenum nitride nanotubes have been obtained by depositing a nitride film on anodic alumina and etching away the template with NaOH [218].

Single crystalline nanotubes of Cd_3P_2 and Zn_3P_2 have been obtained by a template process in which Cd and Zn nanorods formed in-situ act as self-sacrificing templates [219]. A mixture of reactants that contain CdS (or ZnS), P, and Mn_3P_2 is heated to 1350 °C in an induction furnace to obtain nanotubes. Single-crystalline, single-walled, $SbPS_{4-x}Se$ nanotubes with a tunable band-gap have been prepared by taking a mixture of stoichiometric amounts of the elements Sb, P, S, for $SbPS_4$, and the corresponding amounts for $SbPS_3Se$, $SbPS_2Se_2$, and $SbPSSe_3$ in fused silica tubes sealed under vacuum, and heated [220].

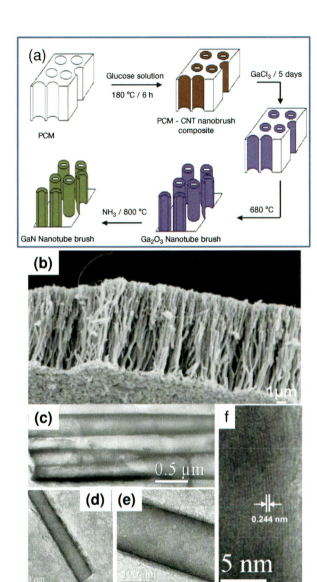

Figure 6.21 a) Schematic showing the formation of nanotube brushes of carbon, Ga$_2$O$_3$, and GaN. b) SEM image of a GaN nanotube brush. c–e) TEM images of GaN nanotubes and f) HRTEM image of the wall of a GaN nanotube. Reproduced with permission from [216]. Copyright 2007, ACS.

6.7
Nanotubes of Carbides and other Materials

SiC nanowires, SiC/SiO$_2$ core–shell nanocables, and SiC nanotubes have been synthesized simultaneously by directly heating Si powder and multiwall carbon nanotubes (MWCNTs) [221]. Single-phase TiC nanotubes are prepared by the reaction of carbon nanotubes with Ti powder at 1300 °C for 30 h [222].

Silicon oxycarbide ceramic nanotubes can be obtained by the pyrolysis of polysilicone nanotubes using a sacrificial AM as a template [223]. In a typical synthesis, the AM templates were immersed in a mixture of low-molecular-weight silicone starting materials, such as $\{CH_3(CH=CH)SiO\}_n$ (n = 3–7), named VC4, and a cyclic-type silicone bearing vinyl groups, and $(CH_3)_3SiO\{CH_3(H)SiO\}_m$ (m = 20), named KF-99 and heated in an oven at two different temperatures, 50 and 200 °C. Large-scale aligned silicon carbonitride nanotube arrays have been synthesized by microwave-plasma-assisted CVD using SiH$_4$, CH$_4$, and N$_2$ as precursors. The nanotubes are 6–7 µm in length and 100–200 nm in diameter [224].

Transition metal halides such as NiCl$_2$ crystallize in the CdCl$_2$ structure, with the metal halide layers held together by weak van der Waals forces. NiCl$_2$ has been shown to form closed cage structures and nanotubes [225]. These were prepared by heating NiCl$_2$·6H$_2$O initially in air to remove the water of crystallization, and heated further at 960 °C under Ar. Zeng et al. [226]. have carried out UV-light-induced fabrication of CdCl$_2$ nanotubes starting from CdSe solid nanocrystals through a Kirkendall effect. In this procedure, CdCl$_2$ nanotubes are obtained starting with a solution of CdSe nanocrystals in o-dichlorobenzene (which generates chlorine free radicals on photolysis).

6.8
Complex Inorganic Nanostructures Based on Nanotubes

In addition to the simple nanotube structures, more complex nanotubular structures of various inorganic materials such as coaxial cables, heterojunctions, and nanotube-nanowire or nanotube-nanoparticle composites have been prepared and characterized by several workers. Typical examples of such nanostructures are the following: Integration of ZnO nanotubes with ordered nanorods [227], conversion of ZnO nanorod arrays into ZnO–ZnS nanocable and ZnS nanotube arrays [228], core–sheath heterostructure CdS–TiO$_2$ nanotube arrays [229], ZnS nanotube–In nanowire core–shell heterostructures [230], ZnO–ZnS core–shall nanotube arrays [231], Cu nanotube–Bi nanowire heterojunctions [232], carbon nanotubes in TiO$_2$ nanotubes [233], TiO$_2$–Pt coaxial nanotube arrays [234], Sn nanowires on TiO$_2$ nanotubes [235], Fe$_2$O$_3$–TiO$_2$ nanorod–nanotube arrays [236], SiO$_2$–Ta$_2$O$_5$ core–shell nanowires and nanotubes [237], multi-walled BCN–carbon nanotube junctions [238], and BN nanotubes with periodic iron nanoparticles [239].

6.9
Outlook

The preceding presentation should suffice to demonstrate the great progress made in the synthesis of inorganic nanotubes. It would appear that today we are able to synthesize almost any inorganic material in tubular form. Not only do we have an arsenal of inorganic nanotubes but also several strategies to synthesize them. There are still many synthetic challenges. One of them relates to the need to synthesize single-walled single-crystalline nanotubes of inorganic materials. There are hardly one or two reports of single-walled inorganic nanotubes at present. With the availability of inorganic nanotubes in sufficient quantities, there is much scope to measure their properties and investigate phenomina associated with them.

This Chapter has been published previously online in
Advanced Materials 3 Aug 2009
Doi: 10.1002/adma.200803720
Copyright © 2009 WILEY-VCH Verlag GmbH & Co. KGaA, Weinheim

References

1 Rao, C.N.R., Muller, A., Cheetham, A.K. (2007) *Nanomaterials Chemistry: Recent Developments*, Wiley-VCH, Weinheim.
2 Rao, C.N.R., Govindaraj, A. (2005) *Nanotubes and Nanowires*, RSC series on Nanoscience, Royal Society of Chemistry, London.
3 Iijima, S. (1991) *Nature*, **354**, 56.
4 Rao, C.N.R. and Nath, M. (2003) *DaltonTrans*, 1.
5 Tenne, R. and Rao, C.N.R. (2004) *Phil. Trans. Royal. Soc. (London)* **362**, 2099.
6 a) Ghadiri, M.R., Granja, J.R., Milligan, R.A., McRee, D.E. and Kazanovich, N. (1993) *Nature*, **366**, 324.
b) Gazit, E. (2007) *Chem. Soc. Rev.*, **36**, 1263.
7 a) Tenne, R. (2002) *Chem. Eur. J*, **8**, 5296, and the, references, there, in.
b) Deepak, F.L. and Tenne, R. (2008) *Cent. Eur. Chem.*, **6**, 373.
8 Brumlik, C.J. and Martin, C.R. (1991) *J. Am. Chem. Soc.*, **113**, 3174.
9 Hulteen, J.C. and Martin, C.R. (1997) *J. Mater. Chem.*, **7**, 1075.
10 Zhang, X., Wang, H., Bourgeois, L., Pen, R., Zhao, D. and Webly, P.A. (2008) *J. Mater. Chem.*, **18**, 463.
11 a) Martin, C.R., Nishizawa, M., Jirage, K. and Kang, M. (2001) *J. Phys. Chem. B*, **105**, 1925. b) Wirtz, M., Martin, C. R. (2003) *Adv. Mater.*, **15**, 455.
12 Kohli, P., Wharton, J.E., Braide, O. and Martin, C.R. (2004) *J. Nanosci. Nanotechnol.*, **4**, 605.
13 Shin, T.-Y., Yoo, S.-H. and Park, S. (2008) *Chem. Mater.*, **20**, 5682.
14 Sachez-Castillo, M.A., Couto, C., Kim, W.B. and Dumesic, J.A. (2004) *Angew. Chem. Int. Ed.*, **43**, 1140.
15 Wang, H.-W., Sheih, C.-F., Chen, H.-Y., Shiu, W.-C., Russo, B. and Cao, G. (2006) *Nanotechnology*, **17**, 2689.
16 Pu, Y.-C., Hwu, J.R., Su, W.-C., Shieh, D.-B., Tzeng, Y. and Yeh, C.-S. (2006) *J. Am. Chem. Soc.*, **128**, 11606.
17 Schwartzberg, A.M., Olson, T.Y., Talley, C.E. and Zhang, J.Z. (2007) *J. Phys. Chem. C*, **111**, 16080.
18 Calvar, M.S., Pacifico, J., Juste, J.P. and Liz-Marzan, L.M. (2008) *Langmiur*, **24**, 9675.
19 Chen, Z., Waje, M., Li, W. and Yan, Y. (2007) *Angew. Chem. Int. Ed.*, **46**, 4060.

20 Takai, A., Yamauchi, Y. and Kuroda, K. (2008) *Chem. Commun.*, 4171.
21 Kijima, T., Yoshimura, T., Uota, M., Ikeda, T., Fujikawa, D., Mouri, S. and Uoyama, S. (2004) *Angew. Chem. Int. Ed.*, **43**, 228.
22 Liang, H.-P., Lawrence, N.S., Wan, L.-J., Jiang, L., Song, W.-G. and Jones, T.G.J. (2008) *J. Phys. Chem. C*, **112**, 338.
23 Han, C.C., Bai, M.Y. and Lee, J.T. (2001) *Chem. Mater.*, **13**, 4260.
24 Bao, J., Tie, C., Xu, Z., Zhou, Q., Shen, D. and Ma, Q. (2001) *Adv. Mater.*, **13**, 1631.
25 Tao, F., Guan, M., Jiang, Y., Zhu, J., Xu, Z. and Xue, Z. (2006) *Adv. Mater.*, **18**, 2161.
26 Cao, H., Wang, L., Qiu, Y., Wu, Q., Wang, G., Zhang, L. and Liu, X. (2006) *ChemPhysChem*, **7**, 1500.
27 Xu, X.J., Yu, S.F., Lau, S.P., Li, L. and Zhao, B.C. (2008) *J. Phys. Chem. C*, **112**, 4168.
28 Narayanan, T.N., Shaijumon, M.M., Ajayan, P.M. and Anantharaman, M.R. (2008) *J. Phys. Chem. C*, **112**, 14281.
29 Kamalakar, M.V. and Raychaudhuri, A.K. (2008) *Adv. Mater.*, **20**, 149.
30 Cheng, C.-L., Lin, J.-S. and Chen, Y.-F. (2008) *Mater. Lett.*, **62**, 1666.
31 Kar, S., Santra, S. and Chaudhuri, S. (2008) *Cryst. Growth. Des.*, **8**, 344.
32 Dou, X., Li, G., Huang, X. and Li, L. (2008) *J. Phys. Chem. C*, **112**, 8167.
33 Li, F.S., Zhou, D., Wang, T., Wang, Y., Song, L.J. and Xu, C.T. (2007) *J. Appl. Phys.*, **101**, 014309.
34 Zhou, D., Cai, L.-H., Wen, F.-S. and Li, F.-S (2007) *Chi. J. Chem. Phys.*, **20**, 821.
35 Liu, L., Zhou, W., Xie, S., Song, L., Luo, S., Liu, D., Shen, J., Zhang, Z., Xiang, Y., Ma, W., Ren, Y., Wang, C. and Wang, G. (2008) *J. Phys. Chem. C*, **112**, 2256.
36 Zhu, Y., Liu, F., Ding, W., Guo, X. and Chin, Y. (2006) *Angew. Chem. Int. Ed.*, **45**, 7211.
37 Gautam, U.K., Nath, M. and Rao, C.N.R. (2003) *J. Mater. Chem.*, **13**, 2845.
38 Gautam, U.K. and Rao, C.N.R. (2004) *J. Mater. Chem.*, **14**, 2530.
39 Xi, G., Xiong, K., Zhao, Q., Zhang, R., Zhang, H. and Qian, Y. (2006) *Cryst. Growth Des.*, **6**, 577.
40 Zhang, H., Yang, D., Ji, Y.J., Ma, X.Y., Xu, J. and Que, D.L. (2004) *J.Phys.Chem. B*, **108**, 1179.
41 Ma, Y.R., Qi, L.M., Ma, J.M. and Cheng, H.M. (2004) *Adv. Mater.*, **16**, 1023.
42 Li, X.M., Li, S.Q., Zhou, W.W., Chu, H.B., Chen, W., Li, I.L. and Tang, Z.K. (2005) *Cryst. Growth Des.*, **5**, 911.
43 Xi, B., Xiong, S., Fan, H., Wang, X. and Qian, Y. (2007) *Cryst. Growth, Des.*, **7**, 1185.
44 Tenne, R., Margulis, L., Genut, M. and Hodes, G. (1992) *Nature*, **360**, 444.
45 Feldman, Y., Wasserman, E., Srolovitch, D.J. and Tenne, R. (1995) *Science*, **267**, 222.
46 Nath, M., Govindaraj, A. and Rao, C.N.R. (2001) *Adv. Mater.*, **13**, 283.
47 Nath, M. and Rao, C.N.R. (2001) *Chem. Commun.*, 2336.
48 Nath, M. and Rao, C.N.R. (2001) *J. Am. Chem. Soc.*, **123**, 4841.
49 Nath, M. and Rao, C.N.R. (2002) *Angew. Chem. Int. Ed.*, **41**, 3451.
50 Nath, M. and Rao, C.N.R. (2002) *Pure Appl. Chem.*, **74**, 1545.
51 a) Deepak, F.L., Margolin, A., Wiesel, I., Bar-Sadan, M., Popovitz-Biro, R. and Tenne, R. (2006) *Nano*, **1**, 167.
b) Margolin, A., Deepak, F.L., Popovitz-Biro, R., Bar-Sadan, M., Feldman, Y. and Tenne, R. (2008) *Nanotechnology*, **19**, 095601.
52 Gordon, J.M., Katz, E.A., Feuermann, D., Yaron, A.A., Levy, M. and Tenne, R. (2008) *J. Mater. Res.*, **18**, 458.
53 Gautam, U.K., Vivekchand, S.R.C., Govindaraj, A., Kulkarni, G.U., Selvi, N.R. and Rao, C.N.R. (2005) *J. Am. Chem. Soc.*, **127**, 3658.
54 Pan, H., Poh, C.K., Zhu, Y., Xing, G., Chin, K.C., Feng, Y.P., Lin, J., Sow, C.H., Ji, W. and Wee, A.T.S. (2008) *J. Phys. Chem. C.*, **112**, 11227.
55 Rao, C.N.R., Govindaraj, A., Deepak, F.L., Gunari, N.A. and Nath, M. (2001) *Appl. Phys. Lett.*, **78**, 1853.
56 Kalyanikutty, K.P., Nikhila, M., Maitra, U. and Rao, C.N.R. (2006) *Chem. Phys. Lett.*, **432**, 190.
57 Wu, C., Yu, S.-H., Chen, S., Liu, G. and Liu, B. (2006) *J. Mater. Chem.*, **16**, 3326.
58 Shen, Q., Jiang, L., Miao, J., Hou, W., Zhu, J.-J. (2008) *Chem. Commun.*, 1683.

59 Zhang, S.-Y., Fang, C.-X., Tian, Y.-P., Zhu, K.-R., Jin, B.-K., Shen, Y.-H. and Yang, J.-X. (2006) *Cryst. Growth Des.*, **6**, 2089.
60 Niu, H. and Gao, M. (2006) *Angew. Chem. Int. Ed.*, **45**, 6462.
61 Ng, C.H.B., Tan, H. and Fan, W.Y. (2006) *Langmuir*, **22**, 9712.
62 Zhang, S.-Y., Fang, C.-X., Wei, W., Jin, B.-K., Tian, Y.-P., Shen, Y.-H., Yang, J.-X. and Gao, H.-W. (2007) *J. Phys. Chem. C*, **111**, 4168.
63 Qin, A., Fang, Y., Tao, P., Zhang, J. and Su, C. (2007) *Inorg. Chem.*, **46**, 7403.
64 Zhu, G., Liu, P., Zhou, J., Bian, X., Wang, X., Li, J. and Chen, B. (2008) *Mater. Lett.*, **62**, 2335.
65 Ye, C.H., Meng, G.W., Jiang, Z., Wang, Y.H., Wang, G.Z. and Zhang, L.D. (2002) *J. Am. Chem. Soc.*, **124**, 15180.
66 Jyoti, R.O. and Suneel, K.S. (2005) *Nanotechnology*, **16**, 2415.
67 Batabyal, S.K., Basu, C., Das, A.R. and Sanyal, G.S. (2006) *Mater. Lett.*, **60**, 2582.
68 Xiao, F., Yoo, B., Lee, K.H. and Myung, N.V. (2007) *J. Am. Chem. Soc.*, **129**, 10068.
69 Li, X.-H., Zhou, B., Pu, L. and Zhu, J.-J. (2008) *Cryst. Growth, Des.*, **8**, 771.
70 Park, K.H., Choi, J., Kim, H.J., Lee, J.B. and Uk Son, S. (2007) *Chem. Mater.*, **19**, 3861.
71 Tong, H., Zhu, Y.-J., Yang, L.-X., Li, L. and Zhang, L. (2006) *Angew. Chem. Int. Ed.*, **45**, 7739.
72 Li, C., Wang, X., Peng, Q. and Li, Y. (2005) *Inorg. Chem.*, **44**, 6641.
73 Bae, C., Yoo, H., Kim, S., Lee, K., Kim, J., Sung, M.M. and Shin, H. (2008) *Chem. Mater.*, **20**, 756.
74 Ogihara, H., Sadakane, M., Nodasaka, Y. and Ueda, W. (2006) *Chem. Mater.*, **18**, 4981.
75 Yuwono, V.M. and Hartgerink, J.D. (2007) *Langmuir*, **23**, 5033.
76 Ji, Q., Iwaura, R. and Shimizu, T. (2007) *Chem. Mater.*, **19**, 1329.
77 Yamanaka, M., Miyake, Y., Akita, S. and Nakano, K. (2008) *Chem. Mater.*, **20**, 2072.
78 Shen, G., Bando, Y. and Golberg, D. (2006) *J. Phys. Chem. B*, **110**, 23170.
79 Zhai, T., Gu, Z., Dong, Y., Zhong, H., Ma, Y., Fu, H., Li, Y. and Yao, J. (2007) *J. Phys. Chem. C*, **111**, 11604.
80 Chen, H., Elabd, Y.A. and Palmese, G.R. (2007) *J. Mater. Chem.*, **17**, 1593.
81 Chen, Y., Li, B., Wu, X., Zhu, X., Suzuki, M., Hanabusa, K. and Yang, Y. (2008) *Chem. Commun.*, 4948.
82 Wu, X., Ruan, J., Ohsuna, T., Terasaki, O. and Che, S. (2007) *Chem. Mater.*, **19**, 1577.
83 Pouget, E., Dujardin, E., Cavalier, A., Moreac, A., Valéry, C., Marchi-Artzner, V., Weiss, T., Renault, A., Paternostre, M. and Artzner, F. (2007), *Nature Mater.*, **6**, 434.
84 He, B., Son, S.J. and Lee, S.B. (2006) *Langmuir*, **22**, 8263.
85 Tuan, H.-Y., Ghezelbash, A. and Korgel, B.A. (2008) *Chem. Mater.*, **20**, 2306.
86 Tan, L.K., Chong, M.A.S. and Gao, H. (2008) *J. Phys. Chem. C*, **112**, 69.
87 Grimes, C.A. (2007) *J. Mater. Chem.*, **17**, 1451.
88 Yoriya, S., Paulose, M., Varghese, O.K., Mor, G.K. and Grimes, C.A. (2007) *J. Phys. Chem. C*, **111**, 13770.
89 Paulose, M., Prakasam, H.E., Varghese, O.K., Peng, L. and Popat, K.C. (2007) *J. Phys. Chem. C*, **111**, 14992.
90 Wang, J. and Lin, Z. (2008) *Chem. Mater.*, **20**, 1257.
91 Albu, S.P., Kim, D. and Schmuki, P. (2008) *Angew. Chem. Int. Ed.*, **47**, 1916.
92 Albu, S.P., Ghicov, A., Macak, J.M. and Schmuki, P. (2007) *Phys. Status Solidi RRL*, **1**, R65.
93 Macak, J.M., Albu, S.P. and Schmuki, P. (2007) *Phys. Status Solidi RRL*, **1**, 181.
94 Allam, N.K. and Grimes, C.A. (2007) *J. Phys. Chem. C*, **111**, 13028.
95 Tang, X. and Li, D. (2008) *J. Phys. Chem. C*, **112**, 5405.
96 Lu, N., Quan, X., Li, J.Y., Chen, S., Yu, H.T. and Chen, G.H. (2007) *J. Phys. Chem. C*, **111**, 11836.
97 Qiu, J., Yu, W., Gao, X. and Li, X. (2007) *Nanotechnology*, **18**, 295604.
98 Bae, C., Kim, S., Ahn, B., Kim, J., Sung, M.M. and Shin, H. (2008) *J. Mater. Chem.*, **18**, 1362.

99 Yue, L., Gao, W., Zhang, D., Guo, X., Ding, W. and Chen, Y. (2006) *J. Am. Chem. Soc.*, **128**, 11042.

100 Yu, A., Lu, G.Q.M., Drennan, J. and Gentle, I.R. (2007) *Adv. Funct. Mater.*, **17**, 2600.

101 Suzuki, M., Nakajima, Y., Sato, T., Shirai, H. and Hanabusa, K. (2006) *Chem. Commun.*, 377.

102 Feng, C., Wang, Y., Jin, Z., Zhang, J., Zhang, S., Wu, Z. and Zhang, Z. (2008) *New J. Chem.*, **32**, 1038.

103 Satishkumar, B.C., Govindaraj, A., Vogl, E.M., Basumallick, L. and Rao, C.N.R. (1997) *J. Mater. Res.*, **12**, 604.

104 Satishkumar, B.C., Govindaraj, A., Nath, M. and Rao, C.N.R. (2000) *J. Mater. Chem.*, **10**, 2115.

105 Eder, D., Kinloch, I.A. and Windle, A.H. (2006) *Chem. Commun.*, 1448.

106 Deng, Q., Wei, M., Ding, X., Jiang, L., Ye, B. and Wei, K. (2008) *Chem. Commun.*, 3657.

107 Miyauchi, M. and Tokudome, H. (2007) *J. Mater. Chem.*, **17**, 2095.

108 Yu, Q., Fu, W., Yu, C., Yang, H., Wei, R., Li, M., Liu, S., Sui, Y., Liu, Z., Yuan, M., Zou, G., Wang, G., Shao, C. and Liu, Y. (2007) *J. Phys. Chem. C*, **111**, 17521.

109 Ku, C.-H. and Wu, J.-J. (2006) *J. Phys. Chem. B*, **110**, 12981.

110 Kar, S. and Santra, S. (2008) *J. Phys. Chem. C*, **112**, 8144.

111 Li, L., Pan, S., Dou, X., Zhu, Y., Huang, X., Yang, Y., Li, G. and Zhang, L. (2007) *J. Phys. Chem. C*, **111**, 7288.

112 Xu, L., Liao, Q., Zhang, J., Ai, X. and Xu, D. (2007) *J. Phys. Chem. C*, **111**, 4549.

113 Tong, Y., Liu, Y., Shao, C., Liu, Y., Xu, C., Zhang, J., Lu, Y., Shen, D. and Fan, X. (2006) *J. Phys. Chem. B*, **110**, 14714.

114 Shen, L., Bao, N., Yanagisawa, K., Domen, K., Grimes, C.A. and Gupta, A. (2007) *J. Phys. Chem. C*, **111**, 7280.

115 Wang, H., Li, G., Jia, L., Wang, G. and Tang, C. (2008) *J. Phys. Chem. C*, **112**, 11738.

116 Zhou, X.F., Hu, Z.L., Chen, Y. and Shang, H.Y. (2008) *Mater. Res. Bull.*, **43**, 2790.

117 Raidongia, K. and Rao, C.N.R. (2008) *J. Phys. Chem. C*, **112**, 13366.

118 Lu, H.B., Liao, L., Li, H., Tian, Y., Wang, D.F., Li, J.C., Fu, Q., Zhu, B.P. and Wu, Y. (2008) *Mater. Lett.*, **62**, 3928.

119 Lee, W., Scholz, R. and Gosele, U. (2008) *NanoLett.*, **8**, 2155.

120 Wang, C.-C., Kei, C.-C., Yu, Y.-W. and Perng, T.-P. (2007) *NanoLett.*, **7**, 1566.

121 Galy, J. (1992) *J. Solid State Chem.*, **100**, 229.

122 Whittingham, M.S., Guo, J., Chen, R., Chirayil, T., Janauer, G. and Zavalji, P.Y. (1995) *Solid State Ionics*, **75**, 257.

123 Spahr, M.E., Bitterli, P., Nesper, R., Müller, M., Krumeich, F., Nissen, H.U. (1998) *Angew. Chem. Int. Ed.*, **37**, 1263; *Angew. Chem*, **110**, 1339.

124 Krumeich, F., Muhr, H.-J., Niederberger, M., Bieri, F., Schnyder, B. and Nesper, R. (1999) *J. Am. Chem. Soc.*, **121**, 8324.

125 Sediri, F., Touati, F. and Gharbi, N. (2007) *Mater. Lett.*, **61**, 1946.

126 Muhr, H.-J., Krumeich, F., Schönholzer, U.P., Bieri, F., Niederberger, M., Gauckler, L.J. and Nesper, R. (2000) *Adv. Mater.*, **12**, 231.

127 Niederberger, M., Muhr, H.-J., Krumeich, F., Bieri, F., Günther, D. and Nesper, R. (2000) *Chem. Mater.*, **12**, 1995.

128 Dobley, A., Ngala, K., Yang, S., Zavalji, P.Y. and Whittingham, M.S. (2001) *Chem. Mater.*, **13**, 4382.

129 Zhou, C., Mai, L., Liu, Y., Qi, Y., Dai, Y. and Chen, W. (2007) *J. Phys. Chem. C*, **111**, 8202.

130 O'Dwyer, C., Navas, D., Lavayen, V., Benavente, E., Santa Ana, M.A., Gonza'lez, G., Newcomb, S.B. and Torres, C.M.S. (2006) *Chem. Mater.*, **18**, 3016.

131 Yan, C. and Xue, D. (2008) *Adv. Mater.*, **20**, 1055.

132 Luo, J., Zhu, H.T., Fan, H.M., Liang, J.K., Shi, H.L., Rao, G.H., Li, J.B., Du, Z.M. and Shen, Z.X. (2008) *J. Phys. Chem. C*, **112**, 12594.

133 Liu, L., Kou, H.-Z., Mo, W., Liu, H. and Wang, Y. (2006) *J. Phys. Chem. B*, **110**, 15218.

134 Zhou, H. and Wong, S.S. (2008) *ACS Nano*, **2**, 944.

135 Bachmann, J., Jing, J., Knez, M., Barth, S., Shen, H., Mathur, S., Gosele, U. and

Nielsch, K. (2007) *J. Am. Chem. Soc.*, **129**, 9554.

136 Yu, X., Cao, C. and An, X. (2008) *Chem. Mater.*, **20**, 1936.

137 Wang, J., Ma, Y. and Watanabe, K. (2008) *Chem. Mater.*, **20**, 20.

138 Wang, T., Wang, Y., Li, F., Xu, C. and Zhou, D. (2006) *J. Phys. Condens. Matter*, **18**, 10545.

139 Geng, F., Zhao, Z., Geng, J., Cong, H. and Cheng, H.-M. (2007) *Mater. Lett.*, **61**, 4794.

140 Lou, X.W., Deng, D., Lee, J.Y., Feng, J. and Archer, L.A. (2008) *Adv. Mater.*, **20**, 258.

141 Shen, X.-P., Miao, H.-J., Zhao, H. and Xu, Z. (2008) *Appl. Phys. A*, **91**, 47.

142 Xu, Y., Wei, J., Yao, J., Fu, J. and Xue, D. (2008) *Mater. Lett.*, **62**, 1403.

143 Malandrino, G., Perdicaro, L.M.S., Fragala, I.L., Nigro, R. Lo, Losurdo, M. and Bruno, G. (2007) *J. Phys. Chem. C*, **111**, 3211.

144 Shi, C., Wang, G., Zhao, N., Du, X. and Li, J. (2008) *Chem. Phys. Lett.*, **454**, 75.

145 Zhao, Z.-G. and Miyauchi, M. (2008) *Angew. Chem. Int. Ed.*, **47**, 7051.

146 Yan, Y., Zhou, L., Zhang, J., Zeng, H., Zhang, Y. and Zhang, L. (2008) *J. Phys. Chem. C*, **112**, 10412.

147 Lu, H.B., Liao, L., Li, H., Tian, Y., Li, J.C., Wang, D.F. and Zhu, B.P. (2007) *J. Phys. Chem. C*, **111**, 10273.

148 Chen, C., Chen, D., Jiao, X. and Wang, C. (2006) *Chem. Commun.*, 4632.

149 Zhong, M., Zheng, M., Ma, L., Li, Y. (2007) *Nanotechnology*, **18**, 465605.

150 Du, N., Zhang, H., Chen, B., Ma, X., Liu, Z., Wu, J. and Yang, D. (2007) *Adv. Mater.*, **19**, 1641.

151 Rajasekharan, V.V. and Buttry, D.A. (2006) *Chem. Mater.*, **18**, 4541.

152 Mao, Y., Huang, J.Y., Ostroumov, R., Wang, K.L. and Chang, J.P. (2008) *J. Phys. Chem. C*, **112**, 2278.

153 Du, N., Zhang, H., Chen, B., Ma, X. and Yang, D. (2008) *Chem. Commun.*, 3028.

154 Wang, N., Cao, X. and Guo, L. (2008) *J. Phys. Chem. C*, **112**, 12616.

155 Lai, M., Martinez, J.A.G., Gratzel, M. and Riley, D.J. (2006) *J. Mater. Chem.*, **16**, 2843.

156 Aoki, Y., Huang, J. and Kunitake, T. (2006) *J. Mater. Chem.*, **16**, 292.

157 Zhao, J., Wang, X., Xu, R., Meng, F., Guo, L. and Li, Y. (2008) *Mater. Lett.*, **62**, 4428.

158 Meng, X., Tan, X., Meng, B., Yang, N. and Ma, Z.-F. (2008) *Mater. Chem. Phys.*, **111**, 275.

159 Han, W.Q., Wu, L.J. and Zhu, Y.M. (2005) *J. Am. Chem. Soc.*, **127**, 12814.

160 Zhou, K., Yang, Z. and Yang, S. (2007) *Chem. Mater.*, **19**, 1215.

161 Inguanta, R., Piazza, S. and Sunseri, C. (2007) *Nanotechnology*, **18**, 485605.

162 Lin, Z.-W., Kuang, Q., Lian, W., Jiang, Z.-Y., Xie, Z.-X., Huang, R.-B. and Zheng, L.-S. (2006) *J. Phys. Chem. B*, **110**, 23007.

163 Bao, N., Shen, L., Srinivasan, G., Yanagisawa, K. and Gupta, A. (2008) *J. Phys. Chem. C*, **112**, 8634.

164 Yang, Y., Wang, X., Zhong, C., Sun, C. and Li, L. (2008) *Appl. Phys. Lett.*, **92**, 122907.

165 Singh, S. and Krupanidhi, S.B. (2007) *Appl. Phys. A*, **87**, 27.

166 Nourmohammadi, A., Bahrevar, M.A. and Hietschold, M. (2008) *Mater. Lett.*, **62**, 3349.

167 Yasuda, K. and Schmuki, P. (2007) *Adv. Mater.*, **19**, 1757.

168 Saupe, G.B., Waraksa, C.C., Kim, H.N., Han, Y.J., Kaschak, D.M., Skinner, D.M. and Mallouk, T.E. (2000) *Chem. Mater.*, **12**, 1556.

169 Du, G., Chen, Q., Yu, Y., Zhang, S., Zhu, W. and Peng, L.M. (2004) *J. Mater. Chem.*, **14**, 1437.

170 Tong, Z., Takagi, S., Shimada, T., Tachibana, H. and Inoue, H. (2006) *J. Am. Chem. Soc.*, **128**, 684.

171 Yu, D., Qian, J., Xue, N., Zhang, D., Wang, C., Guo, X., Ding, W. and Chen, Y. (2007) *Langmuir*, **23**, 382.

172 Yu, J., Wang, A., Tan, J., Li, X., Bokhoven, J.A.V. and Hu, Y. (2008) *J. Mater. Chem.*, **18**, 3601.

173 Fang, Y., Wen, X. and Yang, S. (2006) *Angew. Chem. Int. Ed.*, **45**, 4655.

174 Yan, C., Zhang, T. and Lee, P.S. (2008) *Cryst. Growth, Des.*, **8**, 3144.

175 Wang, Y. and Cao, G. (2007) *J. Mater. Chem.*, **17**, 894.

176 Wang, Z., Zhou, S. and Wu, L. (2007) *Adv. Funct. Mater.*, **17**, 1790.

177 Olson, B.G., Decker, J.J., Nazarenko, S., Yudin, V.E., Otaigbe, J.U., Korytkova, E.N. and Gusarov, V.V. (2008) *J. Phys. Chem. C*, **112**, 12943.

178 Levard, C., Rose, J., Masion, A., Doelsch, E., Borschneck, D., Olivi, L., Dominici, C., Grauby, O., Woicik, J.C. and Bottero, J.-Y. (2008) *J. Am. Chem. Soc.*, **130**, 5862.

179 Yu, T., Park, J., Moon, J., An, K., Piao, Y. and Hyeon, T. (2007) *J. Am. Chem. Soc.*, **129**, 14558.

180 Zhao, Y., Frost, R.L., Martens, W.N. and Zhu, H.Y. (2007) *Langmuir*, **23**, 9850.

181 Yuan, Y., Liu, C., Zhang, Y. and Shan, X. (2008) *Mater. Chem. Phys.*, **112**, 275.

182 Liu, W., Wang, S., Chen, Y., Fang, G., Li, M. and Zhao, X.-Z. (2008) *Sensors Actuators B*, **134**, 62.

183 Fan, H.J., Knez, M., Scholz, R., Nielsch, K., Pippel, E., Hesse, D., Zacharias, M. and Gosele, U. (2006) *Nat. Mater.*, **5**, 627.

184 Ye, C., Bando, Y., Shen, G. and Golberg, D. (2006) *Angew. Chem. Int. Ed.*, **45**, 4922.

185 Song, S., Zhang, Y., Xing, Y., Wang, C., Feng, J., Shi, W., Zheng, G. and Zhang, H. (2008) *Adv. Funct. Mater.*, **18**, 2328.

186 Mukherjee, S., Kim, K. and Nair, S. (2007) *J. Am. Chem. Soc.*, **129**, 6820.

187 Chen, W. and Xia, X.-H. (2007) *Adv. Funct. Mater.*, **17**, 2943.

188 Chen, Y. and Wu, Q. (2008) *Colloids, Surf. A*, **325**, 33.

189 Huang, K.-W., Wang, J.-H., Chen, H.-C., Hsu, H.-C., Chang, Y.-C., Lu, M.-Y., Lee, C.-Y. and Chen, L.-J. (2007) *J. Mater. Chem.*, **17**, 2307.

190 Chen, R., So, M.H., Yang, J., Deng, F., Che, C.-M. and Sun, H. (2006) *Chem. Commun.*, 2265.

191 Loiseau, A., Williame, F., Demonecy, N., Hug, G. and Pascard, H. (1996) *Phys. Rev. Lett.*, **76**, 4737.

192 Chopra, N.G., Luyken, R.G., Cherrey, K., Crespi, V.H., Cohen, M.L., Louie, S.G. and Zettl, A. (1995) *Science*, **269**, 966.

193 Terrones, M., Hsu, W.K., Terrones, H., Zhang, J.P., Ramos, S., Hare, J.P., Castillo, R., Prassides, K., Cheetham, A.K.. Kroto, H.W., Walton, D.R.M. (1996) *Chem. Phys. Lett.*, **259**, 568.

194 a) Yu, D.P., Sun, X.S., Lee, C.S., Bello, I., Lee, S.T., Gu, H.D., Leung, K.M., Zhou, G.W., Dong, Z.F. and Zhang, Z. (1998) *Appl. Phys. Lett.*, **72**, 1966. b) Zhou, G.W., Zhang, Z., Bai, Z.G. and Yu, D.P. (1999) *Solid, State, Commun.*, **109**, 555. c) Terauchi, M., Tanaka, M., Matsuda, H., Takeda, M. and Kimura, K. (1997) *J. Electron Microsc.*, **1**, 75.

195 Laude, T., Matsui, Y., Marraud, A. and Jouffrey, B. (2000) *Appl. Phys. Lett.*, **76**, 3239.

196 a) Golberg, D., Bando, Y., Han, W., Kurashima, K. and Sato, T. (1999) *Chem. Phys. Lett.*, **308**, 337. b) Fowler, P.W., Rogers, K.M., Seifert, G., Terrones, M. and Terrones, H. (1999) *Chem. Phys. Lett.*, **299**, 359.

197 Bengu, E. and Marks, L.D. (2001) *Phys. Rev. Lett.*, **86**, 2385.

198 Tang, C.C., Bando, Y., Sato, T. and Kurashima, K. (2002) *Chem. Commun.*, 1290.

199 Zhi, C.Y., Bando, Y., Tang, C. and Golberg, D. (2005) *Solid State Commun.*, **135**, 67.

200 Deepak, F.L., Vinod, C.P., Mukhopadhyay, K., Govindaraj, A. and Rao, C.N.R. (2002) *Chem. Phys. Lett.*, **353**, 345.

201 Wang, J., Kayastha, V.K., Yap, Y.K., Fan, Z., Lu, J.G., Pan, Z., Ivanov, I.N., Puretzky, A.A. and Geohegan, D.B. (2005) *Nano Lett.*, **5**, 2528.

202 Pal, S., Vivekchand, S.R.C., Govindaraj, A. and Rao, C.N.R. (2007) *J. Mater. Chem.*, **17**, 450.

203 a) Golberg, D., Bando, Y., Tang, C. and Zhi, C. (2007) *Adv. Mater.*, **19**, 2413. b) Golberg, D., Bando, Y., Bourgeois, L., Kurashima, K. and Sato, T. (2000) *Appl. Phys. Lett.*, **77**, 1979.

204 Dai, J., Xu, L., Fang, Z., Sheng, D., Guo, Q., Ren, Z., Wang, K. and Qian, Y. (2007) *Chem. Phys. Lett.*, **440**, 253.

205 Chen, X., Wang, X., Liu, J., Wang, Z. and Qian, Y. (2005) *Appl. Phys. A*, **81**, 1035.

206 Oku, T., Koi, N. and Suganuma, K. (2008) *Diamond Relat. Mater.*, **17**, 1805.
207 Wang, X.Z., Wu, Q., Hu, Z. and Chen, Y. (2007) *Electrochem. Acta*, **52**, 2841.
208 Chen, H., Chen, Y., Liu, Y., Fu, L., Huang, C. and Llewellyn, D. (2008) *Chem. Phys. Lett.*, **463**, 130.
209 Bechelany, M., Bernard, S., Brioude, A., Cornu, D., Stadelmann, P., Charcosset, C., Fiaty, K. and Miele, P. (2007) *J. Phys. Chem. C*, **111**, 13 378.
210 Kim, M.J., Chatterjee, S., Kim, S.M., Stach, E.A., Bradley, M.G., Pender, M.J., Sneddon, L.G. and Maruyama, B. (2008) *NanoLett.*, **8**, 3298.
211 Arenal, R., Stephan, O., Cochon, J.-L. and Loiseau, A. (2007) *J. Am. Chem. Soc.*, **129**, 16 183.
212 Wang, W.L., Bai, X.D., Liu, K.H., Xu, Z., Golberg, D., Bando, Y. and Wang, E.G. (2006) *J. Am. Chem. Soc.*, **128**, 6530.
213 Raidongia, K., Jagadeesan, D., Upadhyay-Kahaly, M., Waghmare, U.V., Pati, S.K., Eswaramoorthy, M. and Rao, C.N.R. (2008) *J. Mater. Chem.*, **18**, 83.
214 Jung, W.-G., Jung, S.-H., Kung, P. and Razeghi, M. (2006) *Nanotechnology*, **17**, 54.
215 Gai, L., Jiang, H., Ma, W., Cui, D., Lun, N., Wang, Q. (2007) *J. Phys. Chem. C*, **111**, 2386.
216 Dinesh, J., Eswaramoorthy, M. and Rao, C.N.R. (2007) *J. Phys. Chem. C*, **111**, 510.
217 Wang, F., Jin, G.-Q. and Guo, X.-Y. (2006) *Mater. Lett.*, **60**, 330.
218 Miikkulainen, V., Suvanto, M. and Pakkanen, T.A. (2008) *Thin Solid Films*, **516**, 6041.
219 Shen, G., Bando, Y., C.Ye, Yuan, X., Sekiguchi, T. and Golberg, D. (2006) *Angew. Chem. Int. Ed.*, **45**, 7568.
220 Malliakas, C.D. and Kanatzidis, M.G. (2006) *J. Am. Chem. Soc.*, **128**, 6538.
221 Li, B., Wu, R., Pan, Y., Wu, L., Yang, G., Chen, J. and Zhu, Q. (2008) *J. Alloys Compounds*, **462**, 446.
222 Taguchi, T. and Yamamoto, H. (2007) *J. Phys. Chem. C*, **111**, 18 888.
223 Wan, C., Guo, G., Zhang, Q. (2008) *Mater. Lett.*, **62**, 2776.
224 Liao, L., Xu, Z., Liu, K.H., Wang, W.L., Liu, S., Bai, X.D., Wang, E.G., Li, J.C. and Liu, C. (2007) *J. Appl. Phys.*, **101**, 114 306.
225 Hacohen, Y.R., Grunbaum, E., Tenne, R., Sloan, J. and Hutchison, J.L. (1998) *Nature*, **395**, 336.
226 Zeng, J., Liu, C., Huang, J., Wang, X., Zhang, S., Li, G. and Hou, J. (2008) *Nano Lett.*, **8**, 1318.
227 Fan, D.H., Shen, W.Z., Zheng, M.J., Zhu, Y.F. and Lu, J.J. (2007) *J. Phys. Chem. C*, **111**, 9116.
228 Yan, C. and Xue, D. (2006) *J. Phys. Chem. B*, **110**, 25 850.
229 Yin, Y., Jin, Z. and Hou, F. (2007) *Nanotechnology*, **18**, 495 608.
230 Gautam, U.K., Fang, X., Bando, Y., Zhan, J. and Golberg, D. (2008) *ACS Nano*, **2**, 1015.
231 Liao, H.-C., Kuo, P.-C., Lin, C.-C. and Chen, S.-Y. (2006) *J. Vac. Sci. Technol.*, **24**, 2198.
232 Yang, D., Meng, G., Zhang, S., Hao, Y., An, X., Wei, Q., Ye, M. and Zhang, L. (2007) *Chem. Commun.*, 1733.
233 Yang, L., Luo, S., Liu, S. and Cai, Q., (2008) *J. Phys. Chem. C*, **112**, 8939.
234 Chen, H., Chen, S., Quan, X., Yu, H., Zhao, H. and Zhang, Y. (2008) *J. Phys. Chem. C*, **112**, 9285.
235 Djenizian, T., Hanzu, I. and Premchand, Y.D. (2008) *Nanotechnology*, **19**, 205 601.
236 Mohapatra, S.K., Banerjee, S. and Misra, M. (2008) *Nanotechnology*, **19**, 315 601.
237 Chueh, Y.-L., Chou, L.-J. and Wang, Z.L. (2006) *Angew. Chem. Int. Ed.*, **45**, 7773.
238 Liao, L., Liu, K., Wang, W., Bai, X., Wang, E., Liu, Y., Li, J. and Liu, C. (2007) *J. Am. Chem. Soc.*, **129**, 9562.
239 Chen, Z.-G., Zou, J., Li, F., Liu, G., Tang, D.-M., Li, D., Liu, C., Ma, X., Cheng, H.-M., Lu, G.Q. and Zhang, Z. (2007) *Adv. Funct. Mater.*, **17**, 3371.

7
Gold Nanoparticles and Carbon Nanotubes: Precursors for Novel Composite Materials
Thathan Premkumar and Kurt E. Geckeler

7.1
Introduction

One of the most stimulating challenges in modern chemistry, materials science, and nanotechnology is the manipulation and assembly at the nanoscale. Carbon nanotubes (CNTs) constitute a novel class of nanomaterials with remarkable applications in diverse domains. The attachment of metal nanoparticles to CNTs represents a new means of obtaining novel hybrid materials with exciting characteristics for a variety of applications such as catalysts and gas sensors, as well as electronic and magnetic devices. These exceptional characteristics, which include excellent electronic properties, good chemical stability, and a large surface area, make CNTs good candidates as support materials for gold nanoparticles in many key applications, ranging from advanced catalytic systems through very sensitive electrochemical sensors and biosensors to highly efficient fuel cells. In this chapter, we focus on the recent progress in this area by exploring the various synthetic approaches and types of assemblies and interactions, in which gold nanoparticles can be attached to CNTs, and also survey the various applications of the resulting composites.

7.2
Gold Nanoparticles

The name of gold (Au) is derived from the Latin *aurum* (meaning "shining dawn"). Gold has the atomic number 79 (electron configuration, [Xe] $4f^{14}$, $5d^{10}$, $6s^{1}$), and is one of the so-called "noble" metals, which are resistant to both corrosion and oxidation. Gold is the most malleable and ductile of the known elements, with characteristic properties of being dense, soft, and shiny. Gold nanoparticles (AuNPs), which are also known as "colloidal gold" or "nanogold", are a colloid of nanosized gold particles in a liquid (generally water). The first description of

Advanced Nanomaterials. Edited by Kurt E. Geckeler and Hiroyuki Nishide
Copyright © 2010 WILEY-VCH Verlag GmbH & Co. KGaA, Weinheim
ISBN: 978-3-527-31794-3

colloidal gold was reported by the philosopher and medical doctor Francisci Antonii in 1618 [1]. Subsequently, in 1676, the chemist Johann Kunckles [2] described "... drinkable gold that contains metallic gold in a neutral, slightly pink solution that exert curative properties for several diseases." Although colloidal gold has been known from ancient times, when it was used as for staining glass, the modern scientific development of colloidal gold did not begin until Michael Faraday's pioneering studies in the nineteenth century. In 1857, Faraday prepared a red-colored solution of colloidal gold in a two-phase system by the reduction of an aqueous solution of chloroaurate ($AuCl_4^-$) using phosphorus in carbon disulfide [3, 4]. The AuNPs are the most stable metal nanoparticles, and have been studied extensively during the past decade owing to their fascinating aspects in materials science, size-related electronics, and optical properties (quantum size effect), as well as their applications in catalysis and biology. This is clearly evident from a literature survey of published articles relating to AuNPs with respect to years [5]. This analysis reveals that reports on AuNPs have increased significantly each year (Figure 7.1a), and indicates that remarkable attention–from both fundamental

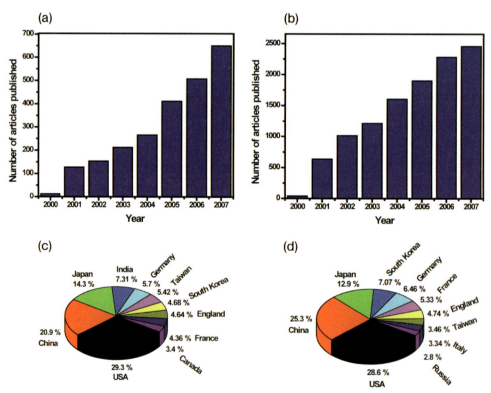

Figure 7.1 Statistical survey of articles published on (a) AuNPs and (b) CNTs. Contributions from the top ten countries in percentage for (c) AuNPs and (d) CNTs.

and applied aspects – has been paid to the synthesis, characterization, and potential application of AuNPs.

When reducing the first, second or third dimensions (1-D, 2-D or 3-D) of bulk material to the nanometer scale, nanometer-thick 2-D layers, 1-D nanowires, or 0-D nanoclusters, respectively, are produced. Nanoparticles have diameters of <100 nm and demonstrate new or enhanced size-dependent properties when compared to larger particles, or to the bulk entity of the same material. Metal nanoparticles – and especially AuNPs – appeared as a new class of materials which were of interest not only in the fields of materials science and chemistry but also in nanotechnology, on the basis of their exceptional catalytic, electronic, and optical properties. Notably, these features related neither to the bulk metal nor to the molecular compounds, but depended heavily on the particle size, the nature of the protecting organic shell, and the shape of the nanoparticle [6]. For example, gold nanocrystals of different shapes possess unique optical scattering responses. Whilst symmetric spherical particles display a single scattering peak, anisotropic shapes such as rods [7], triangular prisms [8], and cubes [9] exhibit multiple scattering peaks in the visible wavelengths due to highly localized charge polarizations at their corners and edges.

With regards to metallic gold, the nucleation and growth of nanoparticles have been most widely achieved using colloidal methods rather than a variety of other approaches [4]. In these techniques, the general approach is to reduce a gold salt precursor in solution (mostly aqueous) in the presence of a stabilizing or protecting agent; this improves the chemical stability of the AuNPs formed by avoiding the aggregation among the particles. The advantages of these methods are: (i) large quantities of nanoparticles can be synthesized; (ii) solution-based processing and assembly can be readily implemented; (iii) no specialized equipment is necessary; (iv) the technique is comparatively low-cost in nature; and (v) the preparation is both facile and user-friendly. The aforesaid reasons are particularly important when considering the real applications of AuNPs; in order to utilize AuNPs as catalytic, electronic, or optical materials a large-scale synthesis and assembly process is required.

7.3
Carbon Nanotubes

Carbon nanotubes, a new carbon allotrope, were discovered by Sumio Iijima in 1991 [10] and have attracted wide and interdisciplinary attention. CNTs are fullerene-related, tube-like structures which may be considered as a rolled-up sheet of graphene. Every carbon atom is covalently bonded to three of its neighbors, and the fourth electron is free to move over the whole structure; that is, it is delocalized (sp^2-hybridization). There are two main types of CNT with high structural perfection: single-walled carbon nanotubes (SWNTs), and multiwalled carbon nanotubes (MWNTs). While the SWNTs consist of a single sp^2-bonded graphene sheet seamlessly wrapped into a hollow cylindrical tube, MWNTs include an array of concentric cylinders (a collection of several concentric SWNTs). Both of these are typically a few nanometers in diameter (ca. between 0.4 and 3 nm for SWNTs, and between 1.4 and 100 nm for MWNTs [11]), and up to several micrometers to millimeters in length. The SWNTs represent a very important type of CNT as they exhibit significant electric properties that are not shared by the MWNT variants. In analogy to the graphite modification of carbon, the unit cells of CNTs are planar hexagons. CNTs may have a variety of helical structures [12], depending on the graphene sheet rolling-up phenomenon. The CNTs form bundles, which are entangled together in the solid state, giving rise to a highly complex network. Depending on the arrangement of the hexagon rings along the tubular surface, CNTs can be either metallic or semiconducting.

The history of CNTs may be traced back to 1976, with the fabrication of very small-diameter (<10 nm) carbon filaments [13, 14]. Nevertheless, until the discovery of fullerene chemistry [15] in 1985, the importance of these as-prepared carbon materials was not well documented. The discovery and development of fullerenes led to an increased interest in carbon materials, and this in turn induced the systematic study of carbon filaments. The CNT structure was first observed by Iijima in 1991, using high-resolution transmission electron microscopy (HR-TEM) [10]. Although the first CNTs to be discovered were MWNTs, subsequently – in 1993 – smaller-diameter SWNTs were independently observed by Iijima [16] and Bethune [17]. Since then, efficient methods for the synthesis of large-quantity [18] and high-quality [19] CNTs, with their exciting electronic, mechanical, and structural properties, as well as their diverse potential applications, have paved the way for global research interest and development activities. Today, these materials form the subject of intensive investigation due not only to their academic and industrial characteristics but also the scope of their application. Indeed, both the number of studies on, and scientific interest in, CNTs have increased dramatically and continue to increase since their discovery – as demonstrated by the huge numbers of articles relating to this topic which are published each year (Figure 7.1b).

Both, SWNTs and MWNTs may generally be synthesized using methods which include arc-discharge [18, 20], laser ablation [19, 21], chemical vapor deposition (CVD) [22], and the gas-phase catalytic process (HiPco) method [23]. Unfortunately, no method has yet been introduced for the synthesis of pure CNTs; rather, all of the methods proposed to date for the synthesis of CNTs lead not only to impure products, such as carbon-coated metal catalysts, carbon-coated metal, and carbon nanoparticles/amorphous carbon, but also to structural defects such as dangling bonds. The impurities may generally be removed using a variety of purification techniques, including: the oxidation of contaminants [24]; flocculation and selective sedimentation [25]; filtration [26]; size-exclusion chromatography [27]; selective interaction with organic polymers [28]; and microwave irradiation [29]. Despite such a variety of techniques, each has been shown to have its own limitations, such that problems persist in the purification of CNTs.

As CNTs can be synthesized with extreme aspect ratios (i.e., length-to-diameter ratios), they are particularly appropriate for applications in electronics and semiconductor devices. The inherent size, hollow geometry and extraordinary electronic properties of CNTs make them promising building blocks for molecular or nanoscale devices. Further, the well-defined geometry, exceptional mechanical properties and remarkable electrical characteristics, in addition to their outstanding physical properties [10] (see Table 7.1), qualify them for potential applications [30] in nanoelectronic circuits, field emitters, nanoelectronic devices, nanotube aquators, batteries, probe tips for scanning probe microscopy, nanotube-reinforced materials, nanoelectromechanical systems (NEMS), and nanorobotic systems [31].

Table 7.1 Important properties of CNTs.

Property	Item	Data	Potential application
Geometric	Layers	Single/multiple	Structures, probes, grippers/tweezers, scissors
	Aspect ratio	10–1000	
	Diameter	~0.4–3 nm (SWNTs)	
		~1.4–100 nm (MWNTs)	
	Length	Several μm to mm	
Mechanical	Young's modulus	~1 TPa (steel: 0.2 TPa)	
	Tensile strength	45 GPa (steel: 2 GPa)	
	Density	~1.33–1.4 g cm^{-3} (Al: 2.7 g cm^{-3})	
	Interlayer friction	Ultra-small	Actuators, bearings, syringes, switches, memories
Electronic	Conductivity	Metallic/semi-conducting	Diodes, transistors, switches, logic gates
	Current carrying capacity	~1 TA cm^{-3} (Cu: 1 GA cm^{-3})	Wires/cables
	Field emission	Activate phosphorus at ~1–3 V	Proximity/position sensors
	Band gap	Eg (eV) ≈ 1 d^{-1} (nm)	
	Electron transport	Ballistic, no scattering	
	Maximum current density	10^{10} A cm^{-2}	
Electromechanical	Piezoresistivity	Positive/negative	Deformation/displacement sensors
Thermal	Heat transmission	>3 kW mK^{-1} (Diamond: 2 kW mK^{-1})	Circuits, sensors, thermal actuators

TPa, Tera-Pascal; GPa, Giga-Pascal; TA, Tera-Ampere; GA, Giga-Ampere.

7.4
CNT–Metal Nanoparticle Composites

In nanocomposites, different materials such as metal nanoparticles, CNTs and clay may be combined with another material, usually a polymer. Nanocomposites have recently attracted considerable interest in both academic and industrial fields due to their unique mechanical, thermal and electronic properties. Nanoscaled particulates such as CNTs have been investigated as fillers in a variety of polymeric matrices to produce enhanced properties. Because of their exceptionally small diameters (several nm), as well as their high Young's modulus (~1 TPa), tensile strength (~200 GPa) and high elongation (10–30%), in addition to a high chemical stability, CNTs represent attractive reinforcement materials for lightweight and high-strength metal matrix composites. Metallic composites (especially with noble metals) containing CNTs would offer distinct advantages over polymeric composites. However, the development of metal matrix composites remains in its infancy, despite its great potential, primarily because of the high fabrication costs involved and difficulties in scaling-up the production process.

It has been predicted, based on computer stimulations, that open nanotubes may be filled with liquid by capillary suction [32]. This has provided much speculation that the filling of extraneous materials into the hollow nanotube cavities might have interesting effects on the physical and electronic properties of the encapsulated materials. The modification of CNTs into composite nanofibers by filling them with molten materials through capillary action has been reported [33, 34], and subsequently metal carbides have been entrapped into the hollow cavities of CNTs by using an arc discharge method [35, 36]. In contrast, in 1994 Ajayan and coworkers [37] reported for the first time that a combination of CNTs and metal nanoparticles could also be achieved by depositing the metal cluster on the surface (outside the tube) of the CNTs; this was termed the "decoration" of CNTs with metal clusters. The method involved the use of SWNTs as a support material for dispersing ruthenium nanoparticles that would then serve as a catalyst in heterogeneous catalysis. As a result, ~0.2% (w/w) of Ru nanoparticles were deposited on the surface of the CNTs. Following these successes, a variety of hybrid composites was introduced and developed with either metal or metal oxide or semiconductor nanoparticles. It is envisaged that these composite materials might find practical uses as nanowires and novel catalysts.

The combination of metal nanoparticles (notably AuNPs) and CNTs may lead to the development of a new class of nanocomposite materials, and lead in turn to the successful integration of the properties of these two components in new hybrid materials that present significant features for catalysis and nanotechnology. The combination of two classes of novel material may be obtained as either particles coated on the surface of the CNTs (exohedral), or encapsulated in the nanotube cavity (endohedral). Here, the CNT surface acts as a template where nanoparticles (either naked or stabilized by protecting agents) are absorbed or, in the case of a functionalized CNT, the nanoparticles may be linked through a functional group of the organic moieties attached to the CNTs. This new type of composite material is important not only for fundamental and academic studies of the interactions between the matrix and the metallic nanoparticles, but also for various applications such as catalysts as well as electronic, optical, and sensor devices.

7.5
CNT–AuNP Composites

A variety of resourceful techniques has been reported for the production of CNT–AuNP composites (in general, the deposition of AuNPs onto the surface of the CNT substrate), each offering different degrees of particle size control and distribution along the CNTs. By changing the size and concentration of the AuNPs deposited/incorporated, the electronic properties of the CNTs can in turn be controlled. Composites of CNTs with AuNPs may be created via different pathways. One approach is to grow and/or incorporate the AuNPs into the hollow cavities of the CNTs, while a second approach is to grow and/or deposit naked AuNPs directly onto the terrace of the CNTs. In yet another process, the AuNPs may be prepared and modified with suitable functional groups that can be connected to the CNT surface via covalent bonding through organic moieties; alternatively, the modified AuNPs may simply be linked to the surface of the CNTs via supramolecular interactions.

7.5.1
Filling of CNTs with AuNPs

In order to synthesize CNTs with predetermined characteristics, it is essential to identify and control the mechanisms that direct CNT growth. The major challenge here is to identify effective ways in which to fill the metal nanoparticles (notably AuNPs) into the hollow CNT cavities, without affecting the latter's individual characteristic properties.

A simple procedure for producing a composite is to fill the MWNTs with AuNPs, simply by mixing an aqueous citric acid solution containing NH_3-treated MWNTs and aqueous auric chloride solution [38]. Heat treatment in NH_3 causes most of the nanotubes to be open, such that functional basic groups are created on their inner walls. Mixing and ultrasonication will then help the citric acid to combine strongly with the basic groups via electrostatic attraction, thus facilitating the *in situ* reduction and subsequent attachment of AuNPs (1–2 nm) inside the nanotubes (Figure 7.2). These hybrid materials may in time become important for investigating and creating a rich variety of electrical and sensor devices.

In another "wet chemistry" technique, a two-step procedure was used to produce a composite material by filling gold metal into the cavities of the CNTs. Here, the nanotubes were opened by oxidation with HNO_3 [39] and then stirred overnight with a concentrated aqueous solution of $AuCl_3$. The resultant CNTs, when separated from excess concentrated solution, were calcined in a furnace under argon

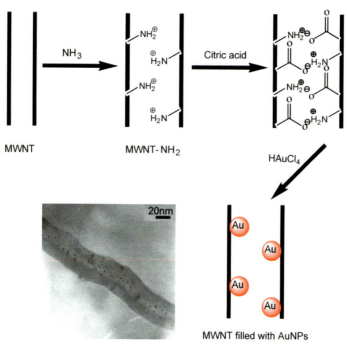

Figure 7.2 Schematic illustration for the attachment of AuNPs to NH$_3$-treated CNTs and the TEM image of AuNP-filled CNTs. Adopted and modified according to Ref. [38]; TEM image reprinted with permission from Ref. [38].

at 600 °C, whereby the AuCl$_3$ was decomposed to produce elemental gold [40]. Most of the gold entrapped within the hollow CNTs was seen to be crystalline in nature, spherical in shape, and to range in size from 1 to 5 nm diameter. A HR-TEM image of a CNT encapsulated with a gold crystal, together with the analytical energy-dispersive spectrum (EDS) of the gold particle, are shown in Figure 7.3. More importantly, by using this method it was possible to produce a relatively high percentage (~70%) of opened nanotubes to be filled with metallic gold [33, 39].

7.5.2
Deposition of AuNPs Directly on the CNT Surface

When depositing AuNPs onto the terrace of the CNTs, it is possible to use gold salts as the precursors for the AuNPs; these salts are produced by a variety of reduction processes, using either reducing agents and/or external energies such as heat (thermal), photochemical, and light, in the presence of the CNTs. The interactions between the AuNPs and CNTs are mostly based on van der Waals forces which, in some cases, appear to be sufficiently strong so as to ensure significant adhesion.

7.5 CNT–AuNP Composites

(a)

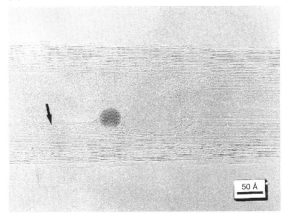

(b)

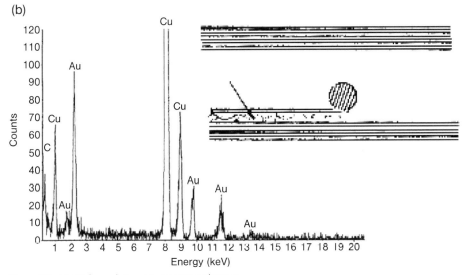

Figure 7.3 (a) High-resolution transmission electron microscopy (HR-TEM) image of a carbon nanotube filled with a spherical Au crystal. The solid arrow shown in the HR-TEM image indicates where there has been intercalation into the gaps where carbon layers are missing; (b) ED spectrum of the Au particle. Reprinted with permission from Ref. [40].

An effective method was introduced by Xue and coworkers for depositing AuNPs onto the walls of CNTs [41]. For this, AuNPs (average size 8 nm) were grown on the surface of the CNTs by the thermal decomposition (400 °C) of gold salts under a hydrogen atmosphere. This synthetic strategy has also been shown to be a generalized process that can easily be extended to the synthesis of different types of metal nanoparticle (Pt, Ag, Pd) onto the CNT surface. It appears that the CNTs play a vital role here, not only as a template for tuning the metal nanoparticles size but also acting as a support material. This finding was supported by the fact that larger particles were observed when metal salts were reduced in the presence of graphite or amorphous carbon. Xue and coworkers suggested that these composite materials might be used as efficient catalysts for certain environmentally advantageous reactions, as well as certain applications in electronic devices.

Raghuveer and coworkers established a novel "eco-friendly" strategy of utilizing microwave irradiation for the rapid introduction of carboxyl, carbonyl, hydroxyl, and allyl terminal groups onto the surface of MWNTs, without using any aggressive oxidants such as HNO_3 and/or ultrasonication [42]. Here, the functional groups served as the preferred nucleation points for reducing gold ions from solution by a microwave-assisted reduction reaction. MWNTs were dispersed in water and added to an aqueous solution containing $HAuCl_4$ and ethylene glycol. After microwave irradiation, the surfaces of the MWNTs were seen to be decorated by uniformly dispersed AuNPs (Figure 7.4) that ranged in size from 3 to 10 nm (average ~6 nm). The MWNTs were derivatized with nanoparticles synthesized by an *in situ* gold-ion reduction during functionalization, all in a single-step process. The notable point here was that the overall tubular structure of the MWNTs remained intact, as was evident from TEM images. This was in great contrast to

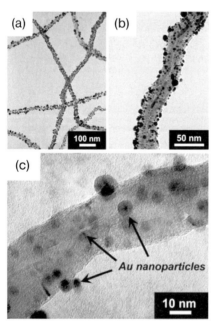

Figure 7.4 (a, b) Low- and (c) high-magnification bright-field TEM images showing the decoration of MWNTs with AuNPs of 3–10 nm diameter. Reprinted with permission from Ref. [42].

the rupture and tube breakage observed during functionalization by aggressive sonication and acid treatment [43].

A different strategy for the formation of AuNPs on the surface of SWNTs, based on the spontaneous reduction of metal ions in solution without the use of a reducing agent, was reported by Choi and coworkers [44]. This approach differed from a typical electroless deposition, which requires either a reducing agent or a catalyst, as a result of direct redox reactions between the ions and nanotubes. Electroless deposition methods rely on a chemical (as opposed to an electrochemical) reduction process, whereby a chemical species with a redox potential suitably lower than that of the metal species being reduced provides the driving force for the reaction [45]. A spontaneous decoration of AuNPs (average size 7 nm) on the sidewalls of the SWNTs was observed following their immersion in $HAuCl_4$ (Au^{3+}) solution for 3 min (Figure 7.5a). The Au^{3+} ion reduction and SWNT oxidation during electroless metal deposition was investigated by measuring the electrical conductance of SWNTs immersed in solutions (Figure 7.5b). When the SWNTs act as electron donors, hole insertion into SWNTs would be expected to cause an increase in the electrical conductance [46] to the already p-type nanotubes due to O_2 doping under ambient conditions [47]. Spontaneous metal deposition onto the SWNTs by an electroless process allows for a facile, efficient, and selective immobilization of metal species on nanotubes, which may be useful for sensor and catalysis applications. However, this selective electroless metal deposition on SWNTs was shown to be effective only for Au or Pt; other metal ions such as Ag^+, Ni^{2+}, and Cu^{2+}, could not be reduced in the same way, perhaps due to their lower redox potentials.

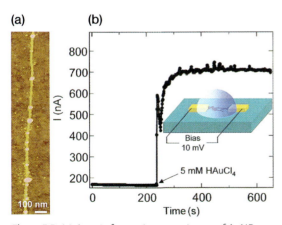

Figure 7.5 (a) Atomic force microscopy image of AuNPs formed on an individual SWNT; (b) Monitoring the change in current across a SWNT during exposure to a 5 mM $HAuCl_4$ solution. The period before the exposure corresponds to the nanotube in a mixture of ethanol and water (1:1). Inset: Schematic for the experimental set-up. Spacing between points = 1 s. Reprinted with permission from Ref. [44].

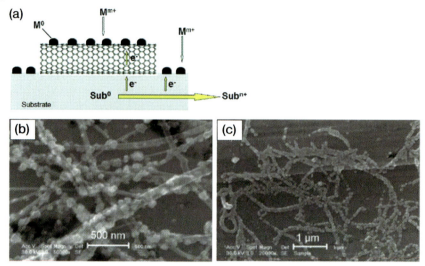

Figure 7.6 (a) Schematic illustration of the metal nanoparticle deposition on CNTs via the substrate-enhanced electroless deposition process; (b, c) Scanning electron microscopy images of (b) MWNTs and (c) SWNTs supported by a copper foil after immersion in an aqueous solution of $HAuCl_4$ (3.8 mM). Reprinted with permission from Ref. [48].

Qu and Dai introduced for the first time another facile, but versatile and effective, method for the electroless deposition of AuNPs on both SWNTs and MWNTs in the absence of any additional reducing agent [48]. Upon immersion of the copper-supported MWNTs or SWNTs into an aqueous solution of $HAuCl_4$ (3.8 mM), the AuNPs were deposited spontaneously onto the terrace of the nanotubes. The advantage of this technique over the previous method is that a large variety of metal nanoparticles (even for metals with a lower redox potential than that of the CNTs, such as Cu and Ag) could be reduced and decorated onto the surface of both MWNTs and SWNTs. The general scheme of the reaction process, together with scanning electron microscopy (SEM) images of AuNPs decorated onto MWNTs and SWNTs, are shown in Figure 7.6.

The deposition of metal nanoparticles is achieved via the redox reaction of a galvanic cell, in which the nanotube acts as a cathode for the metal nanoparticle deposition (M^0, e.g., Au^0) from the reduction of metal ions (M^{m+}, e.g., Au^{3+}) in solution, while the metal substrate (e.g., Cu) serves as an anode where metal atoms (Sub^0) are oxidized into corresponding ions (Sub^{n+}) followed by dissolution [49] (Figure 7.6). This process, which is known as substrate-enhanced electroless deposition (SEED), allows the electroless deposition of many metal nanoparticles onto conducting CNTs, and indicates a great potential for the functionalization of CNTs with various metal nanoparticles.

Geckeler and coworkers have established an unprecedented approach to prepare SWNT–AuNP hybrids in homogeneous phase by using the reaction of gold salts and SWNTs in the presence of a surfactant in aqueous solution, without the addition of a reducing agent [50]. The AuNPs decorated on the sidewalls of the SWNTs were uniform in size and well dispersed (Figure 7.7). Statistical calculations indicated the average AuNP size to be 2.94 ± 0.75 nm, with 7.5×10^{-17} g of gold being coated on an individual nanotube that in turn contained approximately 300 AuNPs. Interestingly, the size of the AuNPs decorated on the surface of the SWNTs could be tailored by altering the concentration of the gold salt solution.

The chemistry of the hybrid material was described by using a frontier-orbital picture. As the relative position of the Fermi level of nanotubes with respect to the mixed metal ion/nanotube highest occupied molecular orbital (HOMO) and lowest unoccupied molecular orbital (LUMO) is suitable for charge transfer, both semiconducting and metallic CNTs may establish attractive interactions with the metal ions, either by four-electron interactions involving two occupied orbitals, or by zero-electron interactions involving two empty orbitals (Figure 7.7d). It was noted that the HOMO level of $AuCl_4^-$ is partly occupied with electrons. This method is easy to scale-up, such that a uniform size of AuNPs may be decorated on the walls of the SWNTs, and size of the AuNPs can also be controlled. More importantly, as the method yields water-soluble composite materials, a much greater variety of applications can be envisaged.

A simple method has been developed recently to prepare hybrid materials from SWNTs and AuNPs, including Pt and Rh nanoparticles. For this, nanoparticles were deposited on the surface of the SWNTs by the mild reduction of metal salts using poly(ethylene glycol)-200 as the reducing agent [51]. The free surface of the nanoparticles attached to the SWNTs was then coated with organic aliphatic molecules such as oleylamine, which enhanced the dispersion of the resulting hybrid material in organic solvents. This method avoids chemical functionalization of the sidewalls and open ends of the SWNTs, and the final hybrid material may be used for the application in the catalysis of organic reactions.

Recently, a simple UV irradiation method was developed to grow uncoated naked AuNPs on carboxy-modified MWNTs by performing UV irradiation on mixed solution containing oxidized MWNTs, $HAuCl_4$, and acetone (acting as a photosensitive agent) at room temperature (Figure 7.8) [52]. The size of AuNPs deposited on the terrace of the MWNTs was found to depend heavily on the diameter of MWNTs and the solution pH. The size of the AuNPs was indirectly proportional to the diameter of MWNTs and the solution pH. As a high catalytic activity (especially for AuNPs) mainly depends on the steps, edges, and corner sites of surface, and also on the electrical interactions between AuNPs and supporting materials, the resultant composites may prove to be advantageous in catalytic reactions.

Although a variety of ingenious strategies to decorate AuNPs onto CNTs is available, the electrodeposition method has its own advantages. Electrochemistry represents a potent technique for the deposition of diverse metals and/or the surface modification of CNTs, being both rapid and facile, and thus allowing the chemist and materials scientist to control with ease the nucleation and growth of the metal nanoparticles [45, 53, 54]. It is very feasible to control the size and distribution of

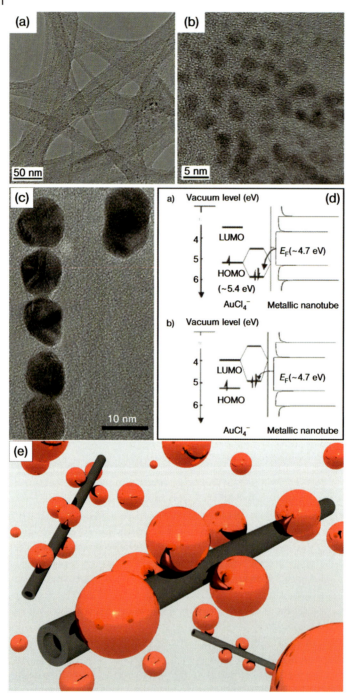

Figure 7.7 (a) Transmission electron microscopy (TEM) image, and (b, c) HR-TEM images of the CNT/AuNPs hybrid; (d) Frontier-orbital picture representation of four-electron interactions (plot a), and zero-electron interactions (plot b) between $AuCl_4^-$ and a metallic CNT. The $AuCl_4^-$ frontier-orbitals are designated as HOMO and LUMO, and the nanotube orbitals are represented by the density of states plots. E_F = Fermi level. (Reprinted with permission from Ref. [50]); (e) Schematic showing decoration of the surface of CNTs with AuNPs. (Kyungjae Lee is gratefully acknowledged for his support in the preparation of this image).

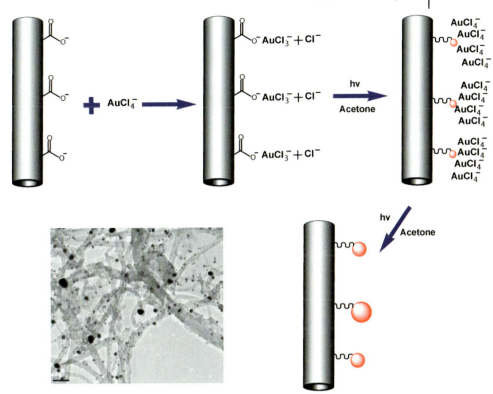

Figure 7.8 Scheme showing the growth of AuNPs on the surface of MWNTs initiated by UV irradiation, and a TEM image of AuNPs attached to the surface of MWNTs at pH 12. Scale bar = 100 nm. Adopted and modified according to Ref. [52]; TEM image reprinted with permission from Ref. [52].

metal nanoparticles simply by varying the deposition potential, time, and substrate. However, the other imaginative methods of depositing nanoparticles onto the CNTs surface involve some tricky, tedious, and time-consuming treatments that allow the impurities in the bath solutions to be included either into the nanoparticles or onto the terrace of the CNTs themselves, thus affecting the optical or catalytic properties of the nanoparticles [54]. Electrochemically deposited nanoparticles – particularly of noble metals such as Au, Pt, or Pd – are often of very high purity, are formed rapidly, and have good adhesion to the CNT substrate [53, 54].

Quinn and coworkers [53] reported a general electrodeposition method for depositing noble metals such as Au, Pt, and Pd onto the surface of the SWNTs by immersing the latter in a solution of the respective metal salts, namely $HAuCl_4$, K_2PtCl_4, and $(NH_4)_2PdCl_4$. Whilst the size of the nanoparticles was tuned by the concentration of the metal precursor salt and the electrochemical deposition parameters, the coverage of the nanoparticles on the surface of the SWNTs was controlled by the nucleation potential. The resultant composite material was

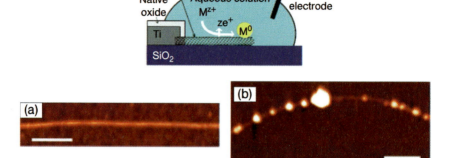

Figure 7.9 General schematic illustration of the SWNT as an electrode used for the electrodeposition of metal nanoparticles, and AFM images of the predeposition (a) and postdeposition (b) of Au. The deposition time was 20 s. Scale bar = 300 nm. Reprinted with permission from Ref. [53].

considered as a metallic wire, as the surface of the SWNTs was thoroughly decorated by the metal nanoparticles (Figure 7.9). It was noted that, under the experimental conditions, both the sidewalls and ends of the SWNTs were coated equally by metal nanoparticles, even though it is believed that the CNTs sidewalls are less reactive than the tube ends. The SWNT serves a dual function in such a way that initially acts as the electrodeposition template, and subsequently as a wire to electrically connect the deposited Au, Pt, and Pd nanoparticles.

The cubic and spherical nanoparticles of gold, with a fairly narrow size distribution, deposited on the surface of the CNTs were produced by immersing copper foil-supported CNTs into an aqueous solution of $HAuCl_4$ with and without $CuCl_2$ at room temperature under different reaction conditions (e.g., different concentration and reaction time) [55, 56]. Both, the shape and size of the AuNPs were found to depend heavily on the gold salt concentration and reaction time, providing considerable room for regulating the morphological features of the resultant nanoparticles. The size of the AuNPs deposited on the surface of CNTs is approximately 60–100 nm, although their size can be tailored by controlling the reaction conditions, especially the deposition time [56]. Gold nanospheres and nanocubes were multisite deposited along the CNT length, with individual nanotubes even threading through the nanoparticles (Figure 7.10). The facile and versatile technique for the shape- and size-controlled syntheses of AuNPs for the site-selective modification of CNTs is very attractive for producing various multicomponent nanoparticle–nanotube hybrid structures that might be useful in a wide range of potential applications, including fuel cells, catalytic, sensing, and optoelectronic systems. Further, in order to investigate and explain the optical response of the CNT–AuNP hybrid from a theoretical standpoint, a 3-D electrodynamic model was built using the finite-difference time-domain [56]. These studies proposed an anisotropic

Figure 7.10 Scanning electron microscopy images of synthesized CNTs that are grafted with gold nanosphere and nanocube systems. Reprinted with permission from Ref. [56].

response, in line with the experimentally observed absorption peaks of such systems in the optical range.

Tello and coworkers introduced a new method, known as solvated metal atom dispersion (SMAD), in combination with CVD, to prepare MWNT and AuNP hybrid materials, and subsequently investigated their thermal stabilities [57]. In the SMAD protocol, a colloid with small and highly reactive gold clusters was prepared by co-evaporation of the gold metal and acetone. The colloid was subsequently condensed into a frozen matrix in a liquid nitrogen atmosphere, and the matrix then allowed to warm to room temperature. The as-prepared gold clusters were reacted with previously incorporated MWNTs in the reactor to yield MWNT–AuNP hybrid materials which were thermally stable up to 400 °C. Moreover, no appreciable changes in particle size were detected as the AuNPs were covered with amorphous carbon after annealing at 200 °C (Figure 7.11).

Interestingly, whilst the average AuNP size was increased approximately from 4 to 20 nm and the nanoparticles were detached from the MWNTs, no damage was induced on the MWNTs when the hybrid material was annealed beyond 600 °C (Figure 7.11c). Surprisingly, further heating (annealing up to 800 °C) induced a severe transformation of the MWNTs (perhaps due to a catalytic activity of the AuNPs) into cylindrical solid carbon nanorods (Figure 7.11d). It would appear that this method provides a new approach to the synthesis of both carbon nanorods and MWNT–AuNP composites that are useful in a variety of key applications.

A new and important approach was developed recently to form naked sub-10-nm AuNPs on individual 2 nm bare CNTs [58]. These assemblies were produced on the terrace of a porous anodic alumina (PAA) template, on which the CNTs (single- or double-walled) were grown by using plasma-enhanced chemical vapor deposition (PECVD). The AuNPs were obtained via an indirect evaporation method using a membrane mask, consisting of a suspended, patterned silicon nitride film; the AuNPs then diffused along the PAA surface into the regions containing CNTs (Figure 7.12a). Three distinct regions were observed on the PAA surface: (i) a semi-continuous gold film composed of multiple grains; (ii) a transition region,

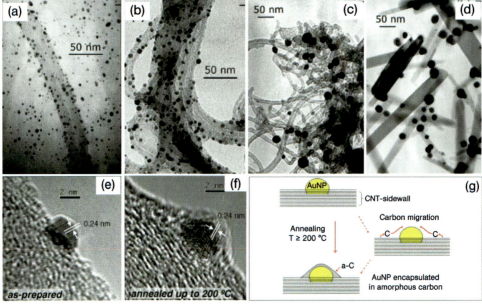

Figure 7.11 Transmission electron microscopy images for the as-prepared AuNP-CNT hybrids (a), and annealed up to 200 °C (b), 600 °C (c), and 800 °C; (d) HR-TEM images of AuNPs anchored on the CNT-walls (e, f); for the as-prepared AuNP–CNT hybrids (e), and annealed up to 200 °C (f). A carbon layer is encapsulating the AuNPs after the thermal process. Lattice fringes on gold particles are consistent with the (111) fcc orientation (0.235 nm); (g) A schematic representation of the process. Reprinted with permission from Ref. [57].

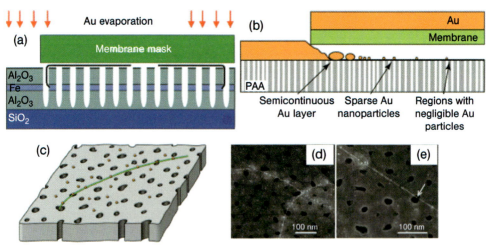

Figure 7.12 (a) Diagram of a cross-section of the experimental set-up; (b) Schematic representation of the three regions with different Au particle distribution on the porous anodic alumina (PAA) film after Au evaporation through a nitride membrane mask; (c) Enlarged view of region sparse AuNPs, showing one CNT emerging from a pore and growing along the PAA surface. The AuNPs diffuse into this region and attach to the CNT; (d) Field-emission scanning electron microscopy images of an area from region sparse AuNPS, showing two crossed CNTs decorated with AuNPs; (e) A sub-5-nm AuNP (indicated by a white arrow) is on a CNT and above the pore in the PAA film. Scale bar = 100 nm. Reprinted with permission from Ref. [58].

in which a sparse coverage of small AuNPs was observed; and (iii) a region where negligible gold was observed. The schematic illustration of the three regions with different gold particle distributions on the PAA film after the evaporation of gold through a nitride membrane mask is shown in Figure 7.12b. The presence of the various regions is attributed to the migration of gold on the surface; as the gold atoms migrate they also aggregate, forming grains (at large surface coverage) and small nanoparticles (at lower surface coverage) [59]. As shown schematically in Figure 7.12c, within region (b) the AuNPs were observed on the PAA surface and on the CNTs. The field-emission scanning electron microscopy (FESEM) image of an area within region (b) shows the presence of a number of well-defined particles with diameters ~5 nm along the CNTs, most likely due to AuNPs (Figure 7.12d). In addition, somewhat diffuse bright regions with the dimensions of 20–40 nm nearer to CNTs were also observed, though these may be due to local charging effects arising during the FESEM imaging. The AuNPs attached strongly to the CNTs, as substantiated by the observations of nanoparticles that were suspended over pores or that moved along with the CNTs (Figure 7.12e). The strong mechanical binding between the AuNPs and the CNTs revealed a comparatively close contact between the two objects, and also showed that this binding energy was larger than that between the cluster and the alumina surface. This behavior is significant for understanding a strongly coupled electronic system. In contrast to most other general methods for the direct evaporation of gold onto CNTs, defect sites on the CNTs are not necessary in this method of creating preformed AuNPs. This approach may provide a new strategy for functionalizing CNTs for chemical or biological sensing and also for fundamental studies of nanoscale contacts to CNTs. Thus, drawbacks such as the complexity in directly functionalizing CNTs, and the inability to obtain individual SWNTs by using common bulk synthesis methods are avoided, which confine the applicability of CNTs as primary elements in sensors.

7.5.3
Interaction Between Modified AuNPs and CNTs

A significant feature of nanoscience and nanotechnology concerns the progress of experimental protocols for the preparation of nanoparticles of diverse chemical compositions, sizes, shapes, and controlled dispersity with a facile approach and no environmental risk [60]. There are many established methods for the synthesis of AuNPs, including conventional chemical reduction, heat-treatment, microwave irradiation, sonochemical, photolytical, seeding growth approaches, and self-reduction using surfactants [4, 61]. It is well known that the reaction medium, reducing agent, and capping or protecting agent are the three key factors for the synthesis and stabilization of metal nanoparticles in general, and for AuNPs in particular. By selecting these factors appropriately, it would be possible to modify the nanoparticles according to their convenient purposes. Therefore, it should also be possible to connect the AuNPs to the surface of the CNTs through either covalent linking or supramolecular (noncovalent) interaction. Hence, the AuNPs prepared are modified with suitable functional groups so as to link the AuNPs to the CNT surface. These connections can be attained by using the functional groups of the modified AuNPs to covalently link with functional groups present on the surface of the CNTs. An alternative method would be simply to adhere the linker onto the CNT surface via supramolecular interactions such as hydrophobic interactions, hydrogen bond linkage, π–π interactions, or electrostatic attraction.

7.5.3.1 Covalent Linkage

As CNTs are chemically inert, the triggering of their surface is a vital prerequisite for linking the metal nanoclusters to them. Chemical functionalization is the most common and widely used way to introduce the linkers, as well as to improve dispersibility of CNTs, which is also significant for the proficient and uniform deposition of nanoparticles. The functionalization of SWNTs by a chemical method (covalent linkage) to enhance its dispersibility was introduced by Chen and coworkers [62].

Azamian and coworkers [63] employed carboxylate chemistry to covalently link the AuNPs to defect sites in controllably oxidized SWNT termini and/or sidewalls. The carboxylic acid groups created on a SWNT were converted to amides by reaction with carbodiimide reagents [N,N'-dicyclohexyl-carbodiimide (DCC) or 1-ethyl-3-(3-dimethyl-amino-propyl)-carbodiimide (EDC)] and 2-aminoethanethiol. Generally, carbodiimide catalyzes the formation of amide bonds between carboxylic acids or phosphates and amines by activating carboxyl or phosphate to form an O-urea derivative. The introduced thiol functionality was consequently linked with well-dispersed gold colloids (Figure 7.13). The attachment of AuNPs to SWNTs sidewalls was corroborated by imaging the pristine SWNTs with atomic force microscopy (AFM) before and after exposure to the coupling reagent and colloidal AuNPs. This technique may be extended to test, with ease, the functionalization of SWNTs with a range of groups.

A simple, direct, solvent-free approach was reported for decorating AuNPs onto the surface of MWNTs functionalized with aliphatic dithiols such as 1,4-butanedithiol, 1,6-hexanedithiol, 1,8-octanedithiol, and 2- aminoethanethiol [64]. While AuNPs (~1.7 nm) with a narrow particle size distribution were produced on the 1,6-hexanedithiol-functionalized MWNTs, the average size of AuNPs was 5.5 nm, obtained on MWNTs derivatized with aminothiol. This difference in the AuNPs size was apparently due to a coalescence phenomenon of AuNPs in aminothiol-functionalized MWNT samples; this could be the result of a nonuniform capping of the aminothiol over the AuNPs surface, avoiding the passivated AuNPs. The main drawback of this approach is that attempts to attach AuNPs to the derivatized MWNTs using water as a solvent medium were not particularly successful, giving rise to a considerable agglomeration of gold over the nanotube bundles, obviously due to the poor CNT dispersibility in water. Instead, the MWNTs nicely decorated with well-dispersed AuNPs can be prepared only when water is substituted by 2-propanol. It is anticipated that this method may be useful for attaching CNTs to gold tips for AFM and scanning tunneling microscopy

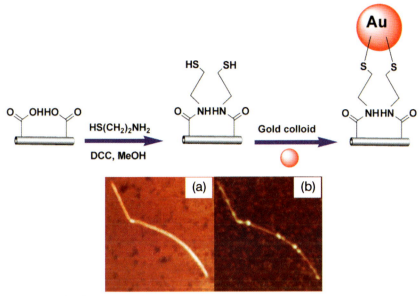

Figure 7.13 The chemistry used to connect the AuNPs to oxidized SWNTs. AFM images of SWNTs before (a) and after (b) exposure to the coupling reagents and colloidal gold particles. DCC = N,N'-dicyclohexyl-carbodiimide. Adopted and modified according to Ref. [63]; AFM images reprinted with permission from Ref. [63].

(STM), and also potentially for monitoring the adsorption and concentration of trace metal ions.

An efficient method was developed to functionalize MWNTs with thiol groups, after which AuNPs were anchored onto them to fabricate new composite materials. Two different simple procedures were followed to produce MWNT–AuNP composite materials, as summarized in the Figure 7.14. The thiol-functionalized MWNTs were stirred with already prepared AuNPs in toluene at room temperature, and in another way AuNPs were produced in the presence of thiol-functionalized MWNTs (*in situ* formation of MWNT–AuNP composite) [65]. The TEM images showed self-assembly of the AuNPs on the MWNTs, where the AuNPs (Figure 7.14a) of 25 nm were dispersed on the MWNTs. The shape and size of the AuNPs obtained using the *in situ* method were different (Figure 7.14b) from those obtained with the former method. This might be explained by the existence of thiol groups surrounded by $AuCl_4^-$ anions, which were reduced to AuNPs in the presence of reducing agents. These AuNPs represent the cores for further growth, as the thiol groups did not cover the whole particles.

Coleman and coworkers used the Bingel reaction to functionalize the SWNTs with a cyclopropane group [66]. For this, the cyclopropane group was tagged using

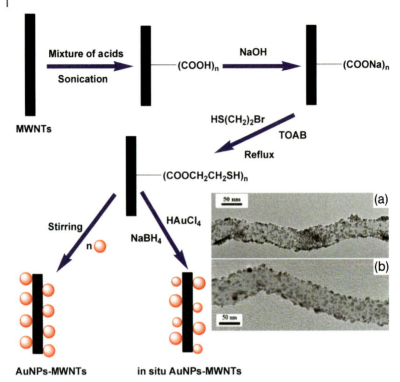

Figure 7.14 Experimental procedure for the fabrication of AuNP–MWNT composites and TEM images of AuNP–MWNT (a) and *in situ* AuNP–MWNT (b). TOAB = Tetraoctylammonium bromide. Adopted and modified according to Ref. [65]; TEM images reprinted with permission from Ref. [65].

preformed ~5 nm gold colloids by exploiting the gold sulfur binding interaction, as shown in the Figure 7.15. AuNPs were observed both on the sidewalls and at the ends of the SWNTs. The Bingel reaction is an example of a [2+1] cycloaddition reaction, and is a popular method in fullerene chemistry.

Recently, a novel DNA biosensor was fabricated for the detection of DNA hybridization based on the layer-by-layer (LBL) self-assembly of MWNTs and AuNPs via covalent linkage, which exhibited an excellent specificity and chemical stability under the DNA-hybridization conditions [67]. The LBL assembly represents one of the simplest ways of producing fundamentally and practically interesting multilayer films with unique mechanical properties; it also provides a precise control over film composition and thickness, which in turn opens up many new opportunities for achieving the ideal model surface, the properties of which are controllable. For this, the cysteamine molecules simply act as a glue to connect activated MWNTs and AuNPs into a 3-D hybrid network on the Au electrode, after which NH_2-ssDNA (ssDNA; single-stranded DNA) was immobilized onto multi-

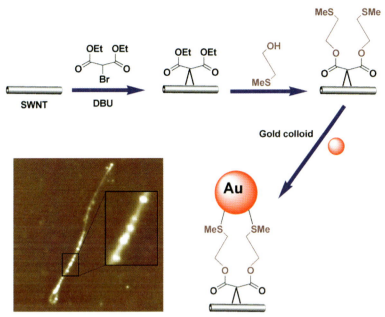

Figure 7.15 Schematic illustration of the chemistry used to connect AuNPs to SWNTs and AFM image of as-formed AuNP–SWNT composites by using the Bingel reaction. DBU = 1,8-diazabicyclo[5.4.0]undec-7-ene. Adopted and modified according to Ref. [66]; AFM image reprinted with permission from Ref. [66].

layer films via the amino link at the 5′ end. Owing to the electron-transfer ability of the CNTs and the catalytic activities of the AuNPs, the sensitivity of DNA biosensors was improved and this DNA biosensor showed an excellent reproducibility and stability under DNA hybridization conditions.

7.5.3.2 Supramolecular Interaction Between AuNPs and CNTs

Another strategy for preparing CNT–AuNP composites is that of the supramolecular or noncovalent functionalization of CNTs, and the subsequent attachment of AuNPs based on noncovalent interaction such as hydrophobic–hydrophobic interaction, weak hydrogen bond linkage, π–π stacking interaction, or electrostatic attraction.

Hydrophobic Interactions and Hydrogen Bonding Hydrophobic interaction is the attractive force between molecules owing to the close positioning of the nonhydrophilic portions of the two or more molecules. Hydrogen bonding is a type of supramolecular interaction or a type of weak attractive (dipole–dipole) interaction between an electronegative atom and a hydrogen atom bonded to another electronegative atom such as nitrogen, oxygen, or fluorine. It can occur between molecules, or within parts of a single molecule. A hydrogen bond tends to be weaker than a covalent or ionic bond, but stronger than van der Waals forces. One method of noncovalent functionalization of CNTs involves their association with amphiphilic molecules through hydrophobic interaction in aqueous medium [68]. Hydrophobic interactions between the ligands forming the monolayer that passivate the metal surface have been used to deposit the metal nanoparticles onto the surface of the CNTs. A novel strategy was reported to link monolayer-protected gold nanoclusters of 1–3 nm diameter to the sidewalls of nonoxidized CNTs through hydrophobic interactions between acetone-activated CNTs and octanethiol-protected gold nanoclusters [69]. The anchorage was provided by the interdigitation of alkyl chains of self-assembled molecular layers protecting the gold nanoclusters and molecular moieties adsorbed on the surface of CNTs (Figure 7.16). These molecularly interlinked hybrid nanoblocks may be significant for exploring and manufacturing a rich variety of molecular nanostructures for potential device applications.

In addition to experimental studies, a theoretical model has been developed to provide a detailed account of the interactions (charge transfer, van der Waals, osmotic, elastic, nonelastic, and covalent) between tetraoctylammonium bromide-stabilized AuNPs and alkyl- and alkylthiol-modified MWNTs, so as to estimate the coverage of AuNPs at the surface of the MWNTs under different experimental conditions [70]. A quantitative description of the interactions between AuNPs and MWNTs was made by comparing between the predictions of the theoretical model and the experimental results. From such a comparison, it was concluded that as the length of the alkyl chains at the surface of the MWNTs increased, coverage of

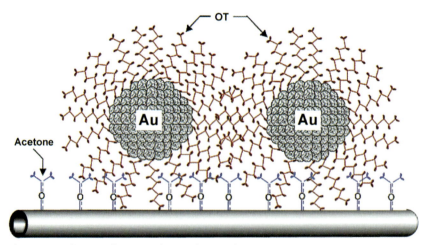

Figure 7.16 Schematic illustrating the attachment of octanethiol (OT)-capped Au nanoclusters to acetone-activated MWNTs. Reprinted with permission from Ref. [69].

the AuNPs decreased (i.e., noncovalent adsorption). In contrast, for alkylthiol-modified MWNTs, coverage of the AuNPs at their surface remained constant, irrespective of the length of the alkyl-thiol chain (covalent adsorption) (Figure 7.17). The theoretical model assumption was in good agreement with the experimental findings, which proved the validity of the predictive model. A significant insight is that, under certain conditions, the coverage of AuNPs is very sensitive to the nature of the MWNT surface modification and the environment, pointing

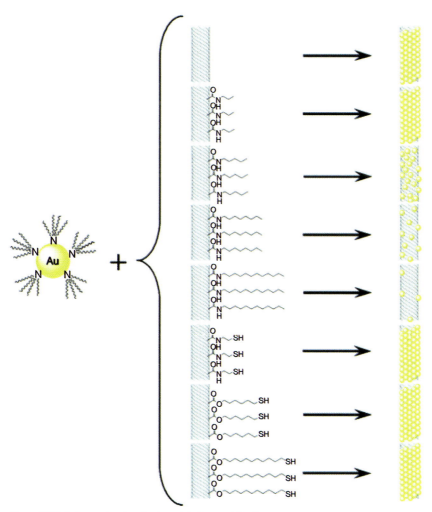

Figure 7.17 Scheme showing the templated assembly of tetraoctylammonium bromide-stabilized AuNPs at the surface of unmodified, alkyl-modified, and alkylthiol-modified MWNTs. Reprinted with permission from Ref. [70].

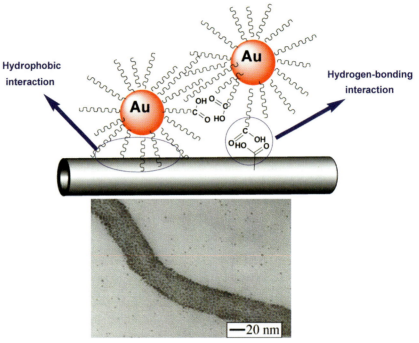

Figure 7.18 Schematic representation of the molecularly mediated assembly of monolayer-capped AuNPs on CNTs, and the characteristic TEM image of the AuNP–CNT composite. Adopted and modified according to Ref. [71]; TEM image reprinted with permission from Ref. [71].

the way to a rational design of functional CNT-based nanoscale devices with potentially widespread applications.

A molecularly mediated assembly of alkanethiolate-capped AuNPs onto the surface of MWNTs through a combination of hydrophobic interactions between the alkyl chains of the capping/linking molecules and the hydrophobic backbones of a nanotube and the hydrogen-bonding interaction between carboxylic groups of the capping/linking molecules and the functional groups on the surface of the nanotube, has been described [71] (Figure 7.18).

Here, the simplicity and effectiveness for assembling alkanethiolate-capped AuNPs of 2–5 nm core size onto CNTs with controllable coverage and spatially isolated character were achieved by adopting such a combination of interfacial chemistry – that is, a combination of hydrophobic and hydrogen-bonding interactions between the organic shells and the CNT surfaces. The benefit of this pathway is that it does not require any difficult and tedious surface modification of CNTs. Rather, the assembly, packing density and distribution of the AuNPs onto the surface of the CNTs depended on the relative concentrations of the AuNPs, CNTs, and capping or linking agents. A thermal treatment to remove any organic capping/linking shells of the AuNPs induced an agglomeration of the AuNPs that attached effectively to the CNTs surface. These composite materials were very stable, with even sonication in hydrophobic solvents being unsuccessful in dissociating the composite material. In time, this approach may find important implications for the design of nanostructured and nanocomposite catalyst and sensory materials. Since it is possible to control the size, shape, loading, and dispersion of the AuNPs on the surface of the conductive CNTs by using a combination of hydrophobic and hydrogen-bonding interactions, this system might also become important for addressing some of the fundamental issues in fuel cell catalysis applications.

Sainsbury and Fitzmaurice [72] reported a smart, noncovalent, self-assembly coverage of dibenzylammonium-cation-modified MWNTs with crown-modified (dibenzo[24]crown) AuNPs, where ammonium cations thread the crown ethers present on the AuNPs surface (Figure 7.19). From the different experimental conditions utilized, it was found that the MWNT-templated self-assembly of a solid nanowire was observed only when uncomplexed cations were present on the surface of the MWNTs and uncomplexed crown was present on the surface of the AuNPs (Figure 7.19a–c). This significant observation led to the perception that the possibility of driving force for the templated nanowire assembly via charge transfer from the crown-modified AuNPs to the cation-modified MWNT could be excluded. Therefore, it was concluded that the templated self-assembly of nanowires was driven by the formation of the surface-confined to pseudorotaxanes that resulted from the electron-poor cation threading the electron-rich crown. In order to initiate the templated self-assembly of a gold nanowire in solution, a dispersion of crown-modified AuNPs in chloroform was added to a freshly sonicated suspension of cation-modified MWNTs in chloroform. The stable dispersion of dodecane-thiol-modified AuNPs was prepared using the well-known method introduced by Brust and coworkers [73]. The nearly size-monodisperse fraction was subsequently modified by the exchange of the adsorbed thiol for thiol-incorporating crown molecules (dodecanethiol incorporating a crown moiety in the terminal position) [74]. The MWNTs were modified by amide-coupling reactions either between the carboxy-modified MWNTs and ethylenediamine, or between the remaining amine of the coupled ethylenediamine and the cation precursor [N-(4-carboxydibenzylamine) carbamate]. This approach may offer the vision of diverse nanoscale components for which many applications can be anticipated.

A facile approach has been developed to prepare a hybrid material of dodecanethiol-protected AuNPs decorated on the surfaces of SWNTs and MWNTs simply by reacting dispersions of both components in dichloromethane [75]. The as-prepared AuNP–SWNT and AuNP–MWNT systems exhibited a strong electronic coupling, which may influence important physico-chemical properties. Although the nature of the electronically coupled state was not clearly understood, it was anticipated that it might involve a charge-transfer character, in which the AuNPs functioned as an electron acceptor, receiving additional electron density from the SWNTs or MWNTs. The well-known absorption features of SWNTs led

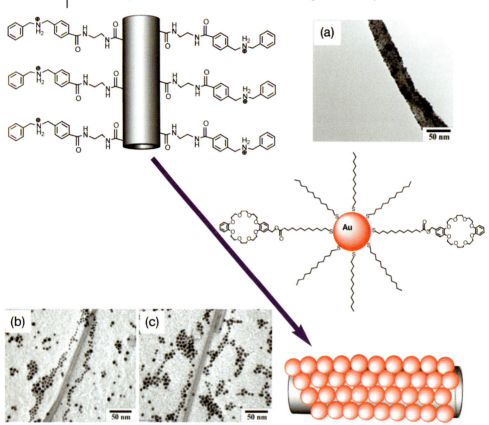

Figure 7.19 A cation-modified MWNT and the noncovalent self-assembly of AuNPs with a crown ether-modified monolayer, and TEM images of (a) crown-modified AuNPs adsorbed at the surface of a cation-modified MWNT. The same AuNPs are not adsorbed at the surface of (b) a cation-precursor-modified MWNTs or (c) an unmodified MWNTs. Adopted and modified according to Ref. [72]; TEM images reprinted with permission from Ref. [72].

to considering the dispersion of AuNP–SWNT instead of the less-explored and less-understood MWNTs characteristics.

In contrast to normal CNTs, Li and coworkers [76] produced bamboo-like carbon nanotubes (BCNTs) that were decorated with a high-density and uniform assembly of AuNPs on the surface, by using a simple and effective method. BCNTs produced by the pyrolysis of iron(II) phthalocyanine were sonicated in toluene for homogeneous dispersion, and then mixed, by gentle shaking, with the toluene solution of AuNPs protected by organic shell. The resultant composite materials, AuNP–BCNTs, were separated by centrifugation and washed repeatedly with toluene. Apart from the hydrophobic interaction between the alkyl chains of the capping/linking molecules and the hydrophobic backbones of the nanotubes, the specific interaction between AuNPs and the nitrogen atoms present on the surface

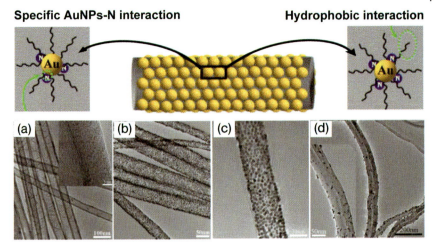

Figure 7.20 Schematic illustration of the direct assembly of AuNPs on BCNTs. (a) TEM image of BCNTs; (b, c) BCNT–AuNP nanohybrids; (d) AuNPs supported on an undoped CNTs. The inset in (a) shows a higher-magnification image. Reprinted with permission from Ref. [76].

of the BCNTs is responsible for the high-density and uniform assembly of AuNPs on BCNTs (Figure 7.20). It is important to note that, compared to the N-doped BCNTs, the pristine undoped CNTs were not efficient for the direct, highly dense and uniform immobilization of AuNPs, due mainly to an absence of specific AuNP–N interactions or other activation processes. As evidently seen from the TEM images, the assembled AuNPs (average size 3.9 nm) were quite uniform, well dispersed, and decorated or packed onto the walls and ends of BCNTs with high density. The unique properties of the BCNTs, combined with this simple and straightforward assembly approach, may facilitate the use of CNTs in a variety of pivotal fields, including biosensors and fuel cell catalysis.

π–π Stacking Interactions This is a type of noncovalent interaction, which is mainly caused by the intermolecular overlapping of p-orbitals in π-conjugated systems (e.g., organic compounds or polymers containing aromatic moieties). The results of different studies [68, 77] have confirmed that organics or polymers with phenyl groups could interact with CNTs through π–π stacking, which is a stronger interaction than the van der Waals forces. Compared to various aromatic compounds, pyrene derivatives are more susceptible to stack on the surface of the CNTs via π–π stacking interactions [78–80]. In this way, the intrinsic electronic properties of the CNTs are preserved, and the surface functional groups on the nanotubes can easily be varied by changing the pyrene derivatives.

By using bifunctionalized pyrene derivatives such as 17-(1-pyrenyl)-13-oxo-hepta-decanethiol (PHT) as an interlinker, AuNPs were decorated on the

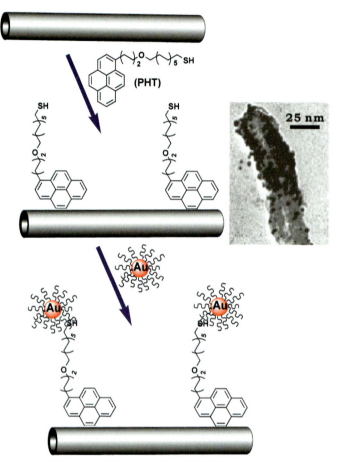

Figure 7.21 Illustration of the connection of AuNPs onto MWNTs through pyrene derivatives as an interlinker, and TEM image of MWNTs decorated with AuNPs.
PHT = 17-(1-pyrenyl)-13-oxo-heptadecanethiol. Adopted and modified according to Ref. [78]; TEM image reprinted with permission from Ref. [78].

surface of the didecylamine-modified MWNTs, where PHT bound to the surface of MWNTs via a π–π stacking interaction between its pyrenyl unit at one end and the sidewall of MWNTs, while the other terminal thiol group was interacted to the surface of AuNPs (Figure 7.21). In this way, AuNPs were self-assembled onto the surface of nanotubes via an organic linker. It should be noted that, while the fluorescence of the resultant composite material was entirely quenched, the Raman response of the CNTs was enhanced considerably. These important consequences imply that there are charge-transfer interactions between the CNTs and AuNPs through the organic interlinker.

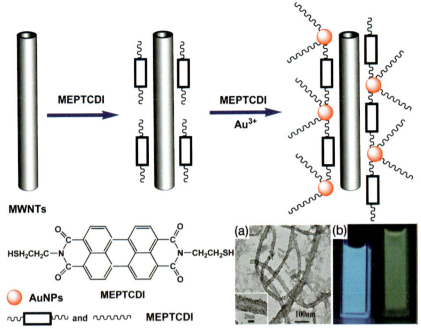

Figure 7.22 Representation for the fabrication processes of the AuNP–MWNT hybrids and molecular structure of MEPTCDI. (a) TEM image of AuNP–MWNT hybrids; (b) Emission from the AuNP–MWNT hybrid with MEPTCDI in water (left) and MEPTCDI in DMF (right) under UV irradiation at 365 nm. MEPTCDI = N,N′-bi(2-mercaptoethyl)-perylene-3,4,9,10-tetracarboxylic diimide. Reprinted with permission from Ref. [81].

Very recently, the *in situ*-solution method was proposed for the synthesis of water-dispersable AuNP–MWNT hybrids with high density and well-distributed AuNPs by using an optoelectronic-active compound of N,N′-bi(2-mercaptoethyl)-perylene-3,4,9,10-tetra-carboxylic diimide (MEPTCDI) as an interlinker and stabilizer for the formation of the AuNP–MWNT hybrids [81]. Here, the MEPTCDI with both phenyl and mercapto groups, played a dual role as an interlinker (Figure 7.22) between the MWNTs and AuNPs (noncovalently wrapping of the MWNTs through π–π stacking) and as a stabilizer for controlling the nucleation and growth of AuNPs on the surface of the MWNTs. This endowed the AuNPs anchored on the surface of the MWNTs with a good stability and dispersability, as well as flexibility of size-control (Figure 7.22a).

AuNP–MWNT hybrids were prepared using an *in situ* procedure in such a way that the MWNTs were dispersed in aqueous solution containing sodium dodecyl benzen sulfonate (SDBS) by sonication and a dimethylformamide (DMF) solution of MEPTCDI was added to the MWNTs suspension, followed by the addition of an aqueous solution of HAuCl$_4$, and stirred for 12 h at room temperature. The nanohybrids formed were further purified by centrifugation and redispersed in water for further characterization. The as-prepared AuNP–MWNT hybrid with

MEPTCDI showed an interesting photoluminescence property under UV irradiation at 365 nm (Figure 7.22b). Based on several blank experiments and reported information [82], it was concluded that the strong luminescence of the hybrids did not result from the atomic Au clusters or from defects of the CNTs or from the organically functionalized CNT hybrids, but rather originated from the AuNP–MWNT hybrids with MEPTCDI, which may be due to energy transfer occurring between the MWNTs and AuNPs or between AuNPs and the MEPTCDI. This property of AuNP–MWNT hybrids in water may expand the potential applications as building blocks to a wider range such as light-emitting sources, biomedical labels, or even tracking materials for drug delivery, and so on.

AuNPs of 2–4 nm diameter were densely decorated on the walls of MWNTs using pyrenealkylamine derivatives such as 1-pyrenemethylamine as the interlinker in aqueous solution [83]. While the pyrene chromophore is noncovalently attached to the terrace of nanotubes through a π–π stacking interaction, the alkylamine substituent of 1-pyrene-methylamine connects to the AuNPs (Figure 7.23). UV-visible absorption and luminescence spectroscopy were employed to monitor the formation of functionalized AuNPs and AuNP–CNT composites. Following surface modification of the AuNPs with 1-pyrenemethylamine, the absorption value of the pyrene chromophore was greatly decreased and the surface plasmon resonance (SPR) of AuNPs showed a red shift from 508 to 556 nm (Figure 7.23a), owing to interparticle plasmon coupling. Further, the photoluminescence property of 1-pyrenemethylamine and its emission intensity were drastically quenched after formation of the AuNP–CNT composites, most likely due to the energy and/or electron transfer from the excited pyrene fluorophore to the AuNPs. This facile strategy for the formation of a high-density assembly of AuNPs onto the surface of CNTs (Figure 7.23b) has been applied to other linking organic molecules with similar structures, such as N-(1-naphthyl)ethylenediamine and phenethylamine, thus demonstrating the general applicability of this approach for preparing AuNP–CNT composites. The resultant composites may become important for applications in low-temperature CO oxidation and other catalytic reactions. Further, the present strategy of depositing spherical AuNPs may be extended to produce CNT–Au nanorod heterojunctions for electronic connection and molecular sensing applications.

Electrostatic Interactions Electrostatic interactions are noncovalent dipole–dipole or induced dipole–dipole interactions that can be either stabilizing or destabilizing. A simple and proficient method has been introduced for the selective anchoring of AuNPs on the surface of nitrogen-doped MWNTs by electrostatic adsorption [84]. Nitrogen-doped MWNTs produced using a pyrolysis method were initially chemically modified by acid treatment (mixture of H_2SO_4 and HNO_3) and the resultant oxidized MWNTs subsequently treated with a cationic polyelectrolyte, namely poly-(diallyldimethylammoniumchloride) (PDDA). The polyelectrolyte was adsorbed onto the surface of the nanotubes by an electrostatic interaction between the carboxyl groups on the chemically oxidized nanotube surface and the polyelectrolyte chains. Negatively charged AuNPs of 10 nm were attached

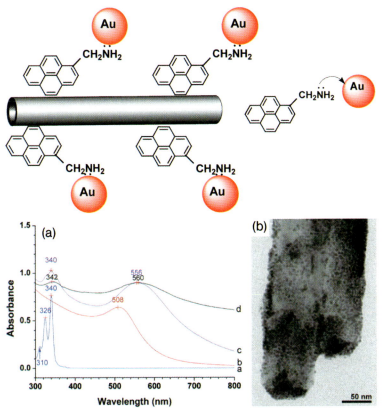

Figure 7.23 AuNP assembly on CNTs through the 1-pyrenemethylamine interlinker. (a) UV-visible absorption spectra of: (curve a) free 1-pyrenemethylamine in ethanol; (curve b) aqueous AuNP solution; (curve c) a mixture of 1-pyrenemethylamine and AuNPs; (curve d) solution containing AuNP–CNT composites; (b) TEM image of the resulting AuNP–CNT composites. Reprinted with permission from Ref. [83].

to the surface of the nanotubes via electrostatic interactions between the polyelectrolyte and AuNPs (Figure 7.24). In this way, novel composites with homogeneously distributed AuNPs on the surface of nitrogen-doped MWNTs were obtained. Therefore, by selecting different types of polyelectrolytes the surfaces of the CNTs can be modified to be either negatively or positively charged, such that accordingly different types of other nanoparticle (e.g., semiconductor nanocrystals, magnetic nanoparticles) can be selectively anchored to the nanotube surfaces. This strategy of decorating nanotubes can be used to identify the location of functional groups on the CNT surfaces, while the resultant composite materials may be used in various domains such as catalytic, electronic, optical, and magnetic applications.

Another example of the electrostatic approach is the formation of AuNP–MWNT composites by the LBL self-assembly technique using polyelectrolytes [85]. To this

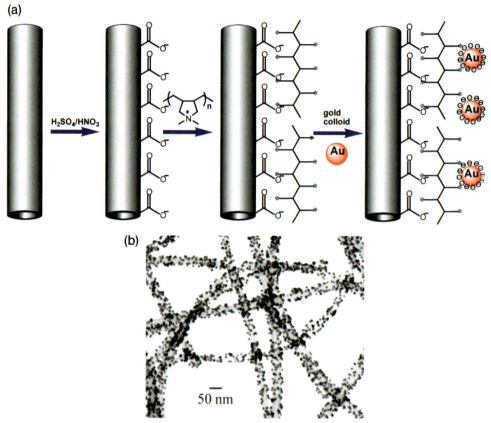

Figure 7.24 (a) Schematic view of the process for anchoring AuNPs onto CNTs; (b) TEM image of AuNP–CNT hybrid structures. Reprinted with permission from Ref. [84].

end, chemically functionalized MWNTs were wrapped with a layer of a positively charged polyelectrolyte (PDDA), followed by a layer of a negatively charged polyelectrolyte such as poly(sodium 4-styrene sulfonate) (PSS). Consequently, positively charged AuNPs prepared by the phase-transfer method [86] were immobilized on negatively charged MWNTs prepared by LBL self-assembly due to electrostatic interactions (Figure 7.25). Positively charged AuNPs also interacted with chemically functionalized MWNTs alone (without polyelectrolyte coating) through electrostatic interaction with the carboxylic acid groups present on MWNTs sidewalls. Hence, it is expected that positively charged AuNPs could be used to detect negatively charged defect sites on CNTs. However, the binding interaction between MWNTs and AuNPs is much more powerful and effective for the LBL approach than the direct approach.

The noncovalent attachment of silica-coated AuNP monolayers and multilayers onto CNT templates was achieved by using the polymer-wrapping technique,

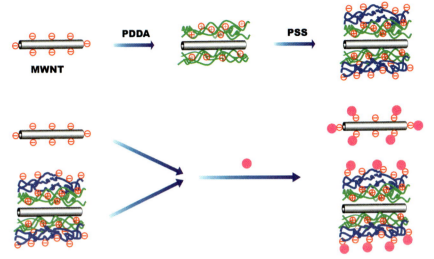

Figure 7.25 Schematic illustration of the LBL self-assembly technique. PDDA = poly(diallyldimethylammoniumchloride); PSS = poly(sodium 4–styrene sulfonate). Adopted and modified according to Ref. [85].

combined with the LBL assembly process [87]. First, a stable dispersion of individual CNTs was obtained by dispersing the MWNTs in water by sonication in the presence of PSS, which acts as the wrapping polymer. PSS has negatively charged sulfonate groups, which serve as primers for the homogeneous adsorption of the cationic polyelectrolyte PDDA. In the LBL approach, the interactions accountable for the assembly are mostly electrostatic, and permit an exploitation of the surface properties of silica to obtain close-packed monolayers and multilayers. As a result, the MWNTs were entirely packed with dense monolayers of silica-coated AuNPs. Such composite nanowires were optically labeled and might have key applications as components of nanoelectronic circuits and waveguides. The repetition of the LBL process allows a good control of the thickness of the nanocomposites by increasing the number of layers deposited. Another strategy for functionalizing SWNT or MWNT surfaces noncovalently with polymer multilayers and AuNPs was also reported by Carrillo and coworkers [88].

Jiang and Gao reported a versatile method for the selective attachment of AuNPs onto the surface of the MWNTs [38]. By using cationic poly(ethyleneimine) (PEI) or anionic citric acid (CA) as the dispersant, the surface properties of the MWNTs were modified to yield a basic or acidic surface. Then, by electrostatic interactions, AuNPs of approximately 10 nm were successfully anchored onto the sidewalls of the MWNTs (Figure 7.26). In the case of the CA-coated MWNTs, the CA forms an adsorption layer around the outer walls of the CNTs and then reduces *in situ* the $HAuCl_4$ to produce the AuNPs. The PEI treatment of the nanotubes and exposure to the $HAuCl_4$ solution generated nitrogen-containing functional groups on

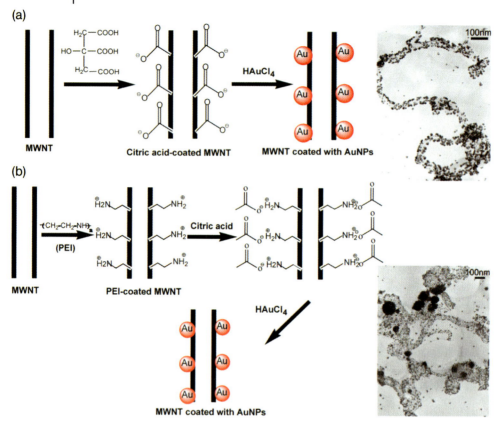

Figure 7.26 The process for attaching AuNPs to: (a) Citric acid-coated CNTs; (b) Poly(ethyleneimine) (PEI)-coated CNTs. The TEM images show the respective products. Adopted and modified according to Ref. [38]; TEM images reprinted with permission from Ref. [38].

the outer walls of the MWNTs. This strategy may be extended to attach other nanoparticles (e.g., semiconductor, electrical and magnetic nanoparticles) onto the terrace of the CNTs by choosing different types of suitable dispersant.

Another straightforward *in situ* reduction procedure for the fabrication of CNT-supported AuNP composite nanomaterials was demonstrated by Hu and coworkers [89]. Here, linear PEI had a dual role as a functionalizing agent for the MWNTs as well as a reducing agent for the formation of AuNPs (Figure 7.27). The driving force involved in the functionalization of PEI on the surface of CNTs is a combination of the electrostatic interaction between the oppositely charged CNTs and the PEI and the physisorption process, which is analogous to the polymer-wrapping process [87]. This synthetic method engages a mild heat-treatment process, which persuades the *in situ* reduction of $HAuCl_4$ on the MWNTs sidewalls and does not require the additional steps of oxidizing the MWNTs with mixed acids and

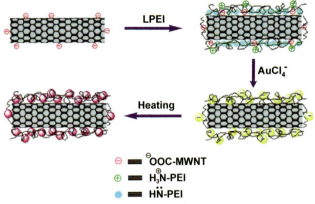

Figure 7.27 Experimental procedure, in which poly(ethyleneimine) possibly interacts with acid-functionalized MWNTs through electrostatic interaction and a physisorption process. Reprinted with permission from Ref. [89].

introducing other reducing agents. More importantly, the dense package of AuNPs on the surface of CNTs was tailored by simply changing the experimental parameters, such as the initial molar ratio of PEI to $HAuCl_4$, the relative concentration of PEI and $HAuCl_4$ to MWNTs, and both the temperature and duration of the heat treatment. This approach is anticipated to become universal for preparing composites of CNTs and other noble metals, as well as heterogeneous CNT–AuNP composite nanowires, and also shows potential for applications such as electronic nanodevices.

Fullam and coworkers have shown that CNTs can be used as templates for the formation of Au nanowires – a new type of composite material which may have the potential for a wide range of applications. The CNT-templated self-assembly of gold nanocrystals and subsequent thermal processing of the nanocrystal assemblies to form continuous polycrystalline gold nanowires extending over many micrometers were described [90]. The AuNPs were decorated on the surface of the CNTs, when mixing the CNTs to a dispersion of tetraoctylammonium bromide-stabilized gold nanocrystals (~6 ± 1 nm) in chloroform. A monolayer coverage of gold nanocrystals on the surface of the CNTs was obtained after precipitation from solution. It was anticipated that the deposition of Au nanocrystals on the CNT surface would be similar to that of phase-transfer catalyst-stabilized gold nanocrystals adsorbed as a monolayer onto the surface of [60]fullerene [91]. Subsequent heat treatment of the gold nanocrystal-coated CNTs yielded a continuous polycrystalline gold nanowire with a CNT core. These findings were of general interest as they showed that nanometer-scale structures self-assembled from nanocrystals might subsequently be processed to yield polycrystalline, nanoscale structures.

In contrast to the use of MWNTs to template the assembly of aligned nanoparticles, which is a relatively common practice, Correa-Duarete and coworkers [92] showed for the first time that the nanorods were assembled with a preferential

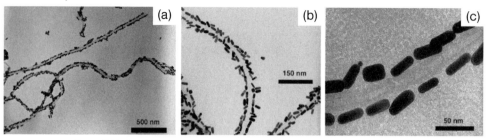

Figure 7.28 (a–c) TEM images of Au nanorods (average aspect ratio 2.94), assembled on MWNTs (average diameter 30 nm) at various magnifications. Reprinted with permission from Ref. [95].

string-like alignment on the surface of MWNTs. The group showed that the formation of uniform electrostatic assembly of Au nanorods on MWNTs occurred in such a way that the nanorods formed strings with end-to-end contacts (Figure 7.28). Such a morphology would result in uniaxial plasmon coupling, and the nanorods could serve as a label to monitor the alignment of CNTs within polymer films or other nanostructured systems, as demonstrated through optical absorption measurements on stretched poly(vinyl alcohol) films loaded with the nanocomposite arrays. Fabrication of the aligned Au nanorods was achieved using the LBL technique. Prior to the assembly of gold nanorods, the MWNTs were wrapped with the negatively charged polyelectrolyte PSS, followed by an electrostatic assembly of the positively charged polyelectrolyte PDDA.

A different approach to fabricate MWNT–AuNPs was proposed via the noncovalent functionalization of MWNTs with surfactants such as sodium dodecyl sulfate (SDS) using the LBL-mediated assembly of a MWNT multilayer film onto indium tin oxide (ITO)-coated glass plates [93]. The supramolecular interactions [94] between the alkyl chains of the SDS and the surface of the MWNTs leads to a good dispersion of MWNTs in water and, on the other hand, creates a distribution of electronic charges at the tube surface. This is very useful for the LBL assembly of uniform MWNT multilayer films on the solid substrate and the fabrication of MWNT–AuNP composites. Hence, it was possible to assemble the MWNT multilayer film onto the ITO plate via the electrostatic interaction of oppositely charged, SDS-modified MWNTs and polyelectrolyte (e.g., PDDA). The same properties of the SDS-functionalized MWNTs were also shown to be useful for mediating the attachment of AuNPs onto the surface of the tubes. This approach to assembling CNT multilayer films and the resultant composites may both find application in the development of bioelectronic nanodevices such as biosensors and biofuel cells.

Biju and coworkers [95] used the functionalization of SWNTs to produce a hierarchical nanoscale composite material of SWNT–AuNPs as well as SWNT–CdSe–ZnS quantum dots (QDs), and studied the photoluminescence properties of the resulting hybrid materials. The excessive sidewall functionalization of the SWNTs into nitro and amino derivatives provided water-solubility to the SWNT

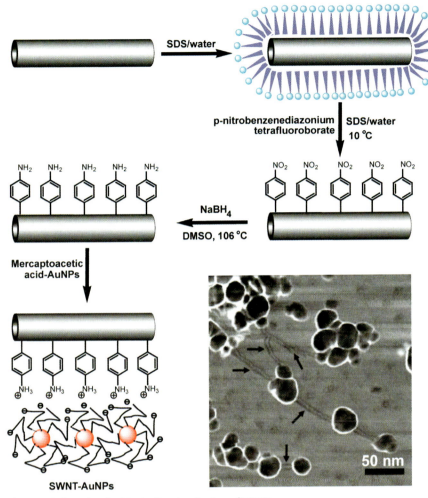

Figure 7.29 Steps involved in the functionalization of SWNTs, including successive conjugation to AuNPs and AFM phase image of SWNT–AuNP conjugates dispersed on a mica plate. Adopted and modified according to Ref. [95]; AFM image reprinted with permission from Ref. [95].

derivatives. The AuNPs modified with mercaptoacetic acid, which provides free surface carboxylic acid groups, were attached to SWNTs through quaternary ammonium salt formation (Figure 7.29). Based on absorption and photoluminescence spectroscopic measurements, it was observed that the conjugation of AuNPs to the SWNT templates induced electronic interactions between the components, which perturbed their energy states, absorption, and photoluminescence properties. The hierarchical nanoscale hybrid structures based on the SWNT and nanoparticle hybrid materials, together with an understanding of their optical properties, were suggested to represent promising building blocks for a light-harvesting system [96] and nanoscale optoelectronic devices.

Another novel method was demonstrated by which to electroporate Gram-negative bacteria (e.g., *Acidothiobacillus ferrooxidans*) through MWNTs [97]. The addition of MWNTs into a solution containing bacteria and AuNPs, and subsequent exposure to microwave radiation, facilitated a rapid transport of AuNPs across the cell wall, without affecting cell viability. In fact, both cell viability and growth provided an insight into the stability of cell under microwave exposure, the reversibility of electroporation via MWNTs, and demonstrated the cells' ability to withstand a high intake of AuNPs. Upon the random addition of MWNTs to a bacteria–AuNPs solution, both the MWNTs and AuNPs remained uniformly suspended, without any strong tendency to accumulate on the bacterial cells. However, upon exposure to microwaves, a large fraction of MWNTs and AuNPs was seen to accumulate on the cell surfaces.

7.6
Applications

It is well known that much attention has been paid to metal nanoparticles based on their pivotal applications as advanced materials in an ample diversity of areas such as catalysis, optical devices, nanotechnology, and the biological sciences [98–100]. Today, many applications of metal nanoparticles are close to profitable implementation, with AuNPs in particular having attracted most attention based on their potential use in applications such as catalysis [101, 102], chemical sensors [103], biological and/or medical areas [104, 105], and the miniaturization of electronic devices due to their unique optical and electrical properties [4]. The applications of thiol-stabilized AuNPs in catalysis comprise asymmetric dihydroxylation reactions [106], carboxylic ester cleavage [107], electrocatalytic reductions [108], and particle-bound ring-opening metathesis polymerization [109], among others [4]. These catalytic applications have in common an exploitation of the carefully designed chemical functionality of the ligand shell, rather than the potential catalytic activity of a nanostructured clean metal surface. From a biological aspect, a color change of AuNP aggregates, from ruby-red to blue, has been exploited in the development of a highly sensitive calorimetric technique for DNA analysis, capable of detecting trace amounts of oligonucleotide sequences. The same effect has also been used to distinguish between perfectly complementary DNA sequences and those exhibiting different degrees of base pair mismatches [110]. One of the most famous potential long-term applications of AuNPs is the fabrication of new (eventually small) electronic devices [111, 112]; for example, a single-electron transistor action has been demonstrated for systems that contain ideally only one particle in the gap between two contacts separated by only few nanometers.

CNTs are of great interest due to their unique mechanical, electronic, and optical properties, as well as their interesting applications [68, 77]. Both, SWNTs and MWNTs have been investigated in numerous promising areas such as field emitters [113], memory elements [114], sensors [46], and nanotube aquators [115]. In order to optimize the applications of CNTs, it is necessary to modify the sidewalls by chemical functionalization and to attach suitable nanoclusters to the nanotubes [90, 116]. In this way, functionalized nanoparticles could be used as versatile building blocks in the construction of nanodevices. An example of this includes the decoration of AuNPs on to the surface of CNT sidewalls, which has shown particular promise towards the development of novel, highly, efficient photoelectrochemical cells, fuel cells, or sensor devices. The noble metal nanoparticles in general – and AuNPs in particular – have become the focal point for many research groups on the basis of their special optical properties [117], their unusual electronic properties, including conductivity by activated electron hopping [118], and their remarkably high catalytic activity [119]. Hybrid nanostructures combining these two types of material might improve these performances, and further extend the applications.

Considering that the broad spectrum of studies connecting metal nanoparticles and CNTs composites are mainly focused on the noble metals (e.g., Pt, Pd, Au, Ag, Ru), it is hardly surprising that the most potential applications of these composite materials are in the field of catalysis. The anchoring of metal nanoparticles in general (and of AuNPs in particular) to CNTs represents another means of producing new types of composite materials with useful properties for gas sensors and catalytic applications [45, 120]. For example, Fasi and coworkers [121] have shown that AuNPs deposited on the surface of the MWNTs were active catalysts in the transformation of methyloxirane (Figure 7.30).

Methyloxirane, which can react in more than one direction and has a versatile functionality, is a favorite of synthetic chemists when creating complicated molecules from relatively simple synthons. It has also been reported that AuNP–CNT composites have been used as catalysts for some environmentally advantageous reactions [41, 71]. Gold-covered MWNTs were proposed as the basic element of a glucose biosensor [122], while the attachment of SWNTs to gold tips not only allowed observations to be made with both AFM and STM [123, 124], but also permitted an investigation of the potential of MWNTs as a solid phase for the adsorption and concentration of true metal ions [125]. A key element when exploring the use of nanoparticle–CNT composites as sensors or catalytic materials is the possibility of an effective and controllable assembly of AuNPs on the surface of CNTs. Such attachment might also serve to detect the presence of certain functional groups on the CNT surface, thus proving that the derivatization had been successful.

7.7
Merits and Demerits of Synthetic Approaches

Several pathways have been introduced to either decorate or attach AuNPs onto the outer walls of CNTs, including processes of reduction, deposition, and chemical linking. By comparing and analyzing these different techniques, it can be

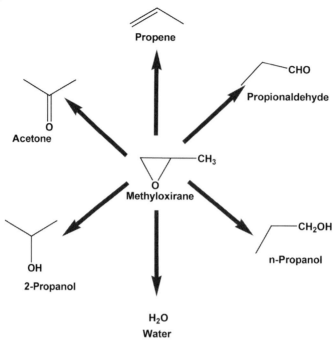

Figure 7.30 Experimentally proven products obtained in the ring-opening reaction of methyloxirane by using AuNP–CNT composite as a catalyst. Reprinted with permission from Ref. [121].

concluded that each and every approach has its merits and demerits. Although, on each occasion the preferred approach can be selected depending on the conditions and specific application purposes, a direct reduction or deposition of AuNPs on the surface of CNTs is generally the easiest and most effective way of ensuring that large quantities of nanoparticles are coated or deposited onto the CNT surfaces. Controlling the shape, size, and distribution of the nanoparticles, as well as the number deposited on the CNT surface, may be very difficult when using this approach, however.

The alternative approach, to link already modified AuNPs with the CNT surface, is a rather complicated, expensive, multistep procedure that typically would include the synthesis and modification of AuNPs, chemical modification of the CNTs, and then linking of the two components. As well as being a very time-consuming process, the formation of composite materials by covalent linking is highly complex, as the number of AuNPs anchored covalently to the surface of the CNTs will be associated with the number of carboxyl groups or other chemical functionalities of the CNT surface. The major advantages of such a linking procedure are that the size and shape of AuNPs can be tuned; likewise, under certain conditions the connection between AuNPs and CNTs is reversible.

The linking of AuNPs to the CNT surface through hydrophobic or electrostatic interactions may be more beneficial, as a fine dispersion with a high density of composite materials can be achieved if previously modified AuNPs are deposited onto the CNT surface in this way. This is especially the case if the CNTs are pre-modified by polymers or surfactants, which enhance such interactions.

7.8
Conclusions

Owing to their unique geometry, exceptional properties and large surface area, CNTs are excellent candidates as support materials for metal nanoparticles in general, and for AuNPs in particular. The blending of these materials to create new composites should in turn lead to the generation of new hybrid materials with applications in catalysis, electronics, and nanotechnology. Moreover, the formation of such composites can be achieved either by the deposition or attachment of AuNPs onto the CNT surface, or by their entrapment within the nanotube cavities.

In this chapter, we have discussed not only the synthetic approaches, assembly and interaction related to AuNP–CNT composites, but also their potential applications. The strategies used to create hybrid materials are variable and depend on the requirements and applications of the composite product. The interaction of CNTs with AuNPs provides the ability to generate significant composite derivatives with numerous potential applications, including advanced catalytic systems, biosensors, electronic nanodevices, probe tips for scanning probe microscopy, electrochemical sensors, and gas sensors. It is perhaps surprising to note that most potential applications for these composites involve catalytic systems, with the high surface area:volume ratio, excellent mechanical properties, thermal and electrical conductivity, chemical stability and inertness, and lack of porosity forming the basis of CNTs being used to support AuNPs. Indeed, with remarkable advances having been made in the area of catalysis, the efficient industrial use of AuNP–CNT composites as catalysts will inevitably attract much interest in the future. Yet, perhaps the most significant point for the future development of these hybrid materials would be to determine influential synthetic protocols and methodologies for the fabrication of composite materials with reproducible properties and performances. Extensive studies are also required with regards to the possible re-use of the composites as catalysts, and consequently attention must be focused on their catalytic reaction mechanisms. Moreover, the spectrum of catalytic activity for AuNP–CNT composites must be extended to the production of industrially important compounds. In addition, simple but effective methods must be identified to deposit gold nanowires and nanorods onto the CNT surface, creating high-density packaging suitable for both electrical and electronic applications. The electron-transfer phenomenon that occurs between deposited AuNPs and CNTs should also be addressed in greater detail, as this may open new opportunities for the development of optoelectronic and light-to-energy-conversion devices. Another future task will be to explore multicomponent systems that contain several different types of structurally cooperating component, including combinations of metal semi-conductor nanoparticles. This may lead in turn to the development of new materials and structures with functional properties that are superior to those of currently available, "conventional" materials. It is envisaged, therefore, that the future will bear witness to a whole range of novel alloys related to AuNP and CNTs, while the list of applications of CNTs that have been functionalized with AuNPs, other metals, metal oxides and semiconductor nanoparticles, should expand significantly in the near future. Undoubtedly, the great attraction of these composite materials will lead to many research groups undertaking increasing numbers of investigations into their properties and practical applications.

In conclusion, it is clear that the extraordinary diversity of structures, properties, and applications of the AuNP–CNT composites will stimulate many investigations into the fundamental properties and applications of these materials. Moreover, such studies will be conducted in association with other molecular, inorganic, and biological areas of research involving chemistry, physics, biology, and medicine.

Acknowledgments

The authors acknowledge financial support from the "Brain Korea 21" project from the Ministry of Education, Science, and Technology, as well as from the World-Class University (WCU), Ministry of Education, Science and Technology (MEST) of Korea (Project No. R31-2008-000-10026-0).

References

1 Antonii, F. (1618) *Panacea Aurea-Auro Potibile*, Bibliopoli Frobeniano, Humberg.
2 Kunckles, J. (1676) *Nuetlicke Observationes oder Anmerkungen von Auro und Argento otabili*, Schutzens, Humburg.
3 Faraday, M. (1857) *Philos. Trans. London*, **147**, 145.
4 Daniel, M.-C. and Astruc, D. (2004) *Chem. Rev.*, **104**, 293 and references there in.
5 Source–ISI web of knowledge. Available at: http://isiknowledge.com
6 Brust, M. and Kiely, C.J. (2002) *Physicochem. Eng. Asp.*, **202**, 186.
7 Yu, Y.Y., Chang, S.S., Lee, C.L. and Wang, C.R.C. (1997) *J. Phys. Chem. B*, **101**, 6661.
8 Shankar, S.S., Rai, A., Ankamwar, B., Singh, A., Ahmad, A. and Sastry, M. (2004) *Nat. Mater.*, **3**, 482.
9 Sun, Y. and Xia, Y. (2002) *Science*, **298**, 2176.
10 Iijima, S. (1991) *Nature*, **354**, 56.
11 Tang, Z.K., Zhang, L., Wang, N., Zhang, X.X., Wen, G.H., Li, G.D., Wang, J.N., Chan, C.T. and Sheng, P. (2001) *Science*, **292**, 2462.
12 Ajayan, P.M. (1999) *Chem. Rev.*, **99**, 1787.
13 Oberlin, A., Endo, M. and Koyama, T. (1976) *J. Cryst. Growth*, **32**, 335.
14 Oberlin, A., Endo, M. and Koyama, T. (1976) *Carbon*, **14**, 133.
15 Kroto, H.W., Heath, J.R., O'Brien, S.C., Curl, R.F. and Smalley, R.E. (1985) *Nature*, **318**, 162.
16 Iijima, S. and Ichihashi, T. (1993) *Nature*, **363**, 603.
17 Bethune, D.S., Kiang, C.H., de Vries, M.S., Gorman, G., Savoy, R., Vazquez, J. and Beyers, R. (1993) *Nature*, **363**, 605.

18. Ebbesen, T.W. and Ajayan, P.M. (1992) *Nature*, **358**, 220.
19. Thess, A., Lee, R., Nikolaev, P., Dai, H.J., Petit, P., Robert, J., Xu, C., Lee, Y.H., Kim, S.G., Rinzler, A.G., Colbert, D.T., Scuseria, G.E., Tomanek, D., Fisher, J.E. and Smalley, R.E. (1996) *Science*, **273**, 483.
20. Journet, C., Maser, W.K., Bernier, P., Loiseau, A., Lamy de la Chapelle, M., Lefrant, A., Denard, P., Lee, R. and Fischer, J.E. (1997) *Nature*, **388**, 756.
21. Rinzler, A.G., Liu, J., Dai, H., Nikolaev, P., Huffman, C.B., Rodriguez-Macias, F.J., Boul, P.J., Lu, A.H., Heymann, D., Colbert, D.T., Lee, R.S., Fischer, J.E., Rao, A.M., Eklund, P.C. and Smalley, R.E. (1998) *Appl. Phys. A*, **67**, 29.
22. Endo, M., Takeuchi, K., Kobori, K., Takahashi, K., Kroto, H.W. and Sarkar, A. (1995) *Carbon*, **33**, 873.
23. Nikolaev, P., Bronikowski, M., Bradley, R., Rohmund, F., Colbert, D.T., Smith, K. and Smalley, R.E. (1999) *Chem. Phys. Lett.*, **313**, 91.
24. Chiang, I.W., Brinson, B.E., Huang, A.Y., Willis, P.A., Bronikowski, M.J., Margrave, J.L., Smalley, R.E. and Hauge, R.H. (2001) *J. Phys. Chem. B*, **105**, 8297.
25. Bonard, J.M., Stora, T., Salvetat, J.-P., Maier, F., Stockli, T., Duschl, C., Forro, L., de Heer, W.A. and Chatelain, A. (1997) *Adv. Mater.*, **9**, 827.
26. Bandow, S., Rao, A.M., Williams, K.A., Thess, A., Smalley, R.E. and Eklund, P.C. (1997) *J. Phys. Chem. B*, **101**, 8839.
27. Duesberg, G.S., Burghard, M., Muster, J., Philipp, G. and Roth, S. (1998) *Chem. Commun.*, 435.
28. Coleman, J.N., Dalton, A.B., Curran, S., Rubio, A., Davey, A.P., Drury, A., McCarthy, B., Lahr, B., Ajayan, P.M., Roth, S., Barklie, R.C. and Blau, W.J. (2000) *Adv. Mater.*, **12**, 213.
29. Vazquez, E., Georgakilas, V. and Prato, M. (2002) *Chem. Commun.*, 2308.
30. Baughman, R.H., Zakhidov, A.A. and de Heer, W.A. (2002) *Science*, **297**, 787.
31. Dong, L., Subramanian, A. and Nelson, B.J. (2007) *Nanotoday*, **2**, 12.
32. Pederson, M.R. and Broughton, J.Q. (1992) *Phys. Rev. Lett.*, **69**, 2689.
33. Ajayan, P.M. and Iijima, S. (1993) *Nature*, **361**, 333.
34. Seshadri, R., Govindaraj, A., Aiyer, H.N., Sen, R., Subbanna, G.N., Raju, A.R. and Rao, C.N.R. (1994) *Curr. Sci.*, **66**, 839.
35. Liu, M. and Cowley, J.M. (1995) *Carbon*, **33**, 225.
36. Liu, M. and Cowley, J.M. (1993) *Carbon*, **33**, 749.
37. Planeix, J.M., Coustel, N., Coq, J., Brotons, V., Kumbhar, P.S., Dutartre, R., Geneste, P., Bernier, P. and Ajayan, P.M. (1994) *J. Am. Chem. Soc.*, **116**, 7935.
38. Jiang, L. and Gao, L. (2003) *Carbon*, **41**, 2923.
39. Tsang, S.C., Chen, Y.K., Harris, P.J.F. and Green, M.L.H. (1994) *Nature*, **372**, 159.
40. Chu, A., Cook, J., Heesom, R.J.R., Hutchison, J.L., Geen, M.L.H. and Sloan, J. (1996) *Chem. Mater.*, **8**, 2751.
41. Xue, B., Chen, P., Hong, Q., Lin, J. and Tan, K.L. (2001) *J. Mater. Chem.*, **11**, 2378.
42. Raghuveer, M.S., Agrawal, S., Bishop, N. and Ramanath, G. (2006) *Chem. Mater.*, **18**, 1390.
43. Chen, J., Rao, A.M., Lyuksyutove, S., Itkis, M.E., Hamon, M.A., Hu, H., Cohn, R.W., Eklund, P.C., Colbert, D.T., Smalley, R.E. and Haddon, R.C. (2001) *J. Phys. Chem. B*, **105**, 2525.
44. Choi, H.C., Shim, M., Bangsaruntip, S. and Dai, H. (2002) *J. Am. Chem. Soc.*, **124**, 9058.
45. Wildgoose, G.G., Banks, C.E. and Compton, R.G. (2006) *Small*, **2**, 182.
46. Kong, J., Franklin, N., Zhou, C., Chapline, M., Peng, S., Cho, K. and Dai, H. (2000) *Science*, **287**, 622.
47. Collins, P.G., Bradley, K., Ishigami, M. and Zettl, A. (2000) *Science*, **287**, 622.
48. Qu, L. and Dai, L. (2005) *J. Am. Chem. Soc.*, **127**, 10806.
49. Wen, X. and Yang, S. (2002) *Nano Lett.*, **2**, 451.
50. Kim, D.S., Lee, T. and Geckeler, K.E. (2006) *Angew. Chem. Int. Ed. Engl.*, **45**, 104.
51. Tzitzios, V., Georgakilas, V., Oikonomou, E., Karakassides, M. and Petridis, D. (2006) *Carbon*, **44**, 848.
52. Zhang, R., Wang, Q., Zhang, L., Yang, S., Yang, Z. and Ding, B. (2008) *Colloid Surf. A*, **312**, 136.

53 Quinn, B.M., Dekker, C. and Lemay, S.G. (2005) *J. Am. Chem. Soc.*, **127**, 6146.
54 Guo, D.J. and Li, H.L. (2004) *J. Electroanal. Chem.*, **573**, 197.
55 Qu, L., Dai, L. and Osawa, E. (2006) *J. Am. Chem. Soc.*, **128**, 5523.
56 Heltzel, A.J., Qu, L. and Dai, L. (2008) *Nanotechnology*, **19**, 245702.
57 Tello, A., Cardenas, G., Haberle, P. and Segura, R.A. (2008) *Carbon*, **46**, 884.
58 Hang, Q., Maschmann, M.R., Fisher, T.S. and Janes, D.B. (2007) *Small*, **3**, 1266.
59 Zhang, Y., Franklin, N.W., Chen, R.J. and Dai, H. (2000) *Chem. Phys. Lett.*, **331**, 35.
60 Geckeler, K.E. and Rosenberg, E. (eds) (2006) *Functional Nanomaterials*, American Scientific Publishers, Valencia, USA.
61 Premkumar, T., Kim, D.S., Lee, K.J. and Geckeler, K.E. (2007) *Gold Bull.*, **40**, 321.
62 Chen, J., Hamon, M.A., Hu, H., Chen, Y.S., Rao, A.M., Ecklund, P.C. and Haddon, R.C. (1998) *Science*, **282**, 95.
63 Azamian, B.R., Coleman, K.S., Davis, J.J., Hanson, N. and Green, M.L.H. (2002) *Chem. Commun.*, 366.
64 Zanella, R., Basiuk, E.V., Santiago, P., Basiuk, V.A., Mireles, E., Puente-Lee, I. and Saniger, J.M. (2005) *J. Phys. Chem. B*, **109**, 16290.
65 Hu, J., Shi, J., Li, S., Qin, Y., Guo, Z.X., Song, Y. and Zhu, D. (2005) *Chem. Phys. Lett.*, **401**, 352.
66 Coleman, K.S., Bailey, S.R., Fogden, S. and Green, M.L.H. (2003) *J. Am. Chem. Soc.*, **125**, 8722.
67 Ma, H., Zhang, L., Pan, Y., Zhang, K. and Zhang, Y. (2008) *Electroanalysis*, **20**, 1220.
68 Tasis, D., Tagmatarchis, N., Bianco, A. and Prato, M. (2006) *Chem. Rev.*, **106**, 1105.
69 Ellis, A.V., Vijayamohanam, K., Goswami, R., Chakrapani, N., Ramanathan, L.S., Ajayan, P.M. and Ramanath, G. (2003) *Nano Lett.*, **3**, 279.
70 Sainsbury, T., Stolarczyk, J. and Fitzmaurice, D. (2005) *J. Phys. Chem. B*, **109**, 16310.
71 Han, L., Wu, W., Kirk, F.L., Luo, J., Maye, M.M., Kariuki, N.N., Lin, Y., Wang, C. and Zhong, C.J. (2004) *Langmuir*, **20**, 6019.
72 Sainsbury, T. and Fitzmaurice, D. (2004) *Chem. Mater.*, **16**, 2174.
73 Brust, M., Walker, M., Buthell, D., Schiffrin, D. and Whyman, R. (1994) *J. Chem. Soc. Chem. Commun.*, 801.
74 Ahern, D., Rao, S. and Fitzmaurice, D. (1999) *J. Phys. Chem. B*, **103**, 1821.
75 Rahman, G.M.A., Guldi, D.M., Zambon, E., Pasquato, L., Tagmatarchis, N. and Prato, M. (2005) *Small*, **1**, 527.
76 Li, X., Liu, Y., Fu, L., Cao, L., Wei, D., Yu, G. and Zhu, D. (2006) *Carbon*, **44**, 3139.
77 Tasis, D., Tagmatarchis, N., Georgakilas, V. and Prato, M. (2003) *Chem. Eur. J.*, **9**, 4000.
78 Liu, L., Wang, T., Li, J., Guo, Z.-X., Dai, L., Zhang, D. and Zhu, D. (2003) *Chem. Phys. Lett.*, **367**, 747.
79 Yang, D.Q., Hennequin, B. and Sacher, E. (2006) *Chem. Mater.*, **18**, 5033.
80 Guldi, D.M., Rahman, G.M.A., Jux, N., Prato, M., Qin, S.H. and Ford, W. (2005) *Angew. Chem. Int. Ed. Engl.*, **44**, 2015.
81 Zhou, R., Shi, M., Chen, X., Wang, M., Yang, Y., Zhang, X. and Chen, H. (2007) *Nanotechnology*, **48**, 485603.
82 Lee, T.H., Gonzalez, J.I., Zheng, J. and Dickson, R.M. (2005) *Acc. Chem. Res.*, **38**, 534.
83 Ou, Y.Y. and Huang, M.H. (2006) *J. Phys. Chem. B*, **110**, 2031.
84 Jiang, K., Eitan, A., Schadler, L.S., Ajayan, P.M., Siegel, R.W., Grobert, N., Mayne, M., Reyes-Reyes, M., Terrones, H. and Terrones, M. (2003) *Nano Lett.*, **3**, 275.
85 Kim, B. and Sigmund, W.M. (2004) *Langmuir*, **20**, 8239.
86 Gittins, D.I. and Caruso, F. (2001) *Angew. Chem. Int. Ed. Engl.*, **40**, 3001.
87 Correa-Duarte, M.A., Sobal, N., Liz-Marzan, L.M. and Giersig, M. (2004) *Adv. Mater.*, **16**, 2179.
88 Carrilo, A., Swartz, J.A., Gamba, J.M., Kare, R.S., Chakrapani, N., Wei, B. and Ajayan, M. (2003) *Nano Lett.*, **3**, 1437.
89 Hu, X., Wang, T., Qu, X. and Dong, S. (2006) *J. Phys. Chem. B*, **110**, 853.
90 Fullam, S., Cottell, D., Rensmo, H. and Fitzmaurice, D. (2000) *Adv. Mater.*, **12**, 1430.

91 Brust, M., Kiely, C.J., Bethell, D. and Schiffrin, D.J. (1998) *J. Am. Chem. Soc.*, **120**, 12367.

92 Correa-Duarte, M.A., Perez-Juste, J., Sanchez-Iglesias, A., Giersig, M. and Liz-Marzan, L.M. (2005) *Angew. Chem. Int. Ed. Engl.*, **44**, 4375.

93 Zhang, M., Su, L. and Mao, L. (2006) *Carbon*, **44**, 276.

94 Geckeler, K.E. (ed.) (2003) *Advanced Macromolecular and Supramolecular Materials and Processes*, Kluwer Academic/Plenum, New York.

95 Biju, V., Itoh, T., Makita, Y. and Ishikawa, M. (2006) *J. Photochem. Photobiol. A-Chem.*, **183**, 315.

96 Ichiya, L.S.H., Basner, B. and Willner, I. (2005) *Angew. Chem. Int. Ed. Engl.*, **44**, 78.

97 Rajas-Chapana, J.A., Correa-Duarte, M.A., Ren, Z., Kempa, K. and Giersig, M. (2004) *Nano Lett.*, **4**, 985.

98 Moreno-Manas, M. and Pleixats, R. (2003) *Acc. Chem. Res.*, **36**, 638.

99 Van Dijk, M.A., Lippitz, M. and Orrit, M. (2005) *Acc. Chem. Res.*, **38**, 594.

100 Han, M., Gao, X., Su, J.Z. and Nie, S. (2001) *Nat. Biotechnol.*, **19**, 631.

101 Haruta, M. (2002) *CATTECH*, **6**, 102.

102 Remediakis, I.N., Lopez, N. and Norskov, J.K. (2005) *Angew. Chem. Int. Ed. Engl.*, **44**, 1824.

103 Zayats, M., Kharitonov, A.B., Pogorelova, S.P., Lioubashevski, O., Katz, E. and Willner, I. (2003) *J. Am. Chem. Soc.*, **125**, 16006.

104 Ghosh, P., Han, G., De, M., Kim, C.K. and Rotello, V.M. (2008) *Adv. Drug Deliv. Rev.*, **60**, 1307.

105 Bhattacharya, R. and Mukherjee, P. (2008) *Adv. Drug Deliv. Rev.*, **60**, 1289.

106 Li, H., Luk, Y.Y. and Mrksich, M. (1999) *Langmuir*, **15**, 4957.

107 Pasquato, L., Rancan, F., Scrimin, P., Mancin, F. and Frigeri, C. (2000) *Chem. Commun.*, 2253.

108 Pietron, J.J. and Murray, R.W. (1999) *J. Phys. Chem. B*, **103**, 4440.

109 Bartz, M., Kuther, J., Scshadri, R. and Tremel, W. (1998) *Angew. Chem. Int. Ed. Engl.*, **37**, 2466.

110 Storhoff, J.J., Elghanian, R., Mucic, R.C., Mirkin, C.A. and Letsinger, R.L. (1998) *J. Am. Chem. Soc.*, **120**, 1959.

111 Sato, T., Ahmed, H., Brown, D. and Johanson, B.F.G. (1997) *J. Appl. Phys.*, **82**, 1007.

112 Person, S.H.M., Olofsson, L. and Hedberg, L. (1999) *Appl. Phys. Lett.*, **74**, 2546.

113 De Heer, W.A., Chatelain, A. and Ugarte, D. (1995) *Science*, **270**, 1179.

114 Rueckes, T., Kim, K., Joselevich, E., Tseng, G.Y., Cheung, C. and Lieber, C.M. (2000) *Science*, **289**, 94.

115 Baughman, R.H., Cui, C.X., Zakhidov, A.A., Lqbal, Z., Barisci, J.N., Spinks, G.M., Wallace, G.G., Mazzoldi, A., De Rossi, D., Rinzler, A.G., Jaschinski, O., Roth, S. and Kertesz, M. (1999) *Science*, **284**, 1340.

116 Chen, Q., Dai, L., Gao, M., Huang, S. and Mau, A. (2001) *J. Phys. Chem. B*, **105**, 618.

117 Dirix, Y., Bastiaansen, C., Caseri, W. and Smith, P. (1999) *Adv. Mater.*, **11**, 233.

118 Terrill, R.H., Postlethwaite, T.A., Chem, C.H., Poon, C.D., Terzis, A., Chen, A., Hutchison, J.E., Clark, M.R., Wignall, G., Landono, J.D., Superfine, R., Falvo, M., Johnson, C.S. Jr., Samulski, E.T. and Murray, R.W. (1995) *J. Am. Chem. Soc.*, **117**, 12537.

119 Haruta, M. and Date, M. (2001) *Appl. Catal. A, General*, **222**, 427.

120 Georgakilas, V., Gournis, D., Tzitzios, V., Pasquato, L., Guldi, D.M. and Prato, M. (2007) *J. Mater. Chem.*, **17**, 2679.

121 Fasi, A., Palinko, I., Seo, J.W., Konya, Z., Hernadi, K. and Kiricsi, I. (2003) *Chem. Phys. Lett.*, **372**, 848.

122 Wang, S.G., Zhang, Q., Wang, R., Yoon, S.F., Ahm, J., Yang, J.D., Tian, J.Z., Li, J.Q. and Zhou, Q. (2003) *Electrochem. Commun.*, **5**, 800.

123 Yang, Y., Zhang, J., Nan, X. and Liu, Z. (2002) *J. Phys. Chem. B*, **106**, 4139.

124 Nishino, T., Ito, T. and Umezawa, Y. (2002) *Anal. Chem.*, **74**, 4275.

125 Liang, P., Liu, Y., Gue, L., Zeng, I. and Lu, H. (2004) *J. Anal. Atomic Spectrom.*, **19**, 1489.

8
Recent Advances in Metal Nanoparticle-Attached Electrodes

Munetaka Oyama, Akrajas Ali Umar, and Jingdong Zhang

8.1
Introduction

Metal nanoparticles (NPs) have attracted vast attention during the past decade on the basis of their unique optical, electronic, magnetic, and catalytic properties. In the case of gold NPs (AuNPs), their assembly, supermolecular chemistry, quantum-size related properties, and applications to biology, catalysis, and nanotechnology have been summarized by Daniel and Astruc [1].

Within the field of electrochemistry, and in particular of electroanalysis, metal NPs have been actively utilized in the modification of electrodes as functional nanomaterials, and this subject has been extensively reviewed [2–4].

In order to utilize metal NPs as unique devices in functional electrodes, they must normally be attached onto the electrode surfaces. Consequently, a wide variety of methods is available for preparing NP-attached functional electrodes with respect to the metal NPs used, the nature of the substrate electrode materials, and the method of attachment. One well-known method of modification is that of electrochemical deposition, which involves the electrochemical reduction of metal ions onto the surfaces of a conducting material, such that the nanoparticles become attached to the surfaces. This procedure is used in the case of AuNPs [5, 6].

A second, "chemical," method is also available for preparing metal NPs. For this, bottom-up-type preparation strategies for metal NPs in solution are well known and well established, again as reported for AuNPs [7–9]. In terms of regularity and uniformity of NP size, the chemical approach is generally superior to electrochemical deposition, although certain difficulties may be encountered when attaching metal NPs which have been formed in solution onto the electrode surfaces. It is possible that the more interesting characteristics of metal NPs might be diminished when they aggregate to form larger clusters, and consequently it is necessary to either fix or attach the NPs while retaining a moderate degree of NP dispersion. The development of such attachment methods with appropriate

---

Advanced Nanomaterials. Edited by Kurt E. Geckeler and Hiroyuki Nishide
Copyright © 2010 WILEY-VCH Verlag GmbH & Co. KGaA, Weinheim
ISBN: 978-3-527-31794-3

degrees of dispersion are, therefore, very important if the best use is to be made of the characteristics of metal NPs for electrode modification. But this situation is also be true for the preparation of nanocomposites, as are used in electronic, sensing, and optical devices.

When attaching AuNPs to the electrode surfaces, a number of peculiar binding molecules that are suitable for connecting AuNPs with substrates have been used. For example, when connecting AuNPs with a glass surface, both 3-mercaptopropyl-trimethoxysilane (MPTMS) and 3-aminopropyl-trimethoxysilane (APTMS) have been adopted; this process utilizes not only the bonding ability of the silanol group towards the glass surfaces but also the affinity of the –SH or –NH_2 group for AuNPs [10–12]. The AuNP-attached surfaces thus formed were successfully applied to surface-enhanced Raman spectroscopic measurements.

Although today, the use of these peculiar binder molecules has become standard strategy when attaching AuNPs to electrode surfaces [13, 14], it is to be expected that the characteristics of metal NPs – notably their conductivity and catalytic ability – might be greatly affected by the chemical reagents that surround or bind them. Hence, although the use of these bridging reagents is clearly effective in some respects (e.g., size-uniformity and rigid-fixation of metal NPs), they may interfere in the process of utilizing the characteristics of the metal NPs.

If it were possible to attach metal NPs to a conducting substrate without the use of a binder molecule, it might be expected that novel conducting nanodispersed materials with unique electrochemical properties involving both the metal NPs and the conducting substrates, could be fabricated. Consequently, our group has recently developed a seed-mediated growth method for surface modification with metal NPs, notably on the surfaces of indium tin oxide (ITO).

In this chapter, we summarize the development and applications of seed-mediated growth methods for the modification of electrode surfaces. Examples include the wet chemical preparation of advanced functional nanomaterials composed of metal NPs and a conducting substrate.

8.2
Seed-Mediated Growth Method for the Attachment and Growth of AuNPs on ITO

During the early stages of the studies, a seed-mediated growth method was used which originally had been developed by Murphy and coworkers to synthesize Au nanorods in aqueous solution, by the chemical reduction of $HAuCl_4$ [15, 16]. This approach was applied to the formation of AuNPs on the ITO surfaces [17].

In the actual procedure, a piece of ITO-coated glass was washed first by sonication in acetone, and then in pure water. The washed ITO substrate was dried in air and prepared for immersion in a seed solution produced by adding 0.5 ml of a cooled pure aqueous solution of 0.1 M $NaBH_4$ to 19.5 ml of a pure aqueous solution of 0.25 mM $HAuCl_4$ and 0.25 mM trisodium citrate, with stirring. When using this procedure, it has been reported that Au nanoseed particles of 3.5 nm were formed in the seed solution [15].

The ITO substrate was then immersed in the seed solution so as to attach the Au nanoseed particles onto the surface. Typically, immersion in the seed solution was maintained for 2 h, and without any particular treatments. The ITO substrate was then removed from the seed solution, and the surface washed carefully by flushing pure water over the surface several times. Any water retained on the surface was removed with tissue paper, simply by touching the edges of the glass.

After blowing the surface of the ITO with N_2 gas until dry, the substrate was immersed in the growth solution. This was prepared by adding 2.5 ml of a pure aqueous solution of 0.01 M $HAuCl_4$, 0.5 ml of a pure aqueous solution of 0.1 M ascorbic acid, and 0.5 ml of a pure aqueous solution of 0.1 M NaOH into 90 ml of a pure aqueous solution of 0.1 M cetyltrimethylammonium bromide (CTAB). Again, the ITO substrate was retained in the growth solution typically for a period of 24 h in order to promote the growth of AuNPs from the attached Au nanoseed particles. Finally, the sample was washed by flushing with pure water.

The real surface nanoscale images were recorded using field emission scanning electron microscopy (FE-SEM). Figure 8.1a shows the FE-SEM image observed

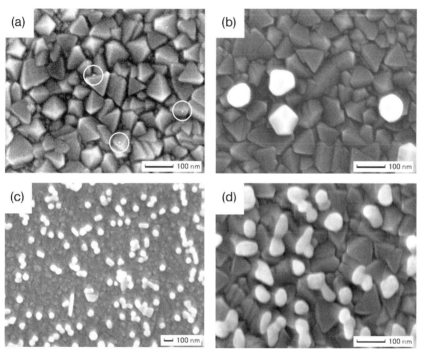

Figure 8.1 FE-SEM images of the ITO surfaces. (a) After immersion of the ITO into the seed solution for 2 h. The white circles indicate areas of attachment of small seed particles; (b) After subsequent immersion of the ITO in the growth solution for 24 h; (c) A low-magnification image of the ITO surface of panel (b); (d) The ITO substrate was modified initially with MPTMS, after which seed-mediated growth was performed in the same manner as for (b). Reproduced with permission from Ref. [17]; © 2005, American Chemical Society.

after seeding, while the images in Figures 8.1b and c were observed after the seed-mediated growth treatment. In the later images, the AuNPs grown on ITO were seen as white particles in the images against a background of ITO crystals. The AuNPs clearly grew to approximately 50–70 nm on the ITO surface from the Au nanoseed particles attached on ITO (Figure 8.1a), following immersion in the growth solution (i.e., by applying the seed-mediated growth method), while retaining a moderate dispersion [17].

The initial attachment of nanoseed particles would be expected to be promoted via the physisorption of ultrasmall NPs which had been dispersed in the seed solution by simply dipping the substrates [18]. The second process should involve crystal growth via the chemical reduction of metal ions in the growth solution, which is inferred to proceed gradually around the nanoseed particles on the substrates.

As shown in Figure 8.1d, the FE-SEM image of AuNPs on the ITO surfaces premodified with MPTMS revealed significant differences in the morphology of the grown AuNPs [17]. Because the crystal-like structures of AuNPs were observed in the absence of MPTMS (Figures 8.1b and c), the seed-mediated growth method was proposed as a unique means of attaching Au nanocrystals onto the ITO surfaces, without the use of binder molecules.

8.3
Electrochemical Applications of AuNP-Attached ITO

The AuNP-directly attached ITO (AuNP/ITO) electrodes were then applied to electrochemical measurements. As inferred by the fact that as-prepared AuNPs on the ITO surface (Figure 8.1c) displayed a different morphology than the MPTMS-linked AuNPs modified on ITO (Figure 8.1d), the AuNP/ITO electrodes provided attractive electrochemical and electrocatalytic properties to promote heterogeneous electron transfer kinetics [19, 20]. In particular, the electrochemical impedance measurements performed to evaluate the effects of AuNPs on the interfacial electron transfer property revealed that the presence of a MPTMS layer caused a significant increase in charge transfer resistance. The relatively low charge transfer resistances of the AuNP/ITO electrodes were clarified from the impedance results, with the typical charge transfer resistance of the AuNP/ITO electrode being approximately one-third that of the AuNP/MPTMS/ITO electrode (Figure 8.2) [20].

The AuNP/ITO electrodes were used for observing the electro-oxidation of uric acid, ascorbic acid, dopamine, norepinephrine (noradrenaline), and epinephrine (adrenaline) [20]. Whilst the AuNP/MPTMS/ITO electrodes caused a depression in selectivity when determining epinephrine in the presence of ascorbic acid, due to the existence of the thiol monolayer, the AuNP/ITO electrodes were able to improve the detection sensitivity while retaining the selectivity [20].

The AuNP/ITO electrodes also provided a biocompatible matrix for the immobilization of hemoglobin (Hb). By using electrochemical impedance

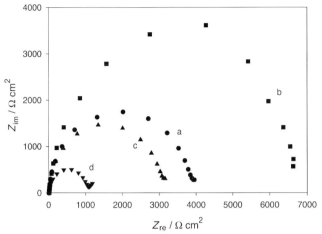

Figure 8.2 Electrochemical impedance spectra of ITO (●), MPTMS/ITO (■), Au/MPTMS/ITO (▲) and Au/ITO (▼) electrodes in 0.1 M phosphate-buffered saline (pH 7.4) containing 1 mM $K_3Fe(CN)_6$ and 1 mM $K_4Fe(CN)_6$. Electrode potential = 0.22 V; Frequency range = 100 kHz to 100 mHz; Voltage amplitude = 5 mV. Reproduced with permission from Ref. [20]; © 2005, Wiley-VCH.

measurements, the modification of AuNPs, followed by the immobilization of Hb on the electrode surfaces, was characterized with the $[Fe(CN)_6]^{3-}/[Fe(CN)_6]^{4-}$ redox probe [21]. Due to the promoted electron transfer of Hb by AuNPs, the Hb-immobilized AuNP/ITO electrodes exhibited an effective catalytic response to the reduction of H_2O_2 with good reproducibility and stability. A linear relationship existed between the catalytic current and the H_2O_2 concentration in the range of 1×10^{-5} to 7×10^{-3} M, with a detection limit [signal-to-noise ratio (SNR) = 3] of 4.5×10^{-6} M [21].

A mediator-free H_2O_2 sensor was also developed along with the application of AuNP/ITO electrodes for the immobilization of myoglobin (Mb) [22]. The Mb-modified AuNP/ITO electrodes showed good reproducibility and stability in pH 7.0 buffer, this being based on the catalytic activity of Mb immobilized on AuNP/ITO towards the reduction of H_2O_2. A linear relationship existed between the catalytic current and the H_2O_2 concentration in the range of 2.5×10^{-6} to 5×10^{-4} M, with a detection limit (SNR = 3) of 4.8×10^{-7} M. In this case, the direct electron transfer of Mb was recorded as a stable and well-behaved voltammetric response, with the Mb-immobilized AuNP/ITO electrodes in buffer solutions [22].

The effects of capping reagents on electron transfer reactions on the AuNP/ITO electrodes were also investigated using cyclic voltammography [23]. In addition to the conventional preparation, with CTAB present in the growth solution, the presence of sodium dodecyl sulfate (SDS) was investigated to determine whether the fabrication of AuNP/ITO was also possible. Systematic cyclic voltammography was then carried out using the AuNP/ITO electrodes prepared with different surfactants – namely cationic CTAB and anionic SDS – for the oxidation of anionic $[Fe(CN)_6]^{4-}$ and reduction of cationic $[Ru(NH_3)_6]^{3+}$. The results showed that the electrochemical responses were significantly improved on both the AuNP/ITO electrodes, but did not depend on the charges of the capping surfactants and the redox species. Thus, the capping of CTAB and SDS on AuNP/ITO was concluded to be quite weak, so that the electron-transfer reactions of both $[Fe(CN)_6]^{4-}$ and $[Ru(NH_3)_6]^{3+}$ were not disturbed [23].

The electrocatalytic activity of a three-dimensional (3-D) monolayer of 3-mercaptopropionic acid (MPA) assembled on AuNPs on the surface of ITO was investigated by using the weak capping of CTAB on the grown AuNPs [24]. The electrochemical behavior of nicotinamide adenine dinucleotide hydrate (NADH) on the 3-D MPA monolayer assembled on the AuNP/ITO electrodes indicated that the MPA layer promoted the electron transfer, which was similar to that of a two-dimensional (2-D) MPA monolayer assembled on planar gold electrodes. However, for the electro-oxidation of ascorbic acid and dopamine, the 3-D MPA monolayer showed an obvious electrocatalytic promotion, while the 2-D MPA monolayer exhibited a blocking effect. The catalytic activity of the 3-D MPA monolayer would be an interesting feature of the proposed AuNP/ITO electrodes.

8.4
Improved Methods for Attachment and Growth of AuNPs on ITO

A number of refined methods for improving or perturbing the seeding processes, namely the attachment of Au nanoseed particles, have been proposed by the present authors' group in order to recognize the significance of the seeding treatment.

The first approach involved an *in situ* reduction method, rather than seeding in the Au colloid solution. Here, it was shown that, when the seeding procedure was modified to be carried out by the impregnation reduction of $AuCl_4^-$ in the presence of the ITO substrate, the density of AuNPs grown directly on the surface was greatly improved [25]. Figure 8.3 shows the FE-SEM images of the AuNP-attached ITO surfaces prepared using the *in situ* reduction method. The initial attachments of the seed particles were recognized on the ITO crystals (Figure 8.3a), after which the AuNP growth could be seen with time on the nanostructured ITO surface. After a 24-h growth period, the smaller AuNPs were found to be attached densely on the ITO surfaces (Figure 8.3c). Repeating the growth treatment was also effective in increasing the size of the AuNPs (Figure 8.3d), although an increasing time for the repeated treatment led to a decrease in the monodispersion properties of the AuNPs on the electrode surface. In other words, the particle size difference became larger, indicating that the AuNPs had not grown uniformly.

The facile heterogeneous electron transfer kinetics resulting from the deposition and growth of the AuNP-arrays was then monitored by using the $[Fe(CN)_6]^{3-}/[Fe(CN)_6]^{4-}$ redox probe for cyclic voltammetry measurements. In addition, the

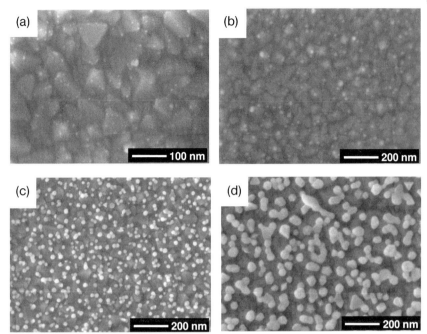

Figure 8.3 FE-SEM images of AuNP-attached ITO surfaces. For seeding, the ITO substrate was immersed into the solution containing $AuCl_4^-$ and citrate ions before the reduction by $NaBH_4$. (a) Observed just after seeding; (b, c) After growth treatment for (b) 15 min and (c) 24 h; (d) After repeating 24-h growth three times. Reproduced with permission from Ref. [25]; © 2005, Elsevier.

AuNP/ITO electrodes thus prepared exhibited a high catalytic activity for the electro-oxidation of nitric oxide (NO), which in turn led to electroanalytical applications for NO sensing [25].

For the second approach, an attempt was made to grow high-density AuNPs on the ITO surfaces by using a "touch" seeding technique rather than the "normal" seeding in the seed-mediated growth procedure [26]. This approach provided a simple and useful strategy to promote the growth of AuNPs on ITO surfaces by simply touching the surface that had already been covered with a drop of Au nanoseed solution with tissue paper.

Subsequent FE-SEM characterization of the growth of AuNPs on two different structured ITOs – namely, rough and smooth structures (Figure 8.4) – has confirmed that this approach was commonly effective and prospective for fostering the growth of high-density AuNPs with a relatively small size (ca. 10–30 nm) of spherical particle [26].

Based on optical absorption measurements of the AuNP-densely attached ITO samples, it was confirmed that the modified ITO substrates could be used for

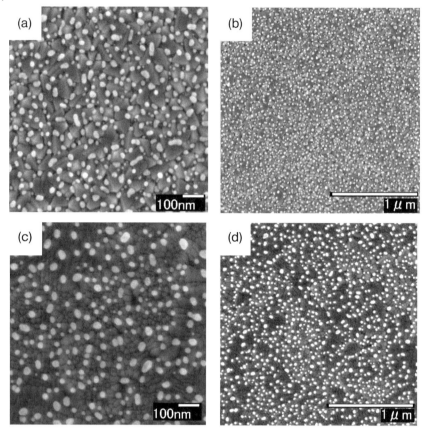

Figure 8.4 FE-SEM images of AuNP-attached ITO surfaces prepared using the touch seed-mediated growth method. (a, c) High-magnification and (b, d) low-magnification images of the surfaces of (a, b) rough ITO and (c, d) smooth ITO. Reproduced with permission from Ref. [26]; © 2005, American Chemical Society.

photoelectrochemical applications, an example being the optically transparent electrode [26]. In addition, the densely attached AuNP/ITO electrodes thus fabricated were applied to observe the electrochemical responses of hydroquinone and p-aminophenol in phosphate buffered solutions [27]. Although the electron-transfer reactions of these compounds were sluggish on bare ITO electrodes, improved electrochemical responses could be remarkably observed on the AuNP/ITO electrodes fabricated with the "touch" seed-mediated growth method [27].

The dense attachment of AuNPs could be also carried out using a cast seed-mediated growth method [28]. Cast seeding which involved three cycles of dropping the seed solution containing Au nanoseed particles and evaporation at 30 °C, followed by treatment in the growth solution containing $HAuCl_4$, CTAB, and ascorbic acid, was found to be suitable for preparing the AuNP-attached ITO

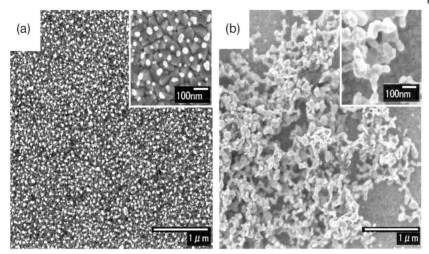

Figure 8.5 FE-SEM images of AuNP-attached ITO surfaces prepared using the cast seed-mediated growth method. The repeated cycles of casting were (a) three times and (b) 10 times. The insets show the higher-magnification images. Reproduced with permission from Ref. [28]; © 2006, Elsevier.

surfaces having a higher density and a narrower size distribution. The FE-SEM image is shown in Figure 8.5a. On the other hand, a 10-cycle cast seeding process formed the connected or networked nanostructures of AuNPs (see Figure 8.5b), while the optical properties were also different from those of the dispersed AuNP-attached ITO [28]. The cast seeding approach provided a facile and practical strategy for attaching AuNPs on the ITO surfaces while controlling the amount of Au loading and without using certain organic binder molecules.

Furthermore, by adjusting the concentration of citrate ions in the seed solution from 1 mM to 50 mM by adding trisodium citrate after the preparation of the Au nanoseed solution, dramatic changes could be observed in the SEM images and in the actual colors of the ITO substrates, which in turn indicated changes in the nanostructures of the AuNPs formed on the ITO surfaces [29]. The attachment of smaller AuNPs with a higher density was observed when 25 mM citrate ions were added in the seed solution (see Figure 8.6). In contrast, larger AuNPs were seen to attach when the concentration of citrate ions was increased to 50 mM. On the basis of this difference and the FE-SEM images observed just after seeding, it was inferred that the citrate ions affected not only the growth process but also the seeding process [29].

Whilst the repulsive power expected from the increased negative charges of citrate ions was not significant, a rather dense attachment was promoted as the particular effect of citrate ions. Such control of the AuNP attachment on ITO would be practically effective because the dense attachment could be achieved by simply changing the composition of the seed solution.

8 Recent Advances in Metal Nanoparticle-Attached Electrodes

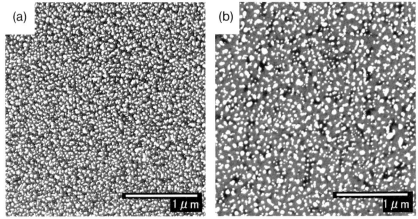

Figure 8.6 FE-SEM images of AuNP-attached ITO surface. The sample was prepared by adding (a) 25 mM and (b) 50 mM citrate ions into the seed solution before seeding, followed by the growth treatment. Reproduced with permission from Ref. [29]; © 2006, Elsevier.

8.5
Attachment and Growth of AuNPs on Other Substrates

The seed-mediated growth method for surface modification with AuNPs could be applicable to the surface treatments of various types of substrate, mainly because that the modification of AuNPs was found to be possible on all of the examined materials, including glassy carbon (GC), mica, stainless, epoxy resin, phenol resin, simple glass, and ZnO film. FE-SEM observations of these materials revealed that the AuNPs attached and grew on the surfaces, although the grown size of the AuNPs and the formation of rod-like particles varied depending on the substrates. Thus, attachment of the Au nanoseed particles by simple immersion into the Au colloid solutions is considered to proceed commonly via physisorption, and not by specific chemical bonding.

As an example, Figure 8.7 shows the FE-SEM images of AuNP-attached GC surfaces observed after seed-mediated growth treatment [30]. Here, although the seeding time was fixed at 2 h for all samples, the growth period was changed from 5 min to 24 h. This resulted in the dense attachment of AuNPs on the GC surfaces, and shorter growth times when compared to the cases where the ITO surfaces were modified. In addition, flat and smaller AuNPs were clearly formed on the GC surfaces (see Figure 8.7a–c). Interestingly, a longer (24 h) treatment in the growth solution promoted the growth of some microcrystals in local areas, as shown in Figure 8.7d. This would be a reflection of ripening in the growth solution, to promote crystal growth and to form the bold bodies. The AuNP-attached

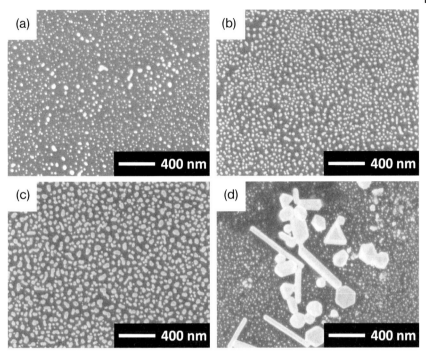

Figure 8.7 FE-SEM images of AuNP-attached GC surfaces. Samples were prepared using the seed-mediated growth method, with seeding for 2 h. The growth period was (a) 5 min, (b) 2 h, (c) 8 h, and (d) 24 h. The observation of (d) was focused on an area where bold crystals were dominantly formed. Reproduced with permission from Ref. [30]; © 2009, Japan Society for Analytical Chemistry.

GC electrodes prepared using the seed-mediated growth method were used for the electrochemical detection of nitrite [31].

The method of attachment used for AuNPs on solid surfaces was applied to the slightly different fabrication of a functional electrode. The technique was used to attach AuNPs onto mesoporous TiO_2 films prepared via a liquid-phase deposition process on GC substrates [32]. Whilst the TiO_2 film strongly inhibited the electron transfer process of the $[Fe(CN)_6]^{3-}/[Fe(CN)_6]^{4-}$ redox couple, electrochemical measurements indicated that the overpotential for the reduction of maleic acid was significantly decreased when the electrode surface was covered with TiO_2 film, due to the electrocatalytic activity. After attaching the AuNPs by the seed-mediated growth method (Figure 8.8), however, the sluggish heterogeneous electron transfer kinetics at the TiO_2 film was effectively improved while the catalytic activity of the TiO_2 film was retained [32]. This example showed the effect of AuNPs on the nature of less-conducting electrode materials.

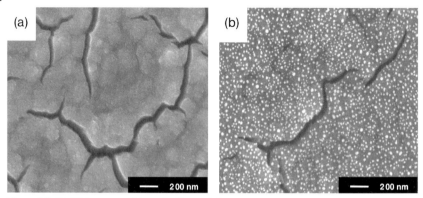

Figure 8.8 FE-SEM images of (a) TiO_2 film and (b) AuNPs attached and grown on the surfaces of TiO_2 film. When the TiO_2 film had been prepared on the GC substrate by liquid-phase deposition, the AuNPs were attached and grown using the seed-mediated growth method. Reproduced with permission from Ref. [32]; © 2005, Electrochemical Society.

8.6
Attachment and Growth of Au Nanoplates on ITO

Although previously, all approaches and trials for preparation have focused on the seeding process, it is likely that a modification of the growth process should, potentially, also be capable of altering the nanostructures of AuNPs on the ITO electrode surfaces.

As a new strategy for attaching Au nanoplates onto the ITO surfaces, a 2-D crystal growth of Au was permitted through a liquid-phase reduction from Au nanoseed particles attached to the ITO surface, using poly(vinylpyrrolidone) (PVP) rather than CTAB as a capping reagent in the growth solution [33]. By controlling the PVP concentration it was possible to form Au nanoplates (see Figure 8.9) with a surface coverage of up to 30%, although variously shaped Au nanocrystals were formed concurrently on the ITO surface. The Au nanoplates were single crystalline in nature, with (111) basal planes and an edge-length of up to approximately ~2 μm, growing parallel to the ITO surface [33]. The concentration of PVP in the growth solution was a key factor in the Au nanoplate formation, as spherical or irregularly shaped AuNPs were formed at either higher or lower concentrations of PVP [33]. The absorption spectra of the Au nanoplate-attached ITO implied anisotropic and specific optical characteristics of the modified ITO glasses.

The Au nanoplate-attached ITO electrode showed some interesting characteristics; for example, an electrochemical response to cytochrome c was observed without using promoter molecules. However, in order to better examine the functions of the Au nanoplates, an increase in their coverage is important. Hence, improved attachment and coverage methods for Au nanoplates are currently under investigation.

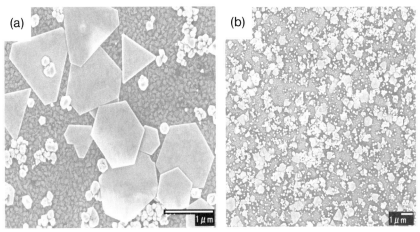

Figure 8.9 FE-SEM images of Au nanoplates attached and grown on ITO surfaces. (a) High-magnification and (b) low-magnification images. Reproduced with permission from Ref. [33]; © 2006, American Chemical Society.

8.7
Attachment and Growth of Silver Nanoparticles (AgNPs) on ITO

By applying the same seed-mediated growth method, silver nanosphere and nanorod particles were successfully attached to the ITO surfaces [34]. As with AuNPs, the attachment of AgNPs could be performed without using a bridging reagent (such as MPTMS), by a simple two-step immersion into first the seed solution, and second into the growth solution containing $AgNO_3$, CTAB, and varying amounts of ascorbic acid [34].

The formed nanostructures were found to be very sensitive to the ascorbic acid content of the growth solution. For example, with an ascorbic acid concentration of 0.64 mM the AgNPs grew on the ITO surface but retained a moderate dispersion (Figure 8.10a). However, when the concentration was raised to 0.86 mM, Ag nanorod and nanowire formation on the ITO surfaces was much less dispersed (Figure 8.10b).

The attachment of AgNPs onto the ITO surfaces was sufficiently strong for further use, for example as a working electrode, and consequently AgNP/ITO electrodes were used in a number of electrochemical measurements. As a result, it was confirmed that the outer spheres of the Ag nanoparticles involved in the redox reaction showed the typical oxidation and reduction waves of Ag metal [34]. In addition, the redox behavior of $[Fe(CN)_6]^{3-}/[Fe(CN)_6]^{4-}$ was improved on the AgNP/ITO electrode, reflecting the low electron transfer resistance of Ag (see Figure 8.11a). This, in turn, indicated that the AgNPs promoted electron transfer reactions by their presence on the conducting ITO surface. The AgNP/ITO electrode was also investigated for reduction of the methyl viologen dication

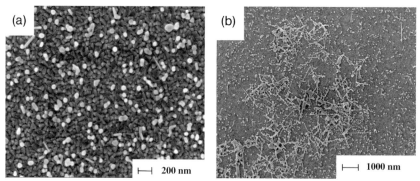

Figure 8.10 FE-SEM images of AgNP-attached ITO surfaces. The ITO substrate was first immersed in the seed solution for 2 h and then into the growth solution containing (a) 0.64 mM or (b) 0.86 mM ascorbic acid for 24 h. Reproduced with permission from Ref. [34]; © 2005, American Chemical Society.

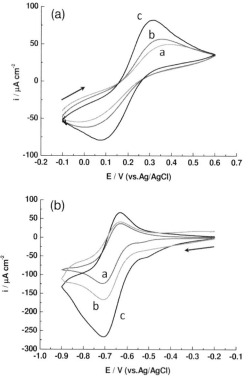

Figure 8.11 (a) Cyclic voltammograms of 0.5 mM [Fe(CN)$_6$]$^{3-}$ and 0.5 mM [Fe(CN)$_6$]$^{4-}$ in 0.1 M phosphate-buffered solution (pH 7.0) recorded using: (curve a) a bare ITO electrode; (curve b) an Ag seed-attached ITO electrode; and (curve c) an AgNP/ITO electrode. Scan rate = 50 mV s^{-1}; (c) Cyclic voltammograms of 0.5 mM methyl viologen dication in 0.1 M Na$_2$SO$_4$ recorded using: (curve a) a bare ITO electrode; (curve b) a conventional Ag electrode; and (curve c) an AgNP/ITO electrode. Scan rate = 100 mV s^{-1}. Reproduced with permission from Ref. [35]; © 2006, American Chemical Society.

(Figure 8.11b), in order to determine the native adsorption features of the fabricated AgNP/ITO electrodes [34].

8.8
Attachment and Growth of Palladium Nanoparticles PdNPs on ITO

Palladium nanoparticles (PdNPs) were also successfully attached and grown on ITO surfaces using the seed-mediated growth method [35]. The FE-SEM images recorded after treating the Pd nanoseed particle-attached ITO substrates in a growth solution for 4 and 12 h, respectively, are shown in Figures 8.12a and b. Crystal growth of PdNPs occurred as the immersion time in the growth solution was increased, with Pd nanocrystals of 60–80 nm being identified after 24 h (Figures 8.12c and d). The major characteristic of the formed nanostructure of Pd was that the nanocrystals tended to adhere to each other. Moreover, such aggregation occurred not only on the surface but was also 3-D in nature; that is, some nanocrystals were seen to grow above the basal nanocrystals. The nanostructure

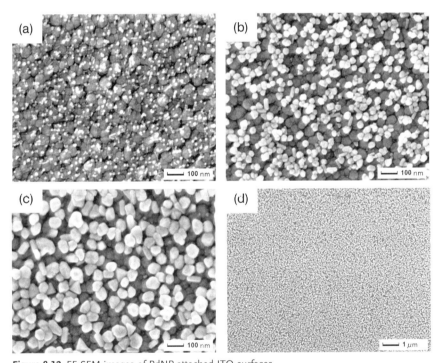

Figure 8.12 FE-SEM images of PdNP-attached ITO surfaces. The ITO substrates were immersed into the seed solution for 2 h and then into the growth solution for: (a) 4 h; (b) 12 h; and (c) 24 h; (d) A low-magnification image of the surface of panel (c). Reproduced with permission from Ref. [35]; © 2006, American Chemical Society.

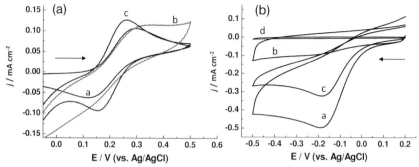

Figure 8.13 (a) Cyclic voltammograms of 1.0 mM [Fe(CN)$_6$]$^{4-}$ in 0.1 M phosphate-buffered solution (pH 7.0) with: (curve a) a bare ITO electrode; (curve b) a bulk Pd electrode; and (curve c) a PdNP/ITO electrode prepared via 24 h growth. Scan rate = 50 mV s^{-1}; (b) Cyclic voltammograms recorded with: (curves a, b) the PdNP/ITO electrode prepared via 24 h growth in (curve a) air-saturated and (curve b) N$_2$-saturated 0.1 M KCl solutions, and with (curve c) a Pd bulk electrode and (curve d) a bare ITO electrode in air-saturated 0.1 M KCl solution. Scan rate = 50 mV s^{-1}. Reproduced with permission from Ref. [35]; © 2006, American Chemical Society.

of Pd grown on ITO was quite different from that of Au and Ag, mainly because the AuNPs and AgNPs tended to form in dispersed states (see above). Hence, the inference was that the identity of the metal changed the aggregation characteristics during the growth process, despite using CTAB as the same capping reagent in all growth procedures.

Due to the dense nanoparticle attachment (see Figure 8.12c), the PdNP/ITO electrodes had a significantly lowered charge transfer resistance compared to that of a bare ITO, and the redox reaction of [Fe(CN)$_6$]$^{3-}$/[Fe(CN)$_6$]$^{4-}$ was observed to be reversible in 0.1 M phosphate-buffered solution (Figure 8.13a) [35]. The electrocatalytic property of PdNPs attached on ITO was confirmed for the reduction of oxygen (Figure 8.13b). In addition, some typical responses were observed in 0.5 M H$_2$SO$_4$ with the PdNP/ITO electrodes, reflecting both the characteristics of the NPs and the thin layer on a nanoscale [35]. The proposed preparation method for PdNP-attached ITO surfaces should be promising for catalytic applications, as well as electrochemical uses.

8.9
Attachment of Platinum Nanoparticles PtNPs on ITO and GC

Although attempts were made to prepare platinum nanoparticles (PtNPs) on ITO surfaces using the seed-mediated growth method, some preliminary studies showed this approach to be difficult [36]. The problems were caused by the appearance of brown precipitates in the growth solution, despite using the same principle (i.e., the mixing of K$_2$PtCl$_4$, CTAB and ascorbic acid), before the seed-attached ITO was immersed.

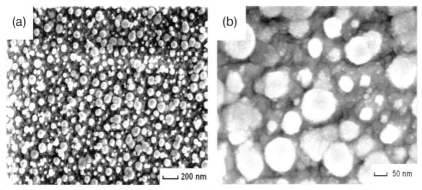

Figure 8.14 FE-SEM images of PtNP-attached ITO surfaces. The ITO substrate was immersed in the growth solution containing 0.25 mM K$_2$PtCl$_4$ and 5 mM ascorbic acid for 24 h. (a) Low-magnification and (b) high-magnification images of the same surface. Reproduced with permission from Ref. [36]; © 2006, American Chemical Society.

However, with further investigation the attachment of PtNPs on ITO was achieved by employing a rather simple method, namely a one-step *in situ* chemical reduction of PtCl$_4^{2-}$ by ascorbic acid, but without using CTAB [36]. The FE-SEM images of the PtNP-attached ITO surfaces prepared via this *in situ* reduction method are shown in Figure 8.14. Here, the attached PtNPs were spherical and showed an agglomerated nanostructure which was composed of small nanoclusters. Based on the morphological changes which were dependent on the growth time, PtNPs were shown to grow via a progressive nucleation mechanism [36].

Characteristically, when PtNP/ITO was used as a working electrode, the charge transfer resistances were found to be significantly lowered due to the PtNP growth. Hence, for the typical redox system of [Fe(CN)$_6$]$^{3-}$/[Fe(CN)$_6$]$^{4-}$, the PtNP/ITO electrodes exhibited electrochemical responses which were similar to that of a bulk Pt electrode [36]. It was also apparent that the PtNP/ITO electrodes had significant electrocatalytic properties for oxygen reduction and methanol oxidation (see Figures 8.15a and b, respectively) [36]. Those PtNPs which demonstrated an agglomerated nanostructure should show promise as a new type of electrode material.

The *in situ* reduction method used to prepare PtNPs was also applied to the modification of GC surfaces. This resulted in a thin continuous Pt film which was composed of small nanoclusters that had a further agglomerated nanostructure of small grains, and could be attached onto the GC surface [37]. FE-SEM images of the Pt nanocluster film (PtNCF) are shown in Figure 8.16. The electrochemical results obtained indicated that the current values for Pt oxidation, Pt oxide reduction and hydrogen-related redox reactions, when recorded with the PtNCF electrode, were almost twice those with the Pt nanocluster dispersedly-attached GC (PtNC/GC) electrode, but this reflected the higher Pt loading (Figure 8.17a) [37]. The electrocatalytic ability of the PtNCF for methanol oxidation was also

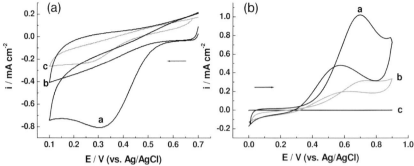

Figure 8.15 (a) Cyclic voltammograms recorded using (curves a, b) a PtNP/ITO electrode prepared via 24 h growth in: (curve a) air-saturated and (curve b) N_2-saturated 0.5 M H_2SO_4 solutions, and (curve c) a Pt bulk electrode for the air-saturated 0.5 M H_2SO_4 solution. Scan rate = 50 mV s^{-1}; (b) Cyclic voltammograms obtained for 0.1 M methanol oxidation in the N_2-saturated 0.5 M H_2SO_4 solution recorded with: (curve a) the prepared PtNP/ITO electrode; (curve b) a Pt bulk electrode; and (curve c) a bare ITO electrode. Scan rate = 50 mV s^{-1}. Reproduced with permission from Ref. [36]; © 2006, American Chemical Society.

Figure 8.16 FE-SEM images of PtNCF attached on GC. The concentration of ascorbic acid for preparation was 5.1 mM. (a) High-magnification and (b) low-magnification images. The left lower corner of (b) was scratched to show the film thickness. Reproduced with permission from Ref. [37]; © 2007, Elsevier.

apparently higher than that of the PtNC/GC or PtNP/ITO electrodes (Figure 8.17b). In addition, the electrocatalytic performance of PtNCF, when expressed in term of Pt content, was clearly superior to that of Pt black formed on GC [37].

Taken together these results indicated that, in spite of the continuous nanostructures, the nanograins of PtNCF functioned effectively for catalytic electrolysis. At present, PtNCF may be regarded as an interesting thin-film material, which can be easily prepared via a one-step chemical reduction.

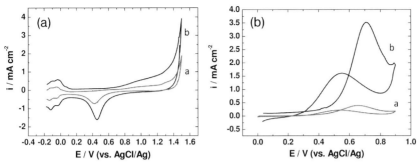

Figure 8.17 (a) Cyclic voltammograms recorded using (a) PtNC/GC and (b) PtNCF electrodes in a solution containing 0.5 M H_2SO_4. Scan rate = 50 mV s^{-1}; (b) Cyclic voltammograms obtained for 1.0 M methanol oxidation in N_2-saturated 0.5 M H_2SO_4 recorded using (a) PtNC/GC and (b) PtNCF electrodes. Scan rate = 50 mV s^{-1}. Reproduced with permission from Ref. [37]; © 2007, Elsevier.

8.10
Electrochemical Measurements of Biomolecules Using AuNP/ITO Electrodes

It has been shown that the AuNP/ITO electrodes can be used for the electrochemical measurements of biomolecules and, in collaboration with Professor Goyal in India, we have acquired a host of data using our AuNP/ITO electrodes, albeit with a differential pulse voltammetry technique. Examples include the electrochemical analysis of uric acid [38], paracetamol (acetaminophen) [39], atenolol [40], nandrolone [41], methylprednisolone acetate [42], and salbutamol [43]. In addition, the simultaneous electrochemical determination of guanosine and guanosine 5′-triphosphate [44], adenosine and adenosine 5′-triphosphate [45], and dopamine and serotonin [46] have been successfully achieved with the AuNP/ITO electrodes.

Although the AuNP/ITO electrodes are not robust, and the modified surfaces are difficult to renew if damaged, Professor Goyal's group used the AuNP/ITO electrodes (which were sent from Japan) with great care and so obtained excellent results. Yet, this may in fact provide evidence of the practical strength of the AuNP/ITO electrodes prepared using the seed-mediated growth method.

8.11
Nonlinear Optical Properties of Metal NP-Attached ITO

Beyond application as electrodes, the metal NP-attached ITO materials exhibit interesting nonlinear optical characteristics, and in order to investigate these properties we are presently collaborating with Professor Kityk in Poland and Professor Ebothé in France. To date, a variety of interesting nonlinear optical properties of AuNPs [47–51] and AgNPs [52, 53], attached either on ITO or glass, have been reported. In addition, PdNPs [54, 55], PtNPs [54], and NiNPs [56] attached to ITO surfaces have also been investigated.

Because the seed-mediated growth technique includes a chemical reduction process in solution, the doping of second ions can be easily carried out. Consequently, the luminescence properties of not only the NPs of pure metals but also of erbium ion-doped AgNPs have been reported [57]. In addition, due to the universal applicability of the seed-mediated method, a ZnO film was modified with AuNPs and the nonlinear optical properties reported [58, 59].

In addition to the nonlinear optical properties of metal NPs, the modification of glasses with metal NPs has the potential for use as a new type of substrate for spectroscopic detection. As an example, AuNP-attached glass plates were described for use in surface plasmon resonance spectroscopy [60].

8.12
Concluding Remarks

In this chapter, we have summarized the wet chemical preparation methods – notably seed-mediated growth – for fabricating various metal NP (or nanostructured crystal)-attached electrodes. It has been shown that small metal seed-NPs can be attached to the surfaces of substrates (typically ITO and GC, but also other materials) without the use of binding reagents, and the seed-NPs can then be grown via chemical reductions while retaining a strong attachment. The simplicity of the preparation method – that is, a two-step immersion first into the seed solution, and second into the growth solution (both of which are aqueous and at ambient temperature) – would be of significant merit in any proposed seed-mediated method for nanofabrication.

The thus-fabricated metal NP-attached electrodes have demonstrated interesting properties in terms of their electrochemical applications, which differ from those of bulk metals. The lower charge transfer resistances should represent a key advantage when constructing electrode materials with metal NPs. Consequently, nanocomposites consisting of metal NPs and base substrates, when prepared using the seed-mediated growth method, should be regarded as advanced-function materials, with potential applications also in nonlinear optics.

References

1 Daniel, M.-C. and Astruc, D. (2004) *Chem. Rev.*, **104**, 293.
2 Hernández-Santos, D., González-García, M.B. and Costa-García, A. (2002) *Electroanalysis*, **14**, 1225.
3 Katz, E., Willner, I. and Wang, J. (2004) *Electroanalysis*, **16**, 19.
4 Welch, C.M. and Compton, R.G. (2006) *Anal. Bioanal. Chem.*, **384**, 601.
5 Finot, M.O., Braybrook, G.D. and McDermott, M.T. (1999) *J. Electroanal. Chem.*, **466**, 234.
6 El-Deab, M.S., Okajima, T. and Ohsaka, T. (2003) *J. Electrochem. Soc.*, **150**, A851.

7 Frens, G. (1973) *Nature*, **241**, 20.
8 Slot, J.W. and Geuze, H.J. (1985) *Eur. J. Cell Biol.*, **38**, 87.
9 Brust, M., Walker, M., Bethell, D., Schiffrin, D.J. and Whyman, R. (1994) *J. Chem. Soc., Chem. Commun.*, 802.
10 Freeman, R.G., Grabar, K.C., Allison, K.J., Bright, R.M., Davis, J.A., Guthrie, A.P., Hommer, M.B., Jackson, M.A., Smith, P.C., Walter, D.G. and Natan, M.J. (1995) *Science*, **267**, 1629.
11 Grabar, K.C., Freeman, R.G., Hommer, M.B. and Natan, M.J. (1995) *Anal. Chem.*, **67**, 735.
12 Chumanov, G., Sokolov, K., Gregory, B.W. and Cotton, T.M. (1995) *J. Phys. Chem.*, **99**, 9466.
13 Cheng, W., Dong, S. and Wang, E. (2002) *Anal. Chem.*, **74**, 3599.
14 Cheng, W., Dong, S. and Wang, E. (2002) *Langmuir*, **18**, 9952.
15 Jana, N.R., Gerheart, L. and Murphy, C.J. (2001) *J. Phys. Chem. B*, **105**, 4065.
16 Busbee, B.D., Obare, S. and Murphy, C.J. (2003) *Adv. Mater.*, **15**, 414.
17 Kambayashi, M., Zhang, J. and Oyama, M. (2005) *Cryst. Growth Des.*, **5**, 81.
18 Bönnemann, H., Braun, G., Brijoux, W., Brinkmann, R., Schulze Tilling, A., Seevogel, K. and Siepen, K. (1996) *J. Organomet. Chem.*, **520**, 143.
19 Zhang, J., Kambayashi, M. and Oyama, M. (2004) *Electrochem. Commun.*, **6**, 683.
20 Zhang, J., Kambayashi, M. and Oyama, M. (2005) *Electroanalysis*, **17**, 408.
21 Zhang, J. and Oyama, M. (2004) *Electrochem. Acta*, **50**, 85.
22 Zhang, J. and Oyama, M. (2005) *J. Electroanal. Chem.*, **577**, 273.
23 Horibe, T., Zhang, J. and Oyama, M. (2007) *Electroanalysis*, **19**, 847.
24 Zhang, J. and Oyama, M. (2007) *Electrochem. Commun.*, **9**, 459.
25 Zhang, J. and Oyama, M. (2005) *Anal. Chim. Acta*, **540**, 299.
26 Ali Umar, A. and Oyama, M. (2005) *Cryst. Growth Des.*, **5**, 599.
27 Ali Umar, A. and Oyama, M. (2005) *Indian J. Chem. A*, **44**, 938.
28 Ali Umar, A. and Oyama, M. (2006) *Appl. Surf. Sci.*, **253**, 2196.
29 Ali Umar, A. and Oyama, M. (2006) *Appl. Surf. Sci.*, **253**, 2933.
30 Oyama, M., Yamaguchi, S. and Zhang, J. (2009) *Anal. Sci.*, **25**, 249.
31 Cui, Y., Yang, C., Zeng, W., Oyama, M., Pu, W. and Zhang, J. (2007) *Anal. Sci.*, **23**, 1421.
32 Zhang, J. and Oyama, M. (2005) *Electrochem. Solid-State Lett.*, **8**, E49.
33 Ali Umar, A. and Oyama, M. (2006) *Cryst. Growth Des.*, **6**, 818.
34 Chang, G., Zhang, J., Oyama, M. and Hirao, K. (2005) *J. Phys. Chem. B*, **109**, 1204.
35 Chang, G., Oyama, M. and Hirao, K. (2006) *J. Phys. Chem. B*, **110**, 20362.
36 Chang, G., Oyama, M. and Hirao, K. (2006) *J. Phys. Chem. B*, **110**, 1860.
37 Chang, G., Oyama, M. and Hirao, K. (2007) *Thin Solid Films*, **515**, 3311.
38 Goyal, R.N., Oyama, M., Sangal, A. and Singh, S.P. (2005) *Indian J. Chem. A*, **44**, 945.
39 Goyal, R.N., Gupta, V.K., Oyama, M. and Bachheti, N. (2005) *Electrochem. Commun.*, **7**, 803.
40 Goyal, R.N., Gupta, V.K., Oyama, M. and Bachheti, N. (2006) *Electrochem. Commun.*, **8**, 65.
41 Goyal, R.N., Oyama, M., Tyagi, A. and Singh, S.P. (2007) *Talanta*, **72**, 140.
42 Goyal, R.N., Oyama, M., Ali Umar, A., Tyagi, A. and Bachheti, N. (2007) *J. Pharm. Biomed. Anal.*, **44**, 1147.
43 Goyal, R.N., Oyama, M. and Singh, S.P. (2007) *J. Electroanal. Chem.*, **611**, 140.
44 Goyal, R.N., Oyama, M. and Tyagi, A. (2007) *Anal. Chim. Acta*, **581**, 32.
45 Goyal, R.N., Oyama, M. and Singh, S.P. (2007) *Electroanalysis*, **19**, 575.
46 Goyal, R.N., Gupta, V.K., Oyama, M. and Bachheti, N. (2007) *Talanta*, **72**, 976.
47 Kityk, I.V., Ali Umar, A. and Oyama, M. (2005) *Physica E*, **27**, 420.
48 Kityk, I.V., Ali Umar, A. and Oyama, M. (2005) *Physica E*, **28**, 178.
49 Kityk, I.V., Ebothé, J., Fuks-Janczarek, I., Ali Umar, A., Kobayashi, K., Oyama, M. and Sahraoui, B. (2005) *Nanotechnology*, **16**, 1687.
50 Kityk, I.V., Plucinski, K.J., Ebothé, J., Ali Umar, A. and Oyama, M. (2005) *J. Appl. Phys.*, **98**, 084304.
51 Kityk, I.V., Ali Umar, A. and Oyama, M. (2005) *Appl. Opt.*, **44**, 6905.

52 Kityk, I.V., Ebothé, J., Chang, G. and Oyama, M. (2005) *Phil. Mag. Lett.*, **85**, 549.

53 Kityk, I.V., Ebothé, J., Ozgad, K., Plucinski, K.J., Chang, G., Kobayashi, K. and Oyama, M. (2006) *Physica E*, **31**, 38.

54 Ebothé, J., Kityk, I.V., Chang, G., Oyama, M. and Plucinski, K.J. (2006) *Physica E*, **35**, 121.

55 Ebothé, J., Kityk, I.V., Nzoghe-Mendon, L., Chang, G., Oyama, M., Sahraoui, B. and Miedzinski, R. (2008) *J. Mod. Opt.*, **55**, 187.

56 Zamorskii, M.K., Nouneh, K., Kobayashi, K., Oyama, M., Ebothé, J. and Reshak, A.H. (2008) *Opt. Laser Technol.*, **40**, 499.

57 Ebothé, J., Ozga, K., Ali Umar, A., Oyama, M. and Kityk, I.V. (2006) *Appl. Surf. Sci.*, **253**, 1626.

58 Ozga, K., Kawaharamura, T., Ali Umar, A., Oyama, M., Nouneh, K., Slezak, A., Fujita, S., Piasecki, M., Reshak, A.H. and Kityk, I.V. (2008) *Nanotechnology*, **19**, 18709.

59 Ozga, K., Kawaharamura, T., Ali Umar, A., Oyama, M., Slezak, A., Fujita, S. and Kityk, I. (2008) *J. Nano Res.*, **2**, 31.

60 Hamamoto, K., Micheletto, R., Oyama, M., Ali Umar, A., Kawai, S. and Kawakami, Y. (2006) *J. Opt. A Pure Appl. Opt.*, **8**, 268.

9
Mesoscale Radical Polymers: Bottom-Up Fabrication of Electrodes in Organic Polymer Batteries

Kenichi Oyaizu and Hiroyuki Nishide

9.1
Mesostructured Materials for Energy Storage Devices

Energy storage is one of the most important worldwide concerns of the twenty-first century, and it is essential that new, low-cost, and environmentally benign (i.e., sustainable) materials for energy conversion and storage purposes are found [1, 2]. Energy-related nanostructured materials have attracted much attention because of the unusual electrical properties that originate from their confined nanoscale dimensions and combined bulk/interfacial properties, which affect their overall behavior [3]. On the other hand, a new trend in materials research has begun to emerge recently, focusing on the fabrication of mesoscale structures [4], which is a broader term (or superordinate concept) of the ordered fine structures, characterized by dimensions larger than those in nanomaterials that typically consist of nanoparticles and top-down nanoarchitectured structures. The exploration of properties of materials on the mesoscale has led to new technological perspectives of electrochemical devices such as batteries and sensors, which rely on ion transport and, thus, are strongly influenced by the size effects of composing materials on mass transfer, transport, and storage. In particular, mesoscale materials are becoming increasingly important for electrochemical energy storage applications, due to the limited distance of diffusion for the mass transfer during charging/discharging processes. A typical example is that of lithium-ion batteries, which consist of a Li-ion intercalating anode (e.g., graphite) and a cathode (e.g., $LiCoO_2$), separated by Li-ion-conducting electrolyte layers such as a solution of $LiPF_6$ in an ethylene carbonate/diethylcarbonate (EC/DEC) mixture (Figure 9.1) [5].

The advantages of ordered mesostructures for these electrode-active materials are the improved accommodation of the lattice transformation resulting from the Li^+ intercalation/deintercalation to improve cycle life, the increased electrode/electrolyte contact area leading to higher charging/discharging rates, and – most importantly – the short path length for electronic and ionic transport which permit operation with low electric conductivity and hence at high power [6].

Advanced Nanomaterials. Edited by Kurt E. Geckeler and Hiroyuki Nishide
Copyright © 2010 WILEY-VCH Verlag GmbH & Co. KGaA, Weinheim
ISBN: 978-3-527-31794-3

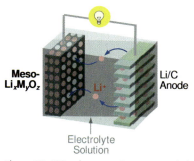

Figure 9.1 Li-ion battery using mesoscale lithium metal oxide cathodes for an improved rate performance.

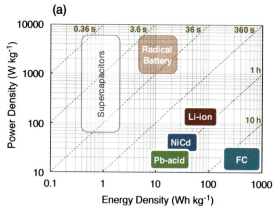
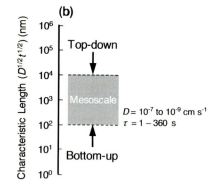

Figure 9.2 (a) Ragone plot for batteries, fuel cells, and supercapacitors. The dashed lines represent the time parameter, τ, which is characteristic of the rate performance; (b) Characteristic length of electrode-active materials defined as the layer thickness of diffusing ions and/or propagating charges during the charging/discharging process, showing the top-down and bottom-up strategies to fabricate the target mesoscale materials.

Ragone plots (Figure 9.2a) of specific power density versus specific energy density for various batteries, displayed with those of fuel cells and supercapacitors as references, show their specific time parameter, τ, that characterizes the period of time required for charging/discharging operations [2]. The diffusion length $(D\tau)^{1/2}$ of ions and propagating charges within the electrode-active materials corresponds to their characteristic dimensional scales, where D is the diffusion coefficient for the mass transfer processes. For cathode-active materials in Li-ion batteries, the lithium cobalt oxide ($LiCoO_2$) is characterized by a relatively large diffusion coefficient of $D = 10^{-7}$ to $10^{-9}\,cm^2\,s^{-1}$ [7], but the recently investigated less expensive and/or less toxic oxides show lower diffusivities for the Li^+ ion, such as those in $Li_{1-\delta}Ni_{0.5}Mn_{0.5-x}Ti_xO_2$ ($x = 0$–0.3, $D = 10^{-9}$ to $10^{-14}\,cm^2\,s^{-1}$) [8], Li_7BiO_6 ($D = 5 \times 10^{-13}\,cm^2\,s^{-1}$) [9], $LiFePO_4$ ($D = 1.8 \times 10^{-14}\,cm^2\,s^{-1}$) [10], and LiV_3O_8 ($D = 2$–$8 \times 10^{-10}\,cm^2\,s^{-1}$) [11]. The resulting dimensional scale, or the characteristic length, is typically in the order of $10^2 < (D\tau)^{1/2} < 10^4$ nm, thus revealing the importance of fabricating mesostructured materials (Figure 9.2b), which are in-between nanoscopic and macroscopic materials in dimension.

Fabrication of the mesostructured oxides has been accomplished by using top-down methods, although mesoscale cathodes are less developed than the mesostructured anodes. For example, a similar approach to the formation of silicon nanopillars has been applied for cathode-active materials. By using suitable templates such as porous alumina and porous polymeric materials, nanopillars of V_2O_5 and $LiMn_2O_4$ have been successfully grown on metal substrates as the current collectors [6]. The mesostructured materials have additional advantages over the conventional electrode-active materials, to accommodate relatively large volumetric changes during charging/discharging cycles, thus improving the overall power-rate performance of the resulting batteries.

Organic molecule-based electrode-active materials are more advantageous than inorganic materials in terms of their potential capability for the molecule-based "bottom-up" fabrication of mesostructures (Figure 9.2b), in addition to their inherent flexibility, safety, and accessibility from unlimited resources [12]. The organic polymers most typically examined for batteries are conducting polymers such as polyacetylene, polyaniline, polypyrrole, polythiophene, and polyphenylene [13–17]. Sulfur and redox polymerization electrodes have also been reported as electroactive materials [13]. However, the low doping levels in conducting polymers result in a small theoretical capacity. Moreover, the charging/discharging processes have often been impeded by the slow kinetics of electrode reactions [13]. Another group of molecules investigated for charge storage is a series of redox polymers with nonconjugated backbones and redox centers localized in pendant groups, such as tetrathiafulvalene [18], ferrocene [19, 20], and carbazole [21, 22]. In this case, the redox centers, rather than the polymer backbones, govern the redox behaviors. Conductivity arises when these redox centers can exchange charge, through for example electron hopping between the redox isolated sites within the polymers (Figure 9.3).

It has been found recently that organic robust radicals are potentially useful as the redox centers due to their rapid one-electron-redox reactions [23, 24]. Moreover, excellent film-forming properties of the organic radical polymers by simple wet processes allow various mesoscale fabrication such as homogeneous layers with precisely controlled thicknesses and ordered mesostructures supported on carbon fiber matrices. This aim of this chapter is to provide a concise overview of recent

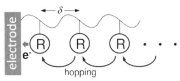

Figure 9.3 Charge propagation within redox polymer layers, swollen in electrolyte solutions, by an electron self-exchange (hopping) mechanism.

322 | *9 Mesoscale Radical Polymers: Bottom-Up Fabrication of Electrodes in Organic Polymer Batteries*

approaches to fabricate mesoscale electrode-active materials, with emphasis placed on the organic polymers for the radical polymer battery [25], which is characterized by an excellent rate performance and capability of fabricating flexible, paper-like, and (semi)transparent rechargeable energy-storage devices.

9.2
Mesoscale Fabrication of Inorganic Electrode-Active Materials

The top-down fabrication of inorganic electrode-active materials has been motivated by the possibility of producing a film-like, bendable lithium-ion battery, which depends in turn on the development of soft electrode-active materials such as those composed of nanoparticles and nanocoatings of metal oxides for the cathodes, and lithium foil or nanocarbon materials for the anodes [1, 26]. The top-down fabrication has also been driven by the recent challenge to maximize transport rates by employing mesostructured inorganic-based materials. The electrode reaction of inorganic-based materials is often rate-determined by the slow kinetics of ion intercalation and migration that accompany lattice transformation [6]. The power-rate properties have been improved by employing the top-down strategy so as to produce a large surface area and suitable path length for the mass transfer processes [27]. Recent interesting approaches have included the virus-enabled synthesis and assembly of cobalt oxide nanowires (Figure 9.4). This has been accomplished by incorporating gold-binding peptides into a filament coat to

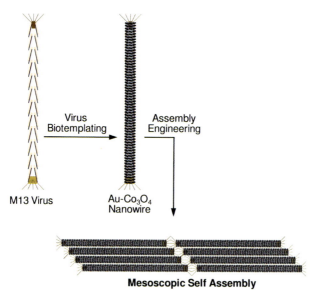

Figure 9.4 Virus-enabled synthesis of Au-Co_3O_4 nanowire and self-assembly strategy to form an ordered monolayer of Co_3O_4 nanowires [28].

form mesoscale gold–cobalt oxide hybrid wires that improve battery capacity as a new cathode-active material [28]. Indeed, engineered electrode mesostructures to enhance electrode kinetics have been a target of intense research, including those based on the electrochemically assisted template growth of Cu nanorods onto a current collector, followed by electrochemical plating of Fe_3O_4 to form the ordered structure at a mesoscale level [27]. The significant effect of the mesoscopic size on transport rates and flexibility has been demonstrated recently by Li-ion batteries made from nanocomposite papers [26, 29]. These are obtained by infiltrating cellulose into carbon nanotubes (CNTs) grown on a silicon substrate and impregnated with the electrolyte, allowing the combination of the cathode (the nanotubes) and the separator (the cellulose) in a single unit [26].

9.3
Bottom-Up Strategy for Organic Electrode Fabrication

9.3.1
Conjugated Polymers for Electrode-Active Materials

Research into organic electrode-active materials was prompted by the discovery of the electric conductivity of doped polyacetylene, which in turn led to the exploration of organic batteries by employing the redox capacity of polyacetylene. This was based on the reversible p-type and n-type doping/de-doping processes for the cathode and the anode, respectively, as the principal electrode reactions [30]. Electrically conducting polymers such as polyaniline, polypyrrole, polythiophene, and their related derivatives, have been similarly examined as electrode-active materials, based on their reversible electrochemical doping behaviors. The limitations of conducting polymers are based on their insufficient doping levels, the resulting low redox capacities, and the chemical instability of the doped states that frequently lead to the self-discharge and degradation of the rechargeable properties of the resulting batteries.

Another concept of using conjugated polymers for batteries is to explore a charge–storage configuration in which the electrolyte layer is sandwiched by thin layers of even less conductive polymers incorporating redox-active groups, with a view to increasing the overall redox capacity. In this case, the polymer backbones provide a matrix and/or a conducting path to interconnect innumerable redox sites for the hopping of electrons by a self-exchange mechanism, and this results in the storage and transport of charge in a homogeneous solid. Along with this concept, polythiophene-based conjugated polymers bearing redox sites such as tetrathiafulvalene derivatives [31] and polypyrrole coupled with redox-active dopants such as indigo carmine derivatives [32–35], have been recently designed. However, the chemical bond cleavage and the formation that are often accompanied by the redox reaction of these closed-shell molecules are frequently electrochemically irreversible or nonreversible, and hence are unfavorable for application as electrode-active materials. Simple methods to fabricate mesostructures by the bottom-up strategies have been also awaited, in order to bring out the performance of the organic polymers.

9.3.2
Mesoscale Organic Radical Polymer Electrodes

A new concept, applicable to the mesostructure fabrication, has been proposed for the organic electrode-active materials, based on the sufficiently large redox capacity of aliphatic redox polymers – that is, organic polymers densely populated with pendant redox-isolated sites. The principal finding to permit the use of the non-conjugated redox polymers is the capability of organic robust radicals, or open-shell molecules, to allow fully reversible, one-electron redox reactions featuring fast electrode kinetics and reactant recyclability. While the chemically inert organic radicals constitute an important and extensively studied group of functional materials, such as organic ferromagnets [36–39], metal-free redox mediators for synthetic applications [40–44], and electron and hole transport materials for organic devices [45, 46], the idea of using them as electroactive materials for rechargeable batteries has never been proposed, except in our recent reports [47–52] and those from other groups [53–64]. Hence, we have focused on organic radicals stabilized thermodynamically by an effective delocalization of the unpaired electron and/or kinetically by bulky substituents to employ them as the redox centers. Radical groups such as the NO-centered nitroxides, N-centered triarylaminium cation radicals, and diphenylpicrylhydrazyl derivatives, O-centered phenoxyls and galvinoxyls, and C-centered trityls and phenalenyls, are typically examined as the pendant group of the radical polymers (Scheme 9.1).

These radicals are persistent at ambient temperatures under air and are characterized by a rapid and reversible one-electron-electrode reactions, typically with heterogeneous electron-transfer rate constants of $k_0 = 10^{-1}\,\text{cm}\,\text{s}^{-1}$ in the

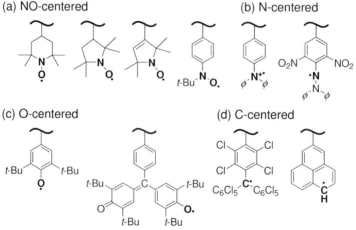

Scheme 9.1 Chemical structures of organic robust radicals in the radical polymers.

case of 2,2,6,6-tetramethylpiperidine-1-oxyl (TEMPO) derivatives [23, 24]. A variety of polymer backbones have been employed to bind the radicals, such as poly(methacrylate)s, polystyrene derivatives, poly(vinyl ether)s, polyethers, and poly(norbornene)s. The unpaired electron density of the polymers, as determined by superconducting quantum interference device (SQUID) measurements, has revealed the presence of radicals substantially per repeating unit. Theoretical redox capacities based on the formula mass of the repeating unit are in the range of 50 to 150 mAh g^{-1}.

For mesoscale electrode fabrication, the radical polymers are conveniently placed on the surface of an electrode or a current collector by a solution-based wet process such as the spin-coating method. The contact stylus profile near a scratched edge provides a layer thickness under dry conditions in the range of 10 to 10^3 nm, according to the concentration of the polymer in the mother liquor and the spinning velocity, and has revealed a flat surface with a roughness of less than several nanometers. The polymer backbones are suitably modified by photo-crosslinking, for example, thus, to be swollen but not to be dissolved in the electrolyte solutions, as typically shown for the TEMPO-substituted polynorbornene **1** in Scheme 9.2 [47].

The capability of forming stable films by the crosslinking method has allowed a novel design of a mesostructure fabrication processes, in which the polymer is coated on the surface of conducting mesoscale matrices as the support. The scanning electron microscopy (SEM) image of the resulting electrode surface, prepared on a Au quartz crystal microbalance (QCM) assembly for simultaneous electrochemical and resonant frequency-based gravimetric analysis (*vide infra*), demonstrates the ordered three-dimensional (3-D) structure successfully fabricated on the mesoscale (Figure 9.5) and which is suitable for ion transport throughout the polymer layer.

The unpaired electron density of the radical polymers amounts to a number of several molar in the bulk of the swollen polymer equilibrated in electrolyte

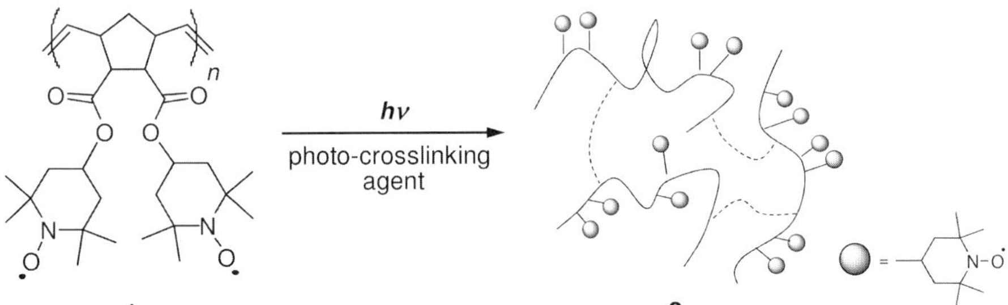

Scheme 9.2 Photo-crosslinking of the TEMPO-substituted polynorbornene **1** to form a polymer layer of **2**, which is insoluble but swellable in electrolyte solutions. The spin density by SQUID measurements amounts to 10^{21} spin per gram of **2**, indicating that most of the radicals survive during the photo-crosslinking reaction.

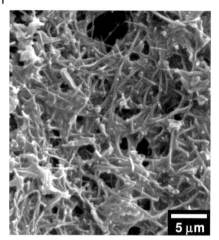

Figure 9.5 Scanning electron microscopy image of the photo-crosslinked polymer **2**/carbon composite electrode coated on the surface of a QCM electrode assembly.

solutions. Under these conditions, charge propagation within the polymer layer is sufficiently fast, leading to high-density charge storage, because the redox sites are so populated that the electron self-exchange reaction goes to completion within a finite distance of the polymer layer attached to the carbon fiber. The concentration gradient-driven charge propagation is accomplished by the electron self-exchange reaction. The apparent diffusion coefficient, D, of the charge produced at the surface of the electrode is determined by the bimolecular rate constant, k_{ex}, for the self-exchange reaction. Thus, when the sites are immobilized in the layer, allowing only diffusional collision of the neighboring sites to undergo an electron self-exchange reaction, this process involves an electron-hopping mechanism with a diffusion coefficient formulated by $D = k_{ex}\delta^2 C^*/6$, where k_{ex} is the bimolecular rate constant, δ is the site distance, and C^* is the site concentration in the polymer layer [65, 66]. The electron self-exchange reaction is sufficiently fast for diffusion-limited outer-sphere redox reactions, resulting in the efficient transfer of charge produced at the surface of the electrode (Figure 9.6).

The electrochemical behavior of the radical polymer layers attached to an electrode have revealed an efficient charge propagation process with diffusion coefficients in the order of 10^{-8} to $10^{-10}\,cm^2\,s^{-1}$. The electrolytes examined are typically tetrabutylammonium or the lithium salts of perchlorate or hexafluorophosphate, dissolved in conventional organic solvents such as CH_3CN and propylene carbonate. The cyclic voltammogram obtained for the electrode (as in Figure 9.5), with a sufficient crosslinking density of 11%, persists without change for several days of continued charging/discharging cycling in CH_3CN, in which the layer appears completely insoluble (Figure 9.7). Wave shapes with a peak-to-peak separation of ΔE_p = ca. 200 mV suggest some contribution from diffusion across the thick layer (Figure 9.7b). In the QCM response, the resonant frequency decreases during the

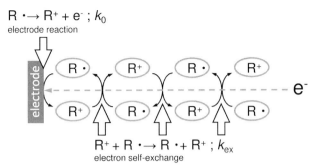

R · → R⁺ + e⁻ ; k_0
electrode reaction

R⁺ + R · → R · + R⁺ ; k_{ex}
electron self-exchange

Figure 9.6 Electrode and chemical reactions related to the charge transfer in radical redox polymers.

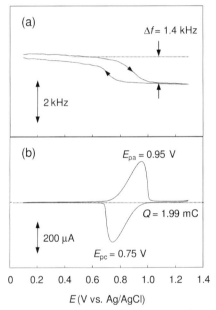

Figure 9.7 (a) QCM response and (b) cyclic voltammogram obtained for the photo-crosslinked polymer **2**/carbon composite electrode coated on the surface of a QCM electrode assembly. The Au electrode surface area A was 0.196 cm². The content of **2** in the composite electrode was 10 wt%. The crosslinking density was 11%. The electrolyte was a solution of 0.1 mol l⁻¹ tetrabutylammonium hexafluorophosphate in CH₃CN. Scan rate = 20 mV s⁻¹.

oxidation and attains its original value after the reduction, indicating reversibility to the mass change (Figure 9.7a). The galvanostatic E–t curves show a plateau region, which agrees with the formal potential. The amount of charge consumed during the potential scan and the constant-current electrolysis both coincide well with the formula weight-based redox capacity, which reveal that almost all of the sites in the layer undergo the redox reaction.

The redox process of the polymer layer is accompanied by the rapid injection and rejection of solvated counterions that compensate the charges produced by the redox reaction. Electroneutrality requires that the removal of each electron from the polymer layer results in the insertion of one electrolyte anion into the layer and/or the Donnan exclusion of the electrolyte cation. The mass change, Δm, associated with the redox reaction is obtained from the resonant frequency change Δf in Figure 9.7a using the Sauerbrey equation ($\Delta f = -C_f \Delta m$), where the sensitive factor C_f is 0.183 Hz cm^2 ng^{-1} for the QCM assembly employed in this study. The redox capacity Q is determined from the area under the waves in Figure 9.7b. The mass change relative to the redox capacity ($\Delta m F/Q \sim 71$ g mol^{-1}) is smaller than the formula mass of the PF$_6^-$ anion, indicating the slight contribution from the nonpermselective incorporation of the electrolyte cation during the mass transfer process.

Nitroxide radicals are potentially reducible to aminoxy anions by a one-electron-transfer process, which corresponds to the n-type doping of the neutral radical polymer (Scheme 9.3).

Attempts have been made to synthesize the n-dopable polymers and to use them as the anode-active material, with a view to employing only nitroxide radicals at both electrodes [48]. It has been demonstrated that suitable electron-withdrawing groups (EWG)s such as trifluoromethyl and cyano substituents, bound proximally to the NO redox center (Scheme 9.4a), shift the n-doping potential positively up to approximately −0.8 V. Acylnitroxides (Scheme 9.4b) are also potentially useful as the anode material. The nitroxide radical polymer can thus be switched from the p-type material for use in cathodes to the n-type anode-active material, by tuning the substituent effect.

Scheme 9.3 Redox couples related to nitroxide radicals.

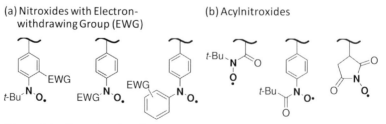

Scheme 9.4 n-Type nitroxide polymers.

The radical polymers act as both cathode- and anode-active materials, because of their capability to tune the redox potentials by the bottom-up strategy based on the molecular design. A couple of polymers, different in redox potentials, are used as the electroactive materials in the organic radical battery. The charging process corresponds to the oxidation of the radical to the cation at the cathode, and the reduction of the radical to the anion at the anode (Figure 9.8). The electromotive force is close to the potential gap between the two redox couples, which typically amounts to 0.5–1.5 V. The radical polymers are reversibly converted to the corresponding polyelectrolytes during the charging process. The rapid electrode reaction of the radicals and the efficient charge propagation within the polymer layer lead to a high rate performance, allowing rapid charging and large discharge currents without any substantial loss of output voltage. The amorphous radical polymers also allow the fabrication of flexible, thin-film devices. A curious feature is the capability of forming a "see-through" battery, due to the absence of significant chromophores in the radical polymer. Here, use could also be made of the slight color changes accompanied by radical redox reactions as an indicator of the charging level.

Efforts have also been directed towards further increasing the theoretical redox capacity of the radical polymer. For this purpose, radical polymers with even smaller formula masses per repeating unit have been designed (Scheme 9.5). The dinitroxide-functionalized polystyrene in Scheme 9.5 with a theoretical capacity of 193 mAh g^{-1} has been typically characterized by a notably high radical density of 4.3×10^{21} unpaired electrons per gram, which leads to the high capacity charging/discharging characteristics of the polymer membrane.

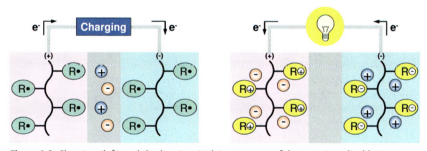

Figure 9.8 Charging (left) and discharging (right) processes of the organic radical battery.

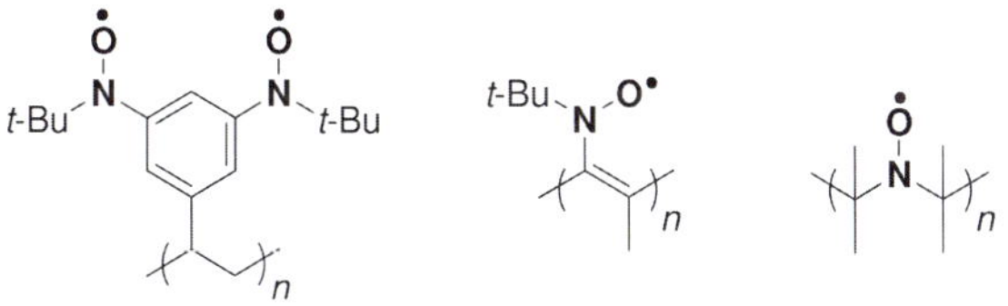

Scheme 9.5 Typical radical polymers designed for high redox capacity.

9.4
Conclusions

Radical polymers have been identified as a new class of redox polymers, which are characterized by a fast kinetics of electrode reactions. The suitably designed polymer backbone allows efficient swelling in conventional electrolyte solutions, which in turn enables an efficient charge propagation by a site-hopping mechanism within the mesostructured polymer layer. The use of radical polymers as electroactive materials leads to the first fabrication of organic electrode-active materials ordered in a mesoscale. The organic radical batteries have several advantages over the Li-ion batteries, such as higher safety, adoptability to wet fabrication processes, easy disposability, and capability of fabrication from less-limited resources. While organic batteries have an intrinsically lower volumetric energy density, this limitation can surely be overcome in the near future so that they can be designed to be compatible with, and installable in, a wide variety of electronic equipment.

References

1 Tarascon, J.-M. and Armand, M. (2001) *Nature*, **414**, 359–367.
2 Service, R.F. (2006) *Science*, **313**, 902.
3 Maier, J. (2005) *Nat. Mater.*, **4**, 805–14.
4 Yamauchi, Y., Takai, A., Nagaura, T., Inoue, S. and Kuroda, K. (2008) *J. Am. Chem. Soc.*, **130**, 5426–5427.
5 Alper, J. (2002) *Science*, **296**, 1224–1226.
6 Aricò, A.S., Bruce, P., Scrosati, B., Tarascon, J.-M. and van Schalkwijk, W. (2005) *Nat. Mater.*, **4**, 366–367.
7 Manthiram, A. and Kim, J. (1998) *Chem. Mater.*, **10**, 2895–2909.
8 Myung, S.-T., Komaba, S., Hosoya, K., Hirosaki, N., Miura, Y. and Kumagai, N. (2005) *Chem. Mater.*, **17**, 2427–2435.
9 Wilkening, M., Mühle, C., Jansen, M. and Heitjans, P. (2007) *J. Phys. Chem. B*, **111**, 8691–8694.
10 Zhang, P., Wu, Y., Zhang, D., Xu, Q., Liu, J., Ren, X., Luo, Z., Wang, M. and Hong, W. (2008) *J. Phys. Chem. A*, **112**, 5406–5410.
11 Yang, G., Wang, G. and Hou, W. (2005) *J. Phys. Chem. B*, **109**, 11186–11196.
12 Nishide, H. and Oyaizu, K. (2008) *Science*, **319**, 737–738.
13 Novák, P., Müller, K., Santhanam, K.S.V. and Haas, O. (1997) *Chem. Rev.*, **97**, 207–281.
14 Coppo, P. and Turner, M.L. (2005) *J. Mater. Chem.*, **15**, 1123–1133.
15 Roncali, J., Blanchard, P. and Frère, P. (2005) *J. Mater. Chem.*, **15**, 1589–1610.
16 Roncali, J. (1997) *J. Mater. Chem.*, **7**, 2307–2321.
17 Roncali, J. (1997) *Chem. Rev.*, **97**, 173–205.
18 Kaufman, F.B., Schroeder, A.H., Engler, E.M., Kramer, S.R. and Chambers, J.Q. (1980) *J. Am. Chem. Soc.*, **102**, 483–488.
19 Iwakura, C., Kawai, T., Nojima, M. and Yoneyama, H. (1987) *J. Electrochem. Soc.*, **134**, 791–795.
20 Hunter, T.B., Tyler, P.S., Smyrl, W.H. and White, H.S. (1987) *J. Electrochem. Soc.*, **134**, 2198–2204.
21 Compton, R.G., Davis, F.J. and Grant, S.C. (1986) *J. Appl. Electrochem.*, **16**, 239–249.
22 Skompska, M. and Peter, L.M. (1995) *J. Electroanal. Chem.*, **383**, 43–52.
23 Yonekuta, Y., Oyaizu, K. and Nishide, H. (2007) *Chem. Lett.*, **36**, 866–867.
24 Suga, T., Pu, Y.-J., Oyaizu, K. and Nishide, H. (2004) *Bull. Chem. Soc. Jpn.*, **77**, 2203–2204.
25 Nishide, H. and Suga, T. (2005) *Electrochem. Soc. Interface*, **14**, 32–36.

26 Scrosati, B. (2007) *Nat. Nanotechnol.*, **2**, 598–599.
27 Taberna, P.L., Mitra, S., Poizot, P., Simon, P. and Tarascon, J.-M. (2006) *Nat. Mater.*, **5**, 567–573.
28 Nam, K.T., Kim, D.-W., Yoo, P.J., Chiang, C.-Y., Meethong, N., Hammond, P.T., Chiang, Y.-M. and Belcher, A.M. (2006) *Science*, **312**, 885–888.
29 Pushparaj, V.L., Shaijumon, M.M., Kumar, A., Murugesan, S., Ci, L., Vajtai, R., Linhardt, R.J., Nalamasu, O. and Ajayan, P.M. (2007) *Proc. Natl Acad. Sci. USA*, **104**, 13574–13577.
30 Nigrey, P.J., MacDiarmid, A.G. and Heeger, A.J. (1979) *J. Chem. Soc., Chem. Commun.*, 594–595.
31 Berridge, R., Skabara, P.J., Pozo-Gonzalo, C., Kanibolotsky, A., Lohr, J., McDouall, J.J.W., McInnes, E.J.L., Wolowska, J., Winder, C., Sariciftci, N.S., Harrington, R.W. and Clegg, W. (2006) *J. Phys. Chem. B*, **110**, 3140–3152.
32 Song, H.-K. and Palmore, G.T.R. (2006) *Adv. Mater.*, **18**, 1764–1768.
33 Song, H.-K. and Palmore, G.T.R. (2005) *J. Phys. Chem. B*, **109**, 19278–19287.
34 Fei, J., Song, H.-K. and Palmore, G.T.R. (2007) *Chem. Mater.*, **19**, 1565–1570.
35 Song, H.-K., Toste, B., Ahmann, K., Hoffman-Kim, D. and Palmore, G.T.R. (2006) *Biomaterials*, **27**, 473–484.
36 Murata, H., Miyajima, D. and Nishide, H. (2006) *Macromolecules*, **39**, 6331–6335.
37 Fukuzaki, E. and Nishide, H. (2006) *Org. Lett.*, **8**, 1835–1838.
38 Kaneko, T., Makino, T., Miyaji, H., Teraguchi, M., Aoki, T., Miyasaka, M. and Nishide, H. (2003) *J. Am. Chem. Soc.*, **125**, 3554–3557.
39 Nishide, H., Kaneko, T., Nii, T., Katoh, K., Tsuchida, E. and Lahti, P.M. (1996) *J. Am. Chem. Soc.*, **118**, 9695–9704.
40 Knoop, C.A. and Studer, A. (2003) *J. Am. Chem. Soc.*, **125**, 16327–16333.
41 Georges, M.K., Lukkarila, J.L. and Szkurhan, A.R. (2004) *Macromolecules*, **37**, 1297–1303.
42 Dijksman, A., Marino-Gonzalez, A., Mairatai Payeras, A., Arends, I.W.C.E. and Sheldon, R.A. (2001) *J. Am. Chem. Soc.*, **123**, 6826–6833.
43 Cameron, N.R. and Reid, A.J. (2002) *Macromolecules*, **35**, 9890–9895.
44 Huang, W., Chiarelli, R., Charleux, B., Rassat, A. and Vairon, J.-P. (2002) *Macromolecules*, **35**, 2305–2317.
45 Pu, Y.-J., Soma, M., Kido, J. and Nishide, H. (2001) *Chem. Mater.*, **13**, 3817–3819.
46 Yonekuta, Y., Susuki, K., Oyaizu, K., Honda, K. and Nishide, H. (2007) *J. Am. Chem. Soc.*, **129**, 14128–14129.
47 Suga, T., Konishi, H. and Nishide, H. (2007) *Chem. Commun.*, 1730–1732.
48 Suga, T., Pu, Y.-J., Kasatori, S. and Nishide, H. (2007) *Macromolecules*, **40**, 3167–3173.
49 Oyaizu, K., Suga, T., Yoshimura, K. and Nishide, H. (2008) *Macromolecules*, **41**, 6646–6652.
50 Takahashi, Y., Hayashi, N., Oyaizu, K., Honda, K. and Nishide, H. (2008) *Polym. J.*, **40**, 763–767.
51 Nishide, H., Iwasa, S., Pu, Y.-J., Suga, T., Nakahara, K. and Satoh, M. (2004) *Electrochim. Acta*, **50**, 827–831.
52 Suga, T., Yoshimura, K. and Nishide, H. (2006) *Macromol. Symp.*, **245-246**, 416–422.
53 Kim, J.-K., Cheruvally, G., Choi, J.-W., Ahn, J.-H., Choi, D.S. and Song, C.E. (2007) *J. Electrochem. Soc.*, **154**, A839–A843.
54 Zhang, X., Li, H., Li, L., Lu, G., Zhang, S., Gu, L., Xia, Y. and Huang, X. (2008) *Polymer.* **49**, 3393–3398.
55 Bugnon, L., Morton, C.J.H., Novak, P., Vetter, J. and Nesvadba, P. (2007) *Chem. Mater.*, **19**, 2910–2914.
56 Suguro, M., Iwasa, S., Kusachi, Y., Morioka, Y. and Nakahara, K. (2007) *Macromol. Rapid Commun.*, **28**, 1929–1933.
57 Yoshikawa, H., Kazama, C., Awaga, K., Satoh, M. and Wada, J. (2007) *Chem. Commun.* 3169–3170.
58 Endo, T., Takuma, K., Takata, T. and Hirose, C. (1993) *Macromolecules*, **26**, 3227–3229.
59 Nakahara, K., Iwasa, S., Iriyama, J., Morioka, Y., Suguro, M., Satoh, M. and Cairns, E.J. (2006) *Electrochim. Acta*, **52**, 921–927.
60 Allgaier, J. and Finkelmann, H. (1993) *Makromol. Chem., Rapid Commun.*, **14**, 267–271.

61 Nakahara, K., Iriyama, J., Iwasa, S., Suguro, M., Satoh, M. and Cairns, E.J. (2007) *J. Power Sources*, **163**, 1110–1113.

62 Nakahara, K., Iwasa, S., Satoh, M., Morioka, Y., Iriyama, J., Suguro, M. and Hasegawa, E. (2002) *Chem. Phys. Lett.*, **359**, 351–354.

63 Nakahara, K., Iriyama, J., Iwasa, S., Suguro, M., Satoh, M. and Cairns, E.J. (2007) *J. Power Sources*, **165**, 398–402.

64 Nakahara, K., Iriyama, J., Iwasa, S., Suguro, M., Satoh, M. and Cairns, E.J. (2007) *J. Power Sources*, **165**, 870–873.

65 Murray, R.W. (ed.) (1992) *Molecular Design of Electrode Surfaces*, John Wiley & Sons, Inc., New York.

66 Bard, A.J. and Faulkner, L.R. (2001) *Electrochemical Methods, Fundamentals and Applications*, 2nd edn, John Wiley & Sons, Inc., New York.

10
Oxidation Catalysis by Nanoscale Gold, Silver, and Copper
Zhi Li, Soorly G. Divakara, and Ryan M. Richards

10.1
Introduction

Catalysis by nanoscale materials is a rapidly growing field which involves the use of nanoparticles as catalysts for a variety of organic and inorganic reactions. Although the application of nanoscale catalysts has a long history in industry, the ability to engineer catalysts on the nanoscale and to explore the related phenomena in a controlled manner has evolved considerably in recent years. Numerous reviews have been published during the past decade on both heterogeneous catalysis in which nanoparticles are supported on solid surfaces (e.g., silica, alumina, MgO) and "soluble heterogeneous" or "quasi-homogeneous" catalysis with colloidal nanoparticles [1–8].

The catalytic performances of nanoparticles can be finely tuned either by their composition, which mediates electronic structure, or by their shape, which determines surface atomic arrangement and coordination. Beyond the exciting potential for tailoring catalysts, nanostructured materials introduce additional challenges to catalyst design. Being small, and with surface atoms of different unsaturated valencies, nanoparticles of specific shape are more liable to change their shape in the harsh medium of chemical reactions; this in turn raises concerns about durability, what the true active species is, and what approaches can be taken to control this phenomena. The surface reconstruction and/or dissolution of active atoms on corners or edges by one or more of the reactants, or even by the solvent, is an additional concern that requires attention. Despite these challenges, several types of chemical reaction have been catalyzed using transition metal nanomaterials, including crosscouplings, electron transfers, hydrogenations, fuel cell electrocatalysis and oxidations. In this chapter, attention will be focused on oxidation reactions, with particular emphasis on the Group 11 metals of the Periodic Table, namely silver, gold, and copper.

The Group 11 metals have a long history for their uses in jewelry, ornaments and as "coinage metals," as well being particularly interesting on the nanoscale for the colors they display as a function of size. However, as research has

Advanced Nanomaterials. Edited by Kurt E. Geckeler and Hiroyuki Nishide
Copyright © 2010 WILEY-VCH Verlag GmbH & Co. KGaA, Weinheim
ISBN: 978-3-527-31794-3

progressed with regards to both size- and shape-related chemical reactivities, the rich chemistry of this group of metals is emerging, and particularly their application as catalysts. Gold, in its bulk state, has long been thought to be far less catalytically active than other transition metals, such as silver, platinum, and palladium. However, many research groups [9–11] have reported pioneering studies to indicate that gold could catalyze hydrogenation, hydrogen exchange, hydrocracking, and carbon monoxide oxidation reactions. These initial findings have inspired numerous investigations into the nature of ultrafine gold particles dispersed on supports, using a wide variety of approaches [12, 13].

For many years, the chemistry of silver and gold was believed to be more similar than is now known to be the case. Silver (Ag) is the best conductor among these metals, and so silver nanoparticles facilitate more electron transfer than do gold (Au) nanoparticles. Silver has a reduction potential of +0.79 V (versus NHE) for an Ag^I (aqueous)/Ag_{metal} system, but for an Ag^I (aqueous)/Ag_{atom} system it is −1.80 V (versus NHE). Among the many different metal nanoparticles under investigation, silver nanoparticles are emerging as one of the most intensively studied, largely because of the broad range of applications that they exhibit. These properties include shape- and size-dependent optical, electronic, and chemical properties, and present many possibilities with respect to technological applications. Currently, copper (Cu) is one of the most widely used materials worldwide, being of major significance in all industries, especially in the electrical sector due to its low cost. Moreover, Cu continues to gain importance on the basis of it being an essential component in future nanodevices due to its excellent conductivity, its good biocompatability, and its surface-enhanced Raman scattering (SERS) activity. Further, copper nanoparticles smaller than 50 nm in size are also considered "superhard" materials, as they do not exhibit the same malleability and ductility as bulk copper.

10.2
Preparations

There are two general paradigms for the preparation of nanoscale materials, namely "top down" and "bottom up". Those preparations characterized by the breaking down of larger starting materials are classified as "top down," while those which are built up from atomic or molecular starting materials are termed "bottom up." Chemical procedures such as alcohol reduction [14–16], hydrogen reduction [17–19], and sodium borohydride reduction [20–22] have in the past been recognized as the most common methods for synthesizing colloidal metal nanoparticles. Other reduction methods such as electrochemical [23, 24], photochemical [25–27], and sonochemical [28, 29] have also been used, but to a smaller extent. Many different stabilizers have been used as capping agents for the synthesis of colloidal metal nanocatalysts, including polymers [30, 31], dendrimers [32, 33], block copolymer micelles [31, 34], and surfactants [35, 36]. Supported metal nanocatalysts have been prepared by the adsorption of colloidal metal nanocatalysts onto supports [37–39], and/or by grafting the nanoparticles onto the support [40]. Supported metal nanocatalysts can also be fabricated lithographically, using electron beam lithography [41, 42].

10.2.1
Silver Nanocatalysts

By using the above-described methods, silver materials with zero-, one-, or two-dimensional nanostructures, including monodisperse nanoparticles, nanowires, nanodisks, nanoprisms, nanoplates, and nanocubes, have each been prepared and are recognized as having great potential for applications in optics, catalysis, and other fields [43–47].

10.2.2
Copper Nanocatalysts

Copper nanoparticles have been synthesized and characterized by different methods. Notably, chemical reduction, pulsed laser ablation, radiolytic reduction, and the reduction of copper ions using supercritical fluids have been developed to synthesize spherical and different-shaped nanoparticles [48–50]. Stability and reactivity are the two important factors that impede the use and development of metal clusters. In contrast to noble metals, such as Ag and Au, pure metallic copper particles usually cannot be obtained via the reduction of simple copper salts (e.g., copper chloride or copper sulfate) in aqueous solution, because the reduction tends to stop at the Cu_2O stage due to the presence of a large number of oxygenous water molecules. However, this problem can be overcome by the addition of other reagents carrying functional groups that can form complexes with copper ions, or by using soluble surfactants as capping agents to prepare copper particles in aqueous solution. Although zero-valent copper forms initially in the solvent, ultimately it can be transformed relatively easily into oxides, in solvents with high dipole moments and under ambient conditions. The use of reverse micelles as microreactors and protecting shells has also helped to overcome some of these complications. Likewise, electrolytic techniques have been used to synthesize a variety of transition metal colloids of either decahedral or isohedral shape, by controlling the electrode potential. The extreme air-sensitivity of copper nanoparticles requires that care be taken during such preparation in order to avoid oxidation.

10.2.3
Gold Nanocatalysts

Beyond preparing nanoscale materials in the form of colloids, considerable effort has been expended in the preparation of supported heterogeneous catalysts, and in particular the nature of the support and the process to immobilize an active metal on the support (mostly metal oxides and active carbon). The use of gold nanocatalysts is cited as an example here when introducing different methods for preparing nanocatalysts.

Due to the lower melting point of gold, and its poor affinity for metal oxides, it is difficult to prepare stable gold catalysts that are well dispersed on metal oxides. The typical impregnation methods that are widely used to prepare supported Pd or Pt catalysts are ineffective in the case of gold, because the presence of chloride ions can cause a significant enhancement in the coagulation of gold particles during the calcination of $HAuCl_4$.

Haruta has summarized the methods used to prepare supported gold catalysts (Table 10.1), and categorized them into four groups [51]:

- The first group includes coprecipitation [54], amorphous alloying [55], and cosputtering [56]. These procedures generally consist of two steps: (i) the preparation of well-mixed gold/metal oxide precursors; and (ii) transformation of the gold precursor into gold particles, normally by calcinations in air above 550 K. Well-mixed precursors and high-temperature calcination are equally

Table 10.1 Preparation techniques for nanoparticulate gold catalysts[51].

Categories	Preparation techniques	Support materials	Reference(s)
Preparation of mixed precursors of Au and the metal component of supports	Coprecipitation (hydroxides or carbonates) (CP)	$Be(OH)_2$, TiO_2, Mn_2O_3, Fe_2O_3, Co_3O_4, NiO, ZnO, In_2O_3, SnO_2	[52–54]
	Amorphous alloy (metals) (AA)	ZrO_2	[55]
	Cosputtering (oxides) in the presence of O_2 (CS)	Co_3O_4	[56]
Strong interaction of Au precursors with support materials	Deposition–precipitation ($HAuCl_4$ in aqueous solution) (DP)	$Mg(OH)_2$, Al_2O_3, TiO_2, Fe_2O_3, Co_3O_4, NiO, ZnO, ZrO_2, CeO_2, Ti-SiO_2	[57]
	Liquid-phase grafting (organogold complex in organic solvents) (LG)	TiO_2, MnOx, Fe_2O_3	[58, 59]
	Gas-phase grafting (organogold complex) (GG)	All types, including SiO_2, Al_2O_3-SiO_2, and activated carbon	[60, 61]
Mixing colloidal Au with support materials	Colloid mixing (CM)	TiO_2, activated carbon	[13]
Model catalysts using single crystal supports	Vacuum deposition (at low temperature) (VD)	Defects are the sites for deposition, MgO, SiO_2, TiO_2	[62–64]

important to ensure a strong contact between the Au particles and the crystalline metal oxides.

- The strategy for the second group is based on the concept of depositing or adsorbing Au compounds onto metal oxide surfaces. Among the three methods referred to here, deposition–precipitation (DP) is widely used to produce active Au catalysts. By controlling the pH and concentration of the $HAuCl_4$ solution, the deposition of $Au(OH)_3$ can be controlled on the surfaces of the support metal oxides so as to prevent precipitation in the liquid phase. Aggregation of the gold nanoparticles, induced by chloride ions, can be prevented by washing the gold compound before drying, and this represents one of the main reasons for the high activity of these catalysts. The primary limitation here is that DP can only be applied to metal oxides with an isoelectric point >5. Although previously it was shown that $Au(OH)_3$ could not be deposited on SiO_2 and active carbon, recent studies have found that this constraint may be overcome by correct surface modification [65].

- In the third group, the procedure involves the direct immobilization of Au colloids on modified metal oxide surfaces. In theory, this method could be applied to all metal oxides, and the catalysts prepared would normally have a good gold particle size distribution. However, there is often a relatively poor contact between the gold particles and the support.

- In the fourth group, vacuum deposition is considered to be an important method for preparing model catalysts that play a critical role when studying reaction mechanisms, and especially the active sites of the supported gold catalysts. Au anion clusters can be deposited with homogeneous dispersion at relatively low temperatures [62, 64] on single crystals of MgO and TiO_2 (rutile). Surface defects or specific surface cages have been suggested as possible sites for stabilizing the Au clusters [62, 63].

10.3
Selective Oxidation of Carbon Monoxide (CO)

10.3.1
Gold Catalysts

During the 1980s, Haruta *et al.* [66] found that gold nanoparticles, when supported on α-Fe_2O_3, were highly active in the oxidation of CO, and especially at very low temperatures, although this surprisingly high activity was not replicated by other metals (Figure 10.1). In a later series of investigations conducted by the same group [52], Au/TiO_2 was found to be an equally effective catalyst, and this in turn led to extensive studies of gold nanocatalysts supported on a variety of metal oxides.

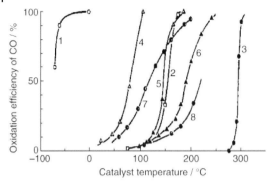

Figure 10.1 CO conversion over various catalysts as a function of temperature. Curve 1, Au/α-Fe$_2$O$_3$ (Au/Fe = 1/19, coprecipitation, 400°C); Curve 2, 0.5 wt% Pd/γ-Al$_2$O$_3$ (impregnation, 300°C); Curve 3, fine Au powder; Curve 4, Co$_3$O$_4$ (carbonate, 400°C); Curve 5, NiO (hydrate, 200°C); Curve 6, α-Fe$_2$O$_3$ (hydrate, 400°C); Curve 7, 5 wt% Au/α-Fe$_2$O$_3$ (impregnation, 200°C); Curve 8, 5 wt% Au/γ-Al$_2$O$_3$ (impregnation, 200°C). Reproduced with permission from Ref. [66]; © 1987, Chemical Society of Japan, Tokyo.

Owing to the possible applications of polymer electrolyte fuel cells to automobiles and also to residential electricity-heat delivery systems, the low-temperature water-gas-shift reaction continues to attract renewed interest. When compared to commercial catalysts that are based on Ni or Cu and operated at 900 K or 600 K, respectively, supported Au catalysts appear to have a clear operational advantage in that they function at temperatures as low as 473 K [67]. During the course of investigating the hydrogenation of CO$_2$ over supported Au catalysts, it was found that Au/TiO$_2$ was selective towards the formation of CO, in that the reverse water-gas-shift reaction could be conducted at a temperature as low as 473 K [67]. Later, Au/TiO$_2$ was confirmed also to be active for the water-gas-shift reaction [68].

The oxidation of CO is a typical reaction for which Au catalysts are extraordinarily active at room temperature, and indeed are much more active than other noble metal catalysts at temperatures below 400 K. One focal point of recent studies has been the elaboration of the mechanism for CO oxidation [69, 70]. Although the available data on this topic are vast—and occasionally contradictory—several pieces of information were identified that were critical to developing an understanding of the mechanism. For example, active catalysts always contain metallic Au particles which produce a CO absorption band at 2112 cm^{-1}, whereas oxidic Au species that produce a CO absorption band at 2151 cm^{-1} are not responsible for steady-state, high catalytic activity [71]. However, as the smooth surfaces of metallic Au do not adsorb CO at room temperature [72], this indicates that CO is adsorbed only on steps, edges, and corner sites. As a consequence, the smaller metallic Au particles are preferable [73].

A theoretical calculation [74] has been used to explain why the smooth surface of Au is noble in the dissociative adsorption of hydrogen. However, when Au is deposited as nanoparticles on metal oxides by means of coprecipitation and DP techniques, it exhibits a surprisingly high catalytic activity for CO oxidation at a

(a)

(b)

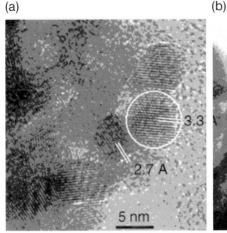
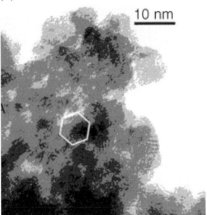

Figure 10.2 High-resolution transmission electron microscopy images of the 2.8% Au/CeO$_2$ sample. (a) The white lines correspond to the (202) Ce$_6$O11 (3.3 Å) and the (200) CeO$_2$ (2.7 Å) lattice spacing; (b) A hexagonal faceted (111) Au crystal is indicated. Reproduced with permission from Ref. [75]; © 2005, WILEY-VCH Verlag GmbH & Co. KGaA, Weinheim.

temperature as low as 200 K [12, 52]. During the 1990s, this finding led to many research groups conducting extensive investigations into the catalysis of Au.

One remarkable study among many was conducted by Corma and coworkers [75], who showed that gold nanoparticles supported on nanocrystalline CeO$_2$, in conjunction with the DP method, proved to be a very active catalyst for CO oxidation (Figure 10.2). Indeed, the catalysts were found to be an order of magnitude more active for CO oxidation than comparable catalysts prepared using a non-nanocrystalline support. The Au/CeO$_2$ catalyst also showed excellent selectivity for CO oxidation in the presence of H$_2$ at 60 °C (close to the operating temperature of fuel cell), where the selectivity of normal Au active catalysts would be negatively affected [76].

Most gold catalysts which have been reported as active for CO oxidation were prepared using the DP method, which provides catalysts with a strong interaction between the gold and the metal oxide matrix. However, a major drawback of this method is that it cannot be used to deposit gold at metal oxides with an isoelectric point (IEP) <5, such as SiO$_2$. Subsequently, Sheng Dai and coworkers [65] successfully deposited gold at the surface of mesoporous SiO$_2$ by using the DP method following a sol–gel surface modification with TiO$_2$. The results showed the gold nanoparticles (0.8–1.0 nm) in the mesopores to be highly active in terms of CO oxidation (Figure 10.3).

Today, whilst it is widely recognized that these supported gold nanocatalysts show a high activity in the low-temperature oxidation of CO, many questions remain unanswered concerning the relatively simple reaction of CO oxidation. The most notable of these are "What is the reaction mechanism?", and "What is the nature of the active site?"

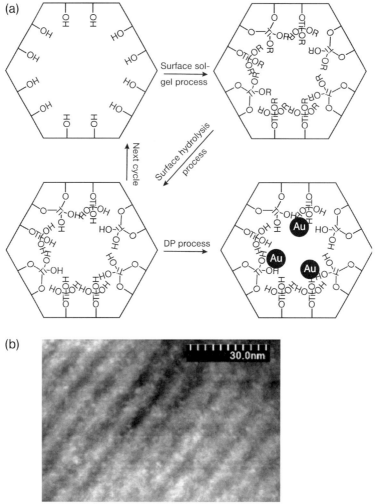

Figure 10.3 (a) Scheme of preparation of gold/mesoporous material catalysts; (b) Z-contrast TEM image of ultrasmall gold nanoparticles on ordered mesoporous materials. The bright spots (0.8–1.0 nm) correspond to gold nanoparticles. Reproduced with permission from Ref. [65]; © 2004, ACS Publications, Washington.

A substantial proportion of the research into CO oxidation catalyzed by supported gold has been motivated by the goal of identifying those catalyst properties which affect the activity. In this respect, two early classes of observation were important in determining the approaches used recently to investigate supported gold catalysts: (i) that various preparation routes lead to catalysts with different activities [52, 77]; and (ii) that catalysts consisting of gold supported on reducible metal oxides (e.g., Fe_2O_3, CeO_2, TiO_2) are typically more active than those

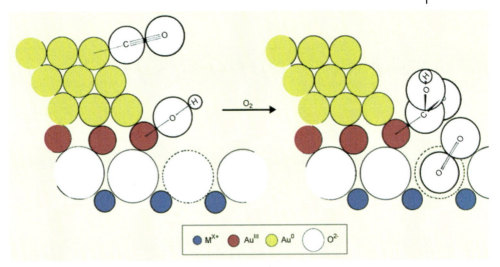

Scheme 10.1 Schematic representation of an active site and possible reaction mechanism for CO oxidation catalyzed by supported gold. Reproduced with permission from Ref. [77]; © 2000, World Gold Council, London.

supported on nonreducible metal oxides (e.g., γ-Al_2O_3, MgO, SiO_2). Such observations led to a wide acceptance of the inferences that the preparation method influenced activity [12], and that the support played a role in the catalysis.

Bond and Thompson [77], in their review of the literature which extended to the year 2000, and in an attempt to reconcile some apparently contradictory hypotheses, proposed a mechanism for CO oxidation that was catalyzed by supported gold (Scheme 10.1). The proposed active site consisted of nanoparticles incorporating both zero-valent and cationic gold, with the latter positioned at the metal–support interface. The suggestion by Bond and Thompson of the presence of cationic gold was based on observations by various authors of vCO infrared (IR) bands that were characteristic of CO bonded to cationic gold. However, evidence was lacking not only of any such species in working catalysts, but also of the suggestion that cationic gold was a "glue" which held the nanoclusters to the support.

In 2002, Haruta [78] presented a review of the literature and proposed, on the basis of measurements of the kinetics of CO oxidation catalyzed by supported gold, that there were three temperature regions, each with different kinetics and activation energies of the CO oxidation reaction. Haruta suggested that, at temperatures below 200 K, the reaction catalyzed by Au/TiO_2 took place at the surfaces of small gold nanoparticles dispersed on the support, but at temperatures above 300 K the reaction occurred at gold atoms at the perimeter sites of the supported gold nanoparticles.

10.3.2
Silver Catalysts

Silver catalysts also have a relatively high activity for the selective oxidation of CO at low temperatures. A silver catalyst was deactivated remarkably following pretreatment in H_2 at high temperatures, but could be reactivated by treatment in oxygen at similarly high temperatures. Interestingly, these changes in activities were mostly reversible. The structures of the silver particles were seen to experience massive changes during the course of various pretreatments, and the existence of subsurface oxygen resulting from an oxygen treatment at high temperatures was shown to be crucial for high selectivity and activity in CO selective oxidation [79, 80]. As CO oxidation is generally claimed to be a structure-sensitive reaction, restructuring of the silver particles is likely to exert an influence on the activity of the catalyst. Yang and Aoyama [81, 82] studied the thermal stability of uniform silver clusters supported on oxidized silicon or aluminum surfaces in both oxidizing and reducing atmospheres, and found the thermal stability of the silver clusters to be significantly lowered under oxidizing conditions. Moreover, heating above 350 °C under oxidizing condition could induce a migration of the silver clusters.

Size selectivity in catalysis was reported for propylene partial oxidation and low-temperature CO-oxidation, with Ag nanoparticles of <5 nm diameter being shown to have equal activity as the Au nanoparticles. In contrast, for ethylene epoxidation only those Ag particles >30 nm could catalyze the reaction [83]. Recently, much attention has been focused on the use of spherical or undetermined-shape nanoparticles for catalyzing reactions. Very few studies have been undertaken in which catalysis was conducted with nanoparticles of known shapes [84], for example, using truncated octahedral Pt nanoparticles to catalyze the electron-transfer reaction, and cubic Pt nanoparticles in the decomposition of the oxalate capping agent. The formation of different oxygen species, depending on the Ag particle size sputtered on the highly ordered pyrolytic graphite (HOPG) surface, resulted in a variation in catalytic activity of CO oxidation using oxygen under (ultra-high vacuum) UHV conditions revealed CO oxidation to be sensitive towards the size of particle [85]. The oxygen uptake of a smaller Ag nanoparticle was seen to be significantly higher than that of a larger particle and a bulk-like Ag which enhanced the reactivity of CO oxidation.

10.3.3
Gold–Silver Alloy Catalysts

The recent progress in polymer electrolyte membrane fuel cells has particularly motivated the search for a highly efficient catalyst for CO selective oxidation at low temperatures. Thus, combinations of metals in the forms of alloys, core–shell and "decorated" surfaces (Pt, Pd, Rh, Ru, Au, Ag, Cu, Co, Fe, In, Ga) with different supports, such as zeolite, Al_2O_3, SiO_2, and activated carbon, have produced active catalysts for the CO reaction. One such alternative catalyst, namely gold–silver

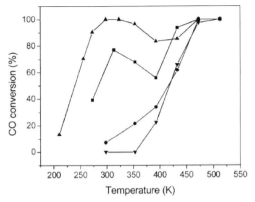

Figure 10.4 CO conversion over reaction temperatures at various molar ratios of Au:Ag. ▲, ratio 3:1; ■, ratio 1:1; ●, ratio 1:0; ▼, ratio 0:1. Reproduced with permission from Ref. [88]; © 2005, ACS Publications, Washington.

alloy nanoparticles deposited on MCM-41, demonstrated an exceptionally high catalytic activity which was comparable to the most active catalysts (Figure 10.4) such as Au/TiO_2 and Au/Fe_2O_3 [86, 87]. The alloying of Au and Ag showed a strong synergistic effect in promoting the low-temperature oxidation of CO [88]. The alloy catalyst activation was shown to depend on the composition (the Ag ratio was crucial), the aluminum content in the support, and the pretreatment conditions.

10.3.4
Copper Catalysts

Copper nanoparticles are also active for the selective oxidation of CO. The majority of studies with copper particles have been performed with finely dispersed copper on various supports, and have demonstrated high catalytic activities for CO oxidation. Likewise, CuO mixed with ZnO or with CeO_2 have also shown promise as catalysts. The results of a recent density functional theory (DFT) study showed that gold and copper had a lower barrier for CO oxidation than for H_2 oxidation.

Ceria has a promoting effect on the activity of the Au/Al_2O_3 catalyst in CO oxidation [89]. The addition of Li_2O and/or CeO_x to copper, silver, and gold catalysts of 3 nm size on γ-Al_2O_3 for the preferential oxidation of CO in a hydrogen atmosphere [90], have shown significant changes in the conversion. The nanoscale metal particles or metal complexes in polymer matrices show quite interesting chemical and catalytic reactivity towards a variety of small gas molecules under relatively mild conditions that differ from those of the corresponding free transition metal complexes, or from those in inorganic oxide-supported systems. The incorporation of copper nanoparticles into cellulose acetate, and the subsequent oxidation of small gas molecules (e.g., CO, H_2, D_2, O_2, NO, and olefins) over a temperature range of 25 to 160 °C, has also been examined [91]. Nanoparticles of various sizes prepared by different routes and hosted in the channels of SBA-15, exhibited a

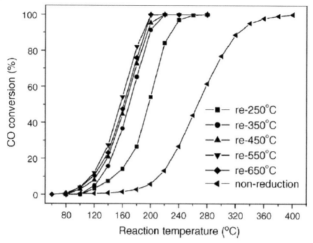

Figure 10.5 CO conversions versus reaction temperature over Cu/SBA-15 (post grafted) calcined at 500 °C and reduced at different temperatures. Reproduced with permission from Ref. [92]; © 2006, Elsevier B.V., Amsterdam.

high catalytic activity for CO oxidation, with complete conversion at 190 °C (Figure 10.5) [92]. Such high catalytic activity was mainly influenced by the size and dispersion of the Cu particles.

10.4
Epoxidation Reactions

10.4.1
Gold Catalysts

Since the first recognition by Hayashi [93] that Au supported on TiO_2 could catalyze the epoxidation of propylene in the gas phase containing O_2 and H_2, the catalytic properties of Au/TiO_2 and related systems have attracted interest not only from the chemical industries but also from academia. Today, propylene oxide (PO) is recognized as one of the world's most important bulk chemicals, and is used in the production of polyurethane and polyols. The current industrial processes utilize two-staged chemical reactions, using either Cl_2 or organic peroxides to yield the byproducts stoichiometrically.

From both environmental and economic points of view, the direct synthesis of PO by using molecular oxygen has long been a major academic challenge, although supported noble metal catalysts such as Ag/carbonates and/or titanates, Pd/TS-1, Pd–Pt/TS-1 [94–96], and Au/TiO_2 (Figure 10.6) [93, 97], have each been reported to be active in this process.

In 1998, using the DP technique, Haruta and coworkers [97] produced nanoscale gold catalysts that showed a high activity towards CO oxidation, based on the

(a)

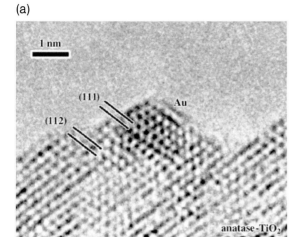

(b)

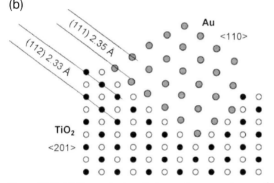

Figure 10.6 (a) Transmission electron microscopy image of the Au/TiO_2 contact interface; (b) A schematic representation of the interface. Reproduced with permission from Ref. [97]; © 1998, VSP, Leiden.

strong contact with the support at highly dispersed isolated tetrahedral Ti^{4+} sites. Moreover, this catalyst also provided a very high PO selectivity (>99%) in a gas phase containing O_2 and H_2, at ambient pressure. The same group also demonstrated the formation of PO over an Au-based catalyst to be a typical structure-sensitive reaction [97], with the selective production of PO being catalyzed only by hemispherical Au particles of a suitable size (2 nm < diameter Au nanoparticle < 10 nm). In more recent investigations, these authors also showed an organically modified mesoporous titanosilicate to be an efficient support [98], providing a reasonably efficient H_2 consumption, high yields, and PO selectivities in excess of 90% [99].

These studies initiated a growing research interest from both industry and academia [100–103], with one of the more remarkable investigations being conducted by Hutchings and coworkers [102]. This group used catalytic amounts of peroxides to initiate the oxidation of alkenes with O_2, and making it unnecessary to sacrifice H_2 in order to activate the O_2. Here, Au/graphite was found to be very active in catalyzing the expoxidation of cyclohexene, styrene, *cis*-stilbene, and cyclooctene, even in a solvent-free system. Although the selectivity could be increased by using the correct solvent (e.g., toluene), the environment-friendly expoxidation in solvent-free systems would be much more attractive to the chemical industry and is certain to become another hot topic of research in the near future.

10.4.2
Silver Catalysts

Silver is considered to be an almost uniquely effective catalyst for heterogeneous epoxidation reactions, and the mechanism of the epoxidation of ethylene with oxygen over a silver catalyst has been the subject of extensive investigation [104–111]. However, despite such numerous studies and its wide use, a number of questions remain unanswered regarding this catalytic system, including: "How do the supports and promoters affect the reaction?"; "What is the mechanism of the primary and secondary reactions?"; and "What relationship exists between the electron and structure factors?" The interaction of oxygen with metal surfaces has been suggested as one of the most important elementary steps in heterogeneous catalysis, and several reviews of oxygen adsorption, active oxygen species, promoter effects and reaction mechanisms on silver catalysts have been produced [112]. The oxygen species on silver were found to play a key role in ethylene epoxidation, and extensive studies have been performed to establish details of the interaction of oxygen with silver surfaces, namely whether the chemisorbed oxygen is atomic or molecular [113]. Details of surface molecular, surface atomic, subsurface atomic, and bulk atomic oxygen species have each been reported in the literature.

Significant efforts to improve selectivity have included the use of different silver precursors, of different preparation techniques, and of different promoters. Such information is gathered following the continuous addition of a chlorine-containing hydrocarbon species to the gaseous reactants as a moderator, which also acts to depress the overall reaction rates. Campbell reported that small amounts of promoters (e.g., Cl) would increase the ethylene oxide selectivity [114, 115]. For an industrial oxidation of ethylene, alkali metal ions are an important additive when using a silver catalyst, and alkali or alkaline earth promoters (e.g., cesium) can provide further substantial improvements [116–118]. Campbell [116] reported the role of a cesium promoter in silver catalysts for the selective oxidation of ethylene. The oxidation of ethylene in solution, catalyzed by polymer-protected silver colloids, and the promotion effect by alkali metal ions on colloidal silver catalysts, have also been studied [119]. Colloidal dispersions of silver nanoclusters, when protected by poly(sodium acrylate), caused increases in the rate of oxidation, the reaction temperature and also the catalytic activity following the addition of Cs(I) and Re(VII) ions [120].

The influence of silver nanoparticle size on catalytic activity is due not only to an enhanced surface area but also to particular electronic properties, which differ from those of bulk silver. The effect of silver particle size on the reaction rate is a well-known property of supported silver catalysts [121, 122]. For example, when monitoring the distribution of Ag supported on alumina, silver particles of 30–70 Å on alumina or silica showed higher activities. Previously, small silver clusters had been shown to be the most effective, probably due to the electronic and other properties of silver (atomic environment, electron work function, electric conductance, etc.), and differed considerably from bulk silver in this respect. Although the majority of studies have used bulk silver samples, uncertainty remains as to the nature of the active sites for ethylene epoxidation on a commercial catalyst. Enlargement of the silver particles has been found to decrease the amount of subsurface oxygen, and result in the appearance of nucleophilic oxygen. These findings have been used to provide a possible explanation for the size effect in ethylene epoxidation over Ag/Al_2O_3 catalysts (Figure 10.7) [123–126]. The kinetic study and shape-controlled catalytic epoxidation of olefins by these nanoparticles on several supports such as α-Al_2O_3, $CaCO_3$ and spherical particles of TiO_2 (all obtained using the Stöber method) were investigated for a non-allylic olefin (e.g., styrene) and for an allylic olefin (e.g., propene), using molecular oxygen and N_2O as oxidants.

Silver supported on titania was found to be active for propene epoxidation using hydrogen/oxygen mixtures at 50 °C [127, 128]. The direct aerobic oxidation of various alkenes catalyzed by $H_5PV_2Mo_{10}O_{40}$ polyoxometalate-stabilized silver nanoparticles supported on Al_2O_3 showed a higher selectivity towards epoxide in the liquid phase [129]. Styrene represents a useful alkene model for studying the reaction mechanism of terminal alkene epoxidation; the size and morphology of the nanoparticles was reported to affect the catalytic behavior of silver catalysts supported on α-Al_2O_3 and MgO, in the selective oxidation of styrene in the gas phase [130–134]. The epoxidation of styrene to its oxide by molecular oxygen was studied using a Cs-loaded silver nanowire catalyst, and resulted in the desired product with greater selectivity.

10.5
Selective Oxidation of Hydrocarbons

The selective oxidation of alkanes with molecular oxygen represents a major challenge [135, 136], as the production of the more valuable oxidized products relative to the low cost of the raw materials is of economic interest. The chemical inertness of hydrocarbons makes the activation of C–H bonds especially difficult, and usually requires dramatic reaction conditions such as high temperature and pressure. The oxidation of cyclohexane is of special interest industrially, because the process produces KA-oil as an intermediate; this is a mixture of cyclohexanone and cyclohexanol that is important in the petroleum chemical industry. KA-oil is used for the production of adipic acid and ε-caprolactam, which are key materials in the respective manufacture of 6,6-nylon and 6-nylon [137].

Modern industrial methods typically require both high pressure and temperature when using a soluble cobalt catalyst. A high selectivity for cyclohexanone and cyclohexanol can only be achieved at a low conversion, as these products are

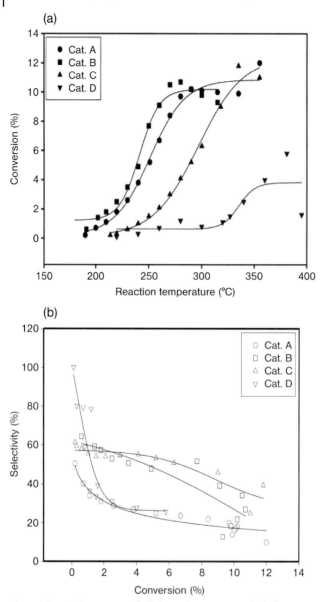

Figure 10.7 (a) Conversion of ethylene; (b) Selectivity and conversion of Ag on γ-Al_2O_3 with catalysts prepared by different methods: A, precipitation; B, modified precipitation; C, water–alcohol; D, microemulsion silver loading (7.5–8.6%). Reproduced with permission from Ref. [126]; © 2003, Elsevier B.V., Amsterdam.

substantially more reactive than the cyclohexane reactant. Thus, under mild conditions it is difficult to achieve, simultaneously, a high conversion and selectivity.

Whilst it is desirable to identify a good catalyst in order to activate the reaction between oxygen and cyclohexane, the technology by which cyclohexane is oxidized by O_2 to produce KA-oil has not yet been improved [138]. In an effort to reduce the use of environmentally harmful elements, heterogeneous catalysts prepared by immobilizing Mn^{III}, Co^{III}, Cr^{III} and Fe^{III} ions on metal oxides have been developed for the oxidation of cyclohexane [139–141], although these systems were subsequently found to suffer from leaching under the reaction conditions used. Mesoporous materials were reported as efficient support materials because in the pores, cyclohexane is oxidized more readily than cyclohexanol, such that the selectivity is enhanced.

10.5.1
Gold Catalysts

Zhao and coworkers were the first to apply gold nanoparticles for this application, and reported that a supported gold catalyst could activate cyclohexane at 150 °C, with selectivities in the region of 90% [142, 143]. Under similar reaction conditions, Kake Zhu and coworkers found that gold nanoparticles immobilized by a variety of methods in the channels of SBA-15 showed a good performance in catalyzing the aerobic oxidation of cyclohexane in a solvent-free system (Figure 10.8) [144], with the highest conversion being reported as 32%. In order to enhance selectivity, the aerobic oxidation of cyclohexane was performed below 100 °C and catalyzed by supported Au, Pt, and Pd catalysts [145]. Although, the selectivity for cylohexanone and cyclohexanol was found to decline rapidly with the enhanced conversion and longer reaction times, the gold catalysts provided an identical performance to the Pt and Pd catalysts.

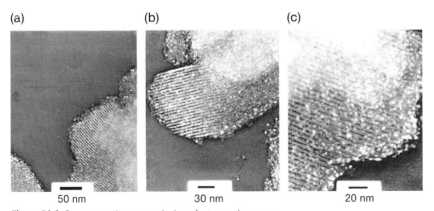

Figure 10.8 Representative transmission electron microscopy images. (a) Au–APS/SBA-15; (b, c) Au–SH/SBA-15 at low (b) and high (c) magnification. Reproduced with permission from Ref. [144]; © 2005, Springer, Netherlands.

10.5.2
Silver Catalysts

Silver catalysts have also found use in hydrocarbon oxidation reactions. Since the granting of the first patent in 1931 for the manufacture of ethylene oxide with an Ag catalyst [146], the industrial production of ethylene oxide by direct oxidation in gas phase has become widely used, with ethylene being converted into ethylene glycol or a variety of other derivatives. In fact, the selective oxidation of ethylene over a supported material represents one of the few uses of silver as an industrially important catalyst, whilst also providing fundamental interests in the surface sciences [112, 147].

Silver has, however, rarely been considered as a catalyst for the selective oxidation of saturated hydrocarbons. Yet, nanoscale silver supported on MCM-41 for the liquid-phase oxidation of cyclohexane was found to be an effective catalyst in the absence of a solvent [148], with a higher turnover number and improved selectivity compared to other support systems of Ag/TS-1 and Ag/Al$_2$O$_3$.

10.5.3
Copper Catalysts

Copper nanoparticles supported on natural zeolites of different structure and origin have been utilized for the complete oxidation of hydrocarbons over a temperature range of 170 to 250 °C. Although a complete conversion was reported, this was shown to depend on the Si/Al ratio of the zeolite matrix and the different nanospecies of copper present [149].

10.6
Oxidation of Alcohols and Aldehydes

The oxidations of alcohols and polyols are important processes in industrial chemistry, and it is not surprising that a significant effort is currently being expended within the scientific community to improve present-day technologies, and in particular to create processes that are more "green." The objective of these research investigations has been the development of a catalytic system that would compete against a stoichiometric approach involving concentrated and toxic oxidizing agents. Supported platinum and palladium catalysts are well known as effective catalysts for the oxidation of polyols under acidic or basic conditions. Yet, supported gold nanoparticles were also found to be very effective for the oxidation of alcohols, including diols, in the presence of a base [150–157]. These catalysts could also be used in the oxidation of sugars, glucose, and sorbitol [158, 159]. When using dioxygen as the oxidant, Carrettin et al. reported a 100% selectivity for the oxidation of glycerol to glycerate, this being catalyzed by an Au catalyst supported on graphite under relatively mild conditions, and with yields approaching 60% [160–162]. In these studies the presence of a base was also found to be essential for both activity and selectivity.

10.6.1
Gold Catalysts

It has been commonly observed that the oxidation of alcohols catalyzed by supported gold catalysts show better performances under basic conditions in terms of high selectivities and reasonable activities, although the mechanism behind this chemistry remains unclear. Another interesting observation is that the types of support used for the gold catalysts, which have important roles in the oxidation of CO and epoxidation, appear to have little effect in the oxidation of alcohols. In fact Rossi and coworkers [163], when monitoring the oxidation of glucose to gluconic acid, found that unsupported colloidal AU particles were equally active as Au nanoparticles supported on active carbon, under the same conditions. A subsequent study extended the application of colloidal gold catalysts to the oxidation of 1,2-diols [164, 165]. The disadvantage of an unsupported colloidal gold catalyst is its poor long-term performance compared to a supported gold nanocatalyst, due to problems of aggregation. Recently, gold nanoclusters stabilized by polymers [166] showed good activity in the aerobic oxidation of benzyl alcohol in aqueous media. However, Corma and coworkers [167, 168] proposed that the support of Au/CeO_2 catalysts could help to stabilize a reactive peroxy intermediate from O_2, and thus could enhance the activities of gold catalysts in the selective oxidation of alcohols to aldehydes/ketones, and of aldehydes to acids. One further interesting point here was that the gold catalysts functioned well in solvent-free systems without any additional base, a point which differed considerably from earlier findings.

One of the most significant advances in the field of alcohol oxidation, provided by the group of Hutchings [169], indicated that an Au/Pt alloy supported on TiO_2 was a highly active catalyst for the oxidation of benzyl alcohol, cinnamyl alcohol, and vanillyl alcohol. In particular, for the oxidation of benzyl alcohol, the performance of the Au-Pd/TiO_2 catalyst was remarkably superior to that of Au/TiO_2 and Pd /TiO_2 in terms of both conversion and selectivity (Figure 10.9).

10.6.2
Silver Catalysts

The silver-catalyzed partial oxidation of methanol to formaldehyde, which is an industrially important chemical transformation of alcohols to carbonyls in the gas phase, and is significant in the synthesis of drugs, vitamins, fragrances, and many complex syntheses, was first employed on an industrial scale by BASF AG in 1905 [170, 171]. Since that time, several silver-based catalysts, including bulk and supported systems, have been developed for the oxidation of alcohols. The main problems encountered problems with bulk silver are low catalytic activities at low temperatures, or the products of cracking and/or overoxidation at higher reaction temperatures. Various attempts have been made to improve the catalytic performance of silver-based catalysts, either by adding additives to the bulk silver catalysts or by dispersing the silver particles on supports [172–175]. Supported silver would be expected to enhance the dispersion and stability of silver, and thus also enhance its catalytic activity at relatively low temperatures. The design of a catalyst combin-

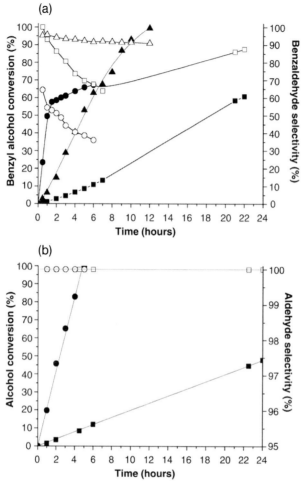

Figure 10.9 (a) Benzyl alcohol conversion and selectivity in benzaldehyde with the reaction time at 373 K and 0.1 MPa pO_2. Squares indicate Au/TiO$_2$; circles indicate Pd/TiO$_2$; triangles indicate Au–Pd/TiO$_2$. Solid symbols indicate conversion, open symbols indicate selectivity; (b) Au–Pd/TiO$_2$-catalyzed reactions at 363 K, 0.1 MPa pO_2, for cinnamyl alcohol (squares) and vanillyl alcohol (circles). Solid symbols indicate conversion, open symbols indicate selectivity to the corresponding aldehydes. Reproduced with permission from Ref. [169]; © 2005, RSC Publishing, Cambridge.

ing the advantages of both bulk silver catalysts (conventional electrolytic silver) and supported silver catalysts, with better performance, remains a major research challenge in the area of alcohol catalytic oxidation [176].

Supported platinum and palladium nanoparticles are generally acknowledged as effective catalysts for the oxidation of polyols [177]. In situ electrolytic nanosilver/zeolite film/copper grid catalysts have shown higher catalytic oxidation properties towards mono, di, and other types of alcohols, and with higher selectivities

[176, 178]. The silver nanoparticles, when generated *in situ*, highly dispersed on the zeolite film and then precoated on a copper grid, demonstrated excellent activities towards polyhydric alcohol at low temperature. Such performance effectively avoids the problems of both overoxidation and C–C bond cracking at high temperatures, or mild oxidation at low temperatures (as in the case of the practical electrolytic silver catalyst). Au–Ag alloy clusters (size range 1.6–2.2 nm) with various Ag contents (5–30%) and prepared using a coreduction method in the presence of poly(N-vinyl-2-pyrrolidone), were investigated in the aerobic oxidation of *p*-hydroxybenzyl alcohol as a model reaction to understand the effect of Ag on the catalytic activity of Au clusters [179]. The rate constants per unit surface area for the Au–Ag:PVP clusters with a small Ag content (<10%) were larger than those of monometallic Au:PVP clusters of comparable size. Spherical nanoparticles anchored on the external walls of a multiwalled carbon nanotube (Ag/MWNT) composite electrode exhibited a high catalytic activity for the electro-oxidation of methanol [180].

10.7
Direct Synthesis of Hydrogen Peroxide

As noted above, there has been much recent interest in the design of new heterogeneous catalysts for selective oxidation under ambient conditions, and these typically use hydrogen peroxide as the oxidant [96]. At present, hydrogen peroxide is produced by the sequential hydrogenation and oxidation of alkyl anthraquinone, with the annual global production approximating 1.9×10^6 tons. However, a number of problems are associated with the anthraquinone route, including the cost of the quinone solvent system and a periodic need to replace the anthraquinone because of hydrogenation. In view of this, considerable interest has been expressed in the direct manufacture of hydrogen peroxide from the catalyzed reaction of hydrogen and oxygen. At present, a degree of success has been achieved using Pd as a catalyst, especially when halides are used as promoters. Typically, dilute solutions of hydrogen peroxide are produced, with earlier studies indicating that the Pd catalyst could be combined with an oxidation catalyst, TS-1, such that the hydrogen peroxide produced could be used *in situ*. To date, however, no commercial process exists for the direct manufacture of hydrogen peroxide. Hutchings and coworkers were the first to show that Au/Al_2O_3 catalysts were effective for the direct reaction, and thus far the best catalysts identified have been Pt–Au alloy supported catalysts (Table 10.2); these showed a better performance than pure Pd or Au catalysts [181, 182]. The suggestion was made that the enhanced activity observed by the addition of Pd to Au was due to an enhanced activation of hydrogen. However, if too much Pd was added, the decomposition activity of the hydrogen peroxide was also enhanced, such that the rate declined.

Subsequently, Ishihara *et al.* [183] have shown that Au/SiO_2 and $Au–Pd/SiO_2$ catalysts are also effective for this reaction at only 10 °C. In recent studies, Hutchings and coworkers have shown that the selectivity for H_2 utilization could be

Table 10.2 Formation of H_2O_2 from the reaction of H_2/O_2 over Au and Pd catalysts[182].

Catalyst	Solvent[a]	Temperature (°C)	Pressure (MPa)	O_2/H_2 (molar ratio)	H_2O_2 mmol g (catalyst)$^{-1}$ h^{-1}
Au/Al$_2$O$_3$	CH$_3$OH	2	3.7	1.2	1530
Au:Pd (1:1)/ Al$_2$O$_3$	CH$_3$OH	2	3.7	1.2	4460
Pd/Al$_2$O$_3$	CH$_3$OH	2	3.7	1.2	370
Au/ZnO	scCO$_2$	35	9.2	1.0	9
Au:Pd (1:3)/ ZnO	scCO$_2$	35	9.2	1.1	7
Au:Pd (1:1)/ ZnO	scCO$_2$	35	9.2	0.8	12
Au:Pd (3:1)/ ZnO	scCO$_2$	35	9.2	0.9	8
Pd/ZnO	scCO$_2$	35	9.2	1.3	0

a   scCO$_2$ = supercritical CO$_2$.

significantly enhanced when Fe$_2$O$_3$ and TiO$_2$ were used as supports [184, 185]. Indeed, with short reaction times the selectivity may exceed 95% for the reaction of dilute H$_2$/O$_2$ mixtures (1:1; 5 vol%) diluted with CO$_2$ (95 vol%). Very high rates of reaction were observed with noncalcined Au–Pd/TiO$_2$ catalysts, but these proved to be unstable as they lost both Au and Pd during the reaction and could not be successfully reused. However, if the catalysts were calcined at 400 °C prior to use, very stable reusable materials were obtained. Detailed structural investigations of these active stable catalysts, using X-ray photoelectron spectroscopy (XPS) and transmission electron microscopy (TEM), showed that the catalysts have a core–shell structure with a gold-rich core and a palladium-rich shell. It was concluded that the Au was acting as an electronic promotor for the Pd-rich surface of the Au–Pd nanocrystals.

10.8
Conclusions

Although the catalytic processes described in this chapter have provided a good overview of the oxidation catalysis properties of Group 11 metals, this is a far from comprehensive survey of the field. However, this group of metals has provided the basis for some of the most interesting size- and shape-dependent catalytic phenomena on the nanoscale. Whilst copper and silver have shown interesting and employable properties, the chemistry exhibited by gold on the nanoscale has truly sparked a whole subfield of research that has regularly yielded exciting and unexpected results. As the research groups continue to unravel the chemistry of these nanomaterials, it is likely that revolutionary technologies will follow.

References

1 Burda, C., Chen, X.B., Narayanan, R. and El-Sayed, M.A. (2005) Chemistry and properties of nanocrystals of different shapes. *Chem. Rev.*, **105** (4), 1025–1102.

2 Min, B.K. and Friend, C.M. (2007) Heterogeneous gold-based catalysis for green chemistry: low-temperature CO oxidation and propene oxidation. *Chem. Rev.*, **107** (6), 2709–2724.

3 Daniel, M.C. and Astruc, D. (2004) Gold nanoparticles: assembly, supramolecular chemistry, quantum-size-related properties, and applications toward biology, catalysis, and nanotechnology. *Chem. Rev.*, **104** (1), 293–346.

4 Wieckowski, A., Savinova, E.R. and Vayenas, C.G. (2003) *Catalysis and Electrocatalysis at Nanoparticle Surfaces*, Dekker, New York.

5 Bonnemann, H. and Richards, R.M. (2001) Nanoscopic metal particles – Synthetic methods and potential applications. *Eur. J. Inorg. Chem.*, **10**, 2455–2480.

6 Xu, Y.P., Tian, Z.J. and Lin, L.W. (2004) Nanostructure and catalytic performance of noble metal solid catalysts. *Chin. J. Catal.*, **25** (4), 331–338.

7 Welch, C.W. and Compton, R.G. (2006) The use of nanoparticles in electroanalysis: a review. *Anal. Bioanal. Chem.*, **384** (3), 601–619.

8 Thomas, J.M. and Raja, R. (2001) Nanopore and nanoparticle catalysts. *Chem. Rec.*, **1** (6), 448–466.

9 Bond, G.C. and Sermon, P.A. (1973) Gold catalysts for olefin hydrogenation. *Gold Bull.*, **6** (1), 102–105.

10 Bond, G.C., Sermon, P.A., Webb, G., Buchanan, D.A. and Wells, P.B. (1973) Hydrogenation over supported gold catalysts. *J. Chem. Soc., Chem. Commun.*, (13), 444b–445b.

11 Huber, H., McIntosh, D. and Ozin, G.A. (1977) A metal atom model for the oxidation of carbon monoxide to carbon dioxide. The gold atom-carbon monoxide-dioxygen reaction and the gold atom-carbon dioxide reaction. *Inorg. Chem.*, **16** (5), 975–979.

12 Haruta, M., Tsubota, S., Kobayashi, T., Kageyama, H., Genet, M.J. and Delmon, B. (1993) Low-temperature oxidation of Co over gold supported on TiO_2, alpha-Fe_2O_3, and Co_3O_4. *J. Catal.*, **144** (1), 175–192.

13 Grunwaldt, J.D., Kiener, C., Wogerbauer, C. and Baiker, A. (1999) Preparation of supported gold catalysts for low-temperature CO oxidation via "size-controlled" gold colloids. *J. Catal.*, **181** (2), 223–232.

14 Teranishi, T. and Miyake, M. (1998) Size control of palladium nanoparticles and their crystal structures. *Chem. Mater.*, **10** (2), 594–600.

15 Li, Y., Boone, E. and El-Sayed, M.A. (2002) Size effects of PVP-Pd nanoparticles on the catalytic Suzuki reactions in aqueous solution. *Langmuir*, **18** (12), 4921–4925.

16 Narayanan, R. and El-Sayed, M.A. (2003) Effect of catalysis on the stability of metallic nanoparticles: Suzuki reaction catalyzed by PVP-palladium nanoparticles. *J. Am. Chem. Soc.*, **125** (27), 8340–8347.

17 Ahmadi, T.S., Wang, Z.L., Green, T.C., Henglein, A. and ElSayed, M.A. (1996) Shape-controlled synthesis of colloidal platinum nanoparticles. *Science*, **272** (5270), 1924–1926.

18 Narayanan, R. and El-Sayed, M.A. (2004) Changing catalytic activity during colloidal platinum nanocatalysis due to shape changes: electron-transfer reaction. *J. Am. Chem. Soc.*, **126** (23), 7194–7195.

19 Narayanan, R. and El-Sayed, M.A. (2005) Effect of colloidal nanocatalysis on the metallic nanoparticle shape: the Suzuki reaction. *Langmuir*, **21** (5), 2027–2033.

20 Sau, T.K., Pal, A. and Pal, T. (2001) Size regime dependent catalysis by gold nanoparticles for the reduction of eosin. *J. Phys. Chem. B*, **105** (38), 9266–9272.

21 Crooks, R.M., Zhao, M.Q., Sun, L., Chechik, V. and Yeung, L.K. (2001) Dendrimer-encapsulated metal nanoparticles: synthesis, characterization, and applications to catalysis. *Acc. Chem. Res.*, **34** (3), 181–190.

22 Schulz, J., Roucoux, A. and Patin, H. (2000) Stabilized rhodium(0) nanoparticles: a reusable hydrogenation catalyst for arene derivatives in a biphasic water-liquid system. *Chem. Eur. J.*, **6** (4), 618–624.

23 Reetz, M.T. and Helbig, W. (1994) Size-selective synthesis of nanostructured transition-metal clusters. *J. Am. Chem. Soc.*, **116** (16), 7401–7402.

24 Reetz, M.T. and Quaiser, S.A. (1995) A new method for the preparation of nanostructured metal-clusters. *Angew. Chem. Int. Ed. Engl.*, **34** (20), 2240–2241.

25 Michaelis, M. and Henglein, A. (1992) Reduction of Pd(ii) in aqueous-solution–stabilization and reactions of an intermediate cluster and pd colloid formation. *J. Phys. Chem.*, **96** (11), 4719–4724.

26 Toshima, N., Takahashi, T. and Hirai, H. (1985) Colloidal platinum catalysts prepared by hydrogen-reduction and photo-reduction in the presence of surfactant. *Chem. Lett.*, **8**, 1245–1248.

27 Kurihara, K., Kizling, J., Stenius, P. and Fendler, J.H. (1983) Laser and pulse radiolytically induced colloidal gold formation in water and in water-in-oil microemulsions. *J. Am. Chem. Soc.*, **105** (9), 2574–2579.

28 Caruso, R.A., Ashokkumar, M. and Grieser, F. (2000) Sonochemical formation of colloidal platinum. *Colloids Surf. A Physicochem. Eng. Asp.*, **169** (1–3), 219–225.

29 Fujimoto, T., Terauchi, S., Umehara, H., Kojima, I. and Henderson, W. (2001) Sonochemical preparation of single-dispersion metal nanoparticles from metal salts. *Chem. Mater.*, **13** (3), 1057–1060.

30 Borsla, A., Wilhelm, A.M. and Delmas, H. (2001) Hydrogenation of olefins in aqueous phase, catalyzed by polymer-protected rhodium colloids: kinetic study. *Catal. Today*, **66** (2–4), 389–395.

31 Bronstein, L.M., Chernyshov, D.M., Volkov, I.O., Ezernitskaya, M.G., Valetsky, P.M., Matveeva, V.G. and Sulman, E.M. (2000) Structure and properties of bimetallic colloids formed in polystyrene-block-poly-4-vinylpyridine micelles: catalytic behavior in selective hydrogenation of dehydrolinalool. *J. Catal.*, **196** (2), 302–314.

32 Scott, R.W.J., Datye, A.K. and Crooks, R.M. (2003) Bimetallic palladium-platinum dendrimer-encapsulated catalysts. *J. Am. Chem. Soc.*, **125** (13), 3708–3709.

33 Pittelkow, M., Moth-Poulsen, K., Boas, U. and Christensen, J.B. (2003) Poly(amidoamine)-dendrimer-stabilized Pd(0) nanoparticles as a catalyst for the Suzuki reaction. *Langmuir*, **19** (18), 7682–7684.

34 Lu, Z.H., Liu, G.J., Phillips, H., Hill, J.M., Chang, J. and Kydd, R.A. (2001) Palladium nanoparticle catalyst prepared in poly(acrylic acid)-lined channels of diblock copolymer microspheres. *Nano Lett.*, **1** (12), 683–687.

35 Mevellec, V., Roucoux, A., Ramirez, E., Philippot, K. and Chaudret, B. (2004) Surfactant-stabilized aqueous iridium(0) colloidal suspension: an efficient reusable catalyst for hydrogenation of arenes in biphasic media. *Adv. Synth. Catal.*, **346** (1), 72–76.

36 Lin, Y. and Finke, R.G. (1994) Novel polyoxoanion-stabilized and bu(4)n(+)-stabilized, isolable, and redissolvable, 20–30-angstrom ir-similar-

to(300)-(900) nanoclusters – the kinetically controlled synthesis, characterization, and mechanism of formation of organic solvent-soluble, reproducible size, and reproducible catalytic activity metal nanoclusters. *J. Am. Chem. Soc.*, **116** (18), 8335–8353.

37 Chen, S.L. and Kucernak, A. (2004) Electrocatalysis under conditions of high mass transport rate: oxygen reduction on single submicrometer-sized Pt particles supported on carbon. *J. Phys. Chem. B*, **108** (10), 3262–3276.

38 Liu, Z.L., Ling, X.Y., Lee, J.Y., Su, X.D. and Gan, L.M. (2003) Nanosized Pt and PtRu colloids as precursors for direct methanol fuel cell catalysts. *J. Mater. Chem.*, **13** (12), 3049–3052.

39 Fachini, E.R., Diaz-Ayala, R., Casado-Rivera, E., File, S. and Cabrera, C.R. (2003) Surface coordination of ruthenium clusters on platinum nanoparticles for methanol oxidation catalysts. *Langmuir*, **19** (21), 8986–8993.

40 Chen, C.W., Serizawa, T. and Akashi, M. (1999) Preparation of platinum colloids on polystyrene nanospheres and their catalytic properties in hydrogenation. *Chem. Mater.*, **11** (5), 1381–1389.

41 Eppler, A.S., Rupprechter, G., Anderson, E.A. and Somorjai, G.A. (2000) Thermal and chemical stability and adhesion strength of Pt nanoparticle arrays supported on silica studied by transmission electron microscopy and atomic force microscopy. *J. Phys. Chem. B*, **104** (31), 7286–7292.

42 Grunes, J., Zhu, J., Anderson, E.A. and Somorjai, G.A. (2002) Ethylene hydrogenation over platinum nanoparticle array model catalysts fabricated by electron beam lithography: determination of active metal surface area. *J. Phys. Chem. B*, **106** (44), 11463–11468.

43 Rashid, H. and Mandal, T.K. (2007) Synthesis and catalytic application of nanostructured silver dendrites. *J. Phys. Chem. C*, **111** (45), 16750–16760.

44 Zhang, W., Qiao, X. and Chen, J. (2007) Synthesis and UV-Vis spectral properties of silver nanoparticles. *Rare Metal Mater. Eng.*, **36**, 64–70.

45 Zhang, T.W., Zhang, L., Yang, S.C., Yang, Z.N. and Ding, B.J. (2007) Shape-controlled synthesis and applications of silver nano-particles. *Rare Metal Mater. Eng.*, **36** (8), 1495–1499.

46 Eastoe, J., Hollamby, M.J. and Hudson, L. (2006) Recent advances in nanoparticle synthesis with reversed micelles. *Adv. Colloid Interface Sci.*, **128**, 5–15.

47 Bajpai, S.K., Mohan, Y.M., Bajpai, M., Tankhiwale, R. and Thomas, V. (2007) Synthesis of polymer stabilized silver and gold nanostructures. *J. Nanosci. Nanotechnol.*, **7** (9), 2994–3010.

48 Anandan, S. and Yang, S.H. (2007) Emergent methods to synthesize and characterize semiconductor CuO nanoparticles with various morphologies – an overview. *J. Exp. Nanosci.*, **2** (1–2), 23–56.

49 Aymonier, C., Loppinet-Serani, A., Reveron, H., Garrabos, Y. and Cansell, F. (2006) Review of supercritical fluids in inorganic materials science. *J. Supercrit. Fluids*, **38** (2), 242–251.

50 Salzemann, C., Lisiecki, L., Urban, J. and Pileni, M.P. (2004) Anisotropic copper nanocrystals synthesized in a supersaturated medium: Nanocrystal growth. *Langmuir*, **20** (26), 11772–11777.

51 Haruta, M. (2004) Gold as a novel catalyst in the 21st century: preparation, working mechanism and applications. *Gold Bull.*, **37** (1–2), 27–36.

52 Haruta, M., Yamada, N., Kobayashi, T. and Iijima, S. (1989) Gold catalysts prepared by coprecipitation for low-temperature oxidation of hydrogen and of carbon-monoxide. *J. Catal.*, **115** (2), 301–309.

53 Kageyama, H., Kamijo, N., Kobayashi, T. and Haruta, M. (1989) XAFS studies of ultra-fine gold catalysts supported on hematite prepared from coprecipitated precursors. *Physica B*, **158** (1–3), 183–184.

54 Sanchez, R.M.T., Ueda, A., Tanaka, K. and Haruta, M. (1997) Selective oxidation of CO in hydrogen over gold supported on manganese oxides. *J. Catal.*, **168** (1), 125–127.

55 Shibata, M., Kawata, N., Masumoto, T. and Kimura, H. (1985) CO hydrogenation over an amorphous gold-zirconium alloy. *Chem. Lett.*, **11**, 1605–1608.

56 Kobayashi, T., Haruta, M., Tsubota, S., Sano, H. and Delmon, B. (1990) Thin-films of supported gold catalysts for CO detection. *Sens. Actuators B Chem.*, **1** (1–6), 222–225.

57 Vogel, W., Cunningham, D.A.H., Tanaka, K. and Haruta, M. (1996) Structural analysis of Au/Mg(OH)(2) during deactivation by Debye function analysis. *Catal. Lett.*, **40** (3–4), 175–181.

58 Yuan, Y.Z., Kozlova, A.P., Asakura, K., Wan, H.L., Tsai, K. and Iwasawa, Y. (1997) Supported Au catalysts prepared from Au phosphine complexes and As-precipitated metal hydroxides: characterization and low-temperature CO oxidation. *J. Catal.*, **170** (1), 191–199.

59 Okumura, M. and Haruta, M. (2000) Preparation of supported gold catalysts by liquid-phase grafting of gold acetylacetonate for low-temperature oxidation of CO and of H-2. *Chem. Lett.*, 4, 396–397.

60 Okumura, M., Tanaka, K., Ueda, A. and Haruta, M. (1997) The reactivities of dimethylgold(III)beta-diketone on the surface of TiO_2–a novel preparation method for Au catalysts. *Solid State Ionics*, **95** (1–2), 143–149.

61 Okumura, M., Tsubota, S., Iwamoto, M. and Haruta, M. (1998) Chemical vapor deposition of gold nanoparticles on MCM-41 and their catalytic activities for the low-temperature oxidation of CO and of H-2. *Chem. Lett.*, 4, 315–316.

62 Wallace, W.T. and Whetten, R.L. (2000) Carbon monoxide adsorption on selected gold clusters: highly size-dependent activity and saturation compositions. *J. Phys. Chem. B*, **104** (47), 10964–10968.

63 Kishi, K., Date, M. and Haruta, M. (2001) Effect of gold on the oxidation of the Si(111)-7 × 7 surface. *Surf. Sci.*, **486** (3), L475–L479.

64 Valden, M., Pak, S., Lai, X. and Goodman, D.W. (1998) Structure sensitivity of CO oxidation over model Au/TiO_2 catalysts. *Catal. Lett.*, **56** (1), 7–10.

65 Yan, W.F., Chen, B., Mahurin, S.M., Hagaman, E.W., Dai, S. and Overbury, S.H. (2004) Surface sol-gel modification of mesoporous silica materials with TiO_2 for the assembly of ultrasmall gold nanoparticles. *J. Phys. Chem. B*, **108** (9), 2793–2796.

66 Haruta, M., Kobayashi, T., Sano, H. and Yamada, N. (1987) Novel gold catalysts for the oxidation of carbon-monoxide at a temperature far below 0-degrees-C. *Chem. Lett.*, **2**, 405–408.

67 Sakurai, H. and Haruta, M. (1995) Carbon-dioxide and carbon-monoxide hydrogenation over gold supported on titanium, iron, and zinc-oxides. *Appl. Catal. A Gen.*, **127** (1–2), 93–105.

68 Sakurai, H., Ueda, A., Kobayashi, T. and Haruta, M. (1997) Low-temperature water-gas shift reaction over gold deposited on TiO_2. *Chem. Commun.*, (3), 271–272.

69 Cosandey, F., Zhang, L. and Madey, T.E. (2001) Effect of substrate temperature on the epitaxial growth of Au on TiO_2(110). *Surf. Sci.*, **474** (1–3), 1–13.

70 Kozlov, A.I., Kozlova, A.P., Liu, H.C. and Iwasawa, Y. (1999) A new approach to active supported Au catalysts. *Appl. Catal. A Gen.*, **182** (1), 9–28.

71 Dekkers, M.A.P., Lippits, M.J. and Nieuwenhuys, B.E. (1998) CO adsorption and oxidation on Au/TiO_2. *Catal. Lett.*, **56** (4), 195–197.

72 Boccuzzi, F., Chiorino, A., Manzoli, M., Lu, P., Akita, T., Ichikawa, S. and Haruta, M. (2001) Au/TiO_2 nanosized samples: a catalytic, TEM, and FTIR study of the effect of calcination temperature on the CO oxidation. *J. Catal.*, **202** (2), 256–267.

73 Mavrikakis, M., Stoltze, P. and Norskov, J.K. (2000) Making gold less noble. *Catal. Lett.*, **64** (2–4), 101–106.

74 Hammer, B. and Norskov, J.K. (1995) Why gold is the noblest of all the metals. *Nature*, **376** (6537), 238–240.

75 Guzman, J., Carrettin, S., Fierro-Gonzalez, J.C., Hao, Y.L., Gates, B.C. and Corma, A. (2005) CO oxidation catalyzed by supported gold: cooperation

between gold and nanocrystalline rare-earth supports forms reactive surface superoxide and peroxide species. *Angew. Chem. Int. Ed. Engl.*, **44** (30), 4778–4781.

76. Qiao, B.T. and Deng, Y.Q. (2003) Highly effective ferric hydroxide supported gold catalyst for selective oxidation of CO in the presence of H-2. *Chem. Commun.*, (17), 2192–2193.

77. Bond, G.C. and Thompson, D.T. (2000) Gold-catalysed oxidation of carbon monoxide. *Gold Bull.*, **33** (2), 41–51.

78. Haruta, M. (2002) Catalysis of gold nanoparticles deposited on metal oxides. *CatTech*, **6** (3), 102–115.

79. Qu, Z.P., Cheng, M.J., Dong, X.L. and Bao, X.H. (2004) Kx, CO selective oxidation in H-2-rich gas over Ag nanoparticles – effect of oxygen treatment temperature on the activity of silver particles mechanically mixed with SiO_2. *Catal. Today*, **93–95**, 247–255.

80. Qu, Z., Huang, W., Cheng, M. and Bao, X. (2005) Restructuring and redispersion of silver on SiO_2 under oxidizing/reducing atmospheres and its activity toward CO oxidation. *J. Phys. Chem. B*, **109** (33), 15842–15848.

81. Yang, M.X., Jacobs, P.W., Yoon, C., Muray, L., Anderson, E., Attwood, D. and Somorjai, G.A. (1997) Thermal stability of uniform silver clusters prepared on oxidized silicon and aluminum surfaces by electron beam lithography in oxidizing and reducing ambients. *Catal. Lett.*, **45** (1–2), 5–13.

82. Aoyama, N., Yoshida, K., Abe, A. and Miyadera, T. (1997) Characterization of highly active silver catalyst for NOx reduction in lean-burning engine exhaust. *Catal. Lett.*, **43** (3–4), 249–253.

83. Jin, L., Qian, K., Jiang, Z.Q. and Huang, W.X. (2007) Ag/SiO(2)catalysts prepared via gamma-ray irradiation and their catalytic activities in CO oxidation. *J. Mol. Catal. A Chem.*, **274** (1–2), 95–100.

84. Narayanan, R. and El-Sayed, M.A. (2004) Shape-dependent catalytic activity of platinum nanoparticles in colloidal solution. *Nano Lett.*, **4** (7), 1343–1348.

85. Lim, D.C., Lopez-Salido, I. and Kim, Y.D. (2005) Size selectivity for CO-oxidation of Ag nanoparticles on highly ordered pyrolytic graphite (HOPG). *Surf. Sci.*, **598** (1–3), 96–103.

86. Wang, A.Q., Hsieh, Y., Chen, Y.F. and Mou, C.Y. (2006) Au-Ag alloy nanoparticle as catalyst for CO oxidation: effect of Si/Al ratio of mesoporous support. *J. Catal.*, **237** (1), 197–206.

87. Wang, A.Q., Chang, C.M. and Mou, C.Y. (2005) Evolution of catalytic activity of Au-Ag bimetallic nanoparticles on mesoporous support for CO oxidation. *J. Phys. Chem. B*, **109** (40), 18860–18867.

88. Liu, J.H., Wang, A.Q., Chi, Y.S., Lin, H.P. and Mou, C.Y. (2005) Synergistic effect in an Au-Ag alloy nanocatalyst: CO oxidation. *J. Phys. Chem. B*, **109** (1), 40–43.

89. Arena, F., Famulari, P., Trunfio, G., Bonura, G., Frusteri, F. and Spadaro, L. (2006) Probing the factors affecting structure and activity of the Au/CeO_2 system in total and preferential oxidation of CO. *Appl. Catal. B Environ.*, **66** (1–2), 81–91.

90. Lippits, M.J., Gluhoi, A.C. and Nieuwenhuys, B.E. (2007) A comparative study of the effect of addition of CeOx and Li_2O on gamma-Al_2O_3 supported copper, silver and gold catalysts in the preferential oxidation of CO. *Top. Catal.*, **44** (1–2), 159–165.

91. Shim, I.W., Noh, W.T., Kwon, J., Cho, J.Y., Kim, K.S. and Kang, D.H. (2002) Preparation of copper nanoparticles in cellulose acetate polymer and the reaction chemistry of copper complexes in the polymer. *Bull. Korean Chem. Soc.*, **23** (4), 563–566.

92. Tu, C.H., Wang, A.Q., Zheng, M.Y., Wang, X.D. and Zhang, T. (2006) Factors influencing the catalytic activity of SBA-15-supported copper nanoparticles in CO oxidation. *Appl. Catal. A Gen.*, **297** (1), 40–47.

93. Hayashi, T., Tanaka, K. and Haruta, M. (1998) Selective vapor-phase epoxidation of propylene over Au/TiO_2 catalysts in the presence of oxygen and hydrogen. *J. Catal.*, **178** (2), 566–575.

94. Meiers, R., Dingerdissen, U. and Holderich, W.F. (1998) Synthesis of propylene oxide from propylene, oxygen,

and hydrogen catalyzed by palladium-platinum-containing titanium silicalite. *J. Catal.*, **176** (2), 376–386.

95 Laufer, W. and Hoelderich, W.F. (2001) Direct oxidation of propylene and other olefins on precious metal containing Ti-catalysts. *Appl. Catal. A Gen.*, **213** (2), 163–171.

96 Jenzer, G., Mallat, T., Maciejewski, M., Eigenmann, F. and Baiker, A. (2001) Continuous epoxidation of propylene with oxygen and hydrogen on a Pd-Pt/TS-1 catalyst. *Appl. Catal. A Gen.*, **208** (1–2), 125–133.

97 Haruta, M., Uphade, B.S., Tsubota, S. and Miyamoto, A. (1998) Selective oxidation of propylene over gold deposited on titanium-based oxides. *Res. Chem. Intermed.*, **24** (3), 329–336.

98 Sinha, A.K., Seelan, S., Tsubota, S. and Haruta, M. (2004) A three-dimensional mesoporous titanosilicate support for gold nanoparticles: vapor-phase epoxidation of propene with high conversion. *Angew. Chem. Int. Ed. Engl.*, **43** (12), 1546–1548.

99 Sinha, A.K., Seelan, S., Okumura, M., Akita, T., Tsubota, S. and Haruta, M. (2005) Three-dimensional mesoporous titanosilicates prepared by modified sol-gel method: ideal gold catalyst supports for enhanced propene epoxidation. *J. Phys. Chem. B*, **109** (9), 3956–3965.

100 Mul, G., Zwijnenburg, A., van der Linden, B., Makkee, M. and Moulijn, J.A. (2001) Stability and selectivity of Au/TiO_2 and Au/TiO_2/SiO_2 catalysts in propene epoxidation: an in situ FT-IR study. *J. Catal.*, **201** (1), 128–137.

101 Zwijnenburg, A., Saleh, M., Makkee, M. and Moulijn, J.A. (2002) Direct gas-phase epoxidation of propene over bimetallic Au catalysts. *Catal. Today*, **72** (1–2), 59–62.

102 Hughes, M.D., Xu, Y.J., Jenkins, P., McMorn, P., Landon, P., Enache, D.I., Carley, A.F., Attard, G.A., Hutchings, G.J., King, F., Stitt, E.H., Johnston, P., Griffin, K. and Kiely, C.J. (2005) Tunable gold catalysts for selective hydrocarbon oxidation under mild conditions. *Nature*, **437** (7062), 1132–1135.

103 Stangland, E.E., Stavens, K.B., Andres, R.P. and Delgass, W.N. (2000) Characterization of gold-titania catalysts via oxidation of propylene to propylene oxide. *J. Catal.*, **191** (2), 332–347.

104 Lambert, R.M., Williams, F.J., Cropley, R.L. and Palermo, A. (2005) Heterogeneous alkene epoxidation: past, present and future. *J. Mol. Catal. A Chem.*, **228** (1–2), 27–33.

105 Sachtler, W.M.H., Backx, C. and Vansanten, R.A. (1981) On the mechanism of ethylene epoxidation. *Catal. Rev. Sci. Eng.*, **23** (1–2), 127–149.

106 Sajkowski, D.J. and Boudart, M. (1987) Structure sensitivity of the catalytic-oxidation of ethene by silver. *Catal. Rev. Sci. Eng.*, **29** (4), 325–360.

107 Vansanten, R.A. and Kuipers, H. (1987) The mechanism of ethylene epoxidation. *Adv. Catal.*, **35**, 265–321.

108 Verykios, X.E., Stein, F.P. and Coughlin, R.W. (1980) Oxidation of ethylene over silver–adsorption, kinetics, catalyst. *Catal. Rev. Sci. Eng.*, **22** (2), 197–234.

109 Yong, Y.S. and Cant, N.W. (1990) Ethylene oxidation over silver catalysts–a study of mechanism using nitrous-oxide and isotopically labeled oxygen. *J. Catal.*, **122** (1), 22–33.

110 Tan, S.A., Grant, R.B. and Lambert, R.M. (1987) Pressure-dependence of ethylene oxidation-kinetics and the effects of added CO_2 and Cs–a study on Ag(111) and Ag/alpha-Al_2O_3 catalysts. *Appl. Catal.*, **31** (1), 159–77.

111 Tan, S.A., Grant, R.B. and Lambert, R.M. (1987) Secondary chemistry in the selective oxidation of ethylene–effect of Cl and Cs promoters on the adsorption, isomerization, and combustion of ethylene-oxide on Ag(111). *J. Catal.*, **106** (1), 54–64.

112 Serafin, J.G., Liu, A.C. and Seyedmonir, S.R. (1998) Surface science and the silver-catalyzed epoxidation of ethylene: an industrial perspective. *J. Mol. Catal. A Chem.*, **131** (1–3), 157–168.

113 Nagy, A., Mestl, G., Ruhle, T., Weinberg, G. and Schlogl, R. (1998) The dynamic restructuring of electrolytic silver during the formaldehyde synthesis reaction. *J. Catal.*, **179** (2), 548–559.

114 Campbell, C.T. (1986) Chlorine promoters in selective ethylene epoxidation over Ag(111) – a comparison with Ag(110). *J. Catal.*, **99** (1), 28–38.

115 Campbell, C.T. and Koel, B.E. (1985) Chlorine promotion of selective ethylene oxidation over Ag(110) – kinetics and mechanism. *J. Catal.*, **92** (2), 272–283.

116 Campbell, C.T. (1985) Cs-promoted Ag(111) – model studies of selective ethylene oxidation catalysts. *J. Phys. Chem.*, **89** (26), 5789–5795.

117 Goncharova, S.N., Paukshtis, E.A. and Balzhinimaev, B.S. (1995) Size effects in ethylene oxidation on silver catalysts – influence of support and Cs promoter. *Appl. Catal. A Gen.*, **126** (1), 67–84.

118 Grant, R.B., Harbach, C.A.J., Lambert, R.M. and Tan, S.A. (1987) Alkali-metal, chlorine and other promoters in the silver-catalyzed selective oxidation of ethylene. *J. Chem. Soc., Faraday Trans. I*, **83**, 2035–2046.

119 Shiraishi, Y. and Toshima, N. (1999) Colloidal silver catalysts for oxidation of ethylene. *J. Mol. Catal. A Chem.*, **141** (1–3), 187–192.

120 Shiraishi, Y. and Toshima, N. (2000) Oxidation of ethylene catalyzed by colloidal dispersions of poly(sodium acrylate)-protected silver nanoclusters. *Colloids Surf. A Physicochem. Eng. Asp.*, **169** (1–3), 59–66.

121 Verykios, X.E., Stein, F.P. and Coughlin, R.W. (1980) Influence of metal crystallite size and morphology on selectivity and activity of ethylene oxidation catalyzed by supported silver. *J. Catal.*, **66** (2), 368–382.

122 Lee, J.K., Verykios, X.E. and Pitchai, R. (1989) Support and crystallite size effects in ethylene oxidation catalysis. *Appl. Catal.*, **50** (2), 171–188.

123 Bukhtiyarov, V.I. and Kaichev, V.V. (2000) The combined application of XPS and TPD to study of oxygen adsorption on graphite-supported silver clusters. *J. Mol. Catal. A Chem.*, **158** (1), 167–172.

124 Podgornov, E.A., Prosvirin, I.P. and Bukhtiyarov, V.I. (2000) XPS, TPD and TPR studies of Cs-O complexes on silver: their role in ethylene epoxidation. *J. Mol. Catal. A Chem.*, **158** (1), 337–343.

125 Minahan, D.M., Hoflund, G.B., Epling, W.S. and Schoenfeld, D.W. (1997) Study of Cs-promoted, alpha-alumina-supported silver, ethylene epoxidation catalysts. 3. Characterization of Cs-promoted and nonpromoted catalysts. *J. Catal.*, **168** (2), 393–399.

126 Kim, Y.C., Park, N.C., Shin, J.S., Lee, S.R., Lee, Y.J. and Moon, D.J. (2003) Partial oxidation of ethylene to ethylene oxide over nanosized Ag/alpha-Al_2O_3 catalysts. *Catal. Today*, **87** (1–4), 153–162.

127 de Oliveira, A.L., Wolf, A. and Schuth, F. (2001) Highly selective propene epoxidation with hydrogen/oxygen mixtures over titania-supported silver catalysts. *Catal. Lett.*, **73** (2–4), 157–160.

128 Nijhuis, T.A., Makkee, M., Moulijn, J.A. and Weckhuysen, B.M. (2006) The production of propene oxide: catalytic processes and recent developments. *Ind. Eng. Chem. Res.*, **45** (10), 3447–3459.

129 Maayan, G. and Neumann, R. (2005) Direct aerobic epoxidation of alkenes catalyzed by metal nanoparticles stabilized by the H5PV2Mo10O40 polyoxometalate. *Chem. Commun.*, (36), 4595–4597.

130 Chimentao, R.J., Medina, F., Sueiras, J.E., Fierro, J.L.G., Cesteros, Y. and Salagre, P. (2007) Effects of morphology and cesium promotion over silver nanoparticles catalysts in the styrene epoxidation. *J. Mater. Sci.*, **42** (10), 3307–3314.

131 Chimentao, R.J., Kirm, I., Medina, F., Rodriguez, X., Cesteros, Y., Salagre, P., Sueiras, J.E. and Fierro, J.L.G. (2005) Sensitivity of styrene oxidation reaction to the catalyst structure of silver nanoparticles. *Appl. Surf. Sci.*, **252** (3), 793–800.

132 Chimentao, R.J., Kirm, I., Medina, F., Rodriguez, X., Cesteros, Y., Salagre, P. and Sueiras, J.E. (2004) Different morphologies of silver nanoparticles as catalysts for the selective oxidation of styrene in the gas phase. *Chem. Commun.*, (7), 846–847.

133 Chimentao, R.J., Medina, F., Fierro, J.L.G., Sueiras, J.E., Cesteros, Y. and Salagre, P. (2006) Styrene epoxidation over cesium promoted silver nanowires catalysts. *J. Mol. Catal. A Chem.*, **258** (1–2), 346–354.

134 Chimentao, R.J., Barrabes, N., Medina, F., Fierro, J.L.G., Sueiras, J.E., Cesteros, Y. and Salagre, P. (2006) Synthesis, characterization and catalytic activity of metal nanoparticles in the selective oxidation of olefins in the gas phase. *J. Exp. Nanosci.*, **1** (4), 399–418.

135 Suresh, A.K., Sharma, M.M. and Sridhar, T. (2000) Engineering aspects of industrial liquid-phase air oxidation of hydrocarbons. *Ind. Eng. Chem. Res.*, **39** (11), 3958–3997.

136 Thomas, J.M., Raja, R., Sankar, G. and Bell, R.G. (1999) Molecular-sieve catalysts for the selective oxidation of linear alkanes by molecular oxygen. *Nature*, **398** (**6724**), 227–230.

137 Schuchardt, U., Cardoso, D., Sercheli, R., Pereira, R., de Cruz, R.S., Guerreiro, M.C., Mandelli, D., Spinace, E.V. and Fires, E.L. (2001) Cyclohexane oxidation continues to be a challenge. *Appl. Catal. A Gen.*, **211** (1), 1–17.

138 Sheldon, R.A., Arends, I. and Lempers, H.E.B. (1998) Liquid phase oxidation at metal ions and complexes in constrained environments. *Catal. Today*, **41** (4), 387–407.

139 Nowotny, M., Pedersen, L.N., Hanefeld, U. and Maschmeyer, T. (2002) Increasing the ketone selectivity of the cobalt-catalyzed radical chain oxidation of cyclohexane. *Chem. Eur. J.*, **8** (16), 3724–3731.

140 Yuan, H.X., Xia, Q.H., Zhan, H.J., Lu, X.H. and Su, K.X. (2006) Catalytic oxidation of cyclohexane to cyclohexanone and cyclohexanol by oxygen in a solvent-free system over metal-containing ZSM-5 catalysts. *Appl. Catal. A Gen.*, **304** (1), 178–184.

141 Anand, R., Hamdy, M.S., Gkourgkoulas, P., Maschmeyer, T., Jansen, J.C. and Hanefeld, U. (2006) Liquid phase oxidation of cyclohexane over transition metal incorporated amorphous 3D-mesoporous silicates M-TUD-1 (M = Ti, Fe, Co and Cr). *Catal. Today*, **117** (1–3), 279–283.

142 Zhao, R., Ji, D., Lv, G.M., Qian, G., Yan, L., Wang, X.L. and Suo, J.S. (2004) A highly efficient oxidation of cyclohexane over Au/ZSM-5 molecular sieve catalyst with oxygen as oxidant. *Chem. Commun.*, (7), 904–905.

143 Lu, G.M., Zhao, R., Qian, G., Qi, Y.X., Wang, X.L. and Suo, J.S. (2004) A highly efficient catalyst Au/MCM-41 for selective oxidation cyclohexane using oxygen. *Catal. Lett.*, **97** (3–4), 115–118.

144 Zhu, K.K., Hu, J.C. and Richards, R. (2005) Aerobic oxidation of cyclohexane by gold nanoparticles immobilized upon mesoporous silica. *Catal. Lett.*, **100** (3–4), 195–199.

145 Xu, Y.J., Landon, P., Enache, D., Carley, A., Roberts, M. and Hutchings, G. (2005) Selective conversion of cyclohexane to cyclohexanol and cyclohexanone using a gold catalyst under mild conditions. *Catal. Lett.*, **101** (3–4), 175–179.

146 Lefort, T.E. (1931) Société Française de Catalyse Generalisée, French Patent 729952, p. 562.

147 van Santen, R.A. and Kuipers, H.P.C.E. (1987) *Advances in Catalysis*, Vol. 35, Academic Press, London, pp. 265–321.

148 Zhao, H., Zhou, J.C., Luo, H., Zeng, C.Y., Li, D.H. and Liu, Y.J. (2006) Synthesis, characterization of Ag/MCM-41 and the catalytic performance for liquid-phase oxidation of cyclohexane. *Catal. Lett.*, **108** (1–2), 49–54.

149 Petranovskii, V.P., Pestryakov, A.N., Kazantseva, L.K., Barraza, F.F.C. and Farias, M. (2005) Formation of catalytically active copper nanoparticles in natural zeolites for complete oxidation of hydrocarbons. *Int. J. Mod. Phys. B*, **19** (15–17), 2333–2338.

150 Prati, L. and Rossi, M. (1997) Chemoselective catalytic oxidation of polyols with dioxygen oil gold supported catalysts, in *3rd World Congress on Oxidation Catalysis*, Vol. 110, Elsevier, Amsterdam, pp. 509–516.

151 Prati, L. and Rossi, M. (1998) Gold on carbon as a new catalyst for selective

liquid phase oxidation of diols. *J. Catal.*, **176** (2), 552–560.

152 Prati, L. and Martra, G. (1999) New gold catalysts for liquid phase oxidation. *Gold Bull.*, **32** (3), 96–101.

153 Porta, F., Prati, L., Rossi, M., Coluccia, S. and Martra, G. (2000) Metal sols as a useful tool for heterogeneous gold catalyst preparation: reinvestigation of a liquid phase oxidation. *Catal. Today*, **61** (1–4), 165–172.

154 Porta, F., Prati, L., Rossi, M. and Scari, G. (2002) New Au(0) sols as precursors for heterogeneous liquid-phase oxidation catalysts. *J. Catal.*, **211** (2), 464–469.

155 Bianchi, C.L., Biella, S., Gervasini, A., Prati, L. and Rossi, M. (2003) Gold on carbon: influence of support properties on catalyst activity in liquid-phase oxidation. *Catal. Lett.*, **85** (1–2), 91–96.

156 Biella, S., Porta, F., Prati, L. and Rossi, M. (2003) Surfactant-protected gold particles: new challenge for gold-on-carbon catalysts. *Catal. Lett.*, **90** (1–2), 23–29.

157 Dimitratos, N., Lopez-Sanchez, J.A., Morgan, D., Carley, A., Prati, L. and Hutchings, G.J. (2007) Solvent free liquid phase oxidation of benzyl alcohol using Au supported catalysts prepared using a sol immobilization technique. *Catal. Today*, **122** (3–4), 317–324.

158 Comotti, M., Della Pina, C., Matarrese, R., Rossi, M. and Siani, A. (2005) Oxidation of alcohols and sugars using Au/C catalysts – Part 2. Sugars. *Appl. Catal. A Gen.*, **291** (1–2), 204–209.

159 Beltrame, P., Comotti, M., Della Pina, C. and Rossi, M. (2006) Aerobic oxidation of glucose II. Catalysis by colloidal gold. *Appl. Catal. A Gen.*, **297** (1), 1–7.

160 Carrettin, S., McMorn, P., Johnston, P., Griffin, K. and Hutchings, G.J. (2002) Selective oxidation of glycerol to glyceric acid using a gold catalyst in aqueous sodium hydroxide. *Chem. Commun.*, (7), 696–697.

161 Carrettin, S., McMorn, P., Johnston, P., Griffin, K., Kiely, C.J. and Hutchings, G.J. (2003) Oxidation of glycerol using supported Pt, Pd and Au catalysts. *Phys. Chem. Chem. Phys.*, **5** (6), 1329–1336.

162 Carrettin, S., McMorn, P., Johnston, P., Griffin, K., Kiely, C.J., Attard, G.A. and Hutchings, G.J. (2004) Oxidation of glycerol using supported gold catalysts. *Top. Catal.*, **27** (1–4), 131–136.

163 Comotti, M., Della Pina, C., Matarrese, R. and Rossi, M. (2004) The catalytic activity of "Naked" gold particles. *Angew. Chem. Int. Ed. Engl.*, **43** (43), 5812–5815.

164 Mertens, P.G.N., Bulut, M., Gevers, L.E.M., Vankelecom, I.F.J., Jacobs, P.A. and De Vos, D.E. (2005) Catalytic oxidation of 1,2-diols to alpha-hydroxy-carboxylates with stabilized gold nanocolloids combined with a membrane-based catalyst separation. *Catal. Lett.*, **102** (1–2), 57–61.

165 Mertens, P.G.N., Vankelecom, I.F.J., Jacobs, P.A. and De Vos, D.E. (2005) Gold nanoclusters as colloidal catalysts for oxidation of long chain aliphatic 1,2-diols in alcohol solvents. *Gold Bull.*, **38** (4), 157–162.

166 Tsunoyama, H., Sakurai, H., Negishi, Y. and Tsukuda, T. (2005) Size-specific catalytic activity of polymer-stabilized gold nanoclusters for aerobic alcohol oxidation in water. *J. Am. Chem. Soc.*, **127** (26), 9374–9375.

167 Abad, A., Concepcion, P., Corma, A. and Garcia, H. (2005) A collaborative effect between gold and a support induces the selective oxidation of alcohols. *Angew. Chem. Int. Ed. Engl.*, **44** (26), 4066–4069.

168 Corma, A. and Domine, M.E. (2005) Gold supported on a mesoporous CeO_2 matrix as an efficient catalyst in the selective aerobic oxidation of aldehydes in the liquid phase. *Chem. Commun.*, (32), 4042–4044.

169 Enache, D.I., Edwards, J.K., Landon, P., Solsona-Espriu, B., Carley, A.F., Herzing, A.A., Watanabe, M., Kiely, C.J., Knight, D.W. and Hutchings, G.J. (2006) Solvent-free oxidation of primary alcohols to aldehydes using Au-Pd/TiO_2 catalysts. *Science*, **311** (5759), 362–365.

170 Reuss, G., Disteldorf, D., Grundler, O. and Hilt, A. (1988) *"Formaldehyde"* in *Ullmann's Encyclopedia of Industrial Chemistry*, Vol. A11, 5th edn, VCH Verlag GmbH, Weinheim.

171 Gerberich, H.R. and Seaman, G.C. (1994) Formaldehyde. In *Kirk-Othmer Encyclopedia of Chemical Technology*, Vol. 11, 4th edn, John Wiley & Sons, Inc., New York.

172 Pestryakov, A.N., Petranovskii, V.P., Pfander, N. and Knop-Gericke, A. (2004) Supported foam-copper catalysts for methanol selective oxidation. *Catal. Commun.*, **5** (12), 777–781.

173 Dai, W.L., Yong, C., Ren, L.P., Yang, X.L., Xu, J.H., Li, H.X., He, H.Y. and Fan, K.N. (2004) Ag-SiO$_2$-Al$_2$O$_3$ composite as highly active catalyst for the formation of formaldehyde from the partial oxidation of methanol. *J. Catal.*, **228** (1), 80–91.

174 Pestryakov, A.N. (1996) Modification of silver catalysts for oxidation of methanol to formaldehyde. *Catal. Today*, **28** (3), 239–244.

175 Pestryakov, A., Davydov, A. and Tsyrulnikov, P. (1996) Role of electronic states of silver and copper catalysis in processes of selective or deep oxidation of alcohols and hydrocarbons. *Abst. Papers Am. Chem. Soc.*, **211**, 64-COLL.

176 Shen, J., Shan, W., Zhang, Y.H., Du, J.M., Xu, H.L., Fan, K.N., Shen, W. and Tang, Y. (2006) Gas-phase selective oxidation of alcohols: in situ electrolytic nano-silver/zeolite film/copper grid catalyst. *J. Catal.*, **237** (1), 94–101.

177 Besson, M. and Gallezot, P. (2000) Selective oxidation of alcohols and aldehydes on metal catalysts. *Catal. Today*, **57** (1–2), 127–141.

178 Shen, J., Shan, W., Zhang, Y.H., Du, J.M., Xu, H.L., Fan, K.N., Shen, W. and Tang, Y. (2004) A novel catalyst with high activity for polyhydric alcohol oxidation: nanosilver/zeolite film. *Chem. Commun.*, (24), 2880–2881.

179 Chaki, N.K., Tsunoyama, H., Negishi, Y., Sakurai, H. and Tsukuda, T. (2007) Effect of Ag-doping on the catalytic activity of polymer-stabilized Au clusters in aerobic oxidation of alcohol. *J. Phys. Chem. C*, **111** (13), 4885–4888.

180 Dao Jun Guo, H.L.L. (2005) *Carbon*, **43**, 1259–1264.

181 Landon, P., Collier, P.J., Papworth, A.J., Kiely, C.J. and Hutchings, G.J. (2002) Direct formation of hydrogen peroxide from H-2/O-2 using a gold catalyst. *Chem. Commun.*, (18), 2058–2059.

182 Landon, P., Collier, P.J., Carley, A.F., Chadwick, D., Papworth, A.J., Burrows, A., Kiely, C.J. and Hutchings, G.J. (2003) Direct synthesis of hydrogen peroxide from H-2 and O-2 using Pd and Au catalysts. *Phys. Chem. Chem. Phys.*, **5** (9), 1917–1923.

183 Ishihara, T., Ohura, Y., Yoshida, S., Hata, Y., Nishiguchi, H. and Takita, Y. (2005) Synthesis of hydrogen peroxide by direct oxidation of H-2 with O-2 on Au/SiO$_2$ catalyst. *Appl. Catal. A Gen.*, **291** (1–2), 215–221.

184 Edwards, J.K., Solsona, B.E., Landon, P., Carley, A.F., Herzing, A., Kiely, C.J. and Hutchings, G.J. (2005) Direct synthesis of hydrogen peroxide from H-2 and O-2 using TiO$_2$-supported Au-Pd catalysts. *J. Catal.*, **236** (1), 69–79.

185 Edwards, J.K., Solsona, B., Landon, P., Carley, A.F., Herzing, A., Watanabe, M., Kiely, C.J. and Hutchings, G.J. (2005) Direct synthesis of hydrogen peroxide from H-2 and O-2 using Au-Pd/Fe$_2$O$_3$ catalysts. *J. Mater. Chem.*, **15** (43), 4595–4600.

11
Self-Assembling Nanoclusters Based on Tetrahalometallate Anions: Electronic and Mechanical Behavior

Ishenkumba A. Kahwa

11.1
Introduction

Self-assembling systems are invaluable sources of novel materials with diverse architectures, morphologies, physical and chemical properties, as well as potential applications [1–9]. The relative simplicity of some preparative procedures, and the precision with which thermodynamic and kinetic factors operate to produce such materials, make self-assembly an intensely studied phenomenon [4, 7]. The major aim of the present author's research has been the preparation of novel multi-metal materials that could: (i) efficiently immobilize toxic metals from the environment [10]; (ii) trap and transport metal ions in human and animal bodies for diagnostic, imaging, or therapeutic benefits [11]; or (iii) simply act as dispersion media for metal ions in the efficient synthesis of multicomponent metal oxides with interesting properties, such as the superconducting oxides, $YBa_2Cu_3O_7$ [12]. These studies have covered self-assembling coordination compounds of multiple metal ions of the 3d-, 4f-, some p-, and nearly all of the s-block series [10–21]. This chapter provides a review of the synthesis of multinuclear self-assembling compounds facilitated by crown ethers, their structures, electronic behavior, and their thermally activated mechanical properties [10–21].

Interest in these compounds was inspired by the potential of crown ether chelates to bind metal ions such as M^+, M^{2+}, and M^{3+}, the charge of which can then be counterbalanced by, among others, negatively charged, metalloanions such as $[CuX_4]^{2-}$ and $[Cu_2X_6]^{2-}$ to produce mixed-metal compounds for the preparation of superconducting ceramic oxides such as $YBa_2Cu_3O_7$ and their derivatives [10]. For thallium-based superconducting ceramic oxide systems, the toxicity of thallium and ensuing environmental health concerns were important considerations [10]. Thus, as thallium-based superconducting ceramics demonstrated a superior potential based on a higher transition temperature (T_c) (e.g., T_c = 120–125 K and 100–105 K for $Tl_2Ba_2Ca_2Cu_3O_{10}$ and $Tl_2Ba_2CaCu_2O_8$, respectively) [22] compared to 90 K for $YBa_2Cu_3O_7$) [23], self-assembling systems that could immobilize large amounts of toxic thallium ions achieved greater significance.

Advanced Nanomaterials. Edited by Kurt E. Geckeler and Hiroyuki Nishide
Copyright © 2010 WILEY-VCH Verlag GmbH & Co. KGaA, Weinheim
ISBN: 978-3-527-31794-3

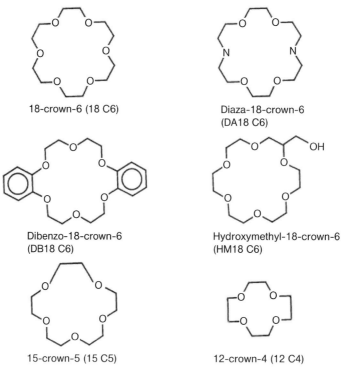

Scheme 11.1 Crown ethers used in these studies.

In this chapter, a review is provided of the studies conducted on the syntheses of metal-rich *Cubic F23* supramolecular complexes [(A(18-Crown-6))$_4$(MX$_4$)] [BX$_4$]$_2 \cdot n$H$_2$O, where A is a monovalent metal or NH$_4^+$ ion; M is a divalent metal ion (normally of 3d element); B is a trivalent metal ion (Tl^{3+} or Fe^{3+}); and X is a halogen (Cl or Br). The 18-crown-6 chelate (Scheme 11.1) in these and other complexes exhibits a thermally activated rotation conformation reorientation motion, which is of interest as a trigger for luminescence and magnetic on/off switching and mechanical nanodevices. The effectiveness of Mn^{2+} to serve as an efficient probe for a variety of coordination environments was also successfully explored, and the results are reported.

11.2
Preparation of Key Compounds

The general preparation of the [(A(18C6))$_4$(MX$_4$)] [BX$_4$]$_2 \cdot n$H$_2$O series has been reported previously [10–21]. The procedure involves a simple refluxing of chelate 18C6 in alcoholic solutions containing the corresponding quantities of metal salts (e.g., RbBr, MnBr$_2$, and TlBr$_3$ when the complex (Rb(18C6))$_4$(MnBr$_4$)][TlBr$_4$]$_2 \cdot n$H$_2$O is desired), and then evaporating off the excess solvent to deposit the crystalline

(1) $Tl^+ + 18\text{-crown-6} \rightleftharpoons [Tl(18\text{-crown-6})]^+$

(2) $[M(H_2O)_6]^{2+} + 4X^- \rightleftharpoons [MX_4]^{2-} + 6H_2O$

(3) $[Tl(18\text{-crown-6})]^+ + 4X^- \rightleftharpoons [TlX_4]^- + 18\text{-crown-6} + 2e^-$

(4) $4[Tl(18\text{-crown-6})]^+ + [MX_4]^{2-} \rightleftharpoons [(Tl(18\text{-crown-6}))_4MX_4]^{2+}$

(5) $[(Tl(18\text{-crown-6}))_4MX_4]^{2+} + 2[TlX_4]^- \rightleftharpoons [(Tl(18\text{-crown-6}))_4MX_4][TlX_4]_2$

Scheme 11.2 Pertinent reactions leading to supramolecular systems: $[(Tl(18\text{-Crown-6}))_4(CuCl_4)][TlCl_4]_2 \cdot nH_2O$. The solvent shells are removed, and reactions occur in air.

materials. For the complex $(Tl(18C6))_4(MX_4)][TlX_4]_2 \cdot nH_2O$, auto-oxidation of Tl^+ to Tl^{3+} or auto-reduction of Tl^{3+} to Tl^+ occurs in the refluxing mixture when the reaction vessels are open to the atmosphere (Scheme 11.2).

Hence, there is no need for both Tl^{3+} and Tl^+ to be added into the reaction mixture. If Fe^{3+} and Tl^+ are used, no auto-oxidation of Tl^+ is observed. Overall, compounds $[(A(18C6))_4\text{-}(MX_4)][BX_4]_2 \cdot nH_2O$ with A = Tl, Na, K, Rb, NH_4, BaX (X = halide or OH); M = Mn, Fe, Co, Ni, Cu, Zn, and B = Tl or Fe have been prepared. Similar results were recently obtained by another group, which demonstrated that for $M^{2+} = Fe^{2+}$ reducing conditions are required; otherwise, oxidation to Fe^{3+} occurs and the cubic compounds are not formed [24]. The crown ethers used in the study are shown in Scheme 11.1.

Changing the crown from 18C6 to DA18C6 or HM18C6 resulted in the formation of compounds similar to those of $[(Rb(18Crown)_4(MnX_4)][TlX_4]_2 \cdot nH_2O$, while DB18C6, 15C5 and 12C4 failed to produce the cubic series. However, the reaction of NaBr, 15C5, and $TlBr_3$ in the presence or absence of $[MX_4]^{2-}$ anions yielded an elegant self-assembling compound $[(Na(15C5))_4Br][TlBr_4]_3$ in which the Br^- anion played the role of concentrating $[Na(Crown)]^+$ cations in a manner similar to that of tetrahedral $[MX_4]^{2-}$ anions $[(A(18C6))_4(MX_4)][BX_4]_2 \cdot nH_2O$.

11.3
Structure of the $[(A(18C6))_4(MX_4)][BX_4]_2 \cdot nH_2O$ Complexes

The structure of compounds of the $[(A(18C6))_4(MX_4)][BX_4]_2 \cdot nH_2O$ is *Cubic F23*, with the $[A(16C6)]^+$ cation perched on the triangular surfaces of the $[MX_4]^{2-}$ anions, as shown in Figure 11.1 for one of the four such links. While the structures remained *Cubic F23* throughout the series $[(A(18C6))_4(MX_4)][BX_4]_2 \cdot nH_2O$, there were subtle differences in the orientation of the 18C6 in some cases, which are temperature-dependent. For instance, for the compound $[(Tl(18C6))_4(CuBr_4)][TlBr_4]_2 \cdot nH_2O$, the room temperature structure is similar to that of the Mn^{2+} analogue, but when it is cooled to 115 K a switch in the O and C positions was found to have occurred (Figure 11.1). The structures of many compounds were studied, and the switch from the low-temperature form (that of $[(Tl(18C6))_4(CuBr_4)][TlBr_4]_2$ at 115 K) to the high-temperature form (that of $[(Tl(18C6))_4(CuBr_4)][TlBr_4]_2$

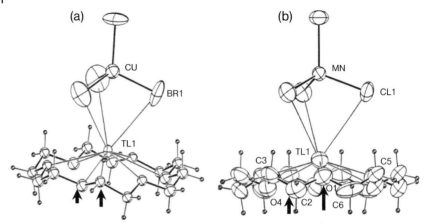

Figure 11.1 Partial structures of the [(Tl(18-Crown-6))$_4$(MX$_4$)]$^{2+}$ supramolecular cations, showing the two orientations. (a) $T = 115$ K, M = Cu, X = Br; (b) Room temperature, M = Mn, X = Cl. The difference is in the 30° rotation for 18C6 ligand resulting in carbon and oxygen switching positions (compare the positions marked by the arrows). Adopted from Ref. [15].

at room temperature) was found to depend on several factors, including the nature of the halide, the presence or absence of crystal water, and the temperature (Table 11.1) [16].

The extended cubic structure of members of the [(A(18C6))$_4$(MX$_4$)] [BX$_4$]$_2$ · nH$_2$O was recently [24] analyzed in detail (Figure 11.2a); the study revealed an M^{2+} nucleus trapped in a tetrahedron of four X$^-$ anions; the resultant [MX$_4$]$^{2-}$ is coordinated on each of its four triangular faces by [A(18C6)]$^+$ cations with resulting [(A(18C6))$_4$(MX$_4$)]$^{2+}$ cations secured inside a cavity made up of a cyclic adamantane-like network of ten [TX$_4$]$^-$ anions constituting six-member rings in chair conformations, as shown in Figure 11.2b [24].

11.4
Structure of the [(Na(15C5))$_4$Br] [TlBr$_4$]$_3$ Complex

Like the [(A(18C6))$_4$(MX$_4$)] [BX$_4$]$_2$ · nH$_2$O, the elegant structure of the [(Na(15C5))$_4$Br] [TlBr$_4$]$_3$ complex reveals that it is stabilized by multiple complexation networks, as shown in Figure 11.3 [18].

11.5
Spectroscopy of the *Cubic F23* [(A(18C6))$_4$(MX$_4$)] [BX$_4$]$_2$·nH$_2$O

The *Cubic F23* structure [(A(18C6))$_4$(MX$_4$)] [BX$_4$]$_2$ · nH$_2$O requires additional support because it requires the [MX$_4$]$^{2-}$ anion, including the [CuX$_4$]$^{2-}$, which is

Table 11.1 18-Crown-6 orientations for structures of [(A(18-Crown-6))₄(MX₄)][TlX₄]₂. The crown orientation is seen to depend on many factors.

A	M	X	T (K)	Orientation
Tl	Cu	Br	295	HT
Tl	Mn	Cl	295	HT
Tl	Cu	Cl	295	HT
K	Fe	Cl	293	HT
Tl	Cu	Br	115	LT
Tl	Cu	Cl (with 0.25 H₂O)	297	LT
K	Mn	Br	295	LT
K	Zn	Br	297	LT

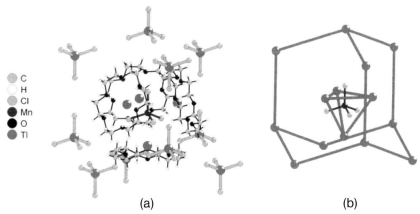

Figure 11.2 (a) The partial extended structure of [(A(18C6))₄(MX₄)] [BX₄]₂ · nH₂O members; (b) The network of ten [TX₄]⁻ anions in an adamantine-type arrangement [24].

highly susceptible to Jahn–Teller distortion [25–27], to occupy a 23T site of perfect T_d symmetry [10]. We thus sought to determine whether the A, M, and B sites were of the indicated oxidation states and coordination geometries by spectroscopic means to complement the X-ray structural evidence. The presence of mixed valence thallium centers in [(Tl(18C6))₄(MCl₄)] [TlCl₄]₂ · nH₂O was established using solid-state nuclear magnetic resonance (NMR) of ^{205}Tl (Figure 11.4), which showed Tl³⁺ with chemical shifts of approximately 2900 ppm relative to Tl(NO₃), while that of Tl⁺ occurred at −782 ppm and −117 ppm for the paramagnetic [(Tl(18C6))₄(CuCl₄)] [TlCl₄]₂ · nH₂O and diamagnetic [(Tl(18C6))₄(ZnCl₄)] [TlCl₄]₂ · nH₂O compounds, respectively. The larger chemical shift and broader resonance of Tl⁺ in [(Tl(18-Crown-6))₄(CuCl₄)] [TlCl₄]₂ · nH₂O, relative to that of [(Tl(18-Crown-6))₄(ZnCl₄)] [TlCl₄]₂, being indicative of the cation's proximity to the paramagnetic Cu²⁺ center.

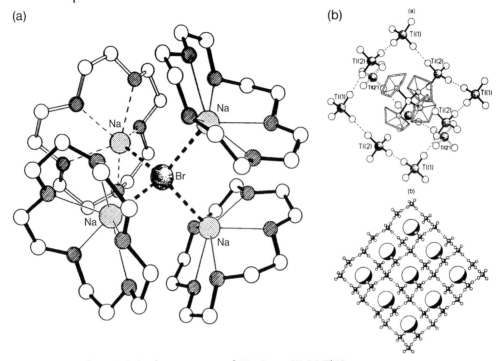

Figure 11.3 An elegant structure of [(Na-Crown-5))$_4$Br] [TlX$_4$]$_3$. (a) The cation [(Na-Crown-5))$_4$Br]$^{3+}$; (b) Upper: the cation in a network of [TlBr$_4$]$^-$ anions. Lower: one layer of stacking egg-tray networks of [TlBr$_4$]$^-$ anions and their [(Na-Crown-5))$_4$Br]$^{3+}$ guest, shown as a partially shaded ball.

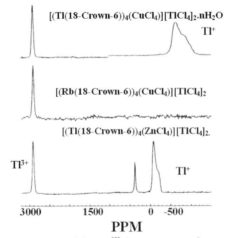

Figure 11.4 Solid-state ^{205}Tl NMR spectra for some members of the [(A(18C6))$_4$(MCl$_4$)] [BCl$_4$]$_2 \cdot n$H$_2$O series, showing the presence of the thallium(I) (Tl$^+$) and thallium(III) (Tl^{3+}) sites [10].

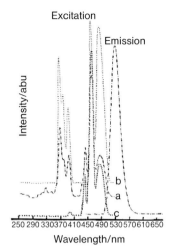

Figure 11.5 Emission and excitation spectra of [(Rb(18C6))$_4$(MnCl$_4$)$_{1-x}$(CuCl$_4$)$_x$] [(TlCl$_4$)$_{1-y}$(FeCl$_4$)$_y$]$_2 \cdot n$H$_2$O. Spectrum a: $x = 0$, $y = 0$; spectrum b: $x = 0.056$, $y = 0$; spectrum c: $x = 0$, $y > 0$) [14].

While the structural details of the copper complexes [(A(18C6))$_4$(MX$_4$)] [BX$_4$]$_2 \cdot n$H$_2$O, M = Cu, were similar to those of M = Co, Mn, Fe, Ni and Zn, it was necessary to determine whether the 23T site is really T_d or is an average of various geometries at room temperature. The deep-blue Co^{2+} cubic compound [(Tl(18C6))$_4$(CoCl$_4$)] [TlCl$_4$]$_2 \cdot n$H$_2$O is consistent with Co^{2+} being in a T_d environment.

The manganese (II) emission from the [(A(18C6))$_4$(MnCl$_4$)] [TlCl$_4$]$_2 \cdot n$H$_2$O (A = Tl or Rb) compounds peaked at about 535 nm, as is typical of the tetrahedral [MnCl$_4$]$^{2-}$ anions [14]; the corresponding excitation spectra were also characteristic of T_d [MnCl$_4$]$^{2-}$ anions (Figure 11.5 for A = Rb). The excitation spectrum of [MnCl$_4$]$^{2-}$ emission in [(Rb(18C6))$_4$(MnCl$_4$)] [TlCl$_4$]$_2 \cdot n$H$_2$O (Figure 11.5a) is attributed to electronic transitions from the $^6A_1$ ground state to accessible higher-energy quartets derived from the crystal field splitting of ^{4}G, ^{4}D, ^{4}P, and ^{4}F levels. Emission spectra from compounds with Cu^{2+} and Fe^{3+} were also typical of $^4T_1(^4G) \rightarrow {}^6A_1$ emission for [MnCl$_4$]$^{2-}$ ions. However, the inner-filter due to charge transfer bands of Cu^{2+} and Fe^{3+} led to significant changes on the excitation spectra (Figure 11.5).

Thus, as required by *Cubic F23* symmetry, the 23T site occupied by the [MX$_4$]$^{2-}$ anions in crystalline [(A(18C6))$_4$(MX$_4$)] [BX$_4$]$_2 \cdot n$H$_2$O compounds is indeed of T_d character. However, the Cu^{2+} charge transfer bands exhibited by the [(A(18C6))$_4$(CuCl$_4$)] [TlCl$_4$]$_2 \cdot n$H$_2$O compounds at 27 000 to 30 000 cm^{-1} (^{2}E ← ^{2}B$_2$) and 21 000 to 23 000 cm^{-1} (^{2}A$_2$ ← ^{2}B$_2$ and ^{2}E ← ^{2}B$_2$) are in regions generally expected of D_{2d} [CuCl$_4$]$^{2-}$ [10]. This ion may be distorted, but this requires additional proof.

11.6
Unusual Luminescence Spectroscopy of Some Cubic [(A(18C6))$_4$(MnX$_4$)] [TlCl$_4$]$_2$·nH$_2$O Compounds

Whereas, the spectral profiles of [(A(18C6))$_4$(MnCl$_4$)] [TlCl$_4$]$_2$ · nH$_2$O (A = Tl or Rb) compounds (Figure 11.5) were typical of Mn^{2+} in T_d environments, the spectra of several other [(A(18C6))$_4$(MnX$_4$)] [TlX$_4$]$_2$ · nH$_2$O compounds exhibited very interesting dynamics and features that depended on the nature of the halide X$^-$ and the large cation A. The spectra of the [(A(18C6))$_4$(MnBr$_4$)] [TlBr$_4$]$_2$ · nH$_2$O compounds with A = K and Rb (Figure 11.6) show the excitation spectra being similar. However, the emission spectrum of rubidium (A = Rb) features a green emission peaking at 535 nm, as is normal for Mn^{2+} in T_d environments, while the potassium (A = K) compound exhibits an unusual orange emission peaking at 575 nm, with the normal green emission being too weak to observe.

The emission spectra indeed do depend on the nature of the A cation, as shown in Figure 11.7, where a progressive spectral shift from the 535 nm to 575 nm emission is observed as the concentration of K$^+$ ions (x) increases from 0 (pure Rb complex) to 1 (pure K complex).

Most interestingly, the ammonium compound [(NH$_4$(18C6))$_4$(MnCl$_4$)] [TlCl$_4$]$_2$ · nH$_2$O, which features temperature-independent spectra, shows both emissions. The normal emission for Mn^{2+} in T_d environments is seen at room temperature, while the strange orange emission dominates at 77 K (Figure 11.8a).

The emission from the bromide complex [(NH$_4$(18C6))$_4$(MnBr$_4$)] [TlBr$_4$]$_2$ · nH$_2$O was normal, but the excitation spectrum featured an unusual broad absorption at 330 nm that was attributed to defects [21]. This led to the conclusion that the

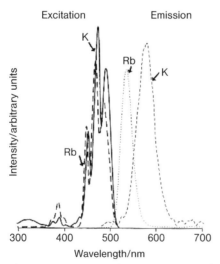

Figure 11.6 Emission and excitation spectra of [(A(18C6))$_4$(MnBr$_4$)] [TlBr$_4$]$_2$·nH$_2$O (A = K or Rb) [16].

11.6 Unusual Spectroscopy of $[(A(18C6))_4(MnX_4)][TlCl_4]_2 \cdot nH_2O$

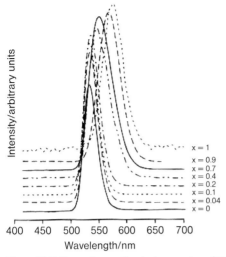

Figure 11.7 Dependence of emission spectra of $[(Rb_{1-x}K_x(18C6))_4(MnBr_4)][TlBr_4]_2 \cdot nH_2O$ on the molar fraction of K [16].

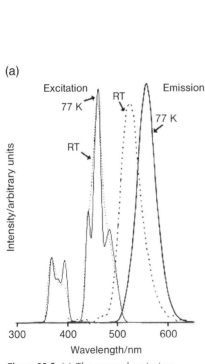

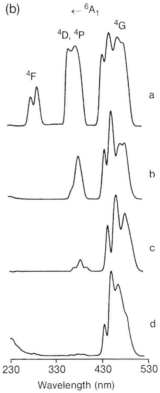

Figure 11.8 (a) The unusual emission behavior of the $[(NH_4(18C6))_4(MnBr_4)][TlBr_4]_2 \cdot nH_2O$ complex [16]; (b) Compared to $[(NH_4(18C6))_2(MnBr_4)]$ (spectrum a), the inner-filter effect of defect sites on the excitation spectral profile of [21]: spectrum b = $(Rb(HM18C6))_4(MnBr_4)][TlBr_4]_2 \cdot nH_2O$; spectrum c = $[(K(18C6))_4(MnBr_4)][TlBr_4]_2 \cdot nH_2O$; spectrum d = $[(Ba(HM18C6))_4(MnBr_4)][TlBr_4]_2 \cdot nH_2O$.

strange orange emission and broad UV absorptions were associated with defect sites. The electronic states of defect sites yielding orange emission are capable of quenching the normal $[MnX_4]^{2-}$ green emission, especially at 77 K. At room temperature, back energy transfer from the defect site to the normal Mn^{2+} seems to be dominant in the case of the $[(NH_4(18C6))_4(MnBr_4)] [TlBr_4]_2 \cdot nH_2O$ compound. Indeed, these defect sites were traced through a variety of compounds of $[MnX_4]^{2-}$ where their ability to act as inner-filters for $[MnBr_4]^{2-}$ UV absorptions is evident (Figure 11.8a).

11.7
Luminescence Decay Dynamics and 18C6 Rotations

The above temperature-dependent structural and luminescence behavior prompted the study of the temperature evolution of luminescence decay dynamics of the $[(A(18C6))_4(MnX_4)] [BX_4]_2 \cdot nH_2O$ compounds. As these fascinating decay dynamics have been investigated and reported in detail elsewhere [14, 16, 21], at this point only the main aspects of our observations and conclusions will be presented.

The luminescence decay rates of $[(A(18C6))_4(MnX_4)] [BX_4]_2 \cdot nH_2O$ compounds are influenced by the nature of the halogen X and metal A for spin–orbit coupling reasons, and also the nature of defect sites, their proximity to the Mn^{2+} site, and the efficiency with which they interact with the emitting $^4T_1(^4G) \to {}^6A_1$ process of Mn^{2+}. Whilst the decay rates change with the composition of the manganese(II) compound, the general features are a temperature-independent component up to approximately 200 K, followed by a rapid decay which involves energy migration over the Mn^{2+} sublattice. The Arrhenius plots (Figure 11.9) show biphasic emission quenching processes, with activation energies of 8–14 kJ mol^{-1} at the start of the emission quenching, followed by a more rapid quenching process with a thermal barrier of 30–50 kJ mol^{-1} [14, 16, 21].

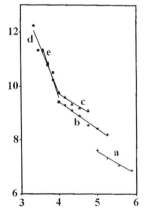

Figure 11.9 Arrhenius plots (ln k_t versus $1/T$) for $170 < T < 300$ K for $[(NH_4(18C6))_4(MnCl_4)]\text{-}[TlCl_4]_2 \cdot nH_2O$. Normal $[MnCl_4]^{2+}$ emission monitored at 510 nm: plots b = 8 kJ mol^{-1}) and e = 35 kJ mol^{-1}. Unusual emission monitored at 590 nm: plot a = 7 kJ mol^{-1}; plot c = 8 kJ mol^{-1}; plot d = 29 kJ mol^{-1}. E (kJ mol^{-1}) ≈ 0.155 T_c (where T_c = transition temperature, i.e., ≈210 K from the above luminescence studies).

The thermal barrier of 9–14 kJ mol^{-1} was attributed to the energy required to bridge the gap between the Stokes shifted exciton-donating $^4T_1(^4G)$ state of [MnX$_4$]$^{2-}$ anions and the energy-accepting $^4T_1(^4G)$ levels of the [MnX$_4$]$^{2-}$ neighbors, this being a key requirement for the energy migration process over the [MnCl$_4$]$^{2+}$ sublattice. The difference between the absorption and emission energies was about 1060 cm^{-1} (or 12.7 kJ mol^{-1}), which was in good agreement with the measured thermal barriers of 9–14 kJ mol^{-1}. The higher thermal barrier of 30–50 kJ mol^{-1} was attributed to the energy required for the onset of the 18C6 rotation–conformation reorientation. These reorientation motions can generate a dynamic electric field at the Mn^{2+} centers, thereby enhancing the electronic transition involved in the energy transfer process from Mn^{2+} to defect sites. The measured thermal barriers of 30–50 kJ mol^{-1} were in good agreement with the Waugh–Fedin approximation for hindered solid-state orientations within the assumptions made in deriving the relationship [16, 25].

The cross-polarization-magic angle spinning (CP-MAS) ^{13}C NMR spectrum of the [(K(18C6))$_4$(ZnBr$_4$)] [TlBr$_4$]$_2$ · nH$_2$O compound showed one –CH$_2$– resonance instead of the pair expected to result from the differential shielding of the up-and-down –CH$_2$– functionalities of 18C6 in D$_{3d}$ symmetry [16]. The single resonance is consistent with –CH$_2$– positions being equilibrated by the rotation–conformational reorientation of the 18C6 ring.

This conclusion is supported by a large body of solid-state NMR data on the compounds of 18C6 and other crown ether ligands in a variety of compounds, and is consistent with the merry-go-round 18C6 motion proposed by Buchanan, Ratcliff *et al.* [26] and supported by Dye *et al.* [27], to account for unusual solid-state NMR behavior. Recent detailed magnetic susceptibility, solid-state NMR and heat capacity measurements on 18C6 sandwich complexes of cesium, [(18C6)Cs(18C6)Cs(18C6)]$^+$ with magnetic [Ni(dmit$_2$)]$^-$ identified 18C6 in these compounds as a rotor (Figure 11.10), the onset motion of which has influence on the physical characteristics of the compounds, including the magnetic behavior of the [Ni(dmit$_2$)]$^-$ anion [28, 29] (Figure 11.9).

11.8
Conclusions

The [(A(18C6))$_4$(MX$_4$)] [BX$_4$]$_2$ · nH$_2$O supramolecular systems have the interesting properties of concentrating and immobilizing toxic metal ions such as thallium(I/III), and are accessible via a relatively simple synthetic procedure. The [(A(18C6))$_4$(MnX$_4$)] [BX$_4$]$_2$ · nH$_2$O compounds hold a total of six thallium ions – four as [Tl(18C6)]$^+$ and two as [TlX$_4$]$^-$. The structural, luminescence, and CP-MAS ^{13}C NMR spectroscopic evidence for 18C6 mechanical activity in the [(A(18C6))$_4$(MnX$_4$)] [BX$_4$]$_2$ · nH$_2$O compounds is compelling. Indeed, when taken together with published reports on heat capacity, magnetic susceptibility and NMR evidence for such mechanical activity in a variety of crown ether compounds, the possibility of developing potentially useful nanodevices becomes of major interest. For example, the

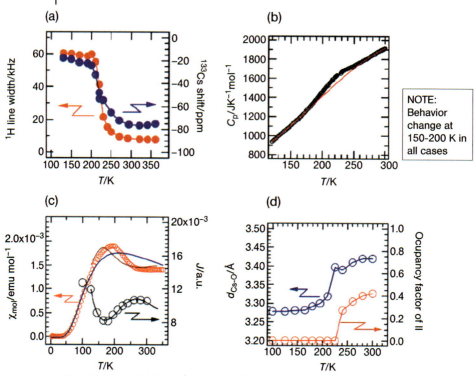

Figure 11.10 (a) Solid-state ^{1}H NMR, (b) heat capacity, (c) magnetic susceptibility, and (d) crystal structural parameters of compounds of [(18C6)Cs(18C6)Cs(18C6)]$^+$ with magnetic [Ni(dmit$_2$)]$^-$ anions, showing behavioral changes at approximately 150–200 K in all cases [28].

changes in physical behavior associated with the onset of rotation–conformation reorientation motion could be used in the form of switches. Indeed, new opportunities for fabricating mechanical nanodevices in which the 18C6 chelate performs useful work as it rotates with –CH$_2$– and –O– moieties flipping up and down can be envisaged. In this regard, it is noteworthy that the behavior of compounds of HM18C6 and 18C6 were similar, as studying the behavior of other substituted 18C6 molecules is important in the search for nanodevices based on the 18C6 motion.

The remarkable sensitivity of Mn^{2+} emission in detecting molecular motion within its environments is also of great interest. It is possible that the defect sites which dominated the emission behavior of [(A(18C6))$_4$(MnX$_4$)] [BX$_4$]$_2 \cdot n$H$_2$O compounds were of manganese(II), in unusual coordination environments. The sensitivity of Mn^{2+} to changing coordination and supramolecular environments was convincingly demonstrated by comparing the luminescence spectroscopy (Figure 11.11) and decay dynamics of compounds featuring Mn^{2+} in coordination numbers 4, 6, 7, and 8. The potential of Mn^{2+} as a luminescence probe for structural and supramolecular environmental deserves further development.

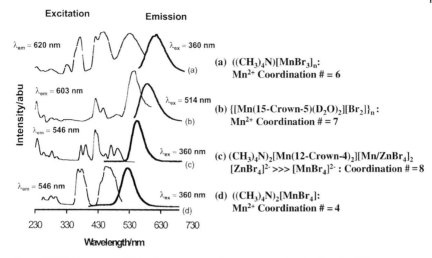

Figure 11.11 Manganese(II) luminescence spectroscopy as a structural probe [19].

Acknowledgments

Contributions made by coauthors of the original publications are greatly appreciated, and especially those of Drs H.O. Reid and N.S. Fender as former students. These studies were funded by the UWI, the Inter-American Development Bank (IADB), and the US NSF.

References

1 Guan, Z. (2007) *Polym. Int.*, **56**, 467.
2 Lehn, J.-M. (2007) *Chem. Soc. Rev.*, **36**, 151.
3 Loeb, S.J. (2007) *Chem. Soc. Rev.*, **36**, 226.
4 Gale, P.A. (ed.) (2007) *Chem. Soc. Rev., Supramolecular Anniversary Issue*, **36**, 125.
5 Yang, Z. and Xu, B. (2007) *J. Mater. Chem.*, **17**, 2385.
6 Schneider, H.-J. and Yatsimirsky, A.K. (2008) *Chem. Soc. Rev.*, **37**, 263.
7 Monti, S. and De Schryver, F. (eds) (2007) *Photochem. Photobiol. Sci., Special Issue*, **6**, 333.
8 Miyata, M., Tohnai, N. and Hisaki, I. (2007) *Acc. Chem. Res.*, **40**, 694.
9 Flood, A.H., Ramirez, R.J.A., Deng, W.-Q., Muller, R.P., Goddard, W.A. and Stoddart, J.F. (2004) *Aust. J. Chem.*, **57**, 301.
10 Kahwa, I.A., Miller, D., Mitchell, M., Fronczek, F.R., Goodrich, R.G., Williams, D.J., O'Mahoney, C.A., Slawin, A.M.Z., Ley, S.V. and Groombridge, C.J. (1992) *Inorg. Chem.*, **31**, 3963.
11 Kahwa, I.A., Folkes, S., Williams, D.J., Ley, S.V., O'Mahoney, C.A. and McPherson, G.L. (1989) *J. Chem. Soc., Chem. Commun.*, 1531.
12 Kahwa, I.A. and Goodrich, R.G. (1989) *J. Mater. Sci. Lett.*, **8**, 755.
13 Kahwa, I.A., Miller, D., Mitchell, M. and Fronczek, F.R. (1993) *Acta Crystallogr. Sect. C, Cryst. Struct. Commun.*, **49**, 320–321.
14 Fairman, R.A., Gallimore, W.A., Spence, K.V.N. and Kahwa, I.A. (1994) *Inorg. Chem.*, **33**, 823–828.
15 Fender, N.S., Finnegan, S.A., Miller, D., Mitchell, M., Kahwa, I.A. and Fronczek,

F.R. (1994) *Inorg. Chem.*, **33**, 4002–4008.

16 Fender, N.F., Fronczek, F.R., John, V., Kahwa, I.A. and McPherson, G.L. (1997) *Inorg. Chem.*, **37**, 5539–5547.

17 Reid, H.O.N., Kahwa, I.A., White, A.J.P. and Williams, D.J. (1998) *Inorg. Chem.*, **37**, 3868–3873.

18 Fender, N.S., Kahwa, I.A., White, A.J.P. and Williams, D.J. (1998) *J. Chem. Soc., Dalton. Trans.*, 1729–1730.

19 Reid, H.O.N., Kahwa, I.A., White, A.J.P. and Williams, D.J. (1999) *Chem. Commun.*, 1565–1566.

20 Reid, H.O.N., Kahwa, I.A., Mague, J.T. and McPherson, G.L. (2001) *Acta Crystallogr.*, **E57**, m3–m4.

21 Fender, N.S., Kahwa, I.A. and Fronczek, F.R. (2002) *J. Solid State Chem.*, **163**, 286–293.

22 Subramanian, M.A., Calabresse, J., Toradi, C.C., Askew, T.R., Flippen, R.B., Morrissey, K.J., Chahudhuri, U. and Sleight, A.W. (1988) *Nature*, **332**, 420.

23 Wu, M.K., Ashburn, J.R., Tomg, C.J., Hor, P.H., Meng, R.L., GAo, L., Huang, Z.J., Wang, Y.Q. and Chu, C.W. (1987) *Phys. Rev. Lett.*, **58**, 908.

24 Rieger, S.F. and Mudring, A.-V. (2005) *Inorg. Chem.*, **44**, 9340.

25 Waugh, J.S. and Fedin, É.I. (1963) *Sov. Phys. Solid State*, **4**, 1633.

26 Ratcliffe, C.I., Ripmester, J.A., Buchanan, G.W. and Denike, J.K. (1992) *J. Am. Chem. Soc.*, **114**, 3294 and references therein.

27 Wagner, M.J., McMills, L.E.H., Ellaboudy, A.S., Elgin, J.L., Dye, J.L., Edward, P.P. and Pyper, N.C. (1992) *J. Phys. Chem.*, **96**, 9656 and references therein.

28 Nishihara, S., Akutagawa, T., Sato, D., Takeda, S., Noro, S.-I. and Nakamura, T. (2005) *J. Am. Chem. Soc.*, **127**, 4397.

29 Takeda, S., Akutagawa, T., Nishihara, S., Nakamura, T. and Saito, K. (2005) *J. Chem. Phys.*, **123**, 044514.

12
Optically Responsive Polymer Nanocomposites Containing Organic Functional Chromophores and Metal Nanostructures

Andrea Pucci, Giacomo Ruggeri, and Francesco Ciardelli

12.1
Introduction

Responsive "smart" materials are certainly one of the most intriguing research areas in modern polymer science and technology, as molecularly designed materials based on macromolecules offer unique opportunities in this connection. Indeed, macromolecules are able to transmit and amplify small signals through involvement of the whole chain, conferring to the material a change in properties of various level and type. We can learn from Nature how these effects may be obtained according to two distinct routes, based either on the covalent bonding of highly responsive molecular species to the chains, or on the nanodispersion of such responsive species within the polymer bulk. In the latter case, those materials based on macromolecules may also affect the behavior of guest low-molecular-weight molecules or noble metal assemblies. In addition to the molecular features of both host and guest, the supramolecular arrangement of the guest can be modulated by external events on the host material. These effects can be identified and conveniently used when easily detected and field-sensitive species are present in the low-molecular-weight component. In this chapter, we review those investigations conducted not only in our laboratory but also by others, that have provided clear examples of the concepts which drive the original idea. In particular, we report our data relating to the effects of external stimuli (i.e., mechanical stretching such as polymer drawing, temperature, and pressure) on the optical properties (absorption and emission) of (nano)composite materials. The (nano)dispersion of active dyes and metal nanoparticles (guest) in inert polymers (host) will then be presented as an example of the effect of the macromolecular environment on respectively the dye or metal atom aggregates (metal nanoparticles), in terms of induced optical properties.

12.2
Organic Chromophores as the Dispersed Phase

12.2.1
Nature of the Organic Dye

Highly sensitive optical techniques based on the luminescence of conjugated aromatic molecular additives dispersed at low concentration (e.g., in Scheme 12.2, less than 1–2 wt%) into the amorphous phase of thermoplastic polymers have been successfully applied to the detection of thermal and mechanical solicitations on plastic films. In fact, a growing interest was devoted to the optical characteristics of macromolecules due to the sensitive response of the photophysical techniques for the study of the dynamic physical properties of macromolecules (i.e., energy transfer, polarization, and trapping phenomena). For example, when luminescent dyes are incorporated into polymers as thermodynamically stable micro-/nano-sized aggregates of a few molecules, emission characteristics are observed which derive mainly from the fluorescence of interacting chromophores through π–π stacking interactions among the planar aromatic backbones. Interactions between the excited state of an aromatic molecule and the ground state of the same molecule actually give rise to the excimer (excited dimer) formation, which represent a powerful diagnostic tool for interacting chromophores (Scheme 12.1) [1–3].

For small molecules separations ($r < 4$ Å), with interactions between ground-state and excited-state molecules, an attractive potential may be obtained due to the configurational interaction between resonance and exciton–resonance states. According to Scheme 12.1, fluorescence from the excimer state will thus be unstructured and at a lower energy than the corresponding monomer emission [2].

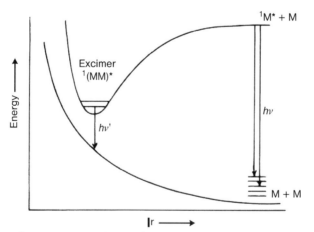

Scheme 12.1 Energy diagram for excimer formation.
M = molecule in the ground state; M* = molecule in the excited state; r = distance between molecules.

Comparisons between the signal intensities of monomeric fluorescence (defined as the contribution of an isolated chromophore covalently attached to the polymer chain or dispersed herein) and the excimer contribution are used very efficiently to obtain accurate information not only on polymer structure and conformation [4–9] but also on mixing at the molecular level in polymer blends [9–16]. Excimer formation in polymer solutions is widely considered to be a diffusion-controlled process, and is influenced by a variety of factors including the solvent, the chromophore microstructure, and the macromolecular conformation [2, 17, 18]. In contrast, in polymer films the formation of aggregates or excimers (static excimers) arises from the structural constraints of the polymer chains. In this case, the static excimers react very sensitively to chromophore aggregation and, as might be expected, also to any variations in the spatial distribution and alignment of molecules within the local environment [19–23].

12.2.2
Polymeric Indicators to Mechanical Stress

12.2.2.1 Oligo(p-Phenylene Vinylene) as Luminescent Dyes

Recently, the possibility of applying the formation of excimers inside polymer matrices, in order to prepare polymers with "built-in" stress–strain deformation sensors, has been effectively demonstrated [24–27]. Weder et al. reported that small amounts (0.01–3 wt%) of excimer-forming oligo(p-phenylene vinylene) synthetic chromophores (Scheme 12.2, **3**) dispersed into a ductile host polymer matrix (e.g., linear low-density polyethylene, LLDPE) as very small (nano)aggregates of dyes may be produced either by guest diffusion or by processing the components in the melt. The authors demonstrated that the phase behavior of the LLDPE/dye blends depended strictly on the supramolecular structure of the dye, which may be easily tuned as a function of the processing conditions and chromophore concentration. On applying a mechanical deformation of the film, a shear-induced mixing between the two phases promoted the break up of the dye's supramolecular

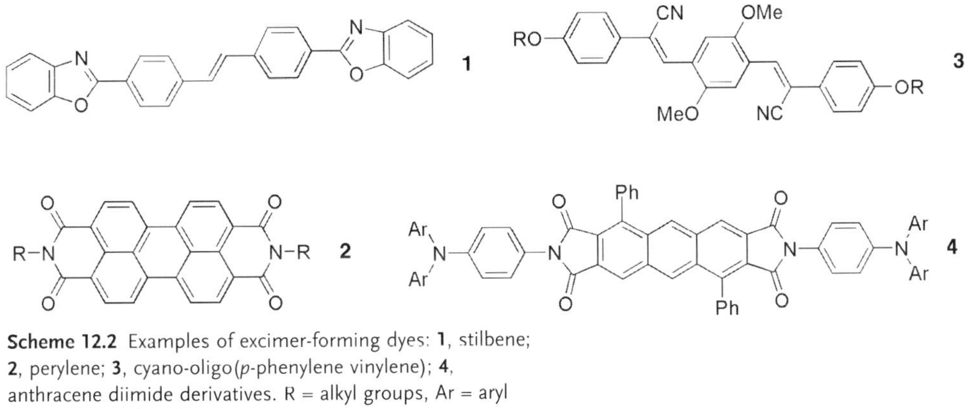

Scheme 12.2 Examples of excimer-forming dyes: **1**, stilbene; **2**, perylene; **3**, cyano-oligo(p-phenylene vinylene); **4**, anthracene diimide derivatives. R = alkyl groups, Ar = aryl groups; Ph = Phenyl.

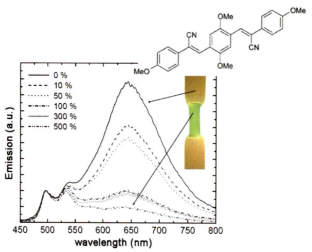

Figure 12.1 Emission spectra of a 0.20 wt% linear low-density polyethylene/1,4-bis(α-cyano-4-methoxystyryl)-2,5-dimethoxybenzene (inset) blend as a function of draw ratio, and image of the same blend after orientation (inset, draw ratio = 500%, $\lambda_{exc.}$ = 365 nm). Adapted with permission from Ref. [27]; © 2003, American Chemical Society.

structure, and led to a mixing of the two components (polymer and dye) and a change in the material's emission properties, from orange-red excimers to the molecularly dissolved green monomers of the dye (Figure 12.1) [26, 27].

According to Weder et al. [28], the extent of the color change observed upon deformation, and thus the ability of the polymer host to break up the dye aggregates, is influenced primarily by three parameters: dye aggregate size; polymer crystallinity; and strain rate. The nucleation rate of dye aggregates, and consequently the size of the aggregates, can be controlled via the structure of the groups attached to the dye. That is, long aliphatic alkyl groups lead to much higher nucleation rates, which in turn promotes the formation of small aggregates. Most importantly, the mechanically induced dispersion of the excimer-forming sensor molecules upon deformation is considered to be related to the plastic deformation process of the polyethylene (PE) crystallites, and thus increases with increasing polymer crystallinity.

In order to determine whether the sensing scheme could be exploited also for elastomers (which change their luminescent color reversibly as a function of the applied strain), Weder et al. recently investigated different polyurethanes that comprised cyano-oligo(p-phenylene vinylene)s (cyano-OPVs) as built-in deformation sensors [29]. The covalent integration of cyano-OPVs into the backbone of the polyurethane led to materials in which large-scale phase separation of the dye was prevented, while the formation of a large population of (presumably very small) ground-state dye aggregates was still allowed. Consequently, these polymers

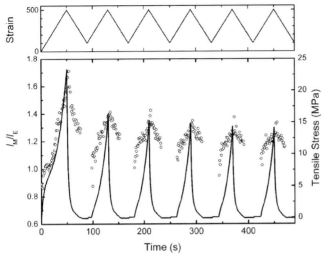

Figure 12.2 Ratio of monomer to excimer emission, I_M/I_E (circles, measured at 540 and 650 nm), and tensile stress (solid line) for a covalented incorporation of cyano-oligo(p-phenylene vinylene)s in polyurethane under a triangular strain cycle between 100% and 500% at a frequency of 0.0125 Hz. Reproduced with permission from Ref. [29]; © 2006, American Chemical Society.

displayed a predominantly excimer emission in the unstretched state, and exhibited a pronounced fluorescence color change upon deformation. The optical change appeared mostly reversible, and reflected the stress–strain behavior of the polymeric material (Figure 12.2).

Recently, properly substituted cyano-OPVs were also discovered as a new class of piezochromic material [30] Liquid-crystalline cyano-OPV dyes showed emission properties which could be switched reversibly and repeatedly from monomer to excimer fluorescence upon compression or quenching of the compounds from a nematic state. In particular, if the powder was either briefly compressed (1 min at 1500 p.s.i. in an IR pellet press) or ground using a mortar and pestle, then a significant bathochromic shift (≥50 nm) occurred and the emission band was seen to broaden. Once compressed, the "excimer form" (turquoise emission) was found to be stable for months if stored under ambient conditions. Rapid heating (ca. 3 min) of the compressed material to 130 °C fully restored the original "monomer" form (green emission).

12.2.2.2 Bis(Benzoxazolyl) Stilbene as a Luminescent Dye

A similar approach to the preparation of stress–strain polymeric indicators was reported by Pucci et al., who developed polypropylene (PP) films that contained different concentrations of the food-grade, luminescent dye bis(benzoxazolyl) stilbene (BBS) (Scheme 12.2, **1**) [31]. Notably, it was clear that the emission

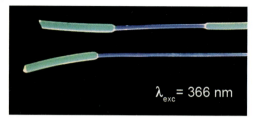

Figure 12.3 Partially stretched linear low-density polyethylene rods prepared in a corotating twin screw extruder (THERMO, Polylab) at a fixed barrel temperature of 180 °C for first zone and 200 °C for the following zones, and with screw speed fixed at 50 r.p.m. with increasing (from top to bottom) amounts of BBS (from 0.2 to 0.5 wt%), examined under UV light.

characteristics of the PP films depended on the BBS concentration and the polymer deformation. A well-defined excimer band was observed with more than 0.2 wt% of BBS, and this conferred to the film a green luminescence. During drawing (130 °C), the PP reorganization caused a break in the BBS excimer-type arrangement, leading to a prevalent blue emission of the single molecules.

In a subsequent phase, the same formulation was obtained by continuous compounding of the components mixture in a co-rotating, twin-screw extruder with a screw diameter of 24 mm and a length-to-diameter (L/D) ratio of 40 [32]. The films obtained through off-line melt-compression molding confirmed the optical responsive properties demonstrated by laboratory-scale, batch-compounded materials (Figure 12.3).

Intelligent films from thermoplastic materials, based on excimer luminescence and responsiveness to mechanical stress, were also obtained through the dispersion (melt-processing) of moderate amounts (0.02–0.2 wt%) of BBS into a thermoplastic, aliphatic, biodegradable polyester [poly(1,4-butylene succinate), PBS] (Figure 12.4) [33].

As reported for PP blends, the PBS morphology reorganization occurred during the drawing process, thus breaking the BBS excimer-type arrangement and leading to a prevalence of blue emission for the single molecules.

12.2.2.3 Perylene Derivatives as Luminescent Dyes

Very recently, analogous thermoplastic films sensitive to mechanical stress were prepared by using a different luminescent probe based on the perylene core (Scheme 12.2, **2**) [34]. Perylene dyes containing different peripheral alkyl chains were synthesized and efficiently dispersed at low loadings (from 0.01 to 0.1 wt%) into LLDPE by processing in the melt. Both perylene bisimides were found to generate supramolecular aggregates promoted by π–π intermolecular interactions between the conjugated planar structure of the dyes, as indicated by spectroscopic investigations on heptane solutions or dispersions into the LLDPE polymer matrix. The occurrence of this phenomenon effectively changed the emission of the dyes from yellow-green (noninteracting dyes) to red (interacting dyes). In particular,

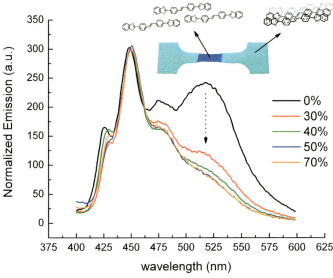

Figure 12.4 Fluorescence emission spectra ($\lambda_{exc.}$ = 277 nm) of 0.1 wt% poly(1,4-butylene succinate)/bis(benzoxazolyl)stilbene film (PBSBBS-0.1), before (0%) and after solid-state drawing (from 30 to 70% elongation). The spectra are normalized to the intensity of the isolated BBS molecules peak (430 nm). Inset: digital image of the uniaxially oriented PBSBBS film containing the 0.1 wt% of BBS molecules, recorded under excitation at 366 nm (50% elongation).

the data (provided by optical experiments and by quantum-mechanical calculations) acquired for dispersion of the dyes into the polymer matrix, revealed that the optical properties and responsiveness to mechanical stimuli were heavily dependent on the compactness of the perylene aggregates provided by the different molecular structures of the dyes.

12.2.3
Polymeric Indicators to Thermal Stress

12.2.3.1 Oligo(p-Phenylene Vinylene) as Luminescent Dyes

Another sensing mechanism may be produced, in contrast, by mixing the cyano-OPVs dyes (or other suitable excimer-forming dyes) with a macromolecular system characterized by a glass transition temperature (T_g) which is above the operating temperature. Hence, molecularly mixed, glassy composites can readily be produced via melt processing and rapid quenching of the melts. In this case, the sensor dyes were kinetically trapped inside the glassy amorphous phase of the host polymers, namely poly(methyl methacrylate) (PMMA) and poly(bisphenyl A carbonate) (PC), with the formation of thermodynamically stable excimers occurring just after annealing above T_g [24]. Subjecting blends of sufficiently high dye concentrations to temperatures above their T_g-values (130 °C for PMMA, 150 °C for PC) led to permanent and significant changes in the materials' emission spectra,

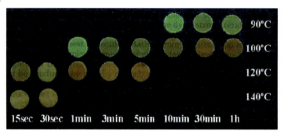

Figure 12.5 Images of initially quenched blends films of 0.9 wt% poly(ethylene terephthalate)/cyano-oligo(p-phenylene vinylene) upon annealing for the time and at the temperature indicated ($\lambda_{exc.}$ = 365 nm). Reproduced with permission from Ref. [25]; © 2006, American Chemical Society.

due to phase separation of the polymer and the excimer-forming dye. This effect appears to bear significant potential for technological applications, including the use of dye/polymer blends as time-temperature indicators (TTIs). In fact, kinetic experiments performed as a function of time, temperature and dye concentration, resulted in some well-fitted, monoexponential growth functions. In particular, plotting the rate of aggregation against the inverse of the annealing temperature indicated the presence of a linear, Arrhenius-type behavior, which suggested that an aggregation of the dye molecules might be predicted when dispersed into a polymer film. Moreover, the derived thermodynamic parameters could easily be determined via luminescence experiments.

Recently, additional data have been acquired by using semicrystalline [poly(ethylene terephthalate); PET] and poly(alkyl methacrylate) -based matrices [25, 35]. Homogeneous blends can be produced by conventional melt-processing protocols that are concluded with a quenching step, with the materials thus prepared displaying emission spectra characteristic of the dyes' monomer emission. For example, annealing above the T_g-value of the polyester (80 °C) led to the formation of excimers and caused significant changes in the materials' emission characteristics. The aggregation processes in the semicrystalline PET/cyano-OPVs blends investigated appeared to be well-described by single-exponential transformation kinetics, while the fluorescence color of these materials was characteristic of their thermal history (Figure 12.5).

In particular, cyano-OPV dyes with a longer, rigid conjugated core showed slower aggregation rates due to their limited mobility within the polymer matrix [35].

An extreme, yet intriguing, application of the multifunctional chromogenic cyano-OPV dyes was recently reported by Weder et al. [36]. The group reported the preparation and characterization of a new shape memory polymer (SMP) with built-in temperature-sensing capabilities, by incorporating a cyano-OPV dye into a semi-crystalline crosslinked poly(cyclo-octene) (PC'O) by guest diffusion. SMPs have the ability to memorize a permanent shape, and so can be manipulated and fixed to a temporary shape under specific conditions of temperature and stress,

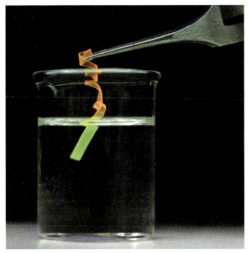

Figure 12.6 Image of the recovery from temporary shape (spiral, orange emission) to the permanent shape (rod, green emission) for a crosslinked poly(cyclo-octene)/cyano-oligo(p-phenylene vinylene) blend. The sample was immersed in silicon oil at ~75 °C (under illumination at 365 nm) [36]. Reproduced with permission of the Royal Society of Chemistry.

yet subsequently relax to the original, stress-free state upon application of an external stimulus [37]. Weder's group described a new SMP/dye system in which a built-in sensor provided additional functionality. The material was based on a chromogenic sensor dye that provided a simple and clear signal indicating if the set/release temperature had been reached. The dye concentration was chosen to allow dye aggregation upon drying, which resulted in the formation of excimers. Exposure of these phase-separated blends to temperatures above the melting point (T_m = 47–48 °C) of the polymer led to dissolution of the dye molecules, yielding a pronounced change in their optical characteristics.

Briefly, the permanent rod-shaped PCO/dye blend was heated to 80 °C, deformed to a temporary shape (spiral), and fixed by cooling to room temperature. The spiral was slowly immersed into a silicon oil bath (ca. 75 °C), allowing recovery of the rod shape which was flanked by a color change from orange to yellow-green (Figure 12.6).

12.2.3.2 Bis(Benzoxazolyl) Stilbene as Luminescent Dye

Polymeric film sensors that were based on excimer luminescence and responsive to temperature stress were also obtained through the dispersion of moderate amounts (0.02–0.2 wt%) of the food-grade dye BBS into the thermoplastic aliphatic biodegradable polyester, PBS, by melt-mixing [33]. Analogous to the cyano-OPVs polymer blends, rapid quenching at 0 °C of the PBS-BBS mixtures from the melt promoted the very fine molecular dispersion of BBS dyes that were kinetically

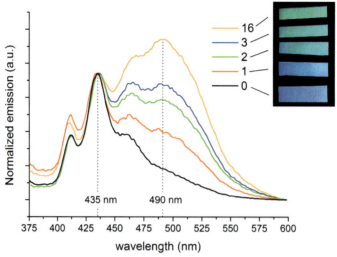

Figure 12.7 Fluorescence emission spectra ($\lambda_{exc.}$ = 277 nm) of initially quenched 0.05 wt% poly(1,4-butylene succinate)/bis(benzoxazolyl)stilbene film, and its color evolution as a function of the annealing time at 65 °C. Inset: The same films, with images recorded under irradiation at 366 nm for specified times (h).

trapped within the PBS matrix, thus avoiding the formation of excimers. The luminescent behavior of PBS-BBS-quenched films was thermally controlled, with BBS aggregation tendencies and emission color changes proportional to the increasing annealing temperature (from 50 to 80 °C). The thermal stress applied to films led to the generation of thermodynamically stable aggregates among BBS dyes, promoting the color change of the material from blue (emission at 435 nm) to green (emission at 490 nm) (Figure 12.7).

The molecular dispersion of BBS dyes into quenched PBS films should allow a prompt optical response to thermal stimuli, that was faster with respect to the isolated dyes incorporated into a glassy amorphous polymer matrix. In the case of PMMA, this occurred at a lower annealing temperature due to the T_g value (−34 °C) of the PBS supporting matrix.

12.2.3.3 Anthracene Triaryl Amine-Terminated Diimide as Luminescent Dye

Another example of thermochromic flexible materials based on polyethylene was recently reported [38] in which the authors described the synthesis and incorporation into PE of a novel anthracene triaryl, amine-terminated diimide fluorescent sensor (Scheme 12.2, 4). When incorporated into the polymer (0.1 wt%, by melt-mixing), the films showed an emission at approximately 620 nm which was progressively red-shifted by increasing the annealing temperature to 543 nm (green color) at 150 °C. Although it is likely that room-temperature emission from the anthracene diimide-doped PE was due to the formation of molecular aggregates, this mechanism cannot fully explain the material's thermochromic behavior in these films. The authors suggested that heating at 150 °C led to disruption of the dye aggregates, while the highly viscous medium of the polymer hampered full conjugation of the molecule, thus providing the same green emission.

12.3
Metal Nanostructures as the Dispersed Phase

12.3.1
Optical Properties of Metal Nanoassemblies

Today, nanoscience represents one of the most rapidly growing research areas, allowing the manipulation of matter at the nanoscale and allowing the controlled fabrication of such systems and devices. Engineered nanoparticles offer potential applications in many areas beneficial for humankind, including sensors, medical imaging, drug delivery systems, sunscreens, cosmetics, and many others [39–41]. Ever since the birth of nanotechnology, it has been clear that the optical properties of nanostructured metal particles depend heavily not only on their dimensions but also their shape [42, 43]. The absorption of visible light by metal nanoparticles was attributed to the induction of a collective oscillation of the free conduction electrons, promoted by their interaction with electromagnetic field. Indeed, when the wavelength of an incident radiation is comparable with the mean free path of the conduction electrons of a metal particle, the electric component of the electromagnetic incident field induces a polarization of the conduction electrons, giving rise to surface plasmon absorption [43, 44].

In particular, noble metal nanoparticles or semi-conducting nanocrystals embedded in bulk polymer matrices demonstrate enhanced optical (absorption, luminescence, and nonlinearity) [42, 43, 45–51] and magnetic properties [52], due to the stabilizing effects of size and aggregation provided by the macromolecular support. When dispersed into polymers in a nonaggregated form, however, those metal nanoparticles with very small diameters (a few nanometers) will allow the preparation of materials with much-reduced light scattering properties for applications as optical filters, linear polarizers, and optical sensors [53, 54].

For example, clusters of noble metals such as gold, silver, or copper, assume a real and natural color due to the absorption of visible light at the surface plasmon resonance (SPR) frequency. The application of composites based on polymeric materials and containing noble metal nanoparticles depends strictly on an ability to control and to modulate their size, shape, and extent of aggregation. As described by the Drude–Lorentz–Sommerfeld theory [42, 43], a decrease in metal particle size leads to a broadening of the absorption band, a decrease in the maximum intensity, and often also to a hypsochromic (blue) shift of the peak. These effects may also depend on cluster topology and packing [44, 55].

In addition, the anisotropic orientation of dipoles in nanoparticles by a uniaxially oriented host polymer matrix leads to the generation of two different excitation modes: (i) with photons polarized along the aggregation direction, leading to a bathochromic (red) shift of the SPR; or (ii) with photons polarized orthogonally to the aggregation direction, resulting in a hypsochromic (blue) shift (Figure 12.8) [53].

The most common procedure for obtaining a dispersion of metal nanoparticles (MNPs) in a polymer matrix is to prepare a colloidal solution of stabilized MNPs, to mix this with the desired polymer in a mutual solvent, and then to cast a film by evaporation from the solution [53]. In contrast, few examples have been reported demonstrating the dispersion of preformed MNPs in a polymer matrix by melt-mixing at high temperature [56, 57].

Usually, a water-soluble metal salt is moved into an organic solvent by using tetra-alkylammonium bromide as phase-transfer agent, followed by successive reduction with sodium borohydride in the presence of an alkylthiol as surface stabilizer to prevent coalescence of the growing nanoparticles [58, 59]. In addition to thiols, a number of different surface stabilizers have been used, including amines [60], poly(vinyl pyrrolidone) (PVP) [61], and sodium poly(acrylate) [62]. By using the above-described colloid chemistry technique, MNPs have been dispersed in ultra-high molecular-weight polyethylene (UHMWPE) [11, 45], high-density polyethylene (HDPE) [57], poly(vinyl alcohol) (PVA) [53, 63, 64], poly(dimethylsiloxane) (PDMS) [65, 66], and poly(styrene-b-ethylene/propylene) [67].

An alternative approach for preparing nanocomposites containing metal nanoparticles involves an *in situ* formation of nanoparticles directly within the

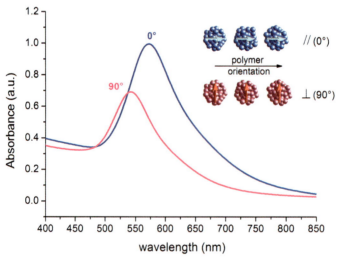

Figure 12.8 Influence of the polarization direction of light on the surface plasmon resonance (SPR) band of uniaxially oriented polymer film containing gold nanoparticles. The polarization direction was either parallel (0°) or perpendicular (90°) to the drawing axis.

polymer matrix [63, 68–80]. This process is relatively straightforward and requires simply a reduction of the metal ions precursors by either a photochemical or a thermally induced process. However, in contrast to nanoparticles prepared via colloid chemistry, controlling the size distribution of these *in situ*-prepared particles is often more difficult due to influential factors such as the polymer matrix composition, and the time and energy density of the photo- or thermo-irradiation process.

12.3.2
Nanocomposite-Based Indicators to Mechanical Stress

12.3.2.1 The Use of Metal Nanoparticles

Nanoparticle dispersions in a polymer matrix can be rendered macroscopically anisotropic, a feature that has allowed their use in nonlinear optical devices and linear absorbing polarizers, for example, in display applications [11, 45, 53, 57, 81–83]. Highly dichroic noble metal nanoparticles are efficiently obtained after mechanical drawing of the polymer matrix. The uniaxial orientation of the macromolecular fibers promotes the anisotropic distribution of both the crystalline and amorphous phases, which then determines the alignment of the metal particles along the direction of drawing [53, 84].

Examples of this include PVA and HDPE film composites with alkyl thiol-coated gold and silver particles which, once uniaxially oriented by stretching, present angular dependencies of the absorption intensity and the color of the transmitted light [53] (Figure 12.9).

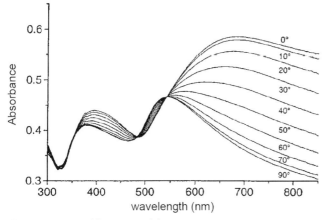

Figure 12.9 UV-visible spectra of drawn nanocomposites comprising high-density polyethylene and gold nanoparticles, taken in linearly polarized light at different angles between the polarization direction of the incident light and the drawing direction [53]. Reproduced with permission of John Wiley & Sons, Ltd.

The absorption of photons is dominated by the excitation of surface plasmons in the metal particles and their aggregates [42, 43]. For example, drawn polyethylene/silver nanocomposites exhibited a strongly polarization-dependent color. Yet, the color of the light transmitted through the oriented nanocomposite shifted from red to yellow when the angle between the polarization axis of an interposed polarizer and the drawing direction of the film was varied from 0° to 90° [57].

Moreover, the optical response of metal nanoparticles can be strongly enhanced through the introduction of photoactive organic molecules, possibly combined with control of the nanoparticle dimensions [85]. The presence of direct electronic interactions between metal and metal-bound chromophores is of particular interest, because it could allow for a fine modulation of the optical properties by inducing an energy transfer from the excited state of the chromophore to the SPR of the metal [86, 87].

For example, the dispersion of gold nanoparticles and gold-binding chromophores in a stretched polymer matrix of PE produces nanocomposites with unusual and anisotropic optical properties [45]. Strongly dichroic, terthiophene-based chromophores, which previously had been used in anisotropic PE dispersions for the preparation of linear polarizers [88], were modified with a thiol group and used for the preparation of gold nanoparticles. These authors demonstrated that the electronic systems of the chromophores were coupled with the gold nanoparticles, and that the polarization of the absorbed radiation could be preserved during energy transfer between a chromophore and a metal particle.

An interesting method for the production of dichroic nanocomposites with a reversible optical response to mechanical stress has been recently reported [65, 66]. Here, Caseri *et al.* [65] based their studies on the difference in the transmitted color of gold and silver elastomeric nanocomposites in dry and in swollen samples, based on the change in interparticle distances which resulted from the swelling process. Caseri's group showed that a given dry gold–polymer nanocomposite could adopt not only one but several colors, and that these colors could be switched through swelling processes, including reversible and irreversible dichroic–monochroic color transitions. A lightly crosslinked rubber such as PDMS was selected as the supporting polymer matrix, as it can take up large quantities of solvent without dissolution, and might also afford reversible dichroic effects upon deformation. The oriented PDMS composite films appeared blue–gray with light polarized parallel to the drawing axis, and red with light in a perpendicular orientation; the dichroism was preserved upon removing the strain. Swelling (with toluene) of the dichroic oriented films caused the specimens to become pink, indicating that the individual particles in the linear assemblies had disconnected and turned blue and nondichroic following evaporation of the toluene.

Recently, nanocomposites based on PVA and poly [ethylene-*co*-vinyl alcohol] polymers and nanostructured gold have been efficiently prepared using a UV photoreduction process [89]. In this case, the polymer matrix based on vinyl alcohol repeating units, acted both as a coreducing agent, as a protective agent against particle agglomeration, and as a macroscopic support. The very rapid process provided dispersed gold nanoparticles with average diameters ranging from 3 to 20 nm, depending on the host polymer matrix and the irradiation time. Uniaxial drawing of the Au/polymer composites promoted anisotropic packing of the embedded gold nanoparticles along the stretching direction of the film, and this resulted in a shift of the SPR of gold well above 30–40 nm (83 nm maximum), thus producing a well-defined polarization-dependent color change from blue to purple [89]. The optical responsiveness to mechanical stimuli appeared similar to that reported previously by Smith et al. for optically anisotropic polyethylene–gold nanocomposites obtained by introducing preformed and annealed alkyl-thiol-protected gold nanoparticles into polyethylene [81]. The UV photogeneration of gold nanoparticles directly into a polymer film precursor represents an easier, faster, and highly competitive technique for the preparation of optically responsive nanocomposites to mechanical stimuli, however.

Among the large number of hybrid organic–inorganic systems investigated, those nanocomposites based on PVA and silver have attracted great interest because of their specific optical, catalytic, electronic, magnetic, and antimicrobial properties [64, 68, 76, 90–97]. Recent developments have been focused in this direction in order to optimize the preparation Ag/PVA nanostructured films by using alternative "in situ" methods such as sun-(UV) [76, 98, 99] or thermally promoted reduction processes [100].

These simple and very rapid methods provide dispersed Ag nanoparticles (<4 wt%) with average diameters that range from 15 to 150 nm, depending on the type of preparation [101]. Thermal annealing and UV irradiation in fact form the basis for these very efficient methodologies, because they take advantage of the formation of a complex between the PVA matrix, while silver nitrate [102]: Ag^+ ions can easily be chelated by the hydroxyl groups of the polymer and then reduced directly in the host matrix (Scheme 12.3).

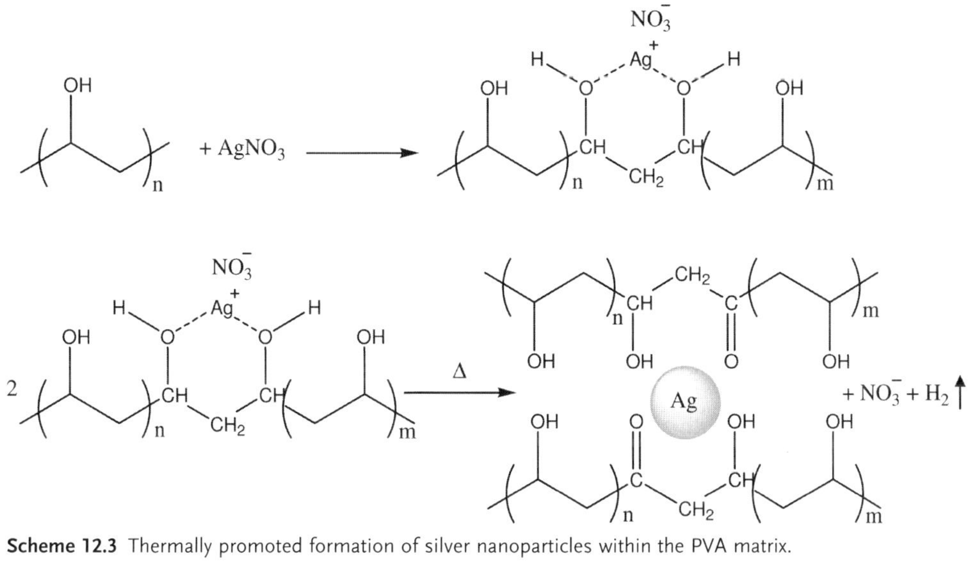

Scheme 12.3 Thermally promoted formation of silver nanoparticles within the PVA matrix.

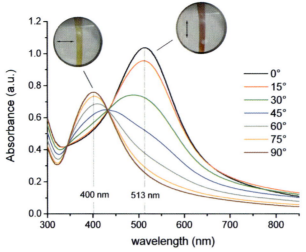

Figure 12.10 UV-visible spectra of oriented sun-promoted Ag/PVA nanocomposite films as a function of the angle between the polarization of light and the drawing direction of the film. The insets show images of the same films under polarized light.

Ag(0) formation was progressively followed using both UV-visible and Fourier transform-infrared (FT-IR) spectroscopies, monitoring the evolution of the SPR of silver (ca. 430 nm) and the carbonyl absorption band of PVA (ca. 1720 cm^{-1}), respectively [101].

After uniaxial orientation, the Ag/PVA nanocomposites showed a pronounced dichroic behavior, based on the anisotropic distribution of the silver assemblies along the stretching direction. In uniaxially oriented samples, the light absorption was seen to depend heavily on the angle between the polarization direction of the incident light and the orientation of the particles embedded onto the host polymeric matrix. Simply by observing oriented samples through a linear polarizer, it is possible to observe that the color of the films depends markedly on the relative orientation between the polarizer and the drawing direction of the film (Figure 12.10).

The maximum shifts measured for the oriented films were more than 100 nm, while the spectra showed a well-defined isosbestic point, thus confirming the existence of two different populations of absorbing nanoparticles [89]. Interestingly, the shift in wavelength of the silver SPR band also occurred at moderate drawings (draw ratio ≥2). The nanocomposite films thus produced were highly optically responsive to mechanical stimuli as uniaxial deformations.

Analogously, thermochromic films based on silver/polystyrene nanocomposites have been reported, these being prepared by the thermal annealing of silver dodecylmercaptide/polystyrene blends at approximately 200 °C [103–105]. It was shown that alkanethiolates of transition metals dissolved in polymers would undergo thermolysis reactions at moderately low temperatures (120–200 °C), producing polymer-embedded metal or metal sulfide clusters. The rapid and reversible color switch (from dark brown to light yellow) that was observed when exposing the nanocomposite films at approximately 80 °C was attributed to a change in the surface-plasmon absorption of silver nanoparticles that follows the variation of inter-particle distances due to polymer matrix expansion [104].

12.3.2.2 The Use of Metal Nanorods

During the past few years, there has been an increasing interest towards the optical properties of elongated nanostructures, as nanorods or nanowires, mainly because their absorption spectrum is characterized by a longitudinal and a transverse SPR, rather than the single resonance, as observed for spherical nanoparticles [40, 106–109]. In other words, despite spherical particles having the greatest symmetry, they have only one plasmon resonance (all modes are degenerate). However, by extending a particle in one dimension, a second, lower-energy resonance band in the longitudinal direction (longitudinal mode) becomes apparent, while the original plasmon resonance (transversal mode) remains unaffected. In general, the number of SPR peaks will increase as the particle symmetry decreases.

In general, those one-dimensional (1-D) structures that range in size from a few hundreds of nanometers to several micrometers, or which have a high aspect ratio (the ratio between the longest and shortest structure axes), are referred to as nanowires, and smaller more elongated structures as nanorods.

In the past, many techniques have been developed for the preparation of 1-D nanostructures, and in order to understand this basic growth process it will first be necessary to consider the spontaneous growth process [110]. In a previous report [44] silver nanorods were prepared via a seed-mediated process and, depending on the surfactant (capping material) used, it was possible to produce shapes other than spheres during seed-mediated growth. Several theories [111] have centered on a preferential adsorption of the surfactant to certain crystal facets on seed particles; in other words, the surfactant binds to the surface radially, but not axially. Unfortunately, however, this behavior blocks crystal growth on these surfaces, such that the particle can grow only in the axial direction.

As reported by the present authors' group and elsewhere [109, 112, 113], in order to produce nanocomposite-containing metal nanostructures which are anisotropic and thus potentially more sensitive and responsive to any mechanical stimuli applied to the polymer matrix, silver nanorods were mixed with an aqueous solution of PVA, such that a homogeneous film was obtained after solvent evaporation.

Following uniaxial stretching, the elongated silver nanostructures (mostly nanorods, but mixed with other spherical assemblies generated during solution casting as a consequence of particle coalescence) that were embedded into the polymer matrix assumed the direction of the drawing, as indicated using transmission electron microscopy (TEM) (Figure 12.11). These dispersed nanorods had diameters between 30 and 35 nm, and a length <180 nm.

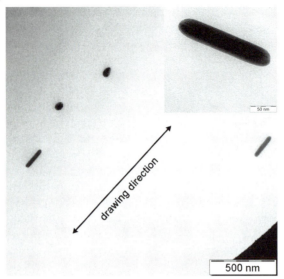

Figure 12.11 Bright-field TEM image of PVA-oriented film containing silver nanostructures [114].

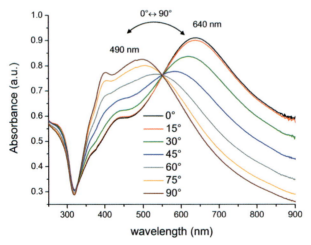

Figure 12.12 UV-visible spectra of oriented Ag nanorods/PVA film as a function of the angle between the light polarization and the drawing direction [114].

As expected, the oriented nanocomposite films showed a strong dichroism when irradiated with a linearly polarized light (Figure 12.12). Changing the relative orientation between the stretching direction of the film and the polarization vector of the incident light induced a suppression of the longitudinal SPR at 640 nm in favor of the transversal SPR at 420 nm. The two isosbestic points, placed at 327

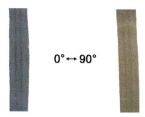

Figure 12.13 Images of oriented Ag nanorods/PVA film with polarization direction of the incident light parallel (blue, 0°) and perpendicular (brown-red, 90°) to the drawing direction [114].

and 550 nm, were originated by the two independent optical modes (transversal and longitudinal) that characterized the optical absorption of silver nanorods. The large, reversible variation (~220 nm) in wavelength of the two absorption maxima was also flanked by an increase in a red component placed at 500 nm and coupled with a strong decreasing of the blue contribution located at 640 nm. Accordingly, the film changed its color under linearly polarized light from blue (parallel configuration, 0°) to brown-red (perpendicular configuration, 90°), with only a moderate drawing (draw ratio ≥2) (Figure 12.13).

Another recently reported and highly interesting example concerned the application of gold nanorods to polymers [115]. This group described the generation of gold nanorod (GNR)–elastin-like polypeptide (ELP) nanoassemblies, the optical responses of which could be manipulated based on their exposure to near-infrared (NIR) light. Cysteine-containing ELPs were actually self-assembled on GNRs mediated by gold–thiol bonds, which led to the generation of GNR–ELP nanoassemblies. Exposure of these assemblies to NIR light at 720 nm resulted in a heating of the GNRs due to SPR (directed in the NIR region of the absorption spectrum at 710 and 820 nm). Subsequent heat transfer from the GNRs led to an increase in the temperature of the self-assembled ELP above its T_g-value (i.e., 33–34 °C), which in turn led to a phase transition and aggregation of the GNR–ELP assemblies and a clear change in the optical density of the system. Based on these findings, it was suggested that polypeptides might be interfaced with GNRs to produce optically responsive nanoasssemblies for sensing and drug delivery applications.

12.4
Conclusions

The information provided in this chapter provides a clear demonstration of the potential offered by controlling the dispersion of luminescent aromatic molecules and noble metal nanoaggregates into a wide range of polymer matrices, down to the nanometer scale. Whilst most of the polymers have been characterized by their excellent thermomechanical properties, a lack of any particular optical response was apparent wavelengths above 200 nm. In contrast, the addition of very small

(<4 wt%) amounts of correctly selected active dyes or noble metal (nano)assemblies allowed the production of flexible or rigid polymeric thin films which demonstrated remarkable optical responses towards external stimuli. Luminescent, excimer-forming dyes have been described as a new class of optical indicators for many thermoplastic and elastomeric polymer matrices for mechanical and thermal uses. However, the greater the interaction and asymmetry of the metal nanostructures, the higher is the dichroic response of the SPR following mechanical stimulation of the nanocomposite film. Thus, these hybrid materials may be regarded as extremely important for future advanced applications, including the possibility of inducing a nanophase organization through external stresses.

Acknowledgments

These studies were supported financially by the FIRB Project RBNE03R78E_005 of the Italian Ministry of University and Research (MIUR).

References

1 Birks, J.B. (1975) *Rep. Prog. Phys.*, **38**, 903–974.
2 Phillips, D. (1985) *Polymer Photophysics: Luminescence, Energy Migration, and Molecular Motion in Synthetic Polymers*, Chapman & Hall Ltd, London.
3 Lakowicz, J.R. (1986) *Principles of Fluorescent Spectroscopy*, Plenum Press, New York.
4 Ceroni, G.B.P., Marchioni, F. and Balzani, V. (2005) *Prog. Polym. Sci.*, **30**, 453–473.
5 Duhamel, J. (2004) *Macromolecules*, **37**, 1987–1989.
6 Vangani, V., Drage, J., Mehta, J., Mathew, A.K. and Duhamel, J. (2001) *J. Phys. Chem. B*, **105**, 4827–4839.
7 Siu, H. and Duhamel, J. (2004) *Macromolecules*, **37**, 9287–9289.
8 Naciri, J. and Weiss, R.G. (1989) *Macromolecules*, **22**, 3928–3936.
9 Morawetz, H. (1999) *J. Polym. Sci., Part A Polym. Chem.*, **37**, 1725.
10 Semerak, S.N. and Frank, C.W. (1984) *Adv. Polym. Sci.*, **54**, 31–85.
11 Pucci, A., Elvati, P., Ruggeri, G., Liuzzo, V., Tirelli, N., Isola, M. and Ciardelli, F. (2003) *Macromol. Symp.*, **204**, 59–70.
12 Frank, C.W., Gashgari, M.A. and Semerak, S.N. (1986) *NATO ASI Ser., Ser. C*, **182**, 523–546.
13 Semerak, S.N. and Frank, C.W. (1984) *Adv. Chem. Ser.*, **206**, 77–100.
14 Semerak, S.N. and Frank, C.W. (1984) *Macromolecules*, **17**, 1148–1157.
15 Semerak, S.N. and Frank, C.W. (1983) *Adv. Chem. Ser.*, **203**, 757–771.
16 Semerak, S.N. and Frank, C.W. (1981) *Polym. Prep. (Am. Chem. Soc., Div. Polym. Chem.)*, **22**, 314–315.
17 Birks, J.B. (1970) *Photophysics of Aromatic Molecules*, John Wiley & Sons, New York.
18 Birks, J.B., Dyson, D.J. and Munro, I.H. (1963) *Proc. Roy. Soc. (London)*, **275**, 575–588.
19 Halkyard, C.E., Rampey, M.E., Kloppenburg, L., Studer-Martinez, S.L. and Bunz, U.H.F. (1998) *Macromolecules*, **31**, 8655–8659.
20 Wilson, J.N., Smith, M.D., Enkelmann, V. and Bunz, U.H.F. (2004) *Chem. Commun.*, 1700–1701.
21 Yang, J., Li, H., Wang, G. and He, B. (2001) *J. Appl. Polym. Sci.*, **82**, 2347–2351.

22 van Hutten, P.F., Krasnikov, V.V., Brouwer, H.J. and Hadziioannou, G. (1999) *Chem. Phys.*, **241**, 139–154.
23 Pucci, A., Biver, T., Ruggeri, G., Itzel Meza, L. and Pang, Y. (2005) *Polymer*, **46**, 11198–11205.
24 Crenshaw, B.R. and Weder, C. (2005) *Adv. Mater.*, **17**, 1471–1476.
25 Kinami, M., Crenshaw, B.R. and Weder, C. (2006) *Chem. Mater.*, **18**, 946–955.
26 Lowe, C. and Weder, C. (2002) *Adv. Mater.*, **14**, 1625–1629.
27 Crenshaw, B.R. and Weder, C. (2003) *Chem. Mater.*, **15**, 4717–4724.
28 Crenshaw, B.R., Burnworth, M., Khariwala, D., Hiltner, A., Mather, P.T., Simha, R. and Weder, C. (2007) *Macromolecules*, **40**, 2400–2408.
29 Crenshaw, B.R. and Weder, C. (2006) *Macromolecules*, **39**, 9581–9589.
30 Kunzelman, J., Kinami, M., Crenshaw, B.R., Protasiewicz, J.D. and Weder, C. (2008) *Adv. Mater.*, **20**, 119–122.
31 Pucci, A., Bertoldo, M. and Bronco, S. (2005) *Macromol. Rapid Commun.*, **26**, 1043–1048.
32 Andreotti, L., Pucci, A., Ruggeri, G., Scatto, M. and Sterner, M. (2007) Monitoring polyethylene films orientation during melt-processing through UV-vis and IR spectroscopy in polarized light. Presented at *European Polymer Congress, Portoroz, Slovenia*, p. 263.
33 Pucci, A., Di Cuia, F., Signori, F. and Ruggeri, G. (2007) *J. Mater. Chem.*, **17**, 783–790.
34 Donati, F., Pucci, A., Cappelli, C., Mennucci, B. and Ruggeri, G. (2008) *J. Phys. Chem. B*, **112**, 3668–3679.
35 Crenshaw, B.R., Kunzelman, J., Sing, C.E., Ander, C. and Weder, C. (2007) *Macromol. Chem. Phys.*, **208**, 572–580.
36 Kunzelman, J., Chung, T., Mather, P.T. and Weder, C. (2008) *J. Mater. Chem.*, **18**, 1082–1086.
37 Lendlein, A. and Kelch, S. (2002) *Angew. Chem. Int. Ed. Engl.*, **41**, 2034–2057.
38 Tyson, D.S., Carbaugh, A.D., Ilhan, F., Santos-Perez, J. and Meador, M.A. (2008) *Chem. Mater.*, **20**, 6595–6596.
39 Baker, C.C., Pradhan, A. and Shah, S.I. (2004) in *Encyclopedia of Nanoscience and Nanotechnology*, American Scientific Publisher, Stevenson Ranch, California, Vol. **5**, pp. 449–473.
40 Jain, P.K., Huang, X., El-Sayed, I.H. and El-Sayed, M.A. (2008) *Acc. Chem. Res.*, **41**, 1578–1586.
41 Nair, L.S. and Laurencin, C.T. (2007) *J. Biomed. Nanotechnol.*, **3**, 301–316.
42 Klabunde, K.J. (2001) *Nanoscale Materials in Chemistry*, John Wiley & Sons, Inc., New York.
43 Kreibig, U. and Genzel, L. (1985) *Surf. Sci.*, **156**, 678–700.
44 Schwartzberg, A.M. and Zhang, J.Z. (2008) *J. Phys. Chem. C*, **112**, 10323–10337.
45 Pucci, A., Tirelli, N., Willneff, E.A., Schroeder, S.L.M., Galembeck, F. and Ruggeri, G. (2004) *J. Mater. Chem.*, **14**, 3495–3502.
46 Beecroft, L.L. and Ober, C.K. (1997) *Chem. Mater.*, **9**, 1302–1317.
47 Gehr, R.J. and Boyd, R.W. (1996) *Chem. Mater.*, **8**, 1807–1819.
48 Godovsky, D.Y. (2000) *Adv. Polym. Sci.*, **153**, 163–205.
49 Trindade, T., O'Brien, P. and Pickett, N.L. (2001) *Chem. Mater.*, **13**, 3843–3858.
50 Caseri, W.R. (2006) *Mater. Sci. Technol.*, **22**, 807–817.
51 Ni, Y., Hao, H., Cao, X., Su, S., Zhang, Y. and Wei, X. (2006) *J. Phys. Chem. B*, **110**, 17347–17352.
52 de la Venta, J., Pucci, A., Pinel, E.F., Garcia, M.A., de Julian Fernandez, C., Crespo, P., Mazzoldi, P., Ruggeri, G. and Hernando, A. (2007) *Adv. Mater.*, **19**, 875–877.
53 Caseri, W. (2000) *Macromol. Rapid Commun.*, **21**, 705–722.
54 Heilmann, A. (2003) *Polymer Films with Embedded Metal Nanoparticles*, Vol. **52**, Springer, Berlin.
55 El-Sayed, M.A. (2001) *Acc. Chem. Res.*, **34**, 257–264.
56 Dirix, Y., Bastiaansen, C., Caseri, W. and Smith, P. (1999) *J. Mater. Sci.*, **34**, 3859–3866.
57 Dirix, Y., Bastiaansen, C., Caseri, W. and Smith, P. (1999) *Adv. Mater.*, **11**, 223–227.
58 Brust, M., Walker, M., Bethell, D., Schiffrin, D.J. and Whyman, R. (1994)

J. Chem. Soc., Chem. Commun., 801–802.
59 Lu, A.H., Lu, G.H., Kessinger, A.M. and Foss, C.A. Jr. (1997) *J. Phys. Chem. B*, **101**, 9139–9142.
60 Leff, D.V., Brandt, L. and Heath, J.R. (1996) *Langmuir*, **12**, 4723–4730.
61 Carotenuto, G. (2001) *Appl. Organomet. Chem.*, **15**, 344–351.
62 Hussain, I., Brust, M., Papworth, A.J. and Cooper, A.I. (2003) *Langmuir*, **19**, 4831–4835.
63 Perez-Juste, J., Rodriguez-Gonzalez, B., Mulvaney, P. and Liz-Marzan, L.M. (2005) *Adv. Funct. Mater.*, **15**, 1065–1071.
64 Mbhele, Z.H., Salemane, M.G., van Sittert, C.G.C.E., Nedeljkovic, J.M., Djokovic, V. and Luyt, A.S. (2003) *Chem. Mater.*, **15**, 5019–5024.
65 Uhlenhaut, D.I., Smith, P. and Caseri, W. (2006) *Adv. Mater.*, **18**, 1653–1656.
66 Pastoriza-Santos, I., Perez-Juste, J., Kickelbick, G. and Liz-Marzan, L.M. (2006) *J. Nanosci. Nanotechnol.*, **6**, 453–458.
67 Bockstaller, M.R. and Thomas, E.L. (2003) *J. Phys. Chem. B*, **107**, 10017–10024.
68 Mallick, K., Witcomb, M.J. and Scurrell, M.S. (2004) *J. Mater. Sci.*, **39**, 4459–4463.
69 Salvati, R., Longo, A., Carotenuto, G., De Nicola, S., Pepe, G.P., Nicolais, L. and Barone, A. (2005) *Appl. Surf. Sci.*, **248**, 28–31.
70 Raikher, Y.L., Stepanov, V.I., Depeyrot, J., Sousa, M.H., Tourinho, F.A., Hasmonay, E. and Perzynski, R. (2004) *J. Appl. Phys.*, **96**, 5226–5233.
71 Dong, S., Tang, C., Zhou, H. and Zhao, H. (2004) *Gold Bull.*, **37**, 187–195.
72 Kaneko, K., Sun, H.-B., Duan, X.-M. and Kawata, S. (2003) *Appl. Phys. Lett.*, **83**, 1426–1428.
73 Henneke, D.E., Malyavanatham, G., Kovar, D., O'Brien, D.T., Becker, M.F., Nichols, W.T. and Keto, J.W. (2003) *J. Chem. Phys.*, **119**, 6802–6809.
74 Callegari, A., Tonti, D. and Chergui, M. (2003) *Nano Lett.*, **3**, 1565–1568.
75 Seibel, M. (2001) In PCT Int. Application (Plasco Ehrich Plasma-Coating GmbH, Germany), WO2001083596, p. 12.
76 Gaddy, G.A., McLain, J.L., Steigerwalt, E.S., Broughton, R., Slaten, B.L. and Mills, G. (2001) *J. Cluster Sci.*, **12**, 457–471.
77 Tanahashi, I. and Kanno, H. (2000) *Appl. Phys. Lett.*, **77**, 3358–3360.
78 Zhou, Y., Wang, C.Y., Zhu, Y.R. and Chen, Z.Y. (1999) *Chem. Mater.*, **11**, 2310–2312.
79 Itakura, T., Torigoe, K. and Esumi, K. (1995) *Langmuir*, **11**, 4129–4134.
80 Porel, S., Singh, S. and Radhakrishnan, T.P. (2005) *Chem. Commun.*, 2387–2389.
81 Dirix, Y., Darribere, C., Heffels, W., Bastiaansen, C., Caseri, W. and Smith, P. (1999) *Appl. Opt.*, **38**, 6581–6586.
82 Crespo, P., Litran, R., Rojas, T.C., Multigner, M., de la Fuente, J.M., Sanchez-Lopez, J.C., Garcia, M.A., Hernando, A., Penades, S. and Fernandez, A. (2004) *Phys. Rev. Lett.*, **93**, 087204.
83 Hao, E., Schatz, G.C. and Hupp, J.T. (2004) *J. Fluoresc.*, **14**, 331–341.
84 Caseri, W. (2003) *Chemistry of Nanostructured Materials*, World Scientific Publishing Co., Mountain View, CA, pp. 359–386.
85 Chandrasekharan, N., Kamat, P.V., Hu, J. and Jones, G. II (2000) *J. Phys. Chem. B*, **104**, 11103–11109.
86 Thomas, K.G. and Kamat, P.V. (2003) *Acc. Chem. Res.*, **36**, 888–898.
87 Thomas, K.G. and Kamat, P.V. (2000) *J. Am. Chem. Soc.*, **122**, 2655–2656.
88 Tirelli, N., Amabile, S., Cellai, C., Pucci, A., Regoli, L., Ruggeri, G. and Ciardelli, F. (2001) *Macromolecules*, **34**, 2129–2137.
89 Pucci, A., Bernabò, M., Elvati, P., Meza, L.I., Galembeck, F., de Paula Leite, C.A., Tirelli, N. and Ruggeri, G. (2006) *J. Mater. Chem.*, **16**, 1058–1066.
90 Kim, J.-H. and Lee, T.R. (2007) *Langmuir*, **23**, 6504–6509.
91 Lin, W.-C. and Yang, M.-C. (2005) *Macromol. Rapid Commun.*, **26**, 1942–1947.
92 Kong, H. and Jang, J. (2006) *Chem. Commun.*, 3010–3012.
93 Khanna, P.K., Singh, N., Charan, S., Subbarao, V.V.V.S., Gokhale, R. and Mulik, U.P. (2005) *Mater. Chem. Phys.*, **93**, 117–121.

94 Karthikeyan, B. (2005) *Physica B*, **364**, 328–332.
95 Cascaval, C.N., Cristea, M., Rosu, D., Ciobanu, C., Paduraru, O. and Cotofana, C. (2007) *J. Optoelectron. Adv. Mater.*, **9**, 2116–2120.
96 Badr, Y. and Mahmoud, M.A. (2006) *J. Mater. Sci.*, **41**, 3947–3953.
97 Fornasiero, D. and Grieser, F. (1987) *Chem. Phys. Lett.*, **139**, 103–108.
98 Gaddy, G.A., Korchev, A.S., McLain, J.L., Slaten, B.L., Steigerwalt, E.S. and Mills, G. (2004) *J. Phys. Chem. B*, **108**, 14850–14857.
99 Gaddy, G.A., McLain, J.L., Korchev, A.S., Slaten, B.L. and Mills, G. (2004) *J. Phys. Chem. B*, **108**, 14858–14865.
100 Clemenson, S., David, L. and Espuche, E. (2007) *J. Polym. Sci., Part A Polym. Chem.*, **45**, 2657–2672.
101 Bernabo, M., Ciardelli, F., Pucci, A. and Ruggeri, G. (2008) *Macromol. Symp.*, **270**, 177–186.
102 Zidan, H.M. (1999) *Polym. Test.*, **18**, 449–461.
103 Carotenuto, G., Nicolais, L. and Perlo, P. (2006) *Polym. Eng. Sci.*, **46**, 1016–1021.
104 Carotenuto, G., La Peruta, G. and Nicolais, L. (2006) *Sens. Actuators B*, **B114**, 1092–1095.
105 Carotenuto, G., Martorana, B., Perlo, P. and Nicolais, L. (2003) *J. Mater. Chem.*, **13**, 2927–2930.
106 Maier, S.A., Kik, P.G., Atwater, H.A., Meltzer, S., Harel, E., Koel, B.E. and Requicha, A.A.G. (2003) *Nat. Mater.*, **2**, 229–232.
107 Gluodenis, M. and Foss, C.A. Jr. (2002) *J. Phys. Chem. B*, **106**, 9484–9489.
108 Mohamed, M.B., Volkov, V., Link, S. and El-Sayed, M.A. (2000) *Chem. Phys. Lett.*, **317**, 517–523.
109 Perez-Juste, J., Rodriguez-Gonzalez, B., Mulvaney, P. and Liz-Marzan, L.M. (2005) *Adv. Funct. Mater.*, **15**, 1065–1071.
110 Cao, G. (2004) *Nanostructures and Nanomaterials, Synthesis, Properties and Applications*, Imperial College Press, London.
111 Gole, A. and Murphy, C.J. (2004) *Chem. Mater.*, **16**, 3633.
112 Wilson, O., Wilson, G.J. and Mulvaney, P. (2002) *Adv. Mater.*, **14**, 1000–1004.
113 Zhou, Y., Yu, S.H., Wang, C.Y., Li, X.G., Zhu, Y.R. and Chen, Z.Y. (1999) *Adv. Mater.*, **11**, 850–852.
114 Bernabo, M. (2009) Synthesis and characterization of optical and magnetic properties of nanostructured metal-polymer systems. PhD Thesis, University of Pisa, Italy.
115 Huang, H.-C., Koria, P., Parker, S.M., Selby, L., Megeed, Z. and Rege, K. (2008) *Langmuir*, **24**, 14139–14144.

13
Nanocomposites Based on Phyllosilicates: From Petrochemicals to Renewable Thermoplastic Matrices

Maria-Beatrice Coltelli, Serena Coiai, Simona Bronco, and Elisa Passaglia

13.1
Introduction

Polymer–phyllosilicate nanocomposites, which are characterized by the presence of fillers that are less than 100 nm in at least one dimension, have been the subject of many reviews and reports during the past decade [1, 2]. The excellent balance between performance and filler content (2–5% by weight of phyllosilicates is sufficient to provide clear improvements in properties) makes these composites especially interesting as innovative materials for a variety of applications. Such improvements, which include high moduli, increased strength and heat resistance, and decreased gas permeability and flammability, can be ascribed to a fundamental feature of polymer nanocomposites, namely that the small size of the fillers leads to a dramatic increase in interfacial area when compared to traditional composites [3].

Many studies have been initiated by considering conventional petrochemical-derived polymer matrices, such as polyolefins, polyamides, and polyesters. Typically, the preparation methods underwent intense examination, the aim being to correlate the structure and morphology of these new nanostructured materials with their ultimate properties.

Today, the development of renewable polymeric materials with excellent properties forms the subject of active research interest worldwide [4], with aliphatic polyesters being among the most promising materials for the production of high-performance, environment-friendly, biodegradable plastics [5]. The need to optimize their properties, in order that in time they can replace present-day commodities, requires the assessment of new, selective, efficient, and sustainable preparation methods. Hence, a comparative review of the preparation methods for both commodities and renewable polyester-based materials should allow comparisons to be made of these two classes of material in terms of their potential and future opportunities.

Advanced Nanomaterials. Edited by Kurt E. Geckeler and Hiroyuki Nishide
Copyright © 2010 WILEY-VCH Verlag GmbH & Co. KGaA, Weinheim
ISBN: 978-3-527-31794-3

13.1.1
Structure of Phyllosilicates

Phyllosilicates [6] are an essential constituent of argillaceous sedimentary rocks and of some low-grade metamorphic rocks. The group includes micas, clays, and serpentine, which are soft minerals with a variable, but generally low, density. Minerals belonging to the phyllosilicate group are characterized by their possessing layers of $[SiO_4]^{4-}$ tetrahedra linked together to form a flat sheet. Rings of tetrahedra are linked by shared oxygens to other rings in a two-dimensional (2-D) plane that produces a sheet-like structure. Typically, the sheets are then connected to each other by layers of cations, together with water or other low molecules devoid of any electrical charge. In some substrates, the sheets are rolled into tubes that produce fibers, as in asbestos serpentine. The silicon:oxygen ratio is generally 1:2.5, because only one oxygen is bonded exclusively to the silicon, while the other three oxygens are half-shared (1.5) with other silicons.

13.1.1.1 Clays

In the case of polymeric composites [7], the most important subcategory is that of clays [8], including kaolinite, montmorillonite/smectite, illite (mica) and chlorite subgroups (Table 13.1). The clay minerals form part of a general, but important, group that contain a high percentage of water trapped between the silicate sheets. Most clays are chemically and structurally analogous to other phyllosilicates, but contain varying amounts of water and so allow more substitution of their cations. Clay minerals tend to form microscopic to submicroscopic crystals, with the percentage of water changing as a function of the humidity change. In the presence of water, clays become plastic and can be molded and formed. The expansion of clay is due to a progressive inclusion of water between the stacked silicate layers. Clays are usually mixed not only with other clays but also with microscopic crystals of carbonates, feldspars, micas, and quartz.

In the case of polymer nanocomposites, the most important fillers belong to the group of smectites. The term "smectite" is used to describe a family of expansible 2:1 phyllosilicate minerals. Specific minerals included in the smectite family [9] include pyrophyllite, talc, vermiculite, sauconite, saponite, nontronite, montmorillonite, and several less common species. Smectites [10] consist of a single octahedral sheet sandwiched between two tetrahedral sheets, with the octahedral sheet sharing the apical oxygens of the tetrahedral sheet. The octahedral sheet may be either dioctahedral or trioctahedral. Some minerals of the smectite groups are neutral (e.g., kaolinite, pyrophillite, talc), but others are characterized by negatively charged layers. The charge arises from substitutions in either the octahedral sheet (typically from the substitution of low-charge species such as Mg^{2+}, Fe^{2+}, or Mn^{2+} for Al^{3+} in dioctahedral species), or in the tetrahedral sheet (where Al^{3+} or occasionally Fe^{3+} substitutes for Si^{4+}), thus producing one negative charge for each such substitution. The charge, which usually expressed as a charge/unit formula, is fundamental to correlate the type of mineral with its final use (Table 13.2).

Table 13.1 Groups, formula, structural description and specific minerals of the clay group.

	Group	Formula	Description	Mineral(s)
Clays	Kaolinite	$Al_2Si_2O_5(OH)_4$	Silicate sheets (Si_2O_5) bonded to aluminum oxide/hydroxide layers ($Al_2(OH)_4$)	Kaolinite Dickite Nacrite
	Smectite	$(Ca, Na, H)(Al, Mg, Fe, Zn)_2(Si, Al)_4O_{10}(OH)_2 - xH_2O$	Silicate layers sandwiching a gibbsite [$Al_2(OH)_4$] (or brucite [$Mg_2(OH)_4$]) layer in between, in an **s-g-s** stacking sequence. Variable amounts of water molecules lie between the **s-g-s** sandwiches	Pyrophyllite *Talc* Vermiculite Sauconite Saponite Nontronite *Montmorillonite*
	Illite (Mica)	$(K, H)Al_2(Si, Al)_4O_{10}(OH)_2 - xH_2O$	Silicate layers sandwiching a gibbsite-like layer in between, in an **s-g-s** stacking sequence. The variable amounts of water molecules would lie between the **s-g-s** sandwiches as well as the potassium ions	Biotite Lepidolite Muscovite Paragonite Phlogopite Zinnwaldite
	Chlorite	$X_{4-6}Y_4O_{10}(OH, O)_8$[a]	Silicate layers sandwiching a brucite or brucite-like layer in between, in an **s-b-s** stacking sequence similar to the above groups. There is an extra weakly bonded brucite layer in between the **s-b-s** sandwiches. This gives the structure an **s-b-s b s-b-s b** sequence. The variable amounts of water molecules would lie between the **s-b-s** sandwiches and the brucite layers	Amesite Baileychlore clinochlore Cookeite corundophilite Daphnite Delessite Gonyerite Nimite Odinite Orthochamosite penninite Pannantite Rhipidolite (prochlore) Sudoite Thuringite etc.

[a] The X represents either aluminum, iron, lithium, magnesium, manganese, nickel, zinc or, rarely, chromium. The Y represents either aluminum, silicon, boron, or iron but mostly aluminum and silicon.

Table 13.2 Charge/unit formula for different phyllosilicates minerals.

	Mineral species	Interlayer cations	Charge/unit formula
Kaolinite-serpentine	Kaolinite, Halloysite	–	0.0
Pyrophillite-talc	Antigorite, Chrysotile	–	0.0
Smectite	Montmorillonite, Beidelite, Nontronite, Saponite, Hectorite	Na, Ca	0.25–0.6
Vermiculite	Dioctahedral-vermiculite Trioctahedral-Vermiculite	Mg	0.6–0.9
Mica	Muscovite, Biotite, Phlogopite	K	1.0

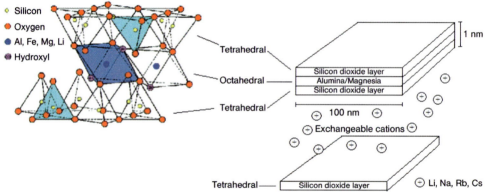

Figure 13.1 Structure of montmorillonite clay.

Montmorillonite (MMT) a member of the smectite group, is commonly used in polymer–clay nanomaterials. The layered structure of MMT (Figure 13.1) consists of two silica tetrahedral (corner-shared) sheets fused to one (edge-shared) octahedral sheet of alumina (aluminosilicate) or magnesia (magnesium silicate). The lateral dimensions of these layers vary from 100 nm to a few microns, and about 1 nm in thickness, depending on the particular silicate. Due to an isomorphic substitution of alumina into the silicate layers (Al^{3+} for Si^{4+}) or magnesium for aluminum (Mg^{2+} for Al^{3+}), each unit cell has a negative charge of between 0.5 and 1.3. The layers are held together with a layer of charge-compensating cations, such as Li^+, Na^+, K^+, and Ca^{2+}; the charge-compensating cations help the intercalation and surface modification of clays, which is required to disperse clays at the nanoscale into polymers. The cation-exchange capacity (CEC), which is indicative of the intercalation capacity, is defined as the number of exchangeable interlayer cations, and has units of mEq $100\,g^{-1}$. Typically, MMT has a CEC ranging from approximately 75 to 115 mEq $100\,g^{-1}$ [11].

Other clays are saponite, a synthetic material that closely resembles MMT, and hectrite, which is a magnesium silicate (CEC 55 mEq $100\,g^{-1}$) and its synthetic equivalent, Laponite®. Typically, sheets of Laponite are 25–30 nm wide, whereas MMT has sheets that are approximately 200 nm in width.

The dispersion of a hydrophilic nanofiller into a hydrophobic polymer matrix requires a preliminary modification of the clay to render it more organophilic. Consequently, a treatment is often carried out with suitable chemicals to replace the inorganic exchange ions in the galleries between the layers with alkylammonium surfactants or other organic cations. Clays treated in this way are referred to as "organoclays". The number of onium ions that can be packed into the galleries depends on the charge density of the clay and the CEC. In fact, at lower charge densities the surfactants pack in monolayers while, as the charge increases, bilayers and trilayers can be formed with different orientations of packed alkyl chains. The intercalation of organophilic cations into the layers can be followed using X-ray diffraction (XRD) analysis, with the distance between a plane in the unit layer and the corresponding plane in the next unit layer being defined as the "basal plane spacing" d_{001}, often known as the "interlayer distance". When a surfactant replaces sodium in the galleries, an increase in interlayer distance can be observed. The organic moiety used for preparing organoclays depends on the host polymer to be used as a matrix for the nanocomposites. Both, the structural and physical characteristics of two common ammonium salts (hexadecyltrimethylammonium bromide and stearyldimethylbenzylammonium chloride) are reported in Table 13.3 [7].

Table 13.3 Structure and physical characterization of two organoclays.

Organic moiety	CEC (mEq kg^{-1})	Moisture content (wt%)	Loss on ignition	Basal Spacing (nm)
HDTM Hexadecyltrimethylammoniumbromide	0.91	1.13	29.98	1.9
SMB Stearyldimethylbenzylammoniumchloride	0.91	1.05	34.19	2.5

The aim of the ion exchange process is to enlarge the interlayer distance so as to promote the accommodation of monomers or polymers, thus affording an organophilic surface and improving wetting between the clays and polymers.

13.1.2
Morphology of Composites

Polymer–clay composites based on layered silicates can be classified into three types, depending on the extent of separation of the silicate layers [12], namely conventional microcomposites, intercalated nanocomposites, and exfoliated nanocomposites (these are shown schematically in Figure 13.2). If the polymer does not enter the galleries, the d_{001} of the clay will remain unchanged, in agreement with the achievement of a microcomposite. If an organic species enters the galleries and causes an increase in d_{001}, but the clay layers remain stacked, then the composite is considered "intercalated." However, if the clay layers are completely pushed apart to create a disordered array, the composite is termed "exfoliated." Usually, a composite for which $d_{001} > 10\,nm$ (a spacing that cannot be determined by conventional XRD analysis) is designated as exfoliated. Recently, Sinha Ray et al. [13] defined another type of nanocomposite within the set of "intercalated nanocomposites" that they termed intercalated-and-flocculated nanocomposites, and which contained aggregates of intercalated silicate layers due to the hydroxylated edge–edge interaction of the silicate layers. In fact, most intercalated tactoids include both single stacks and several connected stacks of clay layers, such that the distinction between flocculation and intercalation will rest on an electron microscopic analysis of the structure. Moreover, the "flocculation" could be attributed to the long molecular chains that intercalate into two or more clay galleries and play a bridging role. Electron microscopy studies have often shown that most nanocomposites are both intercalated and exfoliated, but this cannot be readily deduced from XRD measurements. This situation may occur because the preparation method has not allowed sufficient time for adsorption and penetration

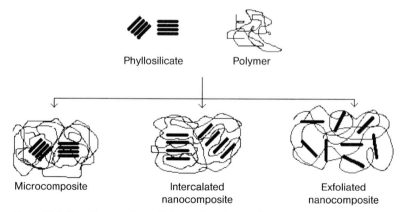

Figure 13.2 Classification of morphologies for polymer–clay composites.

of the galleries to be completed, or because the dispersive mixing has been less effective and that, if more time were available, the polymer–clay system might develop either fully intercalated or exfoliated structures. The extent of clay dispersion in polymer is dependent on the intrinsic properties of the polymer and clay, including the aspect ratio of the clay platelets, the volume fraction of the clay, the nature/structure of polymer, the interactions between the polymer, clay and clay modifier, and the processing conditions used. Fully exfoliated polymer–clay nanocomposites are, in theory, only found for volume fractions of clay less than 3%, as calculated by considering the small size of clay platelets [14].

Three principal methods are available to determine the exfoliated fraction:

- Quantitative XRD, using a strong and independent reflection from an internal standard, can potentially be used to track the decrease in the ratio d_{001} for the clay to the standard peak as exfoliation proceeds.
- A method based on NMR, which was recently established [15].
- Transmission electron microscopy (TEM) provides an indication, but is difficult to use for precise measurement unless very thin specimens are prepared, a large number of images are captured, and quantitative microscopy coupled with stereology is used.

Recently, Luo et al. [16] developed a statistical TEM image analysis methodology to evaluate the dispersion parameters $D_{0.1}$, based on measurement of the free-path spacing distance between the single clay sheets. Drummy et al. [17] reported a morphological characterization of layered silicate nanocomposites by using both electron tomography and small-angle X-ray scattering (SAXS). The latter method has the advantage of providing complementary results with respect to the electron tomography. In fact, it can be used to probe materials with subnanometer resolution while at the same time sampling a large number of nanoparticles, thus providing more than adequate statistics.

Rheological testing [18] might also be developed to measure the degree of exfoliation, but at present this is only used for semi-quantitative analysis. Hence, the morphological characterization of layered silicate nanocomposites requires the use of different techniques in order to achieve a relevance not only in the three-dimensional (3-D) representation of the nanometer-scale part of the sample, but also regarding the average distribution of platelets in the systems.

The characterization of layered silicate polymer blends is complicated by the need to study both the phase and filler distribution. In miscible systems, such as poly(methyl methacrylate) (PMMA)/poly(ethylene oxide) (PEO) [19, 20], the morphological analysis can be carried out as in a general polymeric matrix. However, in immiscible systems, such as poly(propylene) (PP)/poly(styrene) (PS), both XRD patterns and TEM observations have shown that the silicate layers were either intercalated or exfoliated, depending on their interactions with the polymer pair, and were located at the interface between the two polymers [21]. In this case, the compatibilizing action of an organically modified layered silicate resulted in a decrease in the interfacial tension and particle size, and in a remarkable increase in the mechanical properties of the modified immiscible blends. A similar result was obtained by Fang et al. [22], who investigated the morphology of nanocomposites based on 80/20 and 20/80 (w/w) poly(ε-caprolactone) (PCL)/PEO immiscible blends and organophilic layered silicates prepared by melt extrusion. From the TEM analysis, it was observed that the exfoliated silicate platelets were located preferentially at the interface between the two blend phases. However, when the blend-based nanocomposites were prepared via a two-step process, in which the silicates were first premixed with the PEO component or with the PCL component, the silicate layers migrated from the PEO phase or PCL phase to the interface. The emulsifying capability of layered silicates in immiscible blends depends on the structure and physical properties of the couple of polymer components. In fact, in PP/poly(amide) (PA) blends [23], it was shown, using electron microscopy, that in all cases the inorganic filler was enriched in the PA phase, and this resulted in a phase coarsening in comparison with the unfilled PP/PA blend. In contrast, in blends of a poly(vinylidene fluoride)/nylon-6 (PVDF/PA6) 30:70 melt compounded with various organoclays either directly or sequentially [24], the nanocomposite with the best mechanical properties was characterized by a good dispersion of particles throughout the matrix (PA6) and at the PVDF/PA6 interface. The authors ascribed this good result to a suppression of the coalescence of PVDF domains. Moreover, the crystallization of the PVDF domains was suppressed, ultimately creating a blend nanocomposite that was stiffer, stronger, and tougher than the blend without nanoparticles.

In agreement with the latter results, organoclay strongly influences the dynamic evolution of phase morphology, both in partially miscible and immiscible blends. A partially miscible blend was examined by investigating the poly(vinyl methyl ether) (PVME)/poly(styrene) (PS) [25] demixing in the presence of organically modified Laponite. The phase separation of these near-critical blends proceeds by a spinodal decomposition, even with added nanoparticles. However, the presence of nanoparticles slowed the phase-separation kinetics. In immiscible systems, such as poly(ethylene) (PE)/PA6 blends, the presence of an organoclay, preferentially distributed in the polyamide phase, effectively stabilized a co-continuous phase morphology [26] which persisted for more than 500 s. The authors explained this experimental evidence by suggesting that the co-continuous morphology could evolve to a phase-separated morphology in the timescale of their experiments, as the organoclay framework would radically slow down the melt state dynamics of the percolating network formed during the melt processing.

The presence of a compatibilizer [e.g., ethylene–propylene random copolymers (EPM), functionalized with maleic anhydride (EPM-g-MAH), or maleic anhydride-functionalized PP (PP-g-MAH)] in PP/PA6 blends containing organoclay, improves the phase morphology of the blends and promotes the formation of exfoliated nanocomposites, with the nanoplatelets preferentially distributed in the polyamide phase [27, 28]. In particular, atomic force microscopy (AFM) studies [29] carried out after selective chemical or physical etching of PP-g-MAH compatibilized PA6/PP/organoclay polished sample surface, showed that the organoclay was embedded in the PA6-g-PP phase of the PA6/PP blends compatibilized with PP-g-MAH. The preferential location of the clay in the PA6-g-PP phase was attributed to possible chemical interactions between the PA6 and the organic surfactant of the clay. The use of a proper organoclay, modified with the reactive (glycidoxypropyl)trimethoxy silane, in poly(lactic acid) (PLA)/poly(butylene succinate) (PBS) blends [30], allowed an exfoliated blend to be obtain contemporarily, and improved the phase adhesion between the two polymers. The authors ascribed this result to the formation of a grafted hybrid in the interphase region, based on the reaction of a glycidyl unit with the terminal hydroxyl or carboxylic groups of the polyesters.

In general, the phase morphology development and the compatibilization of nanofilled blends requires more studies to be conducted, in order to create a rational scheme concerning the effect of nanoscaled interactions, physical features, and rheological properties on filler and phase distribution, with the aim of modulating the final properties of the composites.

13.1.3
Properties of Composites

The dispersion of phyllosilicates into polymer or blends at nanometer scale allows improvements to be made to the properties of the polymer matrix [31]. In particular:

- the barrier properties to gases are enhanced [32], not only when an exfoliated morphology is obtained [33], but also in an intercalated morphology [34]. In fact, the results of fitting of experimental data with proper models showed that the chain confinement enhanced the barrier properties of the intercalated nanocomposites [35].
- Flame retardancy is improved, based on the dispersion of nanoclays, such that char formation is favored on the burning surface [36, 37].
- Stiffness is improved, with an increase in tensile modulus frequently being observed in phyllosilicate–polymer nanocomposites [21].
- Thermal stability, as degradation temperature, determined via thermogravimetric analysis (TGA), shows an increase of at least 7–8 °C when an exfoliated nanocomposite is tested [38].

Other properties of nanocomposites are currently the object of much research, and in some cases – an example is that of thermal behavior [39, 40] – these investigations have begun to establish new properties that might be exploited such that the nanocomposites can be used for a variety of industrial applications.

13.2
Polyolefin-Based Nanocomposites

The polyolefins (POs) represent one of the most widely used classes of thermoplastic materials, as they offer a unique combination of high thermomechanical properties and good environmental compatibility, at low cost.

Currently, a huge effort is being devoted to extend the field of application for POs and to satisfy their market demand, the latter point being based on these materials having specific and advanced properties, and being adaptable to the environment and, ideally, recyclable. In particular, the preparation of PO nanocomposites by the addition of very small amounts of layered silicate loadings is a very common approach that provides drastic improvements in the elasticity modulus, strength and heat resistance of the materials, as well as a substantial decrease in their gas permeability, when compared to either virgin POs or conventional microcomposites [41–49].

Attempts have been made to prepare either intercalated and exfoliated nanocomposites, through different routes, and extensive studies have been conducted to promote an understanding of the fundamental aspects associated with the clay dispersion, the resultant physical properties, and the inter-relationship between both. Indeed, the properties of polymer–clay nanocomposites depend largely on the synergistic effect of nanoscale structure and interactions occurring between the clay and polymer.

The achievement of a desired level of clay dispersion into the polymer, and interaction between the phases, involves numerous chemical and technological implications. Generally, the chemical modification of one or both system phases is necessary. For this, the hydrophilic clay is often changed to hydrophobic either by the cation exchange or grafting of alkoxysilanes, by exploiting the cation-exchange capability in one case and the reactivity of silanol groups present on the edges of the clay platelets in the other case. Typically, a compatibilizer–that is, a functionalized PO or copolymer–is added to assist in this situation. At present, it is not completely clear which conditions lead to exfoliated layers and which cause only the intercalation of polymer chains into the gallery spacing of the stacked silicate layers. However, what does appear well established is that the interface is not only fundamental for the adhesion between the two phases and stabilization of the attained morphology, but also plays a major role in the bulk properties, due to its large extension [39].

Clearly, the ability to predict and control the morphology of nanocomposites would be of major interest, by virtue of the strict connection between structural control and the final properties of the nanocomposite [49, 50]. Consequently, the main objective of this section is to review the concepts and methods relating to the optimization of layered silicate dispersion in PO matrices by melt intercalation, notably with regard to the use of a compatibilizer and identification of the preferred process conditions.

13.2.1
Overview of the Preparation Methods

Two major synthetic strategies are generally followed for preparing PO/layered silicates nanocomposites: (i) polymer intercalation (including melt intercalation [51–56] and solution intercalation methods [57]); and (ii) *in situ* intercalative polymerization [58–62].

The *in situ* intercalative polymerization method, which is based on the thermo-mechanical forces generated during the process of olefin polymerization inside the interlayers of the silicate, and the energy released during the polymerization, is reported to promote silicate layer dispersion and exfoliation in the nonpolar PO matrix. However, a very poor interaction between the polar silicate layers and the hydrophobic PO matrix makes the nanocomposite structure thermodynamically instable. In fact, it has been demonstrated that melting processes of this type of nanocomposite material cause agglomeration of the silicate layers and collapse of the nanocomposite structure. Even the more recently developed *polymerization filling technique*, which consists of attaching the polymerization catalyst to the surface and into the interlayers of the silicate and polymerizing the olefin *in situ*, although favoring exfoliated structures does not allow thermodynamically stable nanocomposites to be prepared [63]. The method which should be adopted for stabilizing the morphology is to copolymerize the olefin with a functionalized comonomer, thus improving the necessary interaction with the clay. However, many challenges remain for the catalytic polymerization of polar monomers with both traditional Ziegler–Natta and metallocene catalysts, due mainly to the catalyst deactivation.

For *solution intercalation*, the enthalpy of dispersion of the silicate layers in solution must balance the entropy loss of the solvent molecules which intercalate into the silicate layers [64]. Moreover, this method is poorly sustainable because it requires a large amount of organic solvent to ensure a good clay dispersion.

Melt intercalation is the most commonly applied process for preparing PO nanocomposites, because it is sustainable, versatile, and inexpensive. This methodology is limited by the low interaction enthalpy between the nonpolar PO chains and the polar silicate layers. In fact, when using polymer chain intercalation between the silicate layers, an effective and stable morphology is achieved only if the interaction enthalpy balances the entropy loss of the polymer chains confined within the silicate interlayers. However, it has been recognized that the entropy loss due to such polymer confinement is only partially compensated by the entropy gain associated with the increased conformational freedom of the surfactant tails as the interlayer distance increases with polymer intercalation [65, 66], whereas favorable enthalpic interactions are critical for determining the nanocomposite structure [67]. Hence, even if a good dispersion of the clay can be achieved simply by applying strong shear forces during the process, the structure will be unstable and reagglomeration of the particles may occur quite rapidly if an effective interaction between the polymer and clay surface is not ensured [44].

It is somewhat evident that the PO hydrophobicity makes it almost impossible to obtain intercalation without chemical modification of the clay or polymer, or both. In order to overcome the above-described thermodynamic obstacles, and to promote the interaction between the hydrophobic PO and the polar silicate layers, the general approach is to use either: (i) organophilic layered silicates (OLSs), obtained via cation exchange; or (ii) polymer compatibilizers (i.e., functionalized polyolefins).

Whilst the dispersal of layered silicates in polar polymers is relatively straightforward, the achievement of comparable results with hydrophobic POs remains a challenge. Favorable interactions at the inorganic/organic interface are necessary in order to separate the clay monolayers and to successively grant cohesion forces between the exfoliated layers and polymer chains, thus achieving a stable nanostructure.

13.2.2
Organophilic Clay and Compatibilizer: Interactions with the Polyolefin Matrix

Two main methods can be applied for clay modification: (i) to replace the metallic cations that were originally present on the clay surface with large organic cations; and (ii) to graft alkoxysilanes onto the clay by exploiting the reactivity of the silanol groups [64].

The first of these methods has been widely applied, using several types of cation (mostly dialkyldimethyl- or alkyltrimethyl-ammonium cations). The introduction of quaternary ammonium salts with long alkyl chains between the layers (as discussed above) allows the interlamellar space to be expanded, reduces the interaction among the silicate sheets, and also facilitates the diffusion and accommodation of the polymeric matrix. In contrast, the grafting of alkoxysilanes has been used only rarely with layered silicates, due to the small number of reactive-edge silanols present. Moreover, it has been shown recently, by using a combination of morphological analysis and multinuclear solid-state nuclear magnetic resonance (NMR) experiments [68], that the modification of a synthetic Laponite with a combination of organic ammonium cations and alkoxysilanes will result in a disordered arrangement of the clay platelets (which favors interplatelet interactions). This effect was due to self-condensation occurring among clay silanols at the platelet edges, thus limiting dispersion into the polymer matrix.

Several reports have described the use of OLSs modified with alkylammonium surfactants of different nature and length for the preparation of PO/OLS nanocomposites [52, 69]. In particular, the effect of factors such as the length and number of alkyl groups of the cationic surfactant was investigated. It became apparent that the equilibrium structure of the polymer/OLS depended on the nature of the polymer, on the charge density of the layered silicate, and also on the chain length and structure of the cationic surfactant. Balazs et al. [70, 71] showed that an increase in surfactant length improved the separation of the layers by allowing the polymer to adopt more conformational degrees of freedom. Both, exfoliated and intercalated PE/OLS nanocomposites were obtained when the number of methylene units in the alkylammonium chain exceeded 16 [72], whereas it was reported that an excessive density of tethered chains would prevent the formation of intercalated structures. The dispersion level was found to depend strictly on the amount of OLS; in particular, it was shown that the morphology of the PO/OLS systems could be shifted from disordered exfoliated to predominantly intercalated by varying the OLS content from 6 to 36 wt% [73]. Finally, from a thermodynamic point of view, OLSs modified with alkylammonium salts were shown to be adequate in offering a surplus enthalpy for promoting OLS dispersion into a nonpolar matrix, as did a PO. In fact, the surfactant/silicate interactions were more favorable compared to those of PO/silicates. Hence, the synthetic approach was directed towards reducing the enthalpic interactions between the clay and surfactant by using semi-fluorinated surfactants [52] as clay modifiers. Although, in the case of PP/OLS composites this approach led to a significant improvement in miscibility, this type of surfactant was seen to be much too expensive for use in the mass production of PO nanocomposites.

A more general approach for improving the compatibility of a PO with OLS was to introduce into the system a compatibilizer as the coupling agent. In general, the compatibilizer was a functionalized PO, such as a maleic anhydride (MAH) or maleate ester-grafted PO, an oligomeric-functionalized PO, or an ammonium-terminated or OH-terminated PO, as well as a suitable block or random copolymer (Table 13.4). These types of polymer or oligomer reflected some of the theoretical calculations carried out to probe the interactions between polymers and clay sheets, and were used substantially to design the ideal compatibilizer [70, 71]. It became clear that for the simple penetration of polymer chains into the clay gallery, the compatibilizer should contain a fragment which would be highly attracted by the clay surface. In addition, it should incorporate a longer fragment that was not attracted by the sheets, but rather would attempt to gain entropy by pushing the layers apart. Subsequently, when the layers had separated, the surfaces would be sterically hindered from coming into close contact. In this way the compatibilizer could be seen to behave as a steric stabilizer for the PO/OLS colloidal suspension. Thus, the role of the compatibilizer is not only to favor a layered silicate dispersion and to enhance the polarity of the polymer matrix, but also to stabilize the final morphology. The strong interactions between the compatibilizer and clay are not only fundamental for the dispersion step, but also for maintaining the morphology.

Initial attempts to improve the interactions between nonpolar PO and layered silicates were carried out by melt-mixing the polymer and clay with modified oligomers, so as to mediate the polarity [51, 55–57, 74]. Usuki et al. [55] were first to report the preparation of PP/layered silicate nanocomposites using a functional oligomer (PP–OH) with polar telechelic –OH groups as compatibilizer. In this approach, PP–OH was intercalated between the layers of an OLS, after which the PP–OH/OLS was melt-mixed with PP to obtain nanocomposites with an intercalated structure. Another typical example to be reported was that of PP/layered silicate nanocomposites; these were prepared by Toyota, using PP oligomers modified with approximately 10 wt% of MAH groups [51, 56] and clays exchanged with stearyl ammonium cations. Wide-angle X-ray diffraction (WAXD) analyses and TEM observations established the intercalated structure for all of these nanocomposites. Here, the driving force of the intercalation was considered to have originated from the interaction between the MAH group and the oxygen groups of the silicate, through the formation of hydrogen bonding (Figure 13.3).

In general, the compatibilizer is a polymer in which the functional groups are distributed randomly along the backbone, or functionalized at the chain end. Both,

Table 13.4 Some examples of compatibilizers for the preparation of PO/OLS nanocomposites.

Types of compatibilizer for PO/OLS systems	Examples	Main characteristics	Reference(s)
End-functionalized oligomers (i.e., OH-; MAH- terminated)	OH end-functionalized PP oligomer	M_w (by GPC) 20000; OH value 54 mg KOH g^{-1}	[52, 56, 58]
	MAH end-functionalized PP oligomer	M_w (by GPC) 12000–30000. Acid value 7–52 mg KOH g^{-1}	[52, 56, 58]
	MAH end-functionalized PE oligomer	M_n 5700. Acid value; weight fraction of polar comonomer 0.01). This PE-g-MA has a low anhydride content so that, on average, there is one anhydride moiety per two molecules. It can be considered as an end-functionalized PE oligomer.	[64]
End-functionalized polymers (i.e., NH$_3$-, OH-, COOH- terminated)	Linear end-functionalized polyethylene with dimethyl ammonium chloride end groups	Narrow molecular weight distribution	[74, 75]
	Polyethylene with multiple quaternized amine groups along the polymer chain Cl$^-$ $^+$NH Cl$^-$ $^+$NH $^+$NH Cl$^-$ Me Me Me Me Me Me	Narrow molecular weight distribution	[75]

Table 13.4 Continued.

Types of compatibilizer for PO/OLS systems	Examples	Main characteristics	Reference(s)
Block copolymers	Poly(ethylene-block-methacrylic acid) ~~~~CH$_2$CH$_2$CH$_2$~~~~CH$_2$C~~~~ with CH$_3$, C=O, OH groups	Narrow molecular weight distribution	[75]
	Poly(ethylene-block-ethylene glycol) H–[CH$_2$–CH$_2$]$_n$–[O]$_m$–OH	M_n (575 mol g^{-1}); 0.20 molar fraction of polar comonomer; PE-b-PEG is a block-copolymer, the chains of which contain 33 methylene groups and 2.6 ethylene oxide units per molecule on average	[64]
	Poly(propylene- block -methyl methacrylate) (CH$_2$–CH)$_y$ (CH$_2$–C)$_x$ CH$_3$ CH$_3$, C(=O)OCH$_3$	M_w around 200 000; molar fraction of MMA units 1 or 5	[52]
	Styrene-b-ethylene/bu-tylenes-b-styrene block copolymer	A triblock copolymer based on hydrogenated polybutadiene (10 wt% ethyl branches, M_n = 37 000 g mol^{-1}) central block and polystyrene (30 wt%, M_n = 7000 g mol^{-1})	[76]

Table 13.4 Continued.

Types of compatibilizer for PO/OLS systems	Examples	Main characteristics	Reference(s)
Random copolymers	Poly(ethylene-co-vinyl alcohol)	0.69 weight fraction of polar comonomer	[64, 76]
	Poly(ethylene-co-methacrylic acid)	M_n 68 000; 0.11 weight fraction of polar comonomer.	[64]
	PP-r-(PP-OH)$_x$	These copolymers derive from a random polypropylene copolymer which contains 1 mol% p-methylstyrene comonomer. Mw around 200 000; molar fraction of polar units 0.5	[52]
	PP-r-(PP-MA)$_x$		

Table 13.4 Continued.

Types of compatibilizer for PO/OLS systems	Examples	Main characteristics	Reference(s)
Randomly functionalized polyolefins	Maleic anhydride (MAH)-functionalized polyolefins	Functionalization degree ranging between 0.1 and 2% by mol	[32, 48, 51–55, 72, 77–82]
	Diethyl maleate (DEM;-functionalized polyolefins	Functionalization degree ranging between 0.1 and 2% by mol	[32, 79–81, 83–85]
	Maleic anhydride (MAH) and Butyl furanyl propenoate (BFA)-functionalized polypropylene	Functionalized polypropylene having high molecular weight	[86–89]

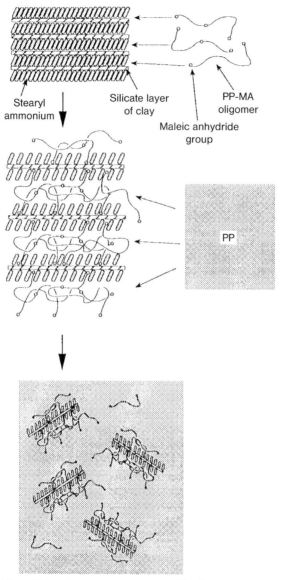

Figure 13.3 Schematic representation of the dispersion process of the organo layered silicate (OLS) in the poly(propylene) (PP) matrix with the aid of a maleic anhydride functionalized PP oligomers. Reproduced with permission from Ref. [51]; © 1997, American Chemical Society.

simulation studies and experimental results [74, 90, 91] produced evidence that end-functionalized polymers – unlike more functionalized polymer systems – assumed a special configuration on the clay surface which allowed complete exfoliation of the clay. In particular, in the case of the PP/OLS nanocomposites, it was shown that by using an ammonium group-terminated PP (PP-NH$_3^+$) as compatibilizer, the terminal hydrophilic NH$_3^+$ groups would be attached by cation exchange on the inorganic surfaces, while the hydrophobic high-molecular-weight PP tails would aid exfoliation of the structure, though this would be maintained even after further mixing with undiluted PP [74]. In contrast, side chain-functionalized PPs or PP block copolymers were seen to form multiple contacts with each of the clay surfaces, which resulted not only in the polymer chains being aligned parallel to the clay surfaces, but also consecutive clay platelets being bridged, thus promoting the production of intercalated structures. This bridging effect was established, for example, by using side chain-functionalized polyolefins characterized by large (>2 wt%) amounts of grafted polar groups [74, 90, 91].

Without doubt, however, the most popular compatibilizers are POs functionalized with MAH by free radical processes. PP functionalized with MAH (PP-g-MAH) was shown to intercalate into the inter-galleries of OLS, much like the analogous oligomers [77]. In fact, a WAXD analysis of the PP/OLS nanocomposites revealed an absence of any peaks representing dispersed OLS in the PP-g-MAH matrix; hence, in this case also, the strong hydrogen bonds between the MAH-grafted groups and the polar clay surface promoted dispersion of the layers. Similar results were obtained for PE–PP rubber (EPM)/OLS nanocomposites which had been prepared by melt blending EPM-g-MAH and OLS modified with C18 alkyl ammonium chains, in a twin-screw extruder [78]. Confirmation of the intense interactions between the functional groups of the compatibilizer and the lamellae of OLS was obtained from experiments with nanocomposites that had been prepared by blending an EPM functionalized with diethylmaleate (DEM) (with 1 mol% grafted functional groups) with increasing amounts (up to 20 wt%) of OLS. The use of a functionalized, rather than virgin, EPM allowed (in the case of 5 wt% OLS nanocomposites) the shift to be made from an intercalated morphology to an almost completely exfoliated morphology [83] (Figure 13.4).

It was hypothesized that the hydrogen bonds between the –OH groups present on the OLS surface and the carbonyl groups of the functionalized EPM allowed both optimization and stabilization of the clay dispersion (Figure 13.5). Most likely, the compatibilizer was placed between the polar silicate layers and contemporarily mixed with the nonpolar polymer chains, thus creating a concentration gradient suitable for providing an exfoliated morphology and avoiding phase separation.

The strong interactions between the OLS and the EPM-g-DEM were confirmed by extracting the nanocomposites with hot toluene; a significant proportion of the rubber proved to be insoluble (up to 55 wt% for 15–20 wt% OLS), and the amount of nonextractable polymer was increased by raising the OLS loading [32, 84]. The occurrence of strong polymer/clay interactions was also proved by comparing the infrared (IR) spectra of the composite and residue (Figure 13.6). The

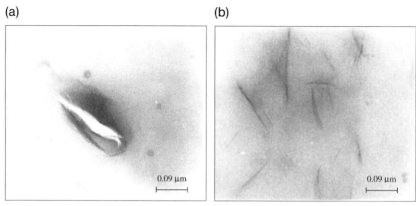

Figure 13.4 Transmission electron microscopy images of (a) ethylene propylene rubber (EPM)/organo-layered silicate (OLS) and (b) EPM functionalized with diethyl maleate (EPM-g-DEM)/OLS 5 wt% nanocomposites. Reproduced with permission from Ref. [84]; © 2005, John Wiley & Sons, Ltd.

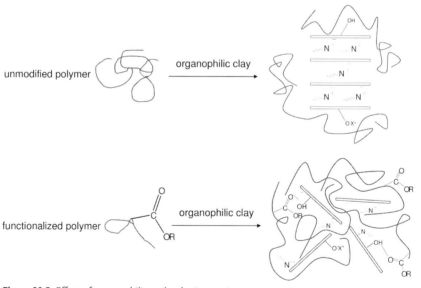

Figure 13.5 Effect of organophilic and polar interactions on the morphologies of ethylene–propylene rubber (EPM) and EPM functionalized with diethyl maleate (EPM-g-DEM)/organo-layered silicate (OLS) composites.

C=O stretching band of the carbonyl ester group of the EPM-g-DEM, when retained in the toluene-extracted residue, was significantly broad and was shifted to lower wavenumbers compared to neat EPM-g-DEM. Moreover, the shoulder at approximately 1715 cm^{-1} suggested a hydrogen-bond interaction between the EPM functional groups and the silicate surface [85].

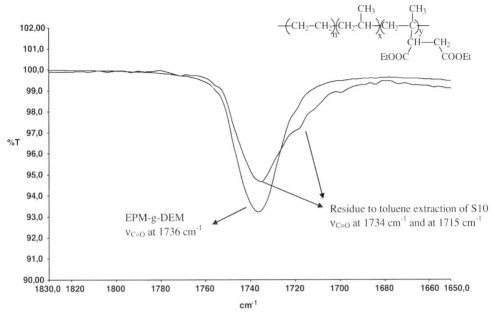

Figure 13.6 Attenuated total reflectance (ATR) spectra of ethylene–propylene rubber functionalized with diethyl maleate (EPM-g-DEM) and residue to toluene extraction of a PE-g-DEM/organo-layered silicate (OLS) nanocomposite 10 wt% OLS. Note the enlargement of the C=O stretching region. Reproduced with permission from Ref. [84]; © 2008, Elsevier Ltd.

Evidence of the strong interaction between the EPM-g-DEM and the layered silicate was also provided with differential scanning calorimetry (DSC) and dielectric relaxation analyses. Initially, the insoluble toluene fraction was seen to have a glass transition temperature (T_g) which was appreciably higher than that of the original composite. Furthermore, the dielectric relaxation analysis of nanocomposites indicated a reduction in both the dielectric strength and α-relaxation frequency as the temperature was increased (Figure 13.7). As the dipolar relaxing units are the side DEM grafted groups, this phenomenon was explained by considering a partial immobilization of such groups on the silicate surface, as consequence of the nanoconfinement of polymer chains. The α-relaxation was shown to disappear for the residue to the toluene extraction, while the cooperative dynamics of the dipolar groups of EPM-g-DEM were strongly reduced as they were interacting with the layers.

The amount of unextractable polymer, the existence of which confirms the polymer/OLS interactions, and can be related to an adsorbed/interacting confined polymeric fraction, was correlated (using a rough geometric model) to the volume occupied by trapped and intercalated macromolecules (Figure 13.8).

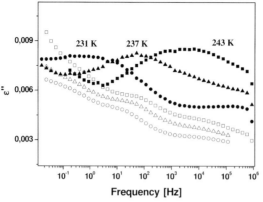

Figure 13.7 Dielectric loss spectra of nanocomposite organo-layered silicate (OLS) 20 wt% (closed symbols) and its residue to solvent extraction (open symbols) above the glass transition temperature. The same symbol type corresponds to the same measurement temperature. Reproduced with permission from Ref. [84]; © 2008, Elsevier Ltd.

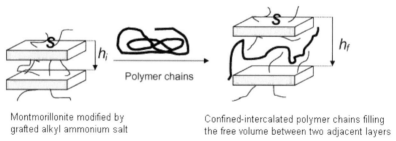

Montmorillonite modified by grafted alkyl ammonium salt

Confined-intercalated polymer chains filling the free volume between two adjacent layers

Figure 13.8 Schematic model of polymer confinement. Reproduced with permission from Ref. [84]; © 2008, Elsevier Ltd.

This comparison between experimental and theoretical data provided evidence of the existence of different interaction levels between the polymer and the organophilic clay, which together caused the formation of an unextractable polymer, namely:

- polymer–surfactant interactions
- polymer–clay surface chemical and physical interactions
- intercalated polymer–bulk polymer interactions.

By applying this model, the PO/OLS nanocomposite can be considered as a system containing at least two phases: (i) the polymer matrix, which does not truly interact with the layers' surface (and thus is extractable with the same T_g of pristine PO); and (ii) a nanostructured phase containing polymer chains that strongly interact and are "confined" (and thus are not extractable, with a higher T_g than pristine PO and the starting composite). Here, the Gordon–Taylor equation was employed to calculate the amounts of both polymer phases by considering the T_g of the polymer in the nanocomposite (as determined by the T_g of the free polymer), and of the confined polymer. This analysis provided results which, in terms of their behavior, were in agreement with experimental data.

Interestingly, it was demonstrated that the unextractable polymer was responsible for reductions in the oxygen permeability of LDPE and EPM/OLS nanocomposites prepared by using functionalized matrices [85]. In fact, by assuming a negligible oxygen permeability through the polymer phase adsorbed onto the surface or trapped inside the galleries of the OLS, these experimental data were finely correlated with the theoretical data predicted from mathematical models.

Besides the nature of the functional groups, it was also clear that the amount of functional groups grafted onto the PO could influence the final morphology of the PO/OLS nanocomposite. In fact, the hydrophilicity of the compatibilizer was seen to increase with the grafting level of MAH, thus improving the level of interaction between the polymer and the clay. In the case of the most common PO-g-MAH compatibilizers, the level of grafting ranged typically between 0.1 and 2 wt% [50, 85, 92, 93]. In the case of MAH-functionalized PO, there was seen to be a limit value for the grafting level (ca. 0.1 wt%), for achieving a good improvement in morphology. In the case of both PE and PP/OLS nanocomposites, if the grafting level was higher than this limit value, a shift or even a disappearance of the basal reflection peak of the OLS on the XRD diffraction pattern could be observed [51, 72, 77, 78, 92].

In general, because of the low grafting percentage of polar groups onto the common commercial compatibilizers, a large amount of functionalized polymer is added to the polyolefin matrix in order to achieve the necessary polarity and hence compatibility with the organoclay. Besides the degree of functionalization (DF), the molecular weight and structure of the compatibilizer, as well as its melt viscosity and rheological properties, are responsible for the composite final morphology [93, 94]. The best results are generally obtained by using a compatibilizer for which the rheological properties are similar to those of the matrix; otherwise, the miscibility between the two polymers might be compromised and the intercalation/exfoliation process inhibited. A typical example is that of PP, for which a high amount of compatibilizer is required to obtain an increase in the OLS basal spacing, but this may cause a deterioration in the properties of the nanocomposite due to the low molecular weight of the commonly used PP-g-MAH samples. Unfortunately, functionalized PPs with high molecular weights are yet not available commercially. The conventional radical functionalization process with peroxide and MAH in the case of PP causes a dramatic decrease in the molecular weight of the polymer, leading to severe damage of its rheological and mechanical properties. Notably, PP is very sensitive to degradation reactions when treated with peroxides above its melting temperature, even in the presence of commonly used maleate functionalizing agents [79–81, 94].

Consequently, conventional PP-g-MAH compatibilizers will in general have a low DF and molecular weight. The problem of obtaining appropriately functionalized PP can be overcome by a new radical functionalization approach involving the use of a furan derivative (BFA). This is added as coagent during the radical functionalization of PP with MAH, which can yield a wide range of PP-g-MAH samples with different DFs and molecular weights, by controlling the macroradical formation and content [86, 87]. These new functionalized PP samples were recently tested in the preparation of PP/OLS nanocomposites, both as a matrix or as a compatibilizer of the system, the aim being to study the effects of both polarity and chain structure/architecture on clay dispersion and the ultimate properties of the corresponding nanocomposites. The results showed that PP-g-MAH with a low molecular weight and a high DF (>2 mol%) had an excellent ability to disperse clay at the nanometric level, especially when used as the matrix of the corresponding nanocomposite. Those samples characterized by a high DF value (>2 mol%) and a branched structure/architecture produced nanocomposites with a lower degree of exfoliation. However, those nanocomposites with a composition of 90/5/5 PP/compatibilizer (the functionalized PP)/organomodified layered silicate, where the sample contained the compatibilizer characterized by a high DF; prepared using a grafting procedure to avoid PP degradation reactions) provided the best performance in terms of morphological and thermomechanical properties. These results not only confirmed the important role of the DF, but also highlighted the fact that the control of molecular weight and structure/architecture during functionalization ensures a good compatibility of the compatibilizer with the PP matrix, which in turn has a positive effect on the ultimate properties of the PP/OLS. In particular, elongation at break point (which usually are poor for similar systems) reached values in excess of 500%, with an excellent reproducibility [88].

13.2.3
The One-Step Process

In addition to a correctly selected OLS and compatibilizer, it is also necessary to utilize an effective mixing protocol (e.g., processing parameters such as mean residence time, screw speed, etc.) capable of promoting both high shear stress and shear rate, in order to assist the OLS dispersion into a PO matrix [82, 95]. It has been shown that the residence time and kneading force applied during extrusion can have a direct effect on the extent of the layered silicate exfoliation [96]. In the case of LDPE/OLS nanocomposites, a correct balance between dispersive and distributive actions, by fine-tuning the screw profile, allowed an optimal nanocomposite morphology [82]. A good dispersion of clay layers was assessed by preparing a master-batch PE-g-MAH/OLS, which highlighted the importance of the shear stress for dispersing, intercalating, and exfoliating lamellae when the correct interface interactions have been provided. A further improvement in dispersion degree was observed during a second processing step, as the melt viscosity of the two polymer components was accurately selected with the aim of improving their compatibility.

Besides the machine parameters, the method selected for feeding the extruder – and therefore also the order of addition for components – can influence the morphology of the final material. In most reports, the best stack delamination was observed when a master batch between the functionalized PO (compatibilizer) and a large proportion of clay (10–50 wt%) was first prepared, followed by dilution in the nonpolar PO matrix (master batch process). The advantage of this method, compared to a direct melt blending of all ingredients together, is that it facilitates the intercalation of the polar polymer chains between the clay stacks, thus obtaining a form of pre-intercalated material.

One rarely discussed aspect is the possibility of obtaining, in a single step, both functionalization of the PO and nanocomposite formation. Such a method would be especially interesting because its successful application would not only avoid selecting the compatibilizer but also save a preparative step. In fact, as noted above, selection of the best compatibilizer may require special care, as this must be sufficiently polar to allow interaction with the clay, but not so hydrophilic as to favor phase separations. Not least, it should have similar rheological properties as the matrix, in order to avoid any deterioration of the composite's ultimate mechanical properties.

Consequently, two different synthetic approaches were very recently compared for the preparation of EPM and PP OLS nanocomposites [89] (Figure 13.9):

- The first approach is a classical procedure where the dispersion of OLS is achieved by using, as the compatibilizer of the system or as a matrix, a previously functionalized PO bearing carboxylate groups on the backbone (two-step process, Method 1). This method involves a first step when the functionalized PO is prepared. Frequently, the traditional approach for the preparation of functionalized PO is by free radical grafting of polar unsaturated molecules.

- The second process is carried out by contemporaneously performing the functionalization of the PO (by using maleic anhydride and/or diethylmaleate and a peroxide as radical initiator) and the dispersion of the filler (one-step process, Methods 2 and 3).

The results obtained using the two-step process (Method 1) indicated that, by using the functionalized PO as matrix, only a fraction of the polar groups on the polyolefin macromolecules would be effective in establishing a bonding with the silicate surface. In order to facilitate the functionalized polymer–clay interactions, it seems convenient to perform the functionalization during mixing of the PO with the clay. In this way, the small polar molecules could more easily penetrate the clay channel, with grafting to the organic macromolecules occurring successively. Comparative experiments were also carried out according to the one-step procedure, following two different routes – Methods 2 and 3 (see Figure 13.9).

In Method 2, the functionalizing reagents and organically modified montmorillonite (OMMT) were premixed and added contemporarily to the molten polymer in a Brabender mixer, such that functionalization and intercalation/exfoliation could occur simultaneously. In Method 3, the functionalizing reagents (initiator,

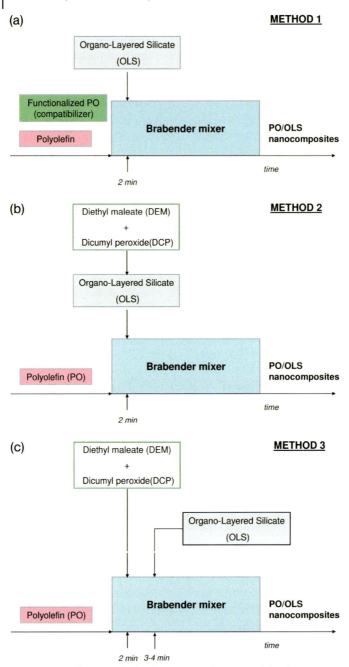

Figure 13.9 Different methods of preparation for compatibilized PO/OLS nanocomposites.

DEM and, occasionally, MAH) were added to the molten polymer, with the OMMT being added after a few minutes. In the former case, the operating conditions the aim essentially was first to have intercalation of the monomer, followed by grafting to the macromolecules. In the latter case, the procedure was essentially similar to the two-step method (Method 1), although for kinetic reasons there was a degree of overlapping within the time scale between the grafting of DEM to PO, and intercalation.

The results obtained indicated that, in terms of the role played by the different parameters (including the polyolefin chemical structure), the process was very complex. In the case of the random ethylene/propylene copolymer (EPM), the simultaneous addition of a functionalizing reagent and OMMT to the molten polymer (Method 2) was the most effective approach, even if an optimal ratio between the DF and clay basal spacing enlargement appeared to exist. In the case of PP, however, the application of Method 2 did not provide the same results, with a better dispersion being obtained via Method 3. Indeed, the polymer functionalization was seen to be favored, and degradation controlled, by the presence of clay. However, the best polymer chain intercalation was obtained when degradation was significant during the functionalization process of PP, indicating that a reduction in molecular weight which favored intercalation would probably be more effective than the presence of a larger amount of grafted polar groups.

13.3
Poly(Ethylene Terephthalate)-Based Nanocomposites

Poly(ethylene terephthalate) (PET), which is one of the most diffused polymers, is used in a wide variety of applications, notably the packaging and textile industries. The need to improve the stiffness, surface properties and/or gas barrier properties of PET has led to a plethora of investigations, in both academia and industry, into PET-based nanocomposites. Within this area, those nanocomposites which contain phyllosilicates are considered superior for providing improved gas barrier properties, due mainly to the strong effect of confinement as the result of a high surface:volume ratio (i.e., reducing chain mobility and permeability [35]), as well as to the enhancement of tortuosity [97] of the path required for small molecules to permeate through a polymer film due to the presence of silicate lamellae. Flame retardancy and gas barrier functions are much-required properties in the area of textiles, the main aim being to develop fabrics with flame self-extinguishing properties [98]. Consequently, many research groups have recently investigated PET/phyllosilicate nanocomposites in an attempt to provide solutions to these important technological problems [99–102].

One especially important field of application is that of recycling post-consumer PET [103]. The recovery of PET from post-consumer packaging products (e.g., beverage bottles) allows flakes to be obtained which have purity levels suitable for reprocessing. Unfortunately, in many countries (e.g., Italy) the recovered material cannot be recycled via a bottle-to-bottle approach, as legislation decrees that post-consumer polymers cannot remain in contact with food. Consequently, there is a need to develop new materials based on post-consumer PET flakes, in order to achieve an effective recycling. At present, blending these flakes with other polymers, especially rubber [104–111], to achieve toughened blends, and/or adding fillers [112], represent the most frequently investigated strategies for producing materials with added value by utilizing post-consumer flakes. This final point is especially important because, if recycling cannot be used to provide materials that perform similarly to commonly used plastic materials, then landfilling or energy recovery using plastic wastes will become inevitable. Nanofillers – and in particular phyllosilicate-based nanocomposites – may well provide the means of modulating the mechanical, thermal, and flame-retardancy properties of post-consumer PET, thus broadening its field of application.

Today, both technological and environmental driving forces continue to attract the attention of scientific research towards PET–phyllosilicate nanocomposites.

13.3.1
In Situ Polymerization

The preparation of PET/MMT nanocomposites by *in situ* polymerization has been the object of several investigations. In particular, Hwang *et al.* [113] reported the preparation of PET nanocomposites containing pristine and organically modified MMT by *in situ* polymerization, since this technique has been shown to produce a homogeneous dispersion of MMT particles in the polymer matrix. The polymerization of PET was carried out through polycondensation [114, 115]. In an esterification tube, ethylene glycol and sodium or organically modified MMT and a catalyst of transesterification (zinc acetate) were subjected to ultrasonication before the ester interchange reaction was initiated. Dimethyl terephthalate was then added to the ethylene glycol slurry to obtain a homogeneously dispersed system. The mixture was then heated to 210 °C to initiate esterification between the silicate layers in the clay. The ester exchange reaction was carried out for 3 h while continuously removing methanol. Finally, polycondensation was performed at reduced pressure, at temperatures between 180 and 285 °C, by adding antimony (III) oxide catalyst as polycondensation catalyst.

The degree of exfoliation was determined by TEM measurements at 20 nm resolution; these showed a clear separation between the silicate layers of the clay, enhancing differences between the composites prepared using either sodium-MMT or OMMT which had been modified with dimethyl, benzyl, hydrogenated tallow, quaternary ammonium. The TEM images revealed the presence in 1% MMT composites of uniformly oriented sodium-MMT particles with complete layer stacks containing an average of 66 silicate sheets per particle, whereas particles in PET/OMMT were largely exfoliated, with a larger proportion of intercalated PET, as evidenced by significantly smaller particles composed of only four to five randomly oriented silicate sheets. Scanning electron microscopy (SEM) images of the nanocomposites confirmed that the organophilic modifier had significantly improved the compatibility between PET and MMT.

DSC analyses showed that the crystallization temperature of PET/sodium-MMT during cooling gradually increased, while the degree of supercooling decreased with clay loading. This suggested that the sodium-MMT had acted as a nucleating agent and accelerated the crystallization rate of the PET matrix. In subsequent heating runs, the nanocomposites exhibited two melting peaks compared to pure PET. The lower melting temperature peak was attributable to the melting of imperfect crystals, which formed as a result of the nucleation effect of sodium-MMT during the cooling run. The higher melting temperature peak corresponded to melting of the melt-reorganized crystal.

DSC curves obtained for the PET/organophilic MMT exhibited an increase in the crystallization peak intensity during the cooling run at 0.5 wt% loading. This peak decreased with increasing clay content, although it maintained a higher intensity than that of neat PET. This indicated that the nucleating effect of organophilic MMT was lower than that of sodium-MMT. This may be attributable to interference by the alkyl groups on the organophilic MMT surfaces with secondary nucleation and diffusion of PET molecules, resulting in a decrease in the DSC crystallization peak.

The dispersion of rigid, exfoliated MMT layers served to enhance the macroscale stiffness of the composite material. Elongation at break values was drastically decreased due to increased stiffness and the formation of microvoids around the clay particles during tensile testing. The main factor contributing to the enhancement of mechanical properties in PET/clay nanocomposites was not the quantity of clay, but rather the degree of dispersion and exfoliation of the clay in the PET matrix.

A very similar *in situ* polymerization procedure was adopted by Chang and Mun [100], who prepared PET nanocomposites by using dodecyltriphenylphosphonium-modified MMT as a reinforcing filler in the fabrication of PET hybrid fibers. The group studied the properties of the composites obtained, for organoclay contents ranging from 0 to 5 wt%, by applying different draw ratios. The modified MMT displayed well-dispersed individual clay layers in the PET matrix, although some particles appeared to be agglomerated at size levels greater than approximately 10 nm.

Ke *et al.* [116] also reported the properties of PET/MMT nanocomposites prepared by *in situ* polymerization of PET onto organophilic-modified MMT. The refined clay, reduced to particles of 40 μm average diameter, was made into a slurry, and formed a solution with an intercalated reagent, which reacted directly with PET monomers in an autoclave [117, 118]. Although, this preparation method led mainly to intercalated nanocomposites, a similar nucleating effect with respect to the report of Chang and Munn was observed.

An alternative *in situ* polymerization method was provided by Lee *et al.* [118], who successfully polymerized ethylene terephthalate cyclic oligomers (ETCOs) to form a high-molecular-weight PET in the presence of OMMT by employing the advantages of the low viscosity of cyclic oligomers and lack of chemical emissions during polymerization (Figure 13.10). Here, the OMMT was prepared via the cation exchange of sodium-MMT with N,N,N-trimethyl octadecylammonium bromide.

The application of ring-opening polymerization (ROP) is usually hindered by the limited efficiencies for synthesizing and isolating cyclic oligomers. Therefore, various types of preparative methods have recently been developed. The present authors followed the direct reaction of acid chlorides with glycols in dilute solution, which provides high yields of ETCs [119]. Subsequently, ETC, OMMT and Ti(O-1-C_3H_7)$_4$ were dissolved in dichloromethane and the solution maintained at room temperature for 12 h. The oligomers, OMMT and Ti(O-1-C3H7)4, dissolved in dichloromethane, were maintained at room temperature for 12 h; the solvent was

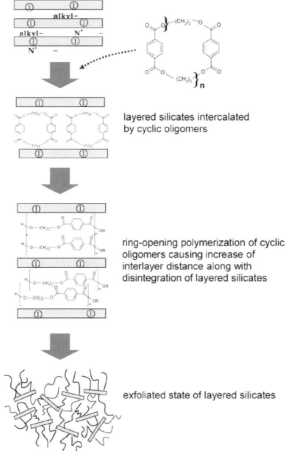

layered silicates intercalated by cyclic oligomers

ring-opening polymerization of cyclic oligomers causing increase of interlayer distance along with disintegration of layered silicates

exfoliated state of layered silicates

Figure 13.10 Schematic representation of nanocomposites formation by ring-opening reaction of cyclic oligomers in-between silicate layers. Reproduced with permission from Ref. [118]; © 2005, Elsevier Ltd.

then removed, dried under reduced pressure and polymerized in a high-vacuum system at temperatures ranging from 240 to 310 °C.

Due to their low molecular weight and low viscosity, the ETCs are successfully intercalated into the clay galleries, as evidenced by XRD analyses which showed a down-shift of the basal plane peak of layered silicate, in good agreement with the results of TEM investigations. The successive ROP of ETCs in-between silicate layers yielded a PET matrix of high molecular weight, along with high disruption of the layered silicate structure and a homogeneous dispersion of the latter in the matrix. However, the coexistence of exfoliation and intercalation states of silicate layers was revealed by morphological investigations. Interestingly, the highest investigated polymerization temperature provided the nanocomposite with the PET having the greatest molecular weight (11 000 mol g^{-1}).

13.3.2
Intercalation in Solution

To date, only one report has been made [120] detailing the preparation of PET/phyllosilicate nanocomposites in solution. Exfoliated PET-layered silicate nanocomposites excluding and including organic modifiers were obtained by solution methods in chloroform/trifluoroacetic acid (TFA) mixtures, with and without solvent–nonsolvent systems, respectively. For this, the PET/OLS/chloroform/TFA solution was added dropwise to the cold methanol to obtain PET nanocomposites excluding an organic modifier; the precipitated materials were then collected by filtration and dried in a vacuum oven (Figure 13.11). In contrast, the PET

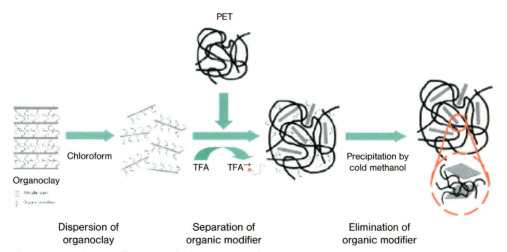

Figure 13.11 Schematic illustration of the preparation of exfoliated PET nanocomposites excluding organic modifier by means of the solvent–nonsolvent system. Reproduced with permission from Ref. [120]; © 2008, John Wiley & Sons, Ltd.

nanocomposite including an organic modifier was prepared by removing both solvents from the prepared PET/OLS/chloroform/TFA solution.

Based on wide-angle X-ray diffraction (WAXRD) and high-resolution TEM analyses, both types of composite were found to have an exfoliated structure that was attributed to a sufficient dispersion of silicate in prepared solvents, regardless of the sample preparation method.

The DSC results indicated that both types of nanocomposite had higher degrees of crystallinity and shorter crystallization half-times than neat PET, since the dispersed silicate layers had acted as nucleating agents in both situations. However, those nanocomposites prepared with the organic modifier exhibited a lower degree of crystallinity and a longer half-time of crystallization than those without an organic modifier. This difference in crystallization behavior between the two nanocomposites was ascribed to the organic modifier, which may have acted as an inhibitor of crystallization.

13.3.3
Intercalation in the Melt

The most frequently reported method of intercalation has been within the melt, mainly because it is significantly more economical and simpler than other methods, and is also compatible with existing processes. In fact, melt processing allows nanocomposites to be formulated directly using ordinary compounding devices such as extruders or mixers, without the need to involve resin production. Consequently, nanocomposite production can be shifted downstream, providing end-use manufacturers with many degrees of freedom with regards to their final product specifications. Notably, melt processing is also environmentally friendly, as no solvents are required. It also enhances the specificity of intercalation of the polymer, by eliminating any competing host–solvent and polymer–solvent interactions.

One topic which has been widely investigated is the influence of an organophilic modification of nanoclay on the morphology of composites. Pegoretti *et al.* [121] compared the effect of a nonmodified natural MMT clay (Cloisite-Na+) and an ion-exchanged clay modified with quaternary ammonium salt (Cloisite 25A) (Table 13.5, row h) in a recycled PET matrix. The Cloisite 25A is a MMT modified with dimethyl, hydrogenated tallow, 2-ethylhexyl quaternary ammonium cations. Following preparation of the composites in an injection-molding machine, subsequent TEM images of the composite fracture surfaces indicated that the particles of Cloisite 25A were much better dispersed in the recycled PET matrix than those of Cloisite-Na+. Moreover, wide-angle X-ray scattering (WAXS) measurements indicated the intercalation of recycled PET between the silicate layers (lamellae) of the clay. In addition, improvements in mechanical properties were more evident in Cloisite 25A nanocomposites than in those containing Cloisite-Na+. More recently, the effects of different modified commercial MMTs on the morphology of PET composites prepared in a twin-screw extruder were investigated [122]. In particular, within a scale of increasing hydrophobicity of

Table 13.5 Montmorillonites used for the preparation of polyesters/MMT nanocomposites by intercalation in the melt.

	Formula of the cation	Commercial name	Reference(s)
(a) Sodium montmorillonite	Na$^+$	Cloisite Na$^+$	[122–124]
(b) Methyl, tallow, bis-2-hydroxyethyl, quaternary ammonium montmorillonite	CH$_3$—N$^+$—T with two CH$_2$CH$_2$OH groups; T is ~tallow (~65% C18; ~30% C16; ~5% C14)	Cloisite 30 B	[123, 125]
(c) 1-[2-(2-hydroxy-3-phenoxy-propoxy)-ethyl]-2,3-dimethyl-3H-imidazolium montmorillonite	[phenoxy-CH$_2$-CH(OH)-CH$_2$-O-CH$_2$CH$_2$-imidazolium(2,3-dimethyl)] Br$^-$	—	[126]
(d) 1,2-dimethyl-3-N-hexadecyl imidazolium montmorillonite	1,2-dimethyl-3-R-imidazolium X$^-$; R = (CH$_2$)$_{15}$CH$_3$; X$^-$ = BF$_4^-$	—	[124]
(e) 1,3-Dioctadecyl imidazolium	H$_3$C—(CH$_2$)$_{17}$—N$^+$(imidazolium)—N—(CH$_2$)$_{17}$—CH$_3$	—	[127]

13.3 Poly(Ethylene Terephthalate)-Based Nanocomposites

Table 13.5 Continued.

	Formula of the cation	Commercial name	Reference(s)
(f) Hexadecyl pyridinium montmorillonite	pyridinium–(CH$_2$)$_{15}$–CH$_3$	–	[127]
(g) Dimethyl, benzyl, hydrogenated tallow, quaternary ammonium montmorillonite	H$_3$C, CH$_2$–Ph, HT attached to N$^+$; HT = hydrogenated tallow	Cloisite 10A Dellite 43B	[123]
(h) Dimethyl, dehydrogenated tallow, 2-ethylhexyl quaternary ammonium montmorillonite	CH$_3$–N$^+$(CH$_3$)(HT)–CH$_2$CH(CH$_2$CH$_3$)CH$_2$CH$_2$CH$_2$CH$_3$	Cloisite 25A	[122, 125]
(i) Dimethyl, dehydrogenated tallow, quaternary ammonium montmorillonite	CH$_3$–N$^+$(CH$_3$)(HT)(HT)	Cloisite 20A (0.95 mg g^{-1} clay) Cloisite 15A (1.25 mg g^{-1} clay) Cloisite 6A (1.4 mg g^{-1} clay) Dellite 72T	[127] [123, 125] [125]
(l) Tributyl hexadecyl phosphonium montmorillonite	(H$_3$C–(CH$_2$)$_3$)$_3$P$^+$–(CH$_2$)$_{15}$–CH$_3$	–	[127]

the cation, sodium-MMT, methyl, tallow, bis-2-hydroxyethyl, quaternary ammonium-MMT (Cloisite 30B) (Table 13.5, row b), dimethyl, benzyl, hydrogenated tallow, quaternary ammonium-MMT (Cloisite 10A) (Table 13.5, row g) and dimethyl, dehydrogenated tallow, quaternary ammonium-MMT (Cloisite 15A) (Table 13.5, row i) were compared.

As revealed by both TEM and WAXD characterization, the shear in the extruder favored the exfoliation process, which resulted in both intercalated and exfoliated morphologies. PET nanocomposites obtained by using clay with polar modifiers showed intercalated and exfoliated morphologies, whereas tactoids were obtained when only apolar modifiers were present. The polymer–clay interactions and the extrusion conditions were sufficient to break the organized arrangement of the natural MMT, so as to disperse it in the polymer matrix. The organic modifier also appeared to sustain the exfoliated clay sheets flat since, in its absence, the intramolecular interactions were stronger and the platelets tended to roll due to their high flexibility. A comparison of these two reports [121, 122] showed contradictory results, however. As the recycled and virgin PET had practically identical structural characteristics (although the recycled form usually had a lower viscosity in the melt), the comparison mainly highlighted the strong effect of shear stresses on the morphology. In fact, the stronger shear stresses developed in the twin-screw extruder were most likely sufficient to achieve an exfoliated/intercalated structure, without using an organophilic modifier. In the study of Pegoretti et al. [121], the lower shear stresses of the injection-molding machine produced the opposite result, as in these conditions the main driving forces were the interactions of the hydrophobic chains with the polymer matrix, which were capable of producing a primary disaggregation of the filler.

The strong effects of processing conditions and rheological properties on morphology were also reported by Krakalik et al. [123], who prepared composites in a co-rotating, twin-screw, microextruder by adding to recycled PET 5% (by weight) of different modified MMTs (Cloisite 6A, 15A, 20A, 25A, and 30B) (Table 13.5). Subsequently, these authors observed the partial exfoliation of Cloisite 25A, 30B, and 10A, which was in good agreement with the results observed by both Pegoretti et al. [121] and Calcagno et al. [122].

Davis et al. [125] investigated the effects of melt-processing conditions on the quality of PET/MMT nanocomposites prepared in a mini twin-screw extruder at 285 °C, and identified the need to use a highly stable ammonium salt in modified MMT. In fact, the use N,N-dimethyl-N,N-dioctadecylammonium-MMT led to the production of black PET nanocomposites as a result of ammonium salt degradation under the processing conditions. The most dispersed, exfoliated PET nanocomposite was achieved by melt-mixing at 21 radians per second for 2 min in a nitrogen atmosphere, after drying the polymer at 120 °C, and the clay at 150 °C. Alternative mixing conditions, longer residence times, and higher screw speeds resulted in poorer-quality nanocomposites. The alternative use of a hexadecyl imidazolium salt, which resulted in an intercalated/exfoliated morphology, allowed improvements in stability, as the imidazolium salt (which has a decomposition temperature of 350 °C) is stable under the conditions of processing. The subject of modified MMT stability was recently investigated by Kim et al. [124], who synthesized an imidazolinium salt with an aromatic group and an hydroxyl group (Table 13.5, row c), which was more stable than a quaternary ammonium salt. Such treatment was found to provide a similar dispersion with respect to Cloisite 15A and 30B, but with the advantage of a higher thermal stability.

More recently, Stoeffler et al. [126] used the surfactants alkyl phosphonium (Table 13.5, row l), alkyl pyridinium (Table 13.5, row f) and dialkyl imidazolium (Table 13.5, row e) as intercalating agents for the preparation of highly thermally stable organophilic MMTs, and compared these to commercial Cloisite 20A (Table 13.5, row i) from the point of view of thermal stability. Although the thermal stability of the former materials was improved with respect to the latter, a mass spectrometric analysis of the volatile products showed the evolution of chloromethane from Cloisite 20A above 200°C, of pyridine from pyridinium-MMT above 250°C, and of tributyl phosphine and/or tributyl phosphine oxide from phosphonium-MMT above 250°C. Based on the toxicity of volatile products, the use of imidazolium or phosphonium derivatives proved unsatisfactory. Rather, the most interesting result obtained was the exfoliation of native sodium-MMT via the establishment of proper extrusion conditions [122].

In only one report [127] was the possibility considered of using a proper compatibilizer, namely a polyester ionomer (Figure 13.12), to improve the compatibility between an OMMT (Cloisite 20A) and PET. Here, the polyester ionomer was efficient in promoting intercalation (or intercalation/exfoliation if the ionomer/OMMT ratio was ≥3) of the OMMT in the PET matrix. Moreover, in terms of rheological behavior, the higher the filler content and/or the degree of intercalation/exfoliation of the OMMT, the more the nanocomposite behaved as a solid. With regards to rheological behavior, the higher the content of filler and/or degree of intercalation/exfoliation of the OMMT, the more the nanocomposite behaved as a solid because of the percolated structure formed by the OMMT layers, and the more the storage and loss modulus, G' and G'', became independent of the frequency at low frequencies.

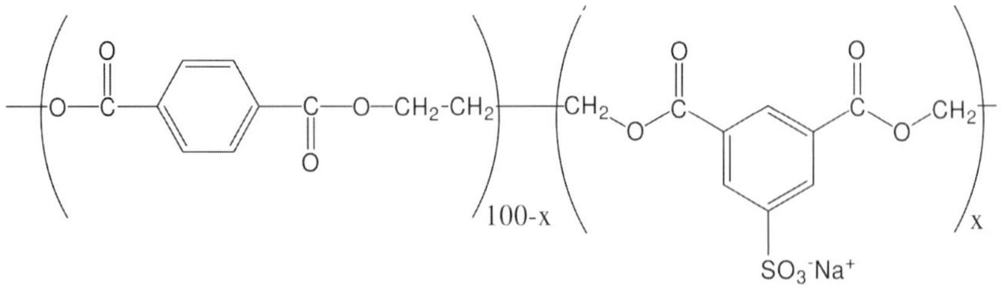

Figure 13.12 Structure of the polyester ionomer (PETi) used by Vidotti et al. [127]. ($x = 0.13$).

13.4
Poly(Lactide) (PLA)-Based Nanocomposites

Today, the development of renewable polymeric materials with excellent properties forms the subject of much active research interest worldwide [4]. Aliphatic polyesters are among the most promising materials for the production of high-performance, environment-friendly, biodegradable plastics [128]. Among these materials, PLA is a renewable and biodegradable polyester, the thermomechanical and gas-barrier properties of which prevent the replacement of traditional polymers in many areas of application. Consequently, innovative methods must be developed to modulate the properties of PLA, such as blending [129], plasticization [130], and the preparation of nanocomposites [131].

13.4.1
Overview of Preparation Methods

13.4.1.1 In Situ Polymerization

The preparation of some biodegradable polyesters, including PLA and PCL, is often achieved via ROP from corresponding lactones [132, 133]. In particular, the synthesis of PLA can be carried out starting either from the cyclic dimer, the lactide (3,6-dimethyl-1,4-dioxane-2,5-dione) by ROP, or by the condensation polymerization of lactic acid or its derivatives.

In terms of molecular weight control, the living ROP of lactide yields a linear relationship between monomer conversion and molecular weight and PLA with a narrow polydispersity index (PDI; defined as the ratio between the weight average and number average molecular weights, M_w/M_n). In contrast, the step-growth condensation polymerization limits the practically accessible range of molecular weights, and leads to a PDI of 2.

The benefits of ROP in conjunction with a "living" method have enabled the controlled synthesis of block, graft, and star polymers [134], and this had led to the present consensus that living ROP represents a powerful and versatile method of addition–polymerization. In particular, coordination–insertion polymerization has been used extensively for the preparation of aliphatic polyesters with well-defined structure and architecture. The most widely used initiators include various aluminum and tin alkoxides and carboxylates. The covalent metal alkoxides or carboxylates with vacant d-orbitals react as coordination initiators by interacting with the oxygen of carbonyl group of the ester. These initiators are capable of producing stereoregular polymers of narrow molecular weight distribution and controlled molecular mass, with well-defined end groups.

Both, Kubies et al. [135] and Lepoittevin et al. [136, 137] studied the ROP of ε-caprolactone (ε-CL) onto modified MMTs. PCL-grafted layered silicate nanohybrids were thus prepared according to a controlled coordination–insertion mechanism. For this purpose, MMT was previously modified by an exchange of the constitutive Na cations by ammonium cations bearing one hydroxy function

((2-hydroxyethyl)dimethylhexadecylammonium), or simply trimethylhexadecylammonium [136]. In a successive study [137], Cloisite 30 B (Table 13.5, row b), modified with methyl, tallow, bis-2-hydroxyethyl, quaternary ammonium, was also employed.

The modified MMT was then treated with aluminum or tin (II or IV) compounds [AlEt$_3$, Bu$_2$Sn(OMe)$_2$ and Sn(2-ethylhexanoate)$_2$] to obtain the respective alkoxides, all of which are known to initiate the controlled polymerization of ε-CL, by creating initiating species onto the MMT layers. The grafting of PCL onto the MMT was achieved in the presence of ε-CL (Figure 13.13).

After polymerization, the molar mass of PCL, as determined by size-exclusion chromatography (SEC) after a reverse ion-exchange reaction with LiCl in tetrahydrofuran (THF), was found to decrease with the increasing content of hydroxy groups available at the clay surface. Indeed, the molar masses were 56 000, 47 000 and 28 000 for OH-deprived organomodified clay and clay containing 50 and 100% monohydroxylated ammonium cations, respectively. This observation suggests, at least qualitatively, that polymerization is initiated by the surface-anchored hydroxy groups activated in Sn(II) alkoxides by reaction with Sn(2-ethylexanoate)$_2$. In the absence of hydroxylated ammonium cations, polymerization is initiated by residual protic impurities (water, silanol, etc.). Polymerization initiation on the clay surface and PCL growth in a "grafting-from" manner each have strong effects on the morphology of the PCL/layered silicate nanocomposites, such that intercalated or exfoliated morphologies were obtained. However, extensive exfoliation occurred only when the silicate sheets were surface-modified with more than 25 wt% monohydroxylated ammonium [136]. Below this value, partially intercalated/partially exfoliated structures were found to coexist, and an intercalated structure was obtained, starting from sodium-MMT.

A good correlation was obtained between morphology and thermal stability, as the composites prepared with the clay that had been modified with hydroxyl groups were more stable than those deprived of them. Moreover, these composites were more stable than those containing sodium-MMT.

More recently, Chrissafis et al. [138] prepared PCL nanocomposites by following a similar approach, but used Ti(OBu)$_4$ as the coordination–insertion catalyst. Moreover, they performed ROPs not only onto sodium-MMT and Cloisite 20A (Table 13.5, row i), but also onto mica and fumed silica. In agreement with the above findings, there was also a decrease in the molecular weight of PCL obtained, compared to the pure material prepared in the absence of fillers. However, the mechanical properties (tensile strength at break, elastic modulus, elongation at break) of the nanocomposites were similar, or even slightly improved, compared to those of the pure PCL.

It appears that only two reports have been made concerning the synthesis of PLA using this method. For example, Paul et al. [139] synthesized PLA/layered aluminosilicate nanocomposites by ROP, using two different types of OMMT (Cloisite 30B and Cloisite 25A; see Table 13.5, rows b and h) for the preparation of nanocomposites. In a typical synthetic procedure, the clay was thoroughly dried and placed in the polymerization vial. The lactide solution in dried THF was then

13.4 Poly(Lactide) (PLA)-Based Nanocomposites

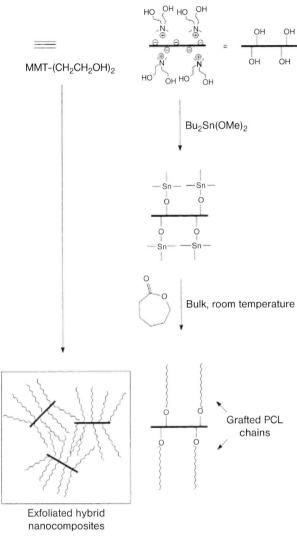

Figure 13.13 Scheme of the synthesis of PCL nanocomposites through *in situ* ROP onto montmorillonite modified with ammonium salt with hydroxyl groups. Reproduced with permission from Ref. [137]; © 2002, American Chemical Society.

transferred to the vial under nitrogen, and the solvent eliminated under reduced pressure. The polymerizations were conducted in bulk at 120 °C for 48 h, after 1 h of clay swelling in the monomer melt. When Cloisite 30B was used, the polymerization was coinitiated by AlEt$_3$, while Sn(2-ethylexanoate)$_2$ was used to catalyze the polymerization of lactide in the presence of Cloisite 25A. The clay Cloisite 30B led to a fully exfoliated structure, whereas Cloisite 25A-based nanocomposites exhibited an intercalated morphology.

In the Cloisite 30B nanocomposites, the growing polymer chains were directly "grafted" onto the clay surface through the hydroxyl-functionalized ammonium cations, yielding exfoliated nanocomposites with an enhanced thermal stability. Notably, with increasing clay content the thermal stability was improved, with a maximum effect at a clay loading of 5 wt%. With a further increase of filler content, a decrease in thermal stability was observed – an effect explained by the relative extent of exfoliation as a function of the amount of organophilic MMT. Nonetheless, a decrease in PLA molecular weight was observed compared to that synthesized in the absence of filler.

A similar result was obtained more recently by Lee [140], who modified the clay surface (Cloisite 30B, Table 13.5 row b) by grafting low-molecular-weight PLA chains (M_n 9400–21600) through the *in situ* polymerization approach, following a preliminary activation of the hydroxyl sites with $Sn(2\text{-ethylexanoate})_2$. The composite obtained was then melt-blended with a high-molecular-weight PLA matrix. This novel clay/PLA nanocomposite showed a high shear-thinning behavior when the molecular weight of the grafted PLA was higher than the critical molecular weight of chain entanglement.

Today, the scientific investigations in this field are ongoing, with the *in situ* polymerization process clearly warranting further investigation. In particular, the ROP of lactide onto clay or organoclay may represent a promising approach for obtaining nanostructured PLA/layered silicate composites, allowing excellent control of the structural features of those materials prepared.

13.4.1.2 Intercalation in Solution

Recently, several studies have focused on the preparation of PLA-layered silicate nanocomposites using intercalation from solution. Initially, Ogata *et al.* [141] dissolved the polymer in hot chloroform in the presence of OMMT, but observed that only microcomposites were formed. Marras *et al.* [142] adopted a similar method but, in contrast, obtained intercalated/exfoliated structures. The differences in experimental procedure here consisted of sonication of the organophilic MMT in chloroform, and of the solution containing both organophilic MMT and PLA. In the latter report, an increase in the d_{001} distance was cited as a function of the organophilic MMT modification, expressed as percentage of the CEC, when investigated over the range of 0 to 250%. The degree of organophilic modification was controlled by an ion-exchange reaction, using controlled amounts of hexadecylamine.

Krikorian and Pochan [143] prepared PLA nanocomposites using dichloromethane (DCM) as the polymer solvent and as the organophilic MMT dispersion medium (Cloisite 30B, 25A and 15A; Table 13.5, rows b, h, and i). These authors obtained intercalated or exfoliated nanocomposites, depending on the type of organophilic MMT used. Exfoliated nanocomposites were formed by using Cloisite 30B; that is, when hydroxyl groups were present in the organic modifier of the clay, due to the favorable enthalpic interaction between the hydroxyl groups and the C–O bonds in the PLA backbone. Cloisite 25A, which showed the higher starting basal interlamellar distance, gave a d_{001} distance in the nanocomposites lower than Cloisite 30B. Hence, the starting basal interlamellar distance, which influences the capability of the organophilic MMT to be dispersed in a polyolefinic matrix, is a key factor but is less important in PLA-based nanocomposites.

Chang et al. [144] reported the preparation of PLA-based nanocomposites with different types of OMMT via solution intercalation using N,N-dimethylacetamide (DMA). The preparation method allowed the production of intercalated/exfoliated nanocomposites, providing material with an improved stiffness and barrier to oxygen compared to pure PLA.

13.4.1.3 Intercalation in the Melt

To date, intercalation in the melt has attracted more attention that other intercalation processes, with different aspects being considered by various research groups with regards to melt blending. These include: the effects of different organophilic modifiers on morphology; the effects of the processing conditions; and the possibility of optimizing the processing and/or properties via blending or plasticization.

Effect of Different Organophilic Modifiers Di et al. [145] mixed two different organophilic MMTs (Cloisite 30B, Table 13.5, row b; and Cloisite 93A) in a PLA matrix by using an internal mixer. Cloisite 93A contains dimethyl-dihydrogenated tallow ammonium as modifier. It was observed that only the Cloisite 30B gave exfoliated nanocomposites, which was in good agreement with results obtained by Krikorian and Pochan [143] for nanocomposites obtained in solution. When Feijo et al. [146] compared two commercial organophilic MMTs, namely Dellite 43B (Table 13.5, row g) and Dellite 72T (Table 13.5, row i), the former material interacted more strongly with PLA, as confirmed by the better dispersion of the organoclay in the PLA matrix when compared to the neat PLA/Dellite 72T composites, which showed aggregates of only micrometric dimensions. Moreover, the PLA/Dellite43B nanocomposite exhibited a slightly higher thermal stability. Hence, replacement in the modifier of a long alkyl chain (hydrogenated tallow) with an aromatic ring renders the modifier more suitable for creating interactions between a modified nanofiller and the PLA polymer matrix. This result could, however, be ascribed to the higher polarity of the Dellite 43B ammonium salt.

Effect of Processing Conditions Pluta [147] investigated the processing of PLA and phyllosilicates by preparing, in a discontinuous laboratory mixer, nanocomposites containing 3% by weight of Cloisite 30B. In particular, the blending time (6.5, 10, 20, and 30 min) and rotor rate (50 and 100 r.p.m.) were varied in order to establish their effects on the morphology and properties of the nanocomposites. Molecular weight changes of the PLA matrices induced by melt compounding were determined using SEC. Whilst the molecular weight was marginally reduced by increasing the blending time, the level of dispersion of the filler into the matrix was improved by increasing the blending time, as revealed by WAXD and TEM characterization.

The behavior of PLA during melt processing depends on many factors, including the grade of PLA used, the processing conditions applied (temperature, rotation speed, residence time, atmosphere, drying efficiency of the components), and the presence of additives and their chemical nature. Although pure PLA degrades slightly during blending, it was noticed that the degradation of PLA was enhanced by the presence of the nanofiller. This suggested important roles for both the shearing forces during compounding and the interaction of PLA with hydroxy groups of the organomodifier in determining the degradation of the PLA matrix. Moreover, the change in rotation speed adopted during nanocomposite preparation, from 50 to 100 r.p.m., caused a decrease in molecular weight but had no effect on the material's morphology and/or properties. It was also found that an increase in the degree of dispersion of the silicate layers led to a pronounced modification of the physical properties of the nanocomposites, via an increase in thermal stability, as revealed by TGA.

The rheological properties of the nanocomposites, as determined during a dynamic frequency sweep, appeared to be very sensitive to the nanostructure evolution. On investigating the rheological behavior of PLA during melt blending with Cloisite 30B, the apparent viscosity (η^*) of the unfilled PLA was unchanged at low frequencies (indicating a Newtonian behavior), but this was followed by a shear-thinning response at a higher-frequency region. The η^* of the nanocomposites showed that, the higher the shear-thinning effect, the better the organoclay dispersion in the PLA matrix. These observed trends allowed an identification of the concentration of organoclay capable of providing a non-negligible interaction between nanoplatelets, based on rheology experiments. Similar results were obtained by Gu et al. [148], who explained the same trend as a function of the organophilic MMT content, by considering the formation of a "percolating network" resulting from reciprocity among the strongly related sheet particles. Because of this "percolating network," the values of entanglement molecular weight (M_e), as calculated from master curves, were lower than that of pure PLA.

During such processing, Lewitus et al. [149] investigated the preparation and dilution of different types of masterbatch. The study results indicated the possibility of preparing PLA/Cloisite 25A nanocomposites via the preliminary preparation of a masterbatch which contained 20% by weight of the nanofiller and PLA. In particular, the most suitable PLA was that with the same grade as the final composite.

Control of Processability and Properties by Blending and Plasticization The addition of a reactive compatibilizer (e.g., an ethylene copolymer functionalized with maleic anhydride) during preparation in a discontinuous mixer of PLA/organophilic MMT nanocomposites was monitored by Pluta et al. [150]. When polylactide-based systems composed of an organoclay (Cloisite 30B; 3–10 wt%) and/or the compatibilizer were investigated, X-ray investigations revealed an exfoliated nanostructure in a 3 wt%-nanocomposite. The degree of exfoliation of the organoclay was noticeably enhanced by compatibilization, due to the combined interactions of the organoclay surfactant with the PLA chains and the MAH groups of the compatibilizer. In the 10 wt%-nanocomposite, mixed (i.e., intercalated and exfoliated) nanostructures were detected due to the high concentration of the filler. Both, the rheological and mechanical properties suggested that a type of silicate network had been formed.

Sinha Ray et al. [151] noted that, although high-molecular-weight PCL was immiscible with PLA, oligomeric PCL was highly miscible (as indicated by a shift in T_g). Consequently, the group used such an α,ω-hydroxy-terminated oligomeric PCL (0.2–3 wt%) as compatibilizer when preparing PLA/octadecylammonium-modified MMT nanocomposites. The incorporation of a compatibilizer into the system led to a better parallel stacking of the silicate layers, and also to a much stronger flocculation as a result of the hydroxylated edge–edge interaction of the silicate layers. Owing to interaction between the clay particles and the PLA matrix with oligomeric PCL, the disk–disk interaction was considered to play an important role in determining the stability of the clay particles, and hence any enhancement of the mechanical properties of such nanocomposites. These systems, which were the first successful intercalated-type PLA/layered silicate nanocomposites, exhibited remarkable improvements in the materials properties of both solid and melt states when compared to the matrix without clay.

Shortly thereafter, Kubies et al. [152] prepared PCL-layered silicate nanocomposite masterbatches by the ROP of ε-CL in the presence of the OMMT, Cloisite 30B. The masterbatch MB30 (28 wt% MMT, PCL with M_w = 1800) had an intercalated structure, while MB8 (7.5 wt% MMT, PCL with M_w = 7500) was exfoliated. Subsequently, high-molecular-weight PLLA nanocomposites containing 0.5–2.5 wt% MMT were obtained by the melt-blending of PLLA (M_w = 4.5 × 10^5 Da) with PCL masterbatches or Cloisite 30B. The MMT particles in PLLA/Cloisite 30B and PLLA/MB30 nanocomposites were intercalated. In contrast to expectation, the exfoliated silicate layers of MB8 were not transferred into the PLLA matrix of the PLLA/MB8 nanocomposites. Due to a low miscibility of PCL and PLLA, MMT remained in the phase-separated masterbatch domains. The stress–strain characteristics of PLLA nanocomposites (i.e., Young's modulus), yield stress and yield strain, were decreased with increasing MMT concentration, associated with an increase in the PCL content (up to 35.4 wt% in PLLA/MB8). The expected stiffening effect of the MMT was low, due to a low aspect ratio of its particles, and was obscured by both the plastifying effect of PCL and the low PLLA crystallinity. Interestingly, in contrast to the neat PLLA, ductility was enhanced in all PLLA/Cloisite 30B materials, and also in the PLLA/masterbatch nanocomposites with low MMT concentrations.

As reported Sinha Ray et al. [151], the miscibility of PCL with the PLA matrix plays a crucial role in its possible use as a compatibilizer. However, Yu et al. [153] more recently prepared PLA/PCL 90/10 blends containing different amounts (1–10 wt%) of organophilic MMTs modified with bis-(2-hydroxyethyl)-methyl-(hydrogenated tallow alkyl) ammonium cations, by using a counter-rotating, discontinuous mixer. The silicate layers of the clay were intercalated and distributed randomly in the matrix. The addition of organophilic MMT to the PLLA/PCL blend led to a significant improvement in the tensile properties and dynamic mechanical properties of the nanocomposites. In contrast, the layered silicate caused a clear improvement in the thermal stability of the PLLA/PCL blends when the organophilic MMT content was less than 5 wt%. SEM images confirmed that the addition of OMMT could reduce the size of the phase-separated particles, causing the material to become more uniform.

Interestingly, it has been suggested that OMMT might play the role of a compatibilizer, and a similar reduction in particle size in phase-separated blends was reported for a triblock PLA-PCL-PLA copolymer used as a compatibilizer in similar blend systems [154]. These authors ascribed this phenomenon to the intercalation of polymer molecules in OMMT, which increases the viscosity ratio and results in the retardation of coalescence of the dispersed phase-separated particles and an enhanced compatibility as a result of the intercalation of both PLLA and PCL molecules into the same OMMT gallery.

Hence, the possible use of a second polymeric phase as a compatibilizer should focus on the miscibility or compatibility between the two polymeric phases, in order to predict the morphology, as this depends not only on the level of dispersion of the lamellae in the polymeric matrix, but also on the phase distribution, which in turn influences the preferential distribution and orientation of the lamellae.

Recently, much interest has been expressed regarding not only the plasticization of PLA [130, 155, 156], but also the possible addition of a plasticizer to PLA/layered silicate nanocomposites. This subject is particularly attractive because it might allow a balance to be achieved between the mechanical and gas-barrier properties, since plasticization will in general provide an improved ductility but reduce the gas-barrier effects. In contrast, adding a nanofiller would improve the stiffness and gas-barrier properties of the PLA. Hence, the plasticization of PLA nanocomposites might provide a more ductile material, without either decreasing and/or improving its gas-barrier properties. Plasticized PLA-based nanocomposites were prepared by Paul et al. [157] by melt-blending the matrix with 20 wt% poly(ethyleneglycol) (PEG; number average molecular weight 1000) and different amounts of MMT, with or without organomodification. For this, four different (organo)clays were dispersed within the plasticized PLA matrix, after which the influence of the interlayer cations on the composite's morphological and thermal properties was studied while maintaining a constant level of inorganic material (3 wt% layered aluminosilicate).

Each of the MMTs studied, including the unmodified sodium-MMT, led to the formation of intercalated nanocomposites, Usually, melt-blending polymer matrices with sodium MMT results in the formation of a microcomposite, as most of the polymers are too highly hydrophobic to migrate into the hydrated Na^+ interlayer space. However, Vaia et al. [158] have reported on the intercalation of PEG between the aluminosilicate layers of an unmodified sodium-MMT that led to an increase in the interlayer distance which was similar to that observed by Paul et al. [157]. Clearly, in the presence of sodium-MMT, the PEG 1000 is able to intercalate preferentially with the interlayer spacing of the clay. This selective PEG intercalation was further confirmed by an inability to form a nanocomposite by melt-blending nonplasticized PLA with sodium-MMT.

At a constant filler level it appears that, among all the clays studied (Cloisite Na+, Cloisite 30B, Cloisite 20A, and Cloisite 15A; see Table 13.5, rows a, b, and i, respectively), the MMT that had been organomodified with bis-(2-hydroxyethyl)-methyl-(hydrogenated tallow alkyl) ammonium cations (Cloisite 30B) brought about a greater effect in terms of improvements in dispersion level and thermal stability of the plasticized nanocomposites, and this was in good agreement with previous reports of PLA/MMT nanocomposites. It was noted, however, by using WAXS and DSC analyses, that there existed a real competition between PEG 1000 and PLA for intercalation into the interlayer spacing of the clay.

As also noted by Pluta *et al.* [159], who investigated the same system, SEC revealed a decrease in the molecular weight of the PLA matrix. This was more consistent at higher filler contents, but essentially independent of the type of organophilic modification of the MMT.

Thellen *et al.* [160] investigated the influence of MMT-layered silicate on plasticized PLA blown films. Here, the plasticized PLA MMT-layered silicate nanocomposites were compounded and blown film-processed, using a co-rotating, twin-screw extruder. The PLA was mixed with 10 wt% acetyltriethyl citrate ester as plasticizer and 5 wt% of an OMMT, at various screw speeds. Both, WAXD and TEM investigations showed that the compounded pellets and the blown-film PLA/OLS nanocomposites had intercalated. The effects of the processing screw speed on the barrier, thermal, mechanical, and biodegradation properties of the nanocomposites were also considered, and compared to the neat polymer. The nanocomposite films showed a 48% improvement in oxygen barrier, and a 50% improvement in water vapor barrier, when compared to PLA. A subsequent TGA revealed an overall 9 °C increase in the decomposition temperature for all of the nanocomposites. A DSC analysis indicated that neither the glass transition, cold crystallization, nor melting point temperature was significantly influenced by the presence of MMT. The mechanical properties of the nanocomposites indicated a 20% increase in the Young's modulus, but that the ultimate elongation of the nanocomposites had not been sacrificed compared to the neat samples. Hence, the use of both plasticizers and nanofillers in a PLA matrix might effectively serve as a successful strategy for providing materials with modulated properties.

13.5
Conclusions

In this chapter we have compared the schemes used to prepare polymer/layered silicate nanocomposites from three classes of material, namely polyolefins, PET and PLA, and have reviewed certain interesting similarities and differences among these materials.

With regards to similarities, analogous synthetic approaches were described for the preparation of nanocomposites, with all three cases – *in situ* polymerization, intercalation from solution, and intercalation in the melt – being investigated. For polyolefins, *in situ* polymerization appears less successful and less sustainable, with synthesis via catalytic polymerization perhaps proving to be a drawback, notably due to poor interactions between the inorganic filler and the polymeric matrix. To overcome this problem, the synthesis should include polyolefins bearing polar groups capable of interacting with the silicate layers. Many challenges remain, however. The catalytic polymerization of polar monomers with both traditional Ziegler–Natta and metallocene catalysts is not feasible, due mainly to catalyst deactivation, while the polymerization of polyesters via ROP onto layered silicates represents an interesting strategy for developing highly compatibilized organic–inorganic hybrid systems. The hydroxyl groups, whether typical of phyllosilicates or purposefully added via modification with hydroxyethyl functional ammonium salts, serve as initiating sites for the polymerization of lactide, or various lactones. Most importantly, polymerizations must be strictly controlled, with the length of the grafted chains being longer than the critical molecular weight of any chain entanglements.

Today, the number of reports describing the preparation of polyolefin-based nanocomposites exceeds that on the preparation of PET and PLA nanocomposites, essentially because preparing nanocomposites from polyolefins is much more difficult due to intrinsic incompatibilities between the polar matrix and the highly polar filler. Whilst for polyolefin-based nanocomposites, extensive data have been acquired by using apolar organophilic layered silicates, in the case of PET, exfoliated silicate nanocomposites were acquired via an extrusion process which utilized a nonmodified sodium-MMT.

It seems that the best structure for a modification agent depends on the polarity of the polymeric matrix. In the case of polyolefins, ammonium salts with long aliphatic chains can permit optimization of the dispersion level. For PLA, the methyl, hydrogenated tallow, bis(hydroxyethyl) ammonium salt seems to be the best dispersing agent for clay, whereas for PET, sodium-MMT, hydrogenated tallow, bis(hydroxyethyl) ammonium-MMT and dimethyl, hydrogenated tallow, benzyl ammonium-MMT, are each capable of providing exfoliated morphologies.

Another crucial point here is the need to use a compatibilizer. When preparing polyolefin/layered silicate nanocomposites, the addition of a compatibilizer is paramount, yet for PLA and PET nanocomposites the inclusion of a compatibilizer is rare. Thus, the polarity of a polymer matrix seems to be the main driving force of the preparation strategy, as summarized in Figure 13.14.

In general, the preparation of polymer/layered silicate nanocomposites requires parallel investigations concerning compatibilizers and processing aids. For polyolefin-based nanocomposites, the challenge resides in setting up and understanding the role of reactive compatibilizers to control and/or tailor the morphologies responsible for the improved material properties. But, for PET the main objective is – and will continue to be – the optimization of thermal stability, as organophilic modifiers are typically unstable at PET processing temperatures. Undoubtedly, improvements in the thermal stability and flame self-extinguishing properties of

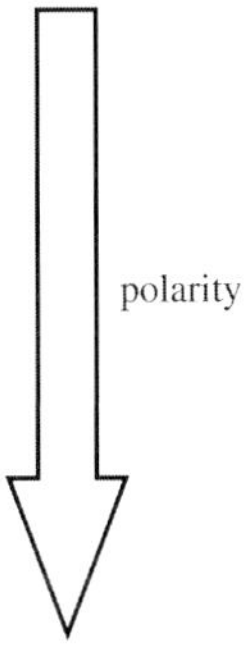

		Best organophilic cations	Number of published papers dealing with compatibilizers
↓ polarity	Polyolefin	With long aliphatic chains [71, 72]	16 [52, 56–58, 68, 76–79, 82–88]
	PLA	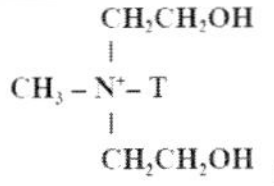 CH_3-N^+-T with CH_2CH_2OH groups [148, 150]	3 [156–158]
	PET	Na^+ [127]; CH_3-N^+-T with CH_2CH_2OH groups [128]; $H_3C-N^+(CH_3)-CH_2-C_6H_5$ with HT [128]	1 [132]

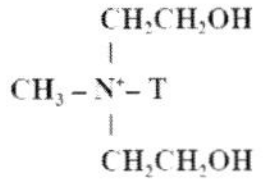

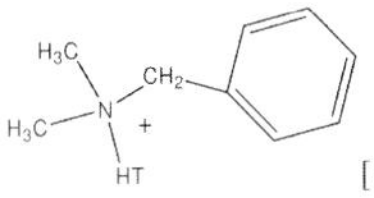

Figure 13.14 Schematic comparison of preparation methods for polymer/layered silicates nanocomposites.

PET will become technological necessities, notably in the development of new fibers and the preparation of nanocomposites. Meanwhile, PET recycling and the need to improve the properties of post-consumer materials remain major challenges for PET/layered silicate nanocomposites.

For PLA, and for biodegradable polyesters in general, new processing aids – developed from renewable resources – should be selected when preparing nanocomposites. Whilst the criteria of nanocomposite preparation are substantially similar to those for petrochemical polymers, the (preferential) need to use naturally derived compounds and polymers will surely lead to innovative methods of production. In particular, replacing commonly used commercial modifiers (ammonium salts) with cations that are not only more sustainable but also biodegradable should prove interesting. In attempts to develop new materials with modulated properties, melt-blending with biodegradable and/or renewable polymers, or perhaps the addition of highly sustainable plasticizers, should allow for the creation of innovative materials that are highly tunable in terms of their mechanical, thermal, gas-barrier, and biodegradation properties. Clearly, the behavior of nanolayers in a complex multiphase system will remain the object of increasing research attention for years to come.

Acknowledgments

The authors thank Prof. Francesco Ciardelli, Stefania Castiello and Francesca Signorini (Dipartimento di Chimica e Chimica Industriale, Università di Pisa) for very helpful discussions. Financial support from MIUR (NANOPACK FIRB 2003

D. D. 2186 grant RBNE03R78E) and from CIPE (Toscana PC-RIPLAS project, 2008–2009) is also gratefully acknowledged.

References

1 Gao, F. (2004) Clay/polymer composites: the story. *Mater. Today*, **7**, 50–5.
2 Zeng, Q.H., Yu, A.B., Lu, G.Q. and Paul, D.R. (2005) Clay-based polymer nanocomposites: research and commercial development. *J. Nanosci. Nanotechnol.*, **5**, 1574–1592.
3 Schadler, L.S., Brinson, L.C. and Sawyer, W.G. (2007) Polymer nanocomposites: a small part of the story. *JOM*, **3**, 53–60.
4 Yu, L., Dean, K. and Li, L. (2006) Polymer blends and composites from renewable resources. *Prog. Polym. Sci.*, **31**, 576–602.
5 Tsutsumi, N., Kono, Y., Oya, M., Sakai, W. and Nagata, M. (2008) Recent development of biodegradable network polyesters obtained from renewable natural resources. *Clean Soil, Air, Water*, **36**, 682–686.
6 Grim, R.E. (1968) *Clay Mineralogy*, 2nd edn, McGraw-Hill, New York, NY, USA.
7 Manocha, L.M. (2006) Composites with nanomaterials, in *Functional Nanomaterials* (eds K.E. Geckeler and E. Rosemberg), American Scientific Publishers, Stevenson Ranch, CA, USA.
8 Dong, H. (2005) Interstratified illite-smectite: a review of contributions of TEM data to crystal chemical relations and reaction mechanisms. *Clay Sci.*, **12** (Suppl. 1), 6.
9 Carnicelli, S., Mirabella, A., Cecchini, G. and Sanesi, G. (1997) Weathering of chlorite to a low charge expandable mineral in a spodosol on the Apennine Mountains, Italy. *Clays Clay Miner.*, **45**, 28.
10 Sakharov, B.A., Lindgreen, H., Salyn, A. and Drits, V.A. (1999) Determination of illite- smectite structures using multispecimen X-ray diffraction profile fitting. *Clays Clay Miner.*, **47** (5), 555–569.
11 Xu, W.B., Ge, M.L. and He, P.S. (2001) Nonisothermal crystallization kinetics of polyoxymethylene/montmorillonite nanocomposites. *J. Appl. Polym. Sci.*, **82**, 2281–2297.
12 Liu, J., Boo, W.-J., Clearfield, A. and Sue, H.-J. (2006) Intercalation and exfoliation: a review on morphology of polymer nanocomposites reinforced by inorganic layer structures. *Mater. Manufact. Process.*, **20**, 143–151.
13 Sinha Ray, S., Okamoto, K. and Okamoto, M. (2003) Structure–property relationship in biodegradable poly(butylene succinate)/layered silicate nanocomposites. *Macromolecules*, **36**, 2355.
14 Chen, B., Evans, J.R.G., Greenwell, H.C., Boulet, P., Coveney, P.V., Bowden, A.A. and Whiting, A. (2008) A critical appraisal of polymer–clay nanocomposites. *Chem. Soc. Rev.*, **37**, 568–594.
15 Samyn, F., Bourbigot, S., Jama, C., Bellayer, S., Nazare, S., Hull, R., Castrovinci, A., Fina, A. and Camino, G. (2008) Crossed characterisation of polymer-layered silicate (PLS) nanocomposite morphology: TEM, X-ray diffraction, rheology and solid-state nuclear magnetic resonance measurements. *Eur. Polym. J.*, **44**, 1642–1653.
16 Luo, Z.P. and Koo, J.H. (2008) Quantification of the layer dispersion degree in polymer layered silicate nanocomposites by transmission electron microscopy. *Polymer*, **49**, 1841–1852.
17 Drummy, L.F., Wang, Y.C., Schoenmakers, R., May, K., Jackson, M., Koerner, O.H., Farmer, B.L., Mauryama, B. and Vaia, R.A. (2008) Morphology of layered silicate- (NanoClay-) polymer nanocomposites by electron tomography and small-angle X-ray scattering. *Macromolecules*, **41**, 2135–2143.
18 Incarnato, L., Scarfato, P., Scatteia, L. and Acierno, D. (2004) Rheological behavior of new melt compounded

copolyamide nanocomposites. *Polymer*, **45** (10), 3487–3496.

19 Lim, S.K., Kim, J.W., Chin, I., Kwon, Y.K. and Choi, H.J. (2002) Preparation and interaction characteristics of organically modified montmorillonite nanocomposite with miscible polymer blend of poly(ethylene oxide) and poly(methyl methacrylate). *Chem. Mater.*, **14**, 1989–1994.

20 Kim, H.B., Choi, J.S., Lee, C.H., Lim, S.T., Jhon, M.S. and Choi, H.J. (2005) Polymer blend/organoclay nanocomposite with poly(ethylene oxide) and poly(methyl methacrylate). *Eur. Polym. J.*, **41**, 679–685.

21 Ray, S.S., Pouliot, S., Bousmina, M. and Utracki, L.A. (2004) Role of organically modified layered silicate as an active interfacial modifier in immiscible polystyrene/polypropylene blends. *Polymer*, **45**, 8403–8413.

22 Fang, Z., Harrats, C., Moussaif, N. and Groeninckx, G. (2007) Location of a nanoclay at the interface in an immiscible poly(ϵ-caprolactone)/poly(ethylene oxide) blend and its effect on the compatibility of the components. *J. Appl. Polym. Sci.*, **106**, 3125–3135.

23 Gahleitner, M., Kretzschmar, B., Pospiech, D., Ingolic, E., Reichelt, N. and Bernreitner, K. (2006) Morphology and mechanical properties of polypropylene/polyamide 6 nanocomposites prepared by a two-step melt-compounding process. *J. Appl. Polym. Sci.*, **100**, 283–291.

24 Vo, L.T. and Giannelis, E.P. (2007) Compatibilizing poly(vinylidene fluoride)/nylon-6 blends with nanoclay. *Macromolecules*, **40**, 8271–8276.

25 Yurekli, K., Karim, A., Amis, E.J. and Krishnamoorti, R. (2003) Influence of layered silicates on the phase-separated morphology of PS-PVME blends. *Macromolecules*, **36**, 7256–7257.

26 Filippone, G., Dintcheva, N.T., Acierno, D. and La Mantia, F.P. (2008) The role of organoclay in promoting co-continuous morphology in high-density poly(ethylene)/poly(amide) 6 blends. *Polymer*, **49**, 1312–1322.

27 Chow, W.S., Mohd Ishak, Z.A., Karger-Kocsis, J., Apostolov, A.A. and Ishiaku, U.S. (2003) Compatibilizing effect of maleated polypropylene on the mechanical properties and morphology of injection molded polyamide 6/polypropylene/organoclay nano composite. *Polymer*, **44**, 7427–7440.

28 Chow, W.S., Abu Bakar, A., Mohd Ishak, Z.A., Karger-Kocsis, J. and Ishiaku, U.S. (2005) Effect of maleic anhydride-grafted ethylene–propylene rubber on the mechanical, rheological and morphological properties of organoclay reinforced polyamide 6/polypropylene nanocomposites. *Eur. Polym. J.*, **41**, 687–696.

29 Chow, W.S., Mohd Ishak, Z.A. and Karger-Kocsis, J. (2005) Atomic force microscopy study on blend morphology and clay dispersion in polyamide-6/polypropylene/organoclay systems. *J. Polym. Sci., Part B: Polym. Phys.*, **43**, 1198–1204.

30 Chen, G.X. and Yoon, J.S. (2005) Thermal stability of poly(L-lactide)/poly(butylene succinate)/clay nanocomposites. *Polym. Degrad. Stab.*, **88**, 206–212.

31 Pavlidou, S. and Papaspyrides, C.D. (2008) A review-layered silicate nanocomposites. *Prog. Polym. Sci.*, **33**, 1190–1198.

32 Passaglia, E., Bertoldo, M., Ceriegi, S., Sulcis, R., Narducci, P. and Conzatti, L. (2008) Oxygen and water vapor barrier properties of MMT nanocomposites from low density polyethylene or EPM with grafted succinic groups. *J. Nanosci. Nanotechnol.*, **8** (4), 1690–1699.

33 Lu, C. and Mai, Y.-W. (2005) Influence of aspect ratio on barrier properties of polymer-clay nanocomposites. *Phys. Rev. Lett.*, **95**, 088303.

34 Xu, B., Zheng, Q., Song, Y. and Shangguan, Y. (2006) Calculating barrier properties of polymer/clay nanocomposites: effects of clay layers. *Polymer*, **47**, 2904–2910.

35 Rittigstein, P., Priestley, R.D., Broadbelt, L.J. and Torkelson, J.M. (2007) Model polymer nanocomposites provide an understanding of confinement effects in real nanocomposites. *Nat. Mater.*, **6** (4), 257–258.

36 Gilman, J.W., Jackson, C.L., Morgan, A.B., Harris, R. Jr., Manias, E., Giannelis, E.P., Wuthenow, M., Hilton, D. and Phillips, S.H. (2000) Flammability properties of polymer-layered-silicate nanocomposites. Polypropylene and polystyrene nanocomposites. *Chem. Mater.*, **12**, 1866–1873.

37 Zanetti, M., Kashiwagi, T., Falqui, L. and Camino, G. (2002) Cone calorimeter combustion and gasification studies of polymer layered silicate nanocomposites. *Chem. Mater.*, **14**, 881–887.

38 Liu, T., Lim, K.P., Tjiu, W.C., Pramoda, K.P. and Chen, Z.-K. (2003) Preparation and characterization of nylon 11/organoclay nanocomposites. *Polymer*, **44**, 3529–3535.

39 Harrats, C. and Groeninckx, G. (2008) Features, questions and future challenges in layered silicates clay nanocomposites with semicrystalline polymer matrices. *Macromol. Rapid. Commun.*, **29**, 14–26.

40 Xu, W., Ge, M. and He, P. (2001) Nonisothermal crystallization kinetics of polyoxymethylene/montmorillonite nanocomposite. *J. Appl. Polym. Sci.*, **82**, 2281–2289.

41 Giannelis, E.P., Krishnamoorti, R. and Manias, E. (1999) Polymer-silicate nanocomposites: model systems for confined polymers and polymer brushes. *Adv. Polym. Sci.*, **138**, 107–147.

42 Alexandre, M. and Dubois, P. (2000) Polymer-layered silicate nanocomposites: preparation, properties and uses of a new class of materials. *Mater. Sci. Eng., R*, **R28**, 1–63.

43 Pinnavaia, T.J. and Beall, G.W. (2000) *Polymer-Clay Nanocomposites*, John Wiley & Sons, Inc., New York.

44 Fischer, H. (2003) Polymer nanocomposites: from fundamental research to specific applications. *Mater. Sci. Eng., C*, **C23**, 763–772.

45 Ray, S.S. and Okamoto, M. (2003) Polymer/layered silicate nanocomposites: a review from preparation to processing. *Prog. Polym. Sci.*, **28**, 1539–1641.

46 Giannelis, E. (1996) Polymer layered silicate nanocomposites. *Adv. Mater.*, **8**, 29–35.

47 Tjong, S.C. (2006) Structural and mechanical properties of polymer nanocomposites. *Mater. Sci. Eng., R*, **R53**, 73–197.

48 Ciardelli, F., Coiai, S., Passaglia, E., Pucci, A. and Ruggeri, G. (2008) Nanocomposites based on polyolefins and functional thermoplastic materials. *Polym. Int.*, **57**, 805–836.

49 Vaia, R.A., Price, G., Ruth, P.N., Nguyen, H.T. and Lichtenhan, J. (1999) Polymer/layered silicate nanocomposites as high performance ablative materials. *J. Appl. Clay. Sci.*, **15**, 67–92.

50 Biswas, M. and Sinha Ray, S. (2001) Recent progress in synthesis and evaluation of polymer-montmorillonite nanocomposites. *Adv. Polym. Sci.*, **155**, 167–221.

51 Kawasumi, M., Hasegawa, N., Kato, M., Usuki, A. and Okada, A. (1997) Preparation and mechanical properties of polypropylene-clay hybrids. *Macromolecules*, **30**, 6333–6338.

52 Manias, E., Touny, A., Wu, L., Strawhecker, K., Lu, B. and Chung, T.C. (2001) Polypropylene/montmorillonite nanocomposites. Review of the synthetic routes and materials properties. *Chem. Mater.*, **13**, 3516–3523.

53 Liu, X.H. and Wu, Q.J. (2001) Polypropylene/clay nanocomposites prepared by grafting-melt. *Polymer*, **42**, 10013–10019.

54 García-López, D., Picazo, O., Merino, J.C. and Pastor, J.M. (2003) Polypropylene-clay nanocomposites: effect of compatibilizing agents on clay dispersion. *Eur. Polym. J.*, **39**, 945–950.

55 Usuki, A., Kato, M., Okada, A. and Kurauchi, T. (1997) Synthesis of polypropylene-clay hybrid. *J. Appl. Polym. Sci.*, **63**, 137–139.

56 Kato, M., Usuki, A. and Okada, A. (1997) Synthesis of polypropylene oligomer-clay intercalation compounds. *J. Appl. Polym. Sci.*, **66**, 1781–1785.

57 Oya, A., Kurokawa, Y. and Yasuda, H. (2000) Factors controlling mechanical properties of clay mineral/polypropylene nanocomposites. *J. Mater. Sci.*, **35**, 1045–1050.

58 Wang, Y., Chen, F.B., Li, Y.C. and Wu, K.C. (2004) Melt processing of

polypropylene/clay nanocomposites modified with maleated polypropylene compatibilizers. *Composites Part B*, **35**, 111–124.
59. Sun, T. and Garcés, M.J. (2002) High-performance polypropylene-clay nanocomposites by in situ polymerization with metallocene/clay catalysts. *Adv. Mater.*, **14**, 128–130.
60. Hwu, J.M. and Jiang, G.J. (2005) Preparation and characterization of polypropylene-montmorillonite nanocomposites generated by in situ metallocene catalyst polymerization. *J. Appl. Polym. Sci.*, **95**, 1228–1236.
61. Ma, J.S., Qi, Z.N. and Hu, Y.L. (2001) Synthesis and characterization of polypropylene/clay nanocomposites. *J. Appl. Polym. Sci.*, **82**, 3611–3617.
62. He, A., Hu, H., Huang, Y., Dong, J.Y. and Han, C.C. (2004) Isotactic poly(propylene)/monoalkylimidazolium-modified montmorillonite nanocomposites: preparation by intercalative polymerization and thermal stability study. *Macromol. Rapid. Commun.*, **25**, 2008–2013.
63. Alexandre, M., Dubois, P., Sun, T., Garces, J.M. and Jerome, R. (2002) Polyethylene-layered silicate nanocomposites prepared by the polymerization-filling technique: synthesis and mechanical properties. *Polymer*, **43**, 2123–2132.
64. Osman, M.A., Rupp, J.E.P. and Suter, U.W. (2005) Effect of non-ionic surfactants on the exfoliation and properties of polyethylene-layered silicate nanocomposites. *Polymer*, **46**, 8202–8209.
65. Vaia, R.A. and Giannelis, E.P. (1997) Lattice model of polymer melt intercalation in organically-modified layered silicates. *Macromolecules*, **30**, 7990–7999.
66. Vaia, R.A. and Giannelis, E.P. (1997) Polymer melt intercalation in organically-modified layered silicates: model predictions and experiment. *Macromolecules*, **30**, 8000–8009.
67. Rogers, K., Takacs, E. and Thompson, M.R. (2005) Contact angle measurement of select compatibilizers for polymer-silicate layer nanocomposites. *Polym. Test.*, **24**, 423–427.
68. Borsacchi, S., Geppi, M., Ricci, L., Ruggeri, G. and Veracini, C.A. (2007) Interactions at the surface of organophilic-modified laponites: a multinuclear solid-state NMR study. *Langmuir*, **23**, 3953–3960.
69. Kàdàr, F., Szàzdi, L., Fekete, E. and Pukànszky, B. (2006) Surface characteristics of layered silicates: influence on the properties of clay/polymer nanocomposites. *Langmuir*, **22**, 7848–7854.
70. Balazs, A.C., Singh, C. and Zhulina, E. (1998) Modeling the interactions between polymers and clay surfaces through self-consistent field theory. *Macromolecules*, **31**, 8370–8381.
71. Balazs, A.C., Singh, C., Zhulina, E. and Lyatskaya, Y. (1999) Modelling the phase behaviour of polymer/clay nanocomposites. *Acc. Chem. Res.*, **32**, 651–657.
72. Wang, K.H., Choi, M.H., Koo, C.M., Choi, Y.S. and Chung, I.J. (2001) Synthesis and characterization of maleated polyethylene/clay nanocomposites. *Polymer*, **42**, 9819–9826.
73. Koo, C.M., Ham, H.T., Kim, S.O., Wang, K.H. and Chung, I.J. (2002) Morphology evolution and anisotropic phase formation of the maleated polyethylene-layered silicate nanocomposites. *Macromolecules*, **35**, 5116–5122.
74. Wang, Z.M., Nakajima, H., Manias, E. and Chung, T.C. (2003) Exfoliated PP/Clay nanocomposites using ammonium-terminated PP as the organic modification for montmorillonite. *Macromolecules*, **36**, 8919–8922.
75. Chrissopoulou, K., Altintzi, I., Anastasiadis, S.H., Giannelis, E.P., Pitsikalis, M., Hadjichristidis, N. and Theophilou, N. (2005) Controlling the miscibility of polyethylene/layered silicate nanocomposites by altering the polymer/surface interactions. *Polymer*, **46**, 12440–12451.
76. Moad, G., Dean, K., Edmond, L., Kukaleva, N., Li, G., Mayadunne, R.T.A., Pfaendner, R., Schneider, A., Simon, G.P. and Wermter, H. (2006) Non-ionic,

poly(ethylene oxide)-based surfactants as intercalants/dispersants/exfoliants for poly(propylene)-clay nanocomposites. *Macromol. Mater. Eng.*, **291**, 37–52.

77 Hasegawa, N., Kawasumi, M., Kato, M., Usuki, A. and Okada, A. (1998) Preparation and mechanical properties of polypropylene-clay hybrids using a maleic anhydride-modified polypropylene oligomer. *J. Appl. Polym. Sci.*, **67**, 87–92.

78 Hasegawa, N., Okamoto, H., Kawasumi, M., Kato, M., Tsukigas, A. and Usuki, A. (2000) Polyolefin–clay hybrids based on modified polyolefins and organoclay. *Macromol. Mater. Eng.*, **280/281**, 76–79.

79 Moad, G. (1999) The synthesis of polyolefin graft copolymers by reactive extrusion. *Prog. Polym. Sci.*, **24**, 81–142.

80 Passaglia, E., Coiai, S., Aglietto, M., Ruggeri, G., Rubertà, M. and Ciardelli, F. (2003) Functionalization of polyolefins by reactive processing: influence of starting reagents on content and type of grafted groups. *Macromol. Symp.*, **198**, 147–160.

81 Ciardelli, F., Aglietto, M., Coltelli, M.B., Passaglia, E., Ruggeri, G. and Coiai, S. (2004) Functionalization of polyolefins in the melt, in *Modification and Blending of Synthetic and Natural Macromolecules* (eds F. Ciardelli and S. Penczek), Kluwer Academic Publishers, The Netherlands.

82 Coiai, S., Scatto, M., Bertoldo, M., Conzatti, L., Andreotti, L., Sterner, M., Passaglia, E., Costa, G. and Ciardelli, F. (2009) Study of the compounding process parameters for morphology control of LDPE/layered silicate nanocomposites. *e-polymers*, accepted.

83 Passaglia, E., Sulcis, R., Ciardelli, F., Malvaldi, M. and Narducci, P. (2005) Effect of functional groups of modified polyolefins on the structure and properties of their composites with lamellar silicates. *Polym. Int.*, **54**, 1549–1556.

84 Passaglia, E., Bertoldo, M., Ciardelli, F., Prevosto, D. and Lucchesi, M. (2008) Evidences of macromolecular chains confinement of ethylene-propylene copolymer in organophilic montmorillonite nanocomposites. *Eur. Polym. J.*, **44**, 1296–1308.

85 Passaglia, E., Bertuccelli, W. and Ciardelli, F. (2001) Composites from functionalized polyolefins and silica. *Macromol. Symp.*, **176**, 299–315.

86 Coiai, S., Passaglia, E., Aglietto, M. and Ciardelli, F. (2004) Control of degradation reactions during radical functionalization of polypropylene in the melt. *Macromolecules*, **37**, 8414–8423.

87 Augier, S., Coiai, S., Gragnoli, T., Passaglia, E., Pradel, J.L. and Flat, J.J. (2006) Coagent assisted polypropylene radical functionalization: monomer grafting modulation and molecular weight conservation. *Polymer*, **47**, 5243–5252.

88 Augier, S., Coiai, S., Pratelli, D., Conzatti, L. and Passaglia, E. (2009) New functionalized polypropylenes as controlled architecture compatibilizers for polypropylene layered silicates nanocomposites. *J. Nanosci. Nanotechnol.*, **9**, 4858–4869.

89 Passaglia, E., Bertoldo, M., Coiai, S., Augier, S., Savi, S. and Ciardelli, F. (2008) Nanostructured polyolefins/clay composites: role of the molecular interaction at the interface. *Polym. Adv. Technol.*, **19**, 560–568.

90 Ginzburg, V.V. and Balazs, A.C. (2000) Calculating phase diagrams for nanocomposites: the effect of adding end-functionalized chains to polymer/clay mixtures. *Adv. Mater.*, **23**, 1805–1809.

91 Sinsawat, A., Anderson, K.L., Vaia, R.A. and Farmer, B.L. (2003) Influence of polymer matrix composition and architecture on polymer nanocomposite formation: coarse-grained molecular dynamics simulation. *J. Polym. Sci., Part B: Polym. Phys.*, **41**, 3272–3284.

92 Marchant, D. and Jayaraman, K. (2002) Strategies for optimizing polypropylene-clay nanocomposite structure. *Ind. Eng. Chem. Res.*, **41**, 6402–6408.

93 Yang, K. and Ozisik, R. (2006) Effects of processing parameters on the preparation of nylon 6 nanocomposites. *Polymer*, **47**, 2849–2855.

94 Perrin-Sarazin, F., Ton-That, M.T., Bureau, M.N. and Denault, J. (2005) Micro- and nano-structure in

95. Cho, J.W. and Paul, D.R. (2001) Nylon 6 nanocomposites by melt compounding. *Polymer*, 42, 1083–1094.
96. Dennis, H.R., Hunter, D.L., Chang, D., Kim, S., White, J.L., Cho, J.W. and Paul, D.R. (2001) Effect of melt processing conditions on the extent of exfoliation in organoclay-based nanocomposites. *Polymer*, 42, 9513–9522.
97. Neilson, L.E. (1967) Models for the permeability of filled polymer systems. *J. Macromol. Sci. Chem.*, 1 (5), 929.
98. Kashiwagi, T., Harris, R.H., Zhang, X., Briber, R.M., Cipriano, B.H., Raghavan, S.R., Srinivasa, R., Awad, W.H. and Shields, J.R. (2004) Flame retardant mechanism of polyamide 6–clay nanocomposites. *Polymer*, 45, 881–891.
99. Jang, K.H., Kim, B.C., Hahm, W.G. and Kikutani, T. (2008) High-speed melt spinning of nanoparticle-filled high molecular weight poly(ethylene terephthalate). *Int. Polym. Process.*, 23 (4), 370–6.
100. Chang, J.-H. and Mun, M.K. (2007) Nanocomposite fibers of poly(ethylene terephthalate) with montmorillonite and mica: thermomechanical properties and morphology. *Polym. Int.*, 56, 57–66.
101. Lange, J. and Wyser, Y. (2003) Recent innovations in barrier technologies for plastic packaging-a review, *Packaging Tech. Sci.*, 16 (4), 149–158.
102. Tzavalas, S., Drakonakis, V., Mouzakis, D.E., Fischer, D. and Gregoriou, V.G. (2006) Effect of carboxy-functionalized multiwall nanotubes (MWNT-COOH) on the crystallization and chain conformations of poly(ethylene terephthalate) PET in PET-MWNT nanocomposites. *Macromolecules*, 39, 9150–9156.
103. Awaja, F. and Pavel, D. (2005) Recycling of PET. *Eur. Polym. J.*, 41, 1453–1477.
104. Coltelli, M.B., Bianchi, S., Savi, S., Liuzzo, V. and Aglietto, M. (2003) Metal catalysis to improve compatibility at PO/PET blends interface. *Macromol. Symp.*, 204 (1), 227–236.
105. Aglietto, M., Coltelli, M.B., Savi, S., Lochiatto, F., Ciardelli, F. and Giani, M. (2004) Post-consumer polyethylene terephthalate (PET)/ polyolefin blends through reactive processing. *J. Mater. Cycles Waste Manage.*, 6, 13–19.
106. Coltelli, M.B., Savi, S., Della Maggiore, I., Liuzzo, V., Aglietto, M. and Ciardelli, F. (2004) A model study of Ti(OBu)$_4$ catalyzed reactions during reactive blending of polyethylene (PE) and poly(ethylene terephthalate). *Macromol. Mater. Eng.*, 289, 400–412.
107. Pracella, M., Pazzagli, F. and Galeski, A. (2002) Reactive compatibilization and properties of recycled poly(ethylene terephthalate)/polyethylene blends. *Polym. Bull.*, 48 (1), 67–74.
108. Coltelli, M.B., Della Maggiore, I., Savi, S., Aglietto, M. and Ciardelli, F. (2005) Modified styrene-*b*-ethylene-*co*-1-butene-*b*-styrene triblock copolymer as compatibiliser precursor in polyethylene/poly(ethylene terephthalate) blends. *Polym. Degrad. Stab.*, 90 (2), 211–223. Erratum (2006) *Polym. Degrad. Stab.*, 91, 987–223.
109. Coltelli, M.-B. and Bianchi, S. (2007) Mauro aglietto, poly(ethyleneterephthalate) (PET) degradation during the Zn catalysed transesterification with dibutyl maleate functionalized polyolefins. *Polymer*, 48, 1276–1286.
110. Coltelli, M.-B., Aglietto, M. and Ciardelli, F. (2008) Influence of the transesterification catalyst structure on the reactive compatibilization and properties of poly(ethylene terephthalate) (PET)/dibutyl succinate functionalized poly(ethylene) blends. *Eur. Polym. J.*, 44, 1512–1524.
111. Coltelli, M.-B., Harrats, C., Aglietto, M. and Groeninckx, G. (2008) Phase morphology development in poly(ethylene terephthalate) (PET)/ low density poly(ethylene) (LDPE) blends: compatibilizer precursors effect. *Polym. Eng. Sci.*, 48, 1424–1433.
112. Pegoretti, A. and Penati, A. (2004) Recycled poly(ethylene terephthalate) and its short glass fibres composites: effects of hygrothermal aging on the thermo-mechanical behavior. *Polymer*, 45, 7995–8004.
113. Hwang, S.Y., Lee, W.D., Lim, J.S., Park, K.H. and Im, S.S. (2008) Dispersibility

of clay and crystallization kinetics for in situ polymerized PET/pristine and modified montmorillonite nanocomposites. *J. Polym. Sci., Part B: Polym. Phys.*, **46**, 1022–1035.

114 Tomita, K. (1973) Studies on the formation of poly(ethylene terephthalate): 2. Rate of transesterification of dimethyl terephthalate with ethylene glycol. *Polymer*, **14**, 55–60.

115 Karayannidis, G.P., Roupakias, C.P., Bikiaris, D.N. and Achilias, D.S. (2003) Study of various catalysts in the synthesis of poly(propylene terephthalate) and mathematical modeling of the esterification reaction. *Polymer*, **44**, 931–942.

116 Ke, Y.C., Long, C.F. and Qi, Z.N. (1999) Crystallization, properties, and crystal and nanoscale morphology of PET–clay nanocomposites. *J. Appl. Polym. Sci.*, **71**, 1139–1146.

117 Ke, Y.C. and Qi, Z.N. (1997) Chinese Patent Application 97104294.6.

118 Lee, S.-S., Ma, Y.T., Rhee, H.-W. and Kim, J. (2005) Exfoliation of layered silicate facilitated by ring-opening reaction of cyclic oligomers in PET–clay nanocomposites. *Polymer*, **46**, 2201–2210.

119 Hubbard, P., Brittain, W.J., Mattice, W.L. and Brunelle, D.J. (1998) Ring-size distribution in the depolymerization of poly(butylene terephthalate). *Macromolecules*, **31**, 1518–1522.

120 Chung, J.W., Son, S.-B., Chun, S.-W., Kang, T.J. and Kwak, S.-Y. (2008) Nonisothermal crystallization behavior of exfoliated poly(ethylene terephthalate)-layered silicate nanocomposites in the presence and absence of organic modifier. *J. Polym. Sci., Part B: Polym. Phys.*, **46**, 989–999.

121 Pegoretti, A., Kolarik, J., Peroni, C. and Migliaresi, C. (2004) Recycled poly(ethylene terephthalate)/layered silicate nanocomposites: morphology and tensile mechanical properties. *Polymer*, **45**, 2751–2759.

122 Calcagno, C.I.W., Mariani, C.M., Teixeira, S.R. and Mauler, R.S. (2007) The effect of organic modifier of the clay on morphology and crystallization properties of PET nanocomposites. *Polymer*, **48**, 966–974.

123 Kráčalík, M., Mikešová, J., Puffr, R., Baldrian, J., Thomann, R. and Friedrich, C. (2007) Effect of 3D structures on recycled PET/organoclay nanocomposites. *Polym. Bull.*, **58**, 313–319.

124 Kim, K.H., Kim, K.H., Huh, J. and Jo, W.H. (2007) Synthesis of thermally stable organosilicate for exfoliated poly(ethylene terephthalate) nanocomposite with superior tensile properties. *Macromol. Res.*, **15** (2), 178–184.

125 Davis, C.H., Mathias, L.J., Gilman, J.W., Schiraldi, D.A., Shields, J.R., Trulove, P., Sutto, T.E. and Delong, H.C. (2002) Effects of melt-processing conditions on the quality of poly(ethylene terephthalate) montmorillonite clay nanocomposites. *J. Polym. Sci., Part B: Polym. Phys.*, **40**, 2661–2666.

126 Stoeffler, K., Lafleur, P.G. and Denault, J. (2008) Thermal decomposition of various alkyl onium organoclays: effect on polyethylene terephthalate nanocomposites' properties. *Polym. Degrad. Stab.*, **93**, 1332–1350.

127 Vidotti, S.E., Chinellato, A.C., Hu, G.-H. and Pessan L.A. (2007) Preparation of poly(ethylene terephthalate)/organoclay nanocomposites using a polyester ionomer as a compatibilizer. *J. Polym. Sci., Part B: Polym. Phys.*, **45**, 3084–3091.

128 Tsutsumi, N., Kono, Y., Oya, M., Sakai, W. and Nagata, M. (2008) Recent development of biodegradable network polyesters obtained from renewable natural resources. *Clean Soil, Air, Water*, **36** (8), 682–86.

129 Signori, F., Coltelli, M.B., Bronco, S. and Ciardelli, F. (2009) Thermal degradation of poly(lactic acid) (PLA) and poly(butylene adipate-*co*-terephthalate) (PBAT) as a consequence of melt processing: effects on pure materials and their blends. *Polym. Degrad. Stab.*, **94**, 74–82.

130 Coltelli, M.B., Della Maggiore, I., Bertoldo, M., Bronco, S., Signori, F. and Ciardelli, F. (2008) Poly(lactic acid) (PLA) properties as a consequence of poly(butylene adipate-*co*-terephthalate)

(PBAT) blending and acetyl tributyl citrate (ATBC) plasticization. *J. Appl. Polym. Sci.*, **110**, 1250–1262.

131 Sinha Ray, S. and Bousmina, M. (2005) Biodegradable polymers and their layered silicate nanocomposites: in greening the 21st century materials world. *Prog. Mater. Sci.*, **50**, 962–1079.

132 Lecomte, P., Riva, R., Jerome, C. and Jerome, R. (2008) Macromolecular engineering of biodegradable polyesters by ring opening polymerization and 'click' chemistry. *Macromol. Rapid. Commun.*, **29** (**12-13**), 982–997.

133 Kamber, N.E., Jeong, W., Waymouth, R.M., Pratt, R.C., Lohmeijer, B.G.G. and Hedrick, J.L. (2007) Organocatalytic ring-opening polymerization. *Chem. Rev.*, **107** (**12**), 5813–5840.

134 Albertsson, A.C. and Varma, I.K. (2003) Recent developments in ring opening polymerization of lactones for biomedical applications. *Biomacromolecules*, **4**, 1466–1486.

135 Kubies, D., Pantoustier, N., Dubois, P., Rulmont, A. and Jerome, R. (2002) Controlled ring-opening polymerization of ε-caprolactone in the presence of layered silicates and formation of nanocomposites. *Macromolecules*, **35**, 3318–3320.

136 Lepoittevin, B., Pantoustier, N., Alexandre, M., Calberg, C., Jerome, R. and Dubois, P. (2002) Polyester layered silicate nanohybrids by controlled grafting polymerization. *J. Mater. Chem.*, **12**, 3528–3532.

137 Lepoittevin, B., Pantoustier, N., Devalckenaere, M., Alexandre, M., Kubies, D., Calberg, C., Jerome, R. and Dubois, P. (2002) Poly(ε-caprolactone)/clay nanocomposites by in-situ intercalative polymerization catalyzed by dibutyltin dimethoxide. *Macromolecules*, **35**, 8385–8390.

138 Chrissafis, K., Antoniadis, G., Paraskevopoulos, K.M., Vassiliou, A. and Bikiaris, D.N. (2007) Comparative study of the effect of different nanoparticles on the mechanical properties and thermal degradation mechanism of in situ prepared poly(ε-caprolactone) nanocomposites. *Compos. Sci. Technol.*, **67**, 2165–2174.

139 Paul, M.-A., Alexandre, M., Degee, P., Calberg, C., Jerome, R. and Dubois, P. (2003) Exfoliated polylactide/ clay nanocomposites by in- situ coordination-insertion polymerization. *Macromol. Rapid Commun.*, **24** (**9**), 561–566.

140 Lee, S., Kim, C.H. and Park, J.-K. (2006) Improvement of processability of clay / polylactide nanocomposites by a combinational method: in situ polymerization of L-lactide and melt compounding of polylactide. *J. Appl. Polym. Sci.*, **101** (**3**), 1664–1669.

141 Ogata, N., Jimenez, G., Kawai, H. and Ogihara, T. (1997) Structure and thermal/mechanical properties of poly(L-lactide)–clay blend. *J. Polym. Sci., Part B: Polym. Phys.*, **35**, 389–396.

142 Marras, S.I., Tsimpliaraki, A., Zuburtikudis, I. and Panayiotou, C. (2007) Surfactant-induced morphology and thermal behavior of polymer layered silicate nanocomposites. *J. Phys.: Conf. Ser.*, **61**, 1366–1370.

143 Krikorian, V. and Pochan, D.J. (2003) Poly (L-lactic acid)/layered silicate nanocomposite: fabrication, characterization, and properties. *Chem. Mater.*, **15**, 4317–4324.

144 Chang, J.-H., Uk-An, Y. and Sur, G.S. (2003) Poly(lactic acid) nanocomposites with various organoclays. I. thermomechanical properties, morphology, and gas permeability. *J. Polym. Sci., Part B: Polym. Phys.*, **41**, 94–103.

145 Di, Y., Iannace, S., Di Maio, E. and Nicolais, L. (2005) Poly(lactic acid)/organoclay nanocomposites: thermal, rheological properties and foam processing. *J. Polym. Sci., Part B: Polym. Phys.*, **43**, 689–698.

146 Feijoo, J.L., Cabedo, L., Gimenez, E., Lagaron, J.M. and Saura, J.J. (2005) Development of amorphous PLA-montmorillonite nanocomposites. *J. Mater. Sci.*, **40**, 1785–1788.

147 Pluta, M. (2006) Melt compounding of polylactide/organoclay: structure and properties of nanocomposites. *J. Polym. Sci., Part B: Polym. Phys.*, **44**, 3392–3405.

148 Gu, S.-Y., Ren, J. and Dong, B. (2007) Melt rheology of polylactide/montmorillonite nanocomposites.

149 Lewitus, D., McCarthy, S., Ophir, A. and Kenig, S. (2006) The effect of nanoclays on the properties of PLLA-modified polymers Part 1: mechanical and thermal properties. *J. Polym. Environ.*, **14**, 171–177.

150 Pluta, M., Jeszka, J.K. and Boiteux, G. (2007) Polylactide/montmorillonite nanocomposites: structure, dielectric, viscoelastic and thermal properties. *Eur. Polym. J.*, **43**, 2819–2835.

151 Sinha Ray, S., Maiti, P., Okamoto, M., Yamada, K. and Ueda, K. (2002) New polylactide/layered silicate nanocomposite. 1. Preparation, characterization, and properties. *Macromolecules*, **35**, 3104–3110.

152 Kubies, D., Scudla, J., Puffr, R., Sikora, A., Baldrian, J., Kovarova, J., Slouf, M. and Rypacek, F. (2006) Structure and mechanical properties of poly(L-lactide)/layered silicate nanocomposites. *Eur. Polym. J.*, **42**, 888–899.

153 Yu, Z., Yin, J., Yan, S., Xie, Y., Ma, J. and Chen, X. (2007) Biodegradable poly(L-lactide)/poly(ε-caprolactone)-modified montmorillonite nanocomposites: preparation and characterization. *Polymer*, **48**, 6439–6447.

154 Dell'Erba, R., Groeninckx, G., Maglio, G., Malinconico, M. and Migliozzi, A. (2001) Immiscible polymer blends of semicrystalline biocompatible components. Thermal properties and phase morphology analysis of PLLA/PCL blends. *Polymer*, **42**, 7831–7840.

155 Piorkowska, E., Kulinski, Z., Galeski, A. and Masirek, R. (2006) Plasticization of semicrystalline poly(L-lactide) with poly(propylene glycol). *Polymer*, **47**, 7178–7188.

156 Ljungberg, N., Colombini, D. and Wesslen, B. (2005) Plasticization of poly(Lactic acid) with oligomeric malonate esteramides: dynamic mechanical and thermal film properties. *J. Appl. Polym. Sci.*, **96**, 992–1002.

157 Paul, M.-A., Alexandre, M., Degeé, P., Henrist, C., Rulmont, A. and Dubois, P. (2003) New nanocomposite materials based on plasticized poly(L-lactide) and organo-modified montmorillonites: thermal and morphological study. *Polymer*, **44**, 443–450.

158 Vaia, R.A., Vasudevan, S., Krawiec, W., Scanlon, L.G. and Giannelis, E.P. (1995) New polymer electrolyte nanocomposites. Melt intercalation of poly(ethylene oxide) in mica-type silicates. *Adv. Mater.*, **7**, 154–156.

159 Pluta, M., Paul, M.-A., Alexandre, M. and Dubois, P. (2006) Plasticized polylactide/clay nanocomposites. I. The role of filler content and its surface organo-modification on the physico-chemical properties. *J. Polym. Sci., Part B: Polym. Phys.*, **44**, 299–311.

160 Thellen, C., Orroth, C., Froio, D., Ziegler, D., Lucciarini, J., Farrell, R., D'Souza, N.A. and Ratto, J.A. (2005) Influence of montmorillonite layered silicate on plasticized poly(L-lactide) blown films. *Polymer*, **46**, 11716–11727.

161 Mainil, M., Alexandre, M., Monteverde, F. and Dubois, P. (2006) Polyethylene organo-clay nanocomposites: the role of the interface chemistry on the extent of clay intercalation/exfoliation. *J. Nanosci. Nanotechnol.*, **6**, 337–344.

162 Lertwimolnun, W. and Vergnes, B. (2005) Influence of compatibilizer and processing conditions on the dispersion of nanoclay in a polypropylene matrix. *Polymer*, **46**, 3462–3471.

163 Lertwimolnun, W. and Vergnes, B. (2006) Effect of processing conditions on the formation of polypropylene/organoclay nanocomposites in a twin screw extruder. *Polym. Eng. Sci.*, **46**, 314–323.

164 Qi, Z.N., Ke, Y.C. and Zhou, Y.Z. (1997) Chinese Patent Application 97104055.9.